MW01676013

PRACTICAL BUSINESS MATH

AN APPLICATIONS APPROACH

ANNOTATED
INSTRUCTOR'S EDITION

PRACTICAL BUSINESS MATH

AN APPLICATIONS APPROACH

EIGHTH EDITION

MICHAEL D. TUTTLE
NORTHWOOD UNIVERSITY

Prentice Hall

Upper Saddle River, New Jersey
Columbus, Ohio

To Cody, Danielle,
Emilie, and Sigurd

Library of Congress Cataloging-in-Publication Data

Tuttle, Michael D.

Practical business math : an applications approach / Michael D. Tuttle.—8th ed.

p. cm.

Includes index.

ISBN 0-13-025667-6 (pbk.)

1. Business mathematics—Problems, exercises, etc. I. Title.

HF5694.T88 2001

650′.01′513—dc21

99-089103

Vice President and Publisher: Dave Garza
Senior Editor: Elizabeth Sugg
Associate Editor: Michelle Churma
Production Editor: Louise N. Sette
Production Supervision: Clarinda Publication Services
Design Coordinator: Robin G. Chukes
Cover Designer: Rod Harris
Cover Art: © Kathy Hanley
Production Manager: Brian Fox
Marketing Manager: Shannon Simonsen

This book was set in Optima by The Clarinda Company. It was printed and bound by Banta Book Group. The cover was printed by Phoenix Color Corp.

Earlier editions copyright © 1994, 1990, 1987, 1983, 1982, 1978 by Wm. C. Brown Communications, Inc.

Copyright © 2001, 1998 by Prentice-Hall, Inc., Upper Saddle River, New Jersey 07458. All rights reserved. Printed in the United States of America. This publication is protected by Copyright and permission should be obtained from the publisher prior to any prohibited reproduction, storage in a retrieval system, or transmission in any form or by any means, electronic, mechanical, photocopying, recording or likewise. For information regarding permission(s), write to: Rights and Permissions Department.

Prentice Hall

10 9 8 7 6 5 4 3 2 1
ISBN: 0-13-025667-6 (Standard Edition)
0-13-025660-9 (Brief Edition)
0-13-026486-5 (Annotated Instructor's Edition)

Contents

TO THE INSTRUCTOR

The eighth edition of *Practical Business Math: An Applications Approach* continues the successful format of previous editions with an added emphasis on the integrated presentation that makes teaching and learning easier. This emphasis is sure to improve student understanding of business math topics. The explanation, example, exercise, assignment, and mastery test format has been continued. This format is enhanced with a complete cross-reference system of chapter objectives, student mastery test answers, essential calculator keystroke solutions, and topic page location. Clearly identified chapter objectives, practical small business examples, and understandable explanations make *Practical Business Math* an excellent learning tool.

We live in a global economy. New and developing businesses are providing challenging opportunities for those prepared to compete in the world of business. *Practical Business Math* prepares the student for the dramatic changes that are taking place and those that are yet unannounced. The practical nature of the presentation, the realistic business scenarios, the problem-solving techniques, and the measurement of success prepare the student for additional courses in business and the world of work.

About This Book

Practical Business Math uses the fundamental tools of basic arithmetic to solve everyday business problems. You may select *Practical Business Math,* Brief Edition (chapters 1–17), or you may select *Practical Business Math,* Standard Edition (chapters 1–23).

Each chapter in *Practical Business Math* consistently follows this format:

Chapter Objectives list the specific goals so students know what is expected. The page reference is provided for easy location of the instructional material.

Sport-A-Merica is a case study that illustrates real-world use of math skills in a business setting. Some specific problems are introduced here and solved in the chapter examples.

Explanation-Example-Exercise format provides clear explanation immediately followed by an illustrative example. Exercises are practice problems designed to reinforce the topics just learned. This material is organized around the chapter objectives. This easy-to-follow format facilitates the learning process.

Assignments check the student's understanding of the concepts covered in the chapter and are divided into skill problems, business application problems, and a challenge problem. All sets of problems are arranged in order of increasing difficulty. Answers to the odd-numbered problems are given in the back of the book and essential keystroke solutions are given in the instructor's edition.

Mastery Test is designed for take-home or in-class examination. The answers to all mastery test problems are in the back of the book.

Chapter 23 provides a review for the student. Fifty problems are keyed to the chapter/page where the concept is introduced.

Writing Across the Curriculum is integrated into assignment and mastery test problems in all chapters. Students are expected to provide answers to some problems in sentence form.

Suggested Classroom Time and Perceived Difficulty bar graphs are included in all chapter openings in the Instructor's Edition to aid in planning. This will assist you in determining the amount of time needed for each of the chapters in the course.

Margin Notations for the instructor and the student are found in all chapters. The notations for the instructor (in the Annotated Instructor's Edition only) will provide additional information that may be useful to introduce topics to the class. The notations to the student (in both the student and annotated instructor's editions) are in the form of alerts, warnings, and reminders to watch for possible errors in solving problems and to clarify misconceptions.

Essential keystrokes are included in the Annotated Instructor's Edition (shown within parentheses in italics). These keystrokes will provide the important values to solve exercises and problems and possibly aid in the instruction process.

Answers to all exercises and problems are shown in the Annotated Instructor's Edition in boldfaced color and italics. Answers to all exercises and odd-numbered Skill and Business Application problems are provided in the back of the student edition. Answers along with essential keystroke solutions and the chapter objective for all end-of-chapter Mastery Tests in the student edition are also shown in the back of the book. The answer to the first exercise is shown along with the essential keystrokes in the student edition.

The Annotated Instructor's Edition of the text places all of the information that you will need for the classroom in a single source. There is no need to carry several items to class and refer from one to the other to lecture, to provide salient comments, and to give detailed solutions to exercises and problems. No more need to fumble around. It is all right here, in this one item!

New to This Edition

All material in each chapter has been updated to include the most recent business practices such as international business scenarios, where applicable, and government regulations, including the minimum wage law changes, payroll tax deductions, and income taxes regulations.

In addition, the chapter on Insurance has been enhanced on the topic of life insurance, including discussion of universal life and variable life. The basis of life insurance rates is explained.

The chapter on Payroll includes an expanded coverage on the topic of F.U.T.A. and state unemployment taxes for the employer.

The chapter on Business Measurements addresses the Euro as a measure of exchange and measure of economic activity.

This edition introduces the term *amount* in the chapter newly titled "Percent." The use of this term will clarify the confusion that surrounds the use of words such as percentage, piece, part, or rate. The term *amount* has been class-tested and has proven to be successful.

The chapter on Installment Loans has been moved ahead of the chapter on Consumer Credit for improved understanding.

Algebriac solutions have been placed in the margin in the chapters on Percent and Markup to aid those who prefer an alternative approach.

A Note about the Sport-A-Merica Case

Though Sport-A-Merica, Inc. is a fictional business, it is based on an actual store operated in the Midwest. The case is designed to provide the student with a scenario of the day-to-day problems faced by small-business owners. The case, introduced chapter by chapter, discusses business problems that pertain specifically to the major emphasis in each chapter. One or more business problems are introduced in each chapter's case discussion. Later, as the chapter material is presented, the problems are solved as part of the worked examples. This will draw the student into the reality of the case and the application of math to business. The chapter can also be presented by itself, without the case.

Supplements for the Instructor

The Annotated Instructor's Edition of *Practical Business Math* provides a complete supplemental package designed to reduce preparatory work. The teaching aids in this package are extensive and will enhance classroom activity and learning.

The Annotated Instructor's Edition provides:

- Answers to all problems at the location of the exercise or problem.
- Essential keystrokes for all exercises, problems, and Mastery Tests at their location to assist in the instruction process.

- Suggested Classroom Time bar graph in each chapter to assist in determining where to spend the most amount of classroom time in the course.
- Perceived Student Difficulty bar graph in each chapter to indicate the areas of the material that have historically been more difficult for students.
- Margin Notations that contain additional information that can be used to enhance the classroom instruction and separately boxed helpful warnings, alerts, and reminders for students.

The Instructor's Resource Section provides:

- A Chapter Interdependencies table that demonstrates the interrelationship of each chapter to the other chapters in the book. This table is very useful for planning purposes, especially for new or adjunct faculty.
- A set of four suggested course outlines that emphasize different areas of study. These outlines will assist in planning which chapters to use in a course outline or syllabus.
- A set of transparency/copy masters that includes the various tables that students would use on tests. These perforated pages can be removed from the book for reproduction purposes. They are ready to copy!
- Mastery Tests A, B, and C in a tear-out ready-to-use form.
- A Final Exam consisting of fifty questions that cover the entire text in a tear-out ready-to-use form.

PH Custom Test for Windows

- A computerized Test Item File (PH Custom Test for Windows) contains 2,200 problems (100 per chapter covering all of the chapter objectives). The problem types include:
  Objective (multiple choice)
  Skill
  Business Application
  Essay

Test Item File

The Test Item File, a printed version of the PH Custom Test for Windows, is also available. This visual tool will assist you in selecting questions you may wish to include on a test. Ask your representative for details.

Companion Website

Invite your students to investigate the Companion Website at www.prenhall.com/tuttle for additional preparation using another vehicle for learning.

Acknowledgments

The creation of a new edition requires the input from many sources. The eighth edition of *Practical Business Math* reflects suggestions and comments from students, adopters, and reviewers.

Additional thanks go to the many instructors who suggested improvements and volunteered valuable suggestions and the many students, especially my own, who provided suggestions to further improve this book.

The four individuals who reviewed this edition deserve a special note of appreciation for their diligent and focused efforts:

Mary Young Bowers, College of Eastern Utah
Dr. James Farris, Golden West College (California)
Arthur Migala
Tom Wiener, Iowa Central Community College

Because I am ever striving to find even the smallest point for clarification, detailed solutions, and accuracy, the final accuracy review by James J. McGinnis, PhD, was extremely valuable. Adopters will appreciate, as I do, his thoroughness.

The editorial staff at Prentice Hall provided the guidance and attention to detail that all authors need. Each member, at every stage of development, was helpful as well as supportive and very organized. Thank you, team!

The revision workload was eased by the valued and extensive contributions of Kathy Tuttle, on her third edition, and, for the fifth time, the proofing skills of Nona Carroll. They have worked to make this the best-ever edition of *Practical Business Math*.

The Sport-A-Merica case involved a lot of work and cooperation from many people. I appreciate the help of the following people in making the story and photos come alive:

Joseph McColly, owner of the Bicyclery
John Molite, manager of Babies•Я•Us
Mary Smith, owner of Wojtusik-Smith & Associates
Donald C. Jacobs, CFP, Prudential
Anthony Marino, Jr., Internal Revenue Service
Decebal Dumitrescu, photo model
Maria Martinez, photo model
Beth Ann Miller, photo model

I know that you will enjoy using this book in your classroom.

To the Student

This book is all about taking very basic math and applying that math to everyday business problems. This is realistic, practical material for future business managers.

If you are one of many people who never felt strong in math, don't feel threatened by this book. This book has been used successfully by business students for more than twenty years. It has a proven track record for making the subjects understandable. The easy-to-follow explanations, worked examples, and exercises work together to help you learn the material without overdoing it! You are first introduced to a bite-size concept, shown how it works, and then you have a chance to put the concept to use. Once you understand that concept, you can move on to the next concept. At the end of the chapter is an assignment section of mixed problems. You can find out how you are doing on the problems at any point by comparing your answers to the answers in the back of the book. There you will also find the essential calculator keystrokes to the Student Mastery Test as well as the objective for the problem. This will allow you to refer back to the part of the chapter that will give you the review you need before you take your test from the instructor. There is even a review chapter to get you ready for the final exam!

As you ready yourself for this course, take the few minutes necessary to learn how this book is designed to make you successful in Business Math.

1. Each chapter opens with a discussion about a business called Sport-A-Merica. In this discussion, some background is provided and a problem is introduced. These are worth reading to give you a feeling of the practical nature of the chapter material.
2. The objectives for the chapter and the page where they appear are shown on the first page of the chapter. This will give you a good cross-reference to use when reviewing the chapter material.
3. Look for the notations in the margin in the chapter. These questions, helpful hints, and warnings will help you solve problems, avoid potential errors, and clarify possible misconceptions. You will also notice business forms located within the chapter. These are business forms of actual companies. They will give you the exposure to the forms that you will use on the job.
4. The chapter is set up to introduce a topic, show an example, and give you an opportunity to practice on some exercises. This piece of the chapter is completed right there in one place.
5. You can assure yourself that you understand the material by checking your answers in the back of the book. You can then go on to the next piece of the chapter.
6. At the end of the chapter you will find the Assignment Section. The problems are arranged in three levels. The Skill problems are similar to those that you completed in the exercises. The Business Application problems are presented in a sentence form. The last problem in the chapter is a Challenge Problem. This problem will be the most difficult problem in the chapter and will challenge you! Answers to the odd-numbered problems are found in the back of the book. Check over your work to see how you are doing.
7. This is a text/workbook. Its design provides adequate room for you to work the exercises and problems in the book. There is no need to carry an extra notebook.

8. At the end of the chapter is a mastery test. This test is a combination of all of the problems in the chapter. It is a good review of the chapter in preparation for the test that your instructor will give you.

This design is used throughout the book. It is a proven approach to learning business math.

Your approach to using this book will make a difference in your academic success. Though your instructor may have some specific instructions for you to follow in the course, you should also consider the following:

1. Read the chapter prior to the class session that will discuss the material. You will have an idea of the topic, a feel for the material, and probably some questions to ask. You may even decide to try a few of the exercises prior to the class session.
2. When your instructor presents the chapter material, you will have a second presentation of the topic. This may clear up the questions that you had when you read the chapter by yourself. If questions remain, now is the time to ask.
3. Do the Assignment Section of the chapter thoroughly. The practice will build confidence and identify any additional questions that you may need to review from the chapter material.
4. Go back to the first page of the chapter. Look at each of the objectives listed on the page. Are you able to do each of the objectives? Do you feel confident that you can pass a test that covers all of the objectives?
5. If you are sure that you understand the chapter material, take the Mastery Test at the end of the chapter without looking back in the chapter for help. You can then check your answers in the back of the book to confirm your understanding.
6. How much time should you spend doing all of this? What is normal? A good place to start is to spend two hours outside of class for every hour that you spend in class. If you do that for yourself, you will probably do the best job academically that you can do.
7. Don't assume that because the arithmetic in this course is very basic that the material is equally easy. This course deals with everyday business activities. Many people who have taken courses similar to this one are doing those activities. They are successful because they learned this material. You can be successful also.

Best wishes to you in Business Math!

Working with Whole Numbers 1

Suggested Classroom Time for This Chapter

hours 1 2 3 4 5

Perceived Student Difficulty with Topical Material

easy moderate difficult

Objectives

After mastering the material in this chapter, you will be able to:

1. Read and write whole numbers. p. 2
2. Round off whole numbers. p. 4
3. Perform the four fundamental arithmetic operations (addition, subtraction, multiplication, and division) of whole numbers with more accuracy and speed. p. 5
4. Work word problems using the four fundamental arithmetic operations. p. 6
5. Understand and use the following terms:
   Place Values
   Addend
   Sum
   Difference
   Minuend
   Subtrahend
   Borrowing
   Multiplicand
   Multiplier
   Product
   Dividend
   Divisor
   Quotient
   Remainder

Sport-A-Merica

Kerry Finn and Hal Balmer are forming Sport-A-Merica, Inc., a retail sports center, specializing in equipment for sports enthusiasts of all ages. Kerry and Hal will form their business as a corporation and will begin in a building located on the main street of the downtown area. The store is near city parking, other retail stores, and city office buildings.

Kerry was first inspired to participate in athletics by her brother Rob. She followed his footsteps in track and cross-country in high school and went on to college on an athletic scholarship in cross-country. Kerry surpassed Rob's performances and competed in many long-distance races, even after college. She was a top finisher in the New York and Boston marathons, in other regional races, and became a hometown heroine.

Kerry first met Hal at a 10-kilometer race. Hal had been active in sports in school and was now a representative of a sportswear firm. Kerry bought a piece of clothing from him. While writing up the sale, they discovered they had attended the same college, although at different times. Hal earned a degree in business and had been in marketing and retail for several years, with most of his experience in sports clothing and athletic shoes.

Following further conversations, Kerry and Hal decided to start their own business in retail sporting goods. Kerry had $50,000 in prize money from some of the races she won. Hal had saved $38,000 from his commissions. They bought a small building for $26,000, purchased merchandise for $41,000, and started their business. Kerry acts as president and makes all the critical decisions—with Hal's advice, of course. She is responsible for sales and public relations. Hal deals with the day-to-day operations of the business. He hires the staff, prices the merchandise, and makes most of the buying decisions. And there are so many decisions to make! With Kerry, Hal, and one sales clerk, Sport-A-Merica is open for business! Kerry and Hal know it will take a great deal of determination to make the business successful.

Kerry and Hal both recognize the importance of good records and the analysis of those records—selling expenses, travel expenses, sales for the month, cost of sales, taxes due, depreciation expenses, checkbook balancing—the list goes on. They realize they need to be on top of the records, as do any employees they may hire. Hal must complete his expense report for the week of October 10, for example. He has the receipts for his meals, travel, and lodging. He expects his reimbursement to be nearly $600. He is also concerned about the check he needs to write to Cindi White for $64. Is there enough money in the checking account? Correct calculations are essential to profits, customer satisfaction, and the success of Sport-A-Merica, Inc.

This chapter will develop your skill in the four fundamental arithmetic operations. These fundamentals are the building blocks for a better understanding of material to be presented later in the text.

Reading and Writing Whole Numbers

place value

Business communications often include numerical information that must be read. To be able to read and write numbers, you must be able to use place values. Each digit in a number, because of its position, has a place value. Once mastered, these **place values** will aid you in reading numbers. The following illustration gives a number with all of its place values labeled.

Objective 1

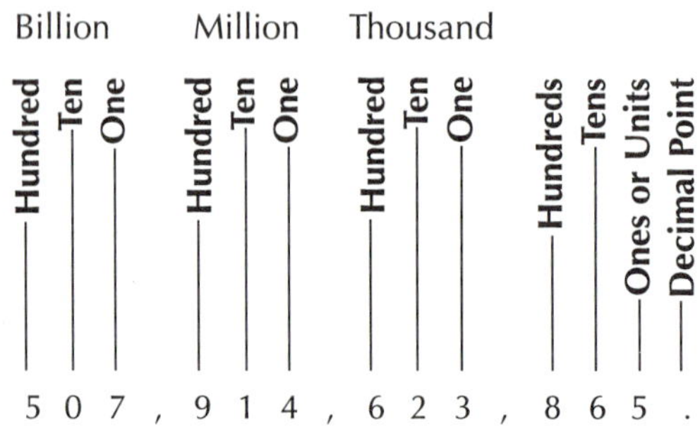

Notice that the number has a comma between each *group* of three digits to the left of the decimal place. The commas break the number into thousands, millions, etc., so that it can be easily read.

The digit in the ten billions place is 0.

The digit in the hundred thousands place is 6.

The digit in the tens place is 6.

The digit in the hundred millions place is 9.

The digit in the ones or units place is 5.

Exercise A

In the number 562,704,831

1. The digit in the hundreds place is 8.
2. The digit in the millions place is *2*.
3. The digit in the tens place is *3*.
4. The digit in the ten thousands place is *0*.
5. The digit in the units place is *1*.

Exercise B

In the number 814,036,725

1. The digit in the thousands place is 6.
2. The digit in the *hundred thousands* place is 0.
3. The digit in the *tens* place is 2.
4. The digit in the *hundreds* place is 7.
5. The digit in the *ten thousands* place is 3.

(Check your progress by turning to the answers at the back of the book.)

With your knowledge of place values, you can read and write numbers using the word description. In the preceding illustration, each group of three digits to the left of the decimal point has a name (i.e., thousands, millions, billions, etc.). To read a number, first read the digit in the leftmost group followed by that group name and continue with this process until all of the digits are read.

The ability to read numbers will enable you to communicate with others in business more successfully. It will also improve your image as a competent student, employee, or manager.

Example 1

The number 46,002 is read: 46 thousand, 2 *or* forty-six thousand, two. Notice that there are no "ands" used in reading the number.

● ***Remind students that no "ands" are used while reading whole numbers.***

Example 2

The number 693,924,806 is read: Six hundred ninety-three million, nine hundred twenty-four thousand, eight hundred six.

Read the following numbers to yourself.

123,066,231	960,012
7,839,850	3,140
7,016	90,008

When reading numbers that are in dollars, add the word "dollars" after the last digit is read.

Example 3

The amount $33,108 is read: 33 thousand, 108 dollars *or* thirty-three thousand, one hundred eight dollars.

Read the following dollar amounts to yourself.

$3,509	$7,750,000
$3,602	$19,436,502,500
$987,503	$51,437

❍ *How many times will you receive or provide monetary values through voice mail or over the phone? This is an important skill!*

Now that you are able to read whole numbers it will be easier for you to write numbers with words. This knowledge will be useful when writing checks. The check form has a place for the amount to be written with words as well as with numbers.

When writing the number, a *comma* is placed between each written portion corresponding to each group of numbers (i.e., thousands, millions, etc.). Values twenty-one through ninety-nine are *hyphenated.*

Example 4

● ***Bankers will use the value written in words if it differs from the value written in numbers on the check. Write carefully and accurately!***

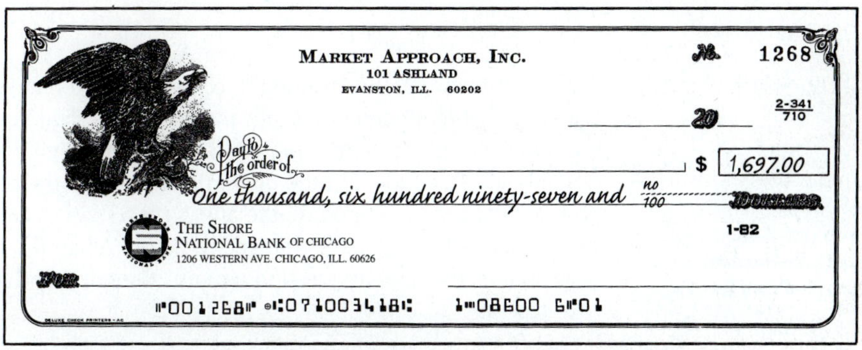

Courtesy of Market Approach, Inc., Evanston IL.

Exercise C

Write in the dollar amounts in the following checks using the value shown on each check. Follow the procedure shown in example 4, check number 1268.

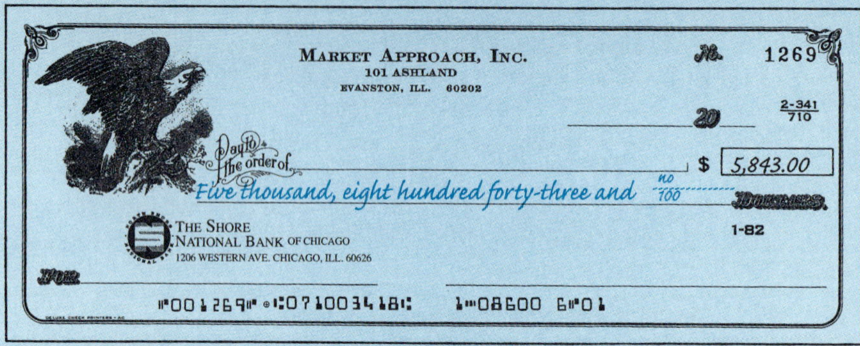

MARKET APPROACH, INC.
101 ASHLAND
EVANSTON, ILL. 60202

No. 1269

20 ___ 2-341/710

Pay to the order of ___ $ 5,843.00

Five thousand, eight hundred forty-three and no/100 DOLLARS

THE SHORE NATIONAL BANK OF CHICAGO
1206 WESTERN AVE. CHICAGO, ILL. 60626

1-82

FOR

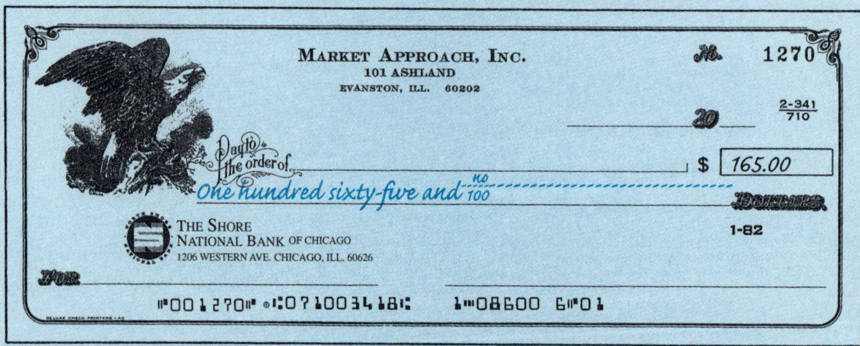

MARKET APPROACH, INC.
101 ASHLAND
EVANSTON, ILL. 60202

No. 1270

20 ___ 2-341/710

Pay to the order of ___ $ 165.00

One hundred sixty-five and no/100 DOLLARS

THE SHORE NATIONAL BANK OF CHICAGO
1206 WESTERN AVE. CHICAGO, ILL. 60626

1-82

FOR

Courtesy of Market Approach, Inc., Evanston IL.

Rounding Off

Objective 2

Accuracy in arithmetic operations is essential and must be a goal of every business student. There are occasions, however, when the answer may need to be rounded off in order for it to be a more useful figure. The rounding off process does not eliminate the need for accurate arithmetic but relies on accuracy in order that the answer may be correctly rounded off. Directions for rounding off usually ask you to round off to the nearest ten, hundred, etc. (See example 5.)

Example 5

"Round off your answer to the nearest ten thousand."

Rounding Rules

- ***Everyday examples of rounding are found in financial statements, sales reports, and income tax returns (no cents).***

If the instructions are to "round off to the nearest ten thousand," you must be concerned with the digit in the ten thousands place and the digit immediately to the right of it, that is, the thousands place. In the number 627,584, the digit 2 is in the ten thousands place. The 2 will be changed to a 3 if the digit immediately to the right of it is a 5 or larger. In this case the digit is a 7; therefore, the answer is 630,000. Notice that all the digits to the right of the rounded number are changed to zeros. When the digit immediately to the right is a 4 or smaller, the digit in question remains the same and the digits to the right are changed to zeros.

Example 6

Digit in question (to be rounded)
Decision-making digit (5 or larger ?)

627,584 = 630,000

If the same number were rounded off to the nearest hundred thousand, then the answer would be 600,000.

Digit in question (to be rounded)
Decision-making digit (5 or larger ?)

627,584 = 600,000

Rounding rules also apply to values on the other side of the decimal point and will be discussed in chapter 3, Working with Decimal Numbers.

Exercise D

Round off the following numbers to the place value indicated.

1. 97,806 to the nearest thousand is 98,000.
2. 2,637 to the nearest ten is *2,640*.
3. 278,901 to the nearest hundred is *278,900*.

Exercise E

1. $1,927,853 rounded to the nearest hundred thousand dollars is $1,900,000.
2. $627,557 rounded to the nearest ten thousand dollars is *$630,000*.
3. $323,306,902 rounded to the nearest ten million dollars is *$320,000,000*.

Addition

Objective 3

sum When adding numbers that have more than one column, you must start with the column farthest to the right, then proceed to the left. The answer is the **sum.**

Example 7

Add the 7 and 4; the sum is 11. Write the 1 under the right column and carry the 10 to the left column. Then add 1 plus 5 plus 9 for a total of 15 tens, giving a **sum,** or answer, of 151.

```
                      1
 57       Addend     57
+94       Addend    +94
---                 ---
          Sum       151
```

(57 + 94 =)

Exercise F

Complete the following problems.

A.

1. 46 + 78 = 124
2. 36 + 17 = ***53***
3. 38 + 74 = ***112***
4. 163 + 56 = ***219***
5. 19 + 71 = ***90***
6. 65 + 41 = ***106***
7. 82 + 17 = ***99***

B.

1. 568 + 204 = ***772***
2. 2,509 + 984 = ***3,493***
3. 8,480 + 6,735 = ***15,215***
4. 735 + 1,294 = ***2,029***
5. 9,450 + 1,205 = ***10,655***

C.

1. $5,432 + 2,684 = ***$8,116***
2. $86,756 + 5,375 = ***$92,131***
3. $9,874 + 6,543 = ***$16,417***
4. $62,747 + 14,154 = ***$76,901***
5. $76,344 + 12,796 = ***$89,140***

D.

1. $3,798 + 6,500 + 47,950 + 56,387 = ***$114,635***
2. $1,699 + 184 + 4,050 + 5,632 = ***$11,565***
3. $450 + 7,699 + 995 + 75,600 = ***$84,744***
4. $14,750 + 6,390 + 2,795 + 4,000 = ***$27,935***
5. $9,950 + 1,795 + 4,350 + 7,885 = ***$23,980***

Horizontal Addition

Occasionally records are kept in such a way that the information can be totaled horizontally as well as vertically. This approach allows for an analysis of the totals for several uses.

Objective 4

Exercise G

Hal has submitted information on the travel expense report for the week of October 10. He has not totaled the amounts for each day, nor for each category. Complete the travel expense report from the information on the form, doing horizontal and vertical addition. How much does Sport-A-Merica owe Hal for his expenses?

SPORT-A-MERICA, INC.
Travel Expense Report

Date: Oct. 10, 1997
For: Hal Balmer

	Monday	Tuesday	Wednesday	Thursday	Friday	Saturday	Sunday	Total
Breakfast		$4	$5	$4				$13
Lunch		3	3					6
Dinner	$18	15	16					49
Travel	186	27	4	136				353
Lodging	38	38	38					114
Tips	3	4	4					11
Misc.				9				9
Total	$245	$91	$70	$149				$555

Horizontal addition is a good way to check your work.

Subtraction

difference
minuend
subtrahend

Even though addition is undoubtedly the most frequently used arithmetic process, in business we must also be concerned with our skill in other fundamental processes. Subtraction is the process of finding the **difference** between two numbers. The larger of the two numbers is termed the **minuend,** and the smaller number is termed the **subtrahend.** The subtrahend is subtracted from the minuend to find the difference.

Example 8

9	Minuend
−5	Subtrahend
4	Difference

(9 − 5 =)

borrowing

The subtraction of larger numbers requires an additional operation, namely, **borrowing.** Borrowing is required when the subtrahend is larger than the minuend in the place value considered. When this occurs, it is necessary to borrow a unit from the place value to the left, adding the unit or ten to the minuend, and subtracting a unit from the value from which it is borrowed. Subtraction then can proceed as in the previous example. An example will help clarify the terms.

Example 9

		2 1	2 1
34	Minuend	~~3~~4	~~3~~4
−18	Subtrahend	−18	−18
	Difference		16

(34 − 18 =)

Checking Addition Answers by Subtraction

In order to be sure of your addition accuracy, it is best to check your answers before proceeding with another problem. The best way to check your answer is by subtraction. After you have found the sum or total of two addends, subtract one addend from the sum or total. Your answer should be the remaining addend.

642	Addend
+927	Addend
1,569	Sum
−927	Addend
642	Addend

Checking Subtraction Answers by Addition

The accuracy of your subtraction can be improved by checking your answers by the use of addition. Although this method is not foolproof, it will help to detect most of the simple errors. This method requires only one step. After you have found the difference between the minuend and the subtrahend, add the difference to the subtrahend. The result should be the minuend.

+ {	854	Minuend
	−687	Subtrahend
	+167	Difference
	854	Minuend (check)

Exercise H

Sport-A-Merica recently fired its accounts payable clerk because she did not keep records in the check register. You are asked, as the new employee, to bring up-to-date some of the records that are not in order. The former employee recorded the amounts of the checks but failed to keep a running balance. Hal, the office manager, explains that the amount of the deposit must be added to the previous balance and that the amount of each check is to be subtracted from the previous balance. The check stubs through check number 1273 are in order. Bring the checkbook up-to-date by following the example shown on check stub number 1268. Is there enough money in the account to pay Cindi White?

Check No.	Date	To	bal. bro't for'd	Deposits	total	amount this check	balance
1268	November 27 19	Thurrel Company	693	260	953	47	906
1269	November 30 19	The Whiting Corp.	906			620	286
1270	December 1 19	Schroeder Supplies	286			100	186
1271	December 2 19	J. J. Originals	186			69	117
1272	December 2 19	Woodman & Assoc.	117			47	70
1273	December 4 19	Cindi White	70	520	590	64	526

Courtesy of Market Approach, Inc., Evanston IL.

Yes, there is enough money to pay Cindi White.

Exercise I

Find the difference in each of the problems below.

A.

1. 65	2. 95	3. 83	4. 71	5. 97	6. 61	7. 98
−46	−27	−59	−29	−69	−45	−36
19	***68***	***24***	***42***	***28***	***16***	***62***

B.

1. \$114	2. \$53	3. \$219	4. \$321	5. \$172	6. \$144
−97	−17	−59	−183	−86	−97
\$ 17	***\$36***	***\$160***	***\$138***	***\$ 86***	***\$ 47***

C.

1. \$653	2. \$247	3. \$1,420	4. \$630	5. \$1,051	6. \$731
−479	−149	−983	−475	−987	−187
\$174	***\$ 98***	***\$ 437***	***\$155***	***\$ 64***	***\$544***

Multiplication

Multiplication is used in business to price stock, compute payroll, and determine the interest due, taxes due, depreciation, and markup amount, along with a variety of other applications. It is really a shortcut approach to addition. Instead of adding a series of values that are all alike, we can multiply the value by the number of repetitions.

multiplicand
multiplier
product

The two numbers used in multiplying are called the **multiplicand** (the number on the top) and the **multiplier** (the number on the bottom). The result of this process is called the **product.**

Example 10

427	Multiplicand
×53	Multiplier
1,281	Partial product (3 × 427)
21 35	Partial product (5 × 427)
22,631	Product (53 × 427)

(427 × 53 =)

Multiplying by 10, 100, 1,000, etc.

Frequently the multiplier or the multiplicand is a 10, 100, or 1,000. The problem can be set up and the product found by the normal multiplication process, but it is much easier and faster to use the following technique.

Example 11

645	Multiplicand
× 100	Multiplier
64,500	Product (645 × 100)

(645 × 100 =)

❍ *You* ***can*** *look good at math!—Note: Count the number of zeros in the multiplier. The same number of additional zeros is found in the product.*

Notice in the example that the product has the same number of zeros in it as the multiplier has.

Example 12

462 × 1,000 = 462,000

(462 × 1,000 =)

Example 13

$98 per share × 100 shares = $9,800 total price

(98 × 100 =)

Exercise J

Complete the following problems using the methods just described.

A.

1.	2.	3.	4.	5.
673 ×100 = 67,300	892 ×100 = ***89,200***	546 ×10 = ***5,640***	891 ×1,000 = ***891,000***	894 ×10 = ***8,940***

B.

1.	2.	3.	4.	5.
819 ×10 = ***8,190***	961 ×100 = ***96,100***	570 ×10 = ***5,700***	469 ×100 = ***46,900***	162 ×1,000 = ***162,000***

C.

1.	2.	3.	4.
$2,871 ×47 = ***$134,937***	$4,712 ×122 = ***$574,864***	$6,534 ×35 = ***$228,690***	$1,927 ×125 = ***$240,875***

D.

1.	2.	3.	4.
$8,965 ×17 = ***$152,405***	$83,935 ×3 = ***$251,805***	$29,171 ×24 = ***$700,104***	$4,782 ×29 = ***$138,678***

Exercise K

George's Hardware Store took inventory on June 30 and found the following merchandise on hand in the paint department. Find the total cost of merchandise on hand in the paint department.

Item Description	Quantity	Cost	Total Cost
Paint			
Pint cans	47	$ 2	$94
Quart cans	93	3	*279*
Gallon cans	89	6	*534*
Stain			
Pint cans	23	2	*46*
Quart cans	36	3	*108*
Brushes			
½-Inch	27	1	*27*
1-Inch	51	2	*102*
2-Inch	40	3	*120*
3-Inch	36	4	*144*
4-Inch	32	6	*192*
Thinner			
Quart cans	41	2	*82*
Gallon cans	26	4	*104*
Rollers			
9-Inch	42	1	*42*
11-Inch	50	2	*100*
Drop Cloths			
9′ × 9′ Plastic	35	2	*70*
12′ × 12′ Plastic	20	3	*60*
12′ × 12′ Cloth	4	16	*64*
12′ × 16′ Cloth	7	18	*126*
Can Holders	25	2	*50*
Cheesecloth	20	1	*20*
Handles	12	1	*12*
Brush Cleaners	16	3	*48*
Wallpapering Tools	12	3	*36*
Paste	17	4	*68*
Total value of merchandise on hand in paint department			*$2,528*

Division

dividend *divisor* *quotient*

Division is the fourth and last of the fundamental operations in arithmetic. Like multiplication is to addition, division is a time-saving approach for subtraction. The number to be divided is called the **dividend,** the number that is used to divide by is called the **divisor,** and the answer or result is the **quotient.**

Example 14

$$\begin{array}{r} 4 \\ 3\overline{)12} \end{array}$$

4 ← Quotient
12 ← Dividend
3 ← Divisor

(12 ÷ 3 =)

The symbol $\overline{)}$ signifies division. $3\overline{)12}$ is read as "3 divided into 12." The division sign (÷) can also be used in the same problem and would be written 12 ÷ 3 or read "12 divided by 3." Notice the key words "divided into" and "divided by." This will become even more important in chapter 3, Working with Decimal Numbers.

Long Division

The problem 5,301 ÷ 57 requires that 57 be first divided into the group of digits farthest to the left of the decimal point. Fifty-seven cannot be divided into 5 or 53, but 57 will divide into 530. Fifty-seven will divide into 530 nine times. The nine is written directly over the last digit to the right in the group (530).

Example 15

```
      9
57)5,301
   5 13
     17
```

(5,301 ÷ 57 =)

remainder Seventeen is left over, or is a **remainder** from dividing 57 into 530. The 1 is brought down as the next digit in the dividend. Now 57 can be divided into 171. We find that 57 divides into 171 three times.

```
     93
57)5,301
   5 13
    171
    171
```

In this example the remainder is zero. If there had been a remainder of, say, 14, the answer would have been stated as the quotient "93 with a remainder of 14."

Example 16

```
   3,076
18)55,385
   54
    1 38
    1 26
      125
      108
       17
```

Answer: 3,076 remainder 17

(55,385 ÷ 18 =)

Dividing by 10, 100, 1,000, etc.

❍ *Now you can look even **better** at math! Note: Count the number of zeros in the divisor. The same number of zeros is removed from the dividend.*

When the divisor is a 10, 100, or 1,000, the technique of multiplication by 10, 100, or 1,000 may be used. The example of 9,853 × 100 was used to show that the zeros are added to the multiplicand in order to find the product 985,300. If the problem was changed to read 985,300 ÷ 100, the answer could be found by removing the two zeros from the divisor 100 and from the dividend and writing the quotient 9,853.

Example 17 64,820,400 ÷ 10 = 6,482,040

(64,820,400 ÷ 10 =)

Example 18 98,800 ÷ 100 = 988

(98,800 ÷ 100 =)

Example 19 1,600,000 ÷ 1,000 = 1,600

(1,600,000 ÷ 1,000 =)

Exercise L

Complete the following problems.

1. $627\overline{)98,439}$ = 157
2. $49\overline{)65,366}$ = *1,334*
3. $801\overline{)94,518}$ = *118*
4. $183\overline{)638,519}$ = *3,489 r 32*
5. $867\overline{)365,007}$ = *421*
6. $37\overline{)75,254}$ = *2,033 r 33*
7. $389\overline{)87,500}$ = *224 r 364*
8. $34\overline{)657}$ = *19 r 11*

Exercise M

Complete the following problems by the shortcut method. A check can always be made by dividing in the normal manner.

1. 793,000 ÷ 100 = 7,930
2. 47,000 ÷ 100 = *470*
3. 90,600 ÷ 100 = *906*
4. 891,000 ÷ 1,000 = *891*
5. 600,500 ÷ 10 = *60,050*

Checking Answers

Division and multiplication problems can be checked easily by using the opposite process to find the answer.

To check a divisor of 7, a dividend of 58, a quotient of 8, and a remainder of 2, multiply the divisor by the quotient and add the remainder to the product.

$$\begin{array}{r} 8 \\ 7\overline{)58} \\ \underline{56} \\ 2 \end{array} \qquad \text{Check} \quad \begin{array}{r} 7 \\ \underline{\times 8} \\ 56 \\ \underline{+2} \\ 58 \end{array}$$

Example 20

$$\begin{array}{r} 6,054 \text{ remainder } 7 \\ 8\overline{)48,439} \end{array} \qquad \text{Check} \quad \begin{array}{r} 6,054 \\ \underline{\times 8} \\ 48,432 \\ \underline{+7} \\ 48,439 \end{array}$$

(48,439 ÷ 8 =)

Multiplication can be checked by dividing the product by the multiplier or the multiplicand. The quotient will be the multiplier or multiplicand, depending on which was used as a divisor.

$$\begin{array}{r} 602 \\ \underline{\times 9} \\ 5,418 \end{array} \qquad \begin{array}{r} 602 \\ 9\overline{)5,418} \\ \underline{5\ 4} \\ 18 \\ \underline{18} \end{array} \qquad \textit{or} \qquad \begin{array}{r} 9 \\ 602\overline{)5,418} \\ \underline{5,418} \end{array}$$

Use of this check method will help you find errors. Accuracy is important!

ASSIGNMENT Chapter 1

Name ______________________ **Date** ______________ **Score** ____________

Skill Problems

In the number 865,903,427

1. The digit in the hundreds place is *4*.

2. The digit in the tens place is *2*.

3. The digit in the units place is *7*.

4. The digit in the millions place is *5*.

In the number 6,423,570,198

5. The digit in the *billions* is 6.

6. The digit in the *hundreds* is 1.

7. The digit in the *hundred thousands* is 5.

8. The digit in the *hundred millions* is 4.

Fill in the dollar values written in words for the following check amounts.

9. $148.00 *One hundred forty-eight and 00/100* DOLLARS.

10. $3,579.00 *Three thousand, five hundred seventy-nine and no/100* DOLLARS.

11. $250,000.00 *Two hundred fifty thousand and no/100* DOLLARS.

Round off the following numbers to the place value indicated.

12. 56,278 to the nearest ten thousand is *60,000*.

13. 5,449 to the nearest ten is *5,450*.

14. 654,729 to the nearest hundred thousand is *700,000*.

Select the answer from the choices below.

A. Place Values
B. Addend
C. Sum
D. Difference
E. Minuend
F. Subtrahend
G. Borrowing
H. Multiplicand
I. Multiplier
J. Product
K. Dividend
L. Divisor
M. Quotient
N. Remainder

15. *D* The result of subtraction.

16. *B* Numbers being added.

17. *G* Using a larger place value to provide enough to subtract from.

18. *C* The amount resulting from the addition of two or more numbers.

19. *I* The bottom number of two numbers that are to be multiplied.

Complete the travel expense report doing horizontal and vertical addition.

	Monday	Tuesday	Wednesday	Thursday	Total
Breakfast	$5	$6	$4	—	**20.** $ *15*
Lunch	—	4	6	—	**21.** *10*
Dinner	19	14	11	$25	**22.** *69*
Travel	257	35	26	281	**23.** *599*
Lodging	65	65	65	—	**24.** *195*
Tips	4	5	4	—	**25.** *13*
Misc.	2	—	3	—	**26.** *5*
Totals	**27.** $ *352*	**28.** $ *129*	**29.** $ *119*	**30.** $ *306*	**31.** $ *906*

Complete the following problems. Check your answers.

32.
```
 11
 532
 +89
 621
 -89
 532
```

33.
```
   1
 9,650
 + 425
10,075
  -425
 9,650
```

34.
```
 71
 $87
 -59
 $28
 +59
 $87
```

35.
```
   91
 1,028
  -584
   444
  +584
 1,028
```

36.
```
     753
  ×1,000
 753,000

            753
 1,000)753,000
```

37.
```
  $206
  × 27
 1,442
  4 12
$5,562

     $206
27)$5,562
    5 4
      16
       0
     162
     162
       0
```

38. 6,603
560
1,450
935
+451
9,999

39. 847 ÷ 62
13 r 41

40. $634
×42
$26,628

41. $658
×42
$27,636

42. $82,861
−17,973
$64,888

43. 42)984
23 r 18

44. $568
3,904
+257
$4,729

45. $82,564
−12,593
$69,971

46. $3,549,560 ÷ 25
$141,982 r $10

47. Round off your answer to the nearest ten.
492 ÷ 6 *= 82 = 80*

48. Round off your answer to the nearest hundred.
5,680
427
11,473
+1,865
19,445 = 19,400

49. Round off your answer to the nearest hundred dollars.
$875
×37
$32,375 = $32,400

50. Write the answer to this problem in words.
$860
×562
$483,320
Four hundred eighty-three thousand, three hundred twenty dollars

51. 239
×16
3,824

52. Find the difference between sixty-two thousand, eight hundred fifty-seven and twenty-nine thousand, six hundred eight.
62,857 − 29,608 = 33,249

53. 472,800 ÷ 10
47,280

54. $6,547
−898
$5,649

55. $568,950
−425,982
$142,968

56. 76)34,656
456

57. $693
×47
$32,571

58. 8)192
24
16
32
32

3
24
× 8
192

59. 64)1,985
31 r 1
1 92
65
64
1

1
64
×31
64
1 92
1,984 + r 1 = 1,985

60. 1,~~000~~)1,900,000,~~000~~
1,900,000

1,000
×1,900,000
1,900,000,000

Business Application Problems

Complete the following problems using the appropriate fundamental operation.

1. The shipping department sent out six packages on Friday. The weights of each of the six packages were 15, 16, 43, 5, 12, and 25 pounds. What was the total weight of the six packages sent?
15 + 16 + 43 + 5 + 12 + 25 = 116 pounds in total

2. The Gift Shop received 40 boxes. Each box contained 144 cards. How many cards did they receive?
40 × 144 = 5,760 cards

3. Amy sold a customer $420 of dishes for use in a restaurant. Each dish was priced at $6. How many dishes were sold? *$420/$6 = 70 dishes sold*

4. Patrick sold 62,431 items the first week of the new year. He also indicated that 1,834 units were returned. How many items were kept by customers?
62,431 − 1,834 = 60,597 units kept

5. Bill estimates the expenses for the month to be as follows for his company: $218 for electricity, $2,200 for rent, $8,950 for labor, $180 for telephone, and $350 for equipment repair. What is the total of the estimated expenses for the month?

$218 + $2,200 + $8,950 + $180 + $350 = $11,898 total expenses

6. A total of 3,592 customers entered The Pet Shop in the month of June. Sales for June totaled $50,288. How many dollars of sales were made for each customer that entered the shop in the month?

$50,288/3,592 = $14 per customer

7. A delivery van used 32 gallons of fuel on Thursday. The van is able to travel 17 miles on each gallon. How many miles did the van travel? *32 × 17 = 544 miles traveled*

8. One of every six units of product produced by Raco, Inc. is destined to be shipped out of the country. Each unit is sold for $23. Raco, Inc. produced 8,412 units in one week. What is the dollar value of the product shipped out of the country in the week?

8,412/6 = 1,402 × $23 = $32,246 value of product

9. There are 27 employees at the Viril Spa. The average wage is $620 a week. What is the total of the wages paid for the week? *$620 × 27 = $16,740 total wages*

10. The voice mail message from the eastern regional sales representative stated that sales were "fifty-eight thousand, seven hundred sixty dollars" for the month of October. Not considered in the message were $985 of merchandise returns by customers. Using words, indicate below the amount of actual sales, after the returns, by "voice-mail" message to the sales representative.

$58,760 − $985 = $57,775 Fifty-seven thousand, seven hundred seventy-five dollars.

11. The income from a design studio must be split evenly among seven partners. The income was $115,486 for the year. What was the share for Patsy, a partner?

$115,486 ÷ 7 = $16,498 for Patsy

12. A lawsuit has been decided in favor of five complainants in the amount of $225,000. Before they can receive their share, the lawyer is entitled to half. What is the amount each complainant will receive after the lawyer's cut?

$225,000 ÷ 2 = $112,500 for the lawyer
$112,500 ÷ 5 = $22,500 each

13. The ABC Party Store had $1,723 in the cash register at the end of a day's business. When they opened for business that morning, there was $437 in the register. Mr. Lee said that $86 of the money was sales tax they had collected for the state. How much did they sell during the day?

$1,723
−437
$1,286 − $86 tax = $1,200 in sales

14. Kerry contributed $50,000 in cash and Hal contributed $38,000 in cash to start Sport-A-Merica. The company bought a building for $26,000 and purchased $41,000 in merchandise, paying cash for both. How much cash does the company have left?

$50,000	*$88,000*
+38,000	*−26,000*
$88,000	*−41,000*
	$21,000 cash left

15. At the beginning of the month, Mr. Harold's business car had 15,975 miles on it. At the end of the month, the odometer read 17,193 miles. He estimated that 147 miles were for personal use. How many miles did he travel on business during the month?

17,193
−15,975
1,218 − 147 personal miles
= 1,071 business miles

16. An employee produced 682 parts in an eleven-hour day. How many parts were produced each hour?

682 ÷ 11 = 62 parts per hour

17. Janie decides to purchase her books for the fall semester. At the College Book Store she finds the prices as follows: math, \$65; English, \$50; history, \$78; and accounting, \$90. She decides that the prices are too high and goes across the street to the Campus Book Store. She finds the prices there as follows: math, \$60; English, \$50; history, \$81; and accounting, \$95. At which store should she buy all her books to get the lowest total price?

College	*Campus*	
\$65	*\$60*	
50	*50*	
78	*81*	
+90	*+95*	
\$283	*\$286*	*College Book Store*

18. The production schedule for the Seldon Manufacturing Corporation calls for 1,645 units to be produced on each shift. The company has a three-shift operation. How many units will be produced in five days of production?

1,645
× 3 shifts
4,935 × 5 days = 24,675 units

19. You are given the task of determining the cost of a banquet for ninety-seven community leaders. The hall will cost \$290. The meal will consist of a steak at an estimated cost of \$10; a baked potato at \$3; vegetable, relish dish, beverage, and roll at \$6. Determine the total cost of the banquet rounded off to the nearest hundred dollars.

\$10
3
+6
\$19 meal × 97 people = \$1,843 food cost
+290 hall
\$2,133 total cost
= \$2,100

20. Alex sold three building lots for \$16,000 each. He also sold a house for \$62,500. What was the total value of the real estate he sold?

\$16,000
×3
\$48,000 + \$62,500 = \$110,500

21. The Municipal Transportation Department bought seven new minibuses at a cost of \$27,300 each. They also bought eleven cars for \$12,200 each. What was the total expenditure?

\$27,300 *\$12,200*
×7 minibuses *×11 cars*
\$191,100 + \$134,200
= \$325,300 total expenditure

22. The Retail Trade Association has decided to create a package to be used by managers for the training of their employees. The cost will be shared equally by each of the 42 members of the association. Each package will consist of 7 CDs at \$14 each; a set of 5 booklets at \$10 each set; stationery supplies at \$1; a hardbound container at \$4; and a set of practice formats at \$42 each. The association has requested that twenty training packages be assembled for each association member. What will be the total cost for twenty training packages?

\$14
×7
\$98 for CDs

\$98 CDs
10 booklets
1 stationery
4 containers
42 formats
\$155 per package × 20 packages = \$3,100 total cost

23. A salesperson must travel a route of 611 miles in order to cover the sales territory. There are twenty-eight stops to make on the route; each stop takes about one hour. Traveling time for the complete route is about eleven hours. If the salesperson wishes to work only eight hours a day, how many days will it take to complete the route?

28 hours for stops
+11 hours for traveling
39 hours for complete route

4 days 7 hours = 5 days
8)39

24. The Stuben Oil Company read the meter on the gasoline storage tank. On Monday, the meter read 4,736,902 gallons; on Wednesday, it read 4,019,456. On Thursday morning a delivery of 550,000 gallons was added to the tank. On Saturday at closing time, the meter read 3,457,995. How many gallons of gas were sold during the week?

4,736,902
−3,457,995
1,278,907 + 550,000 delivery
= 1,828,907 gallons

25. The town of Norden had a population of 47,867 people on January 1 of last year. During the year 427 babies were born; 381 residents died; 188 people, composing 62 families, moved in; and 73 people moved out. What was the population of Norden at the end of the year?

47,867 + 427 + 188 − 381 − 73
= 48,028 persons

26. The inventory of Spectrum, Inc., totaled 842 parts of item J926. The cost of the inventory was \$13,472. What was the cost per unit?

\$13,472 ÷ 842 = \$16 per unit

27. A resort found that 473 clients had been there last year. This year, they lost 239 customers from last year but recruited 357 new customers. How many customers did they have this year?

473
−239
+357
591 customers this year

28. Write a complete sentence in answer to this question: Each of the 49 restaurants in the town of Witon served 167 meals per hour for the five hours that preceded the homecoming game. How many meals were served in Witon in the five-hour period?

167
×49
8,183 × 5 hours = 40,915 meals
There were 40,915 meals served in the town of Witon in a five-hour period.

29. Write a complete sentence in answer to this question: New lounge chairs for a pool area will cost $127 each. The Recreation Department has decided on a purchase of twenty-three chairs. What will be the cost of the purchase?

$127
×23
$2,921 total
The cost of twenty-three chairs will be $2,921.

30. Write a complete sentence in answer to this question: The Food Mart has decided to add 42 items to their existing inventory of 1,219 items. At the same time, they will discontinue 81 items. How many items will they carry in inventory?

1,219 + 42 − 81 = 1,180 items
The number of items in the new inventory will be 1,180.

★ Challenge Problem

At the beginning of the month, the Gene Robins Company had 422 units of a product, "Talson," on hand. The company purchased another 960 pieces during the month at a cost of $1.40 per unit. Forty pieces of the new shipment were found to be defective and were returned to the seller. During the month, the company sold 865 units after having 14 returned by customers because of poor packing. If the company made a profit of $1.23 on each unit sold, (a) how much will be made on the units already sold and (b) how much would be made if all of the acceptable units remaining were sold?

a. *865*
×$1.23
$1,063.95 already sold

b. *422*
+960
− 40
−865
477 × $1.23
= $586.71 yet to be sold

MASTERY TEST Chapter 1

Name ______________________ **Date** ____________ **Score** __________

1. Find the product of the following numbers: 64 and 850.
54,400

2. \$ 1,457
5,800
3,575
\+ 41
\$10,873

3. \$ 42,803
\+ 787,276
\$830,079

4. 18)75,859
4,214 r 7

5. The profit of the Edrunton Company must be evenly distributed between the six department managers. The profit for the quarter was \$64,638. What will each department manager receive as his or her share of the profits?
\$64,638 ÷ 6 = \$10,773 share per manager

6. 549 × 187 = *102,663*

7. 768 ÷ 40 = *19 r 8*

8. The Sales Department of the Wolton Corporation reported the following figures for May: marine equipment—\$6,185,005; automotive equipment—\$54,980; service repair parts—\$7,681; and miscellaneous—\$1,822. What were the total sales for the month?
\$6,185,005
54,980
7,681
\+ 1,822
\$6,249,488 sales for the month

9. Jay Moore produced 891 parts on his machine for the week. One hundred forty-seven of those were determined to be oversize and, therefore, not part of the acceptable lot. Another 63 failed a strength test set up by the Quality Department. How many of the parts were acceptable (to the nearest ten)?
891
−147
− 63
681 = 680 acceptable

10. The answering service tape has recorded a message from the Western Sales Region Manager, A. F. Litona. His statement indicates that sales of the "Aferton" product were "eighty-five thousand, six hundred five dollars," the "Virmonon" product "two hundred thousand, thirty-eight dollars," and the "Olsand" product "forty-one thousand, eighty dollars." What were the total sales for the period?
\$ 85,605
200,038
41,080
\$326,723 total sales

11. Find the total difference between the following: 651 units at \$387 each and 120 units at \$456 each.

\$ 387	*\$ 456*	*\$251,937*
× 651	*× 120*	*− 54,720*
\$251,937	*\$54,720*	*\$197,217 difference*

12. Write the following value using words: \$43,679.
Forty-three thousand, six hundred seventy-nine dollars

13. Subtract and round off your answer to the nearest ten thousand dollars.
\$459,073
− 91,752
\$367,321 = \$370,000

14. Mrs. Jesop knows that the salespeople in her group will be successful on one out of every two sales calls that they make. If each call takes two hours and each successful call develops a \$98 order, what will be the volume of sales from a sales force of twelve people, each working forty hours a week?
40 hours ÷ 2 = 20 calls;
20 calls ÷ 2 = 10 successful calls per person
× \$98 order
\$980 per person
× 12 people
\$11,760 sales per week

15. Round off 685,224 to the nearest hundred.
685,200

(If you missed problems 2, 3, and 8, you need to review addition; problems 9, 11, or 13, you need to review subtraction; problems 1, 6, or 14, you need to review multiplication; problems 4, 5, or 7, you need to review division; problems 10 or 12, you need to review reading and writing whole numbers; and problems 13 or 15, you need to review rounding off.) Record your score on the Record of Performance on Mastery Tests in the back of the book.

Working with Fractions

2

Suggested Classroom Time for This Chapter

hours 1 2 3 4 5

Perceived Student Difficulty with Topical Material

easy moderate difficult

Objectives

After mastering the material in this chapter, you will be able to:

1. Identify fraction types. p. 22
2. Change mixed numbers to improper fractions, and improper fractions to mixed numbers. p. 23
3. Reduce fractions to their lowest terms. p. 24
4. Perform the basic fundamental processes (addition, subtraction, multiplication, and division) with accuracy in problems dealing with fractions. p. 25
5. Develop a ratio from data provided. p. 34
6. Understand and use the following terms:
   Denominator
   Numerator
   Common Fraction
   Improper Fraction
   Mixed Numbers
   Prime Numbers
   Common Denominator
   Lowest Common Denominator (LCD)
   Ratio

Sport-A-Merica

The selection of the store Kerry and Hal agreed to purchase took into consideration the many types of equipment Sport-A-Merica will handle. Hal knew there would need to be space for seasonal merchandise. Other space would have to be set aside for equipment, end-of-season sale merchandise, an office area, a receiving area, and an employee lounge.

When Kerry saw the first building Hal located, she asked, "How much space will we have for baseball equipment?" Kerry knew that there were a lot of baseball leagues in the immediate area. By knowing that one-fifth of the space will be set aside for baseball equipment, Hal was able to give her an answer. As a result, they turned down the first building because the floor space would not allow enough room for baseball equipment once the other space needed was also taken into consideration. Knowing that three-fourths of the baseball business would involve the baseball leagues, this building would not have been a good choice.

Kerry and Hal looked further. Hal found the right place—one that offered 1,000 square feet, the proper space to allow for each area needed in the store. Hal is happy that he had considered the floor space and had applied fractions to the planning. Now he must determine the number of square feet to be used for baseball leagues. He will also want to know the ratio of office floor space to sales floor. Kerry is sure to ask.

Description and Types

Objective 1

A fraction can be described as a part of a whole. This description fits many business situations. Mastery of fractions and their applications is therefore required.

Example 1

The fraction $\frac{3}{4}$ states that the whole is divided into four equal parts. Three of the four parts make up the portion that the fraction is describing. The fraction can be presented graphically in a variety of forms, each describing the same part of a whole, $\frac{3}{4}$.

3/4 = $\frac{3}{4}$ =

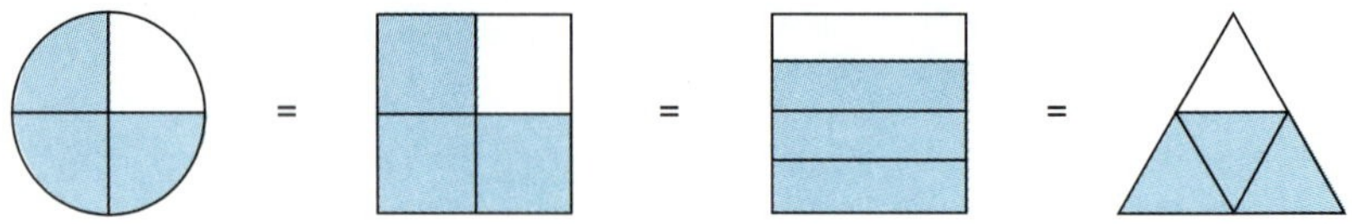

denominator *numerator*

The bottom portion of the fraction that describes the number of equal parts is called the **denominator.** The top portion that specifies the number of equal parts that the fraction is describing is called the **numerator.**

A fraction can be read in two different ways. The fraction $\frac{7}{8}$ can be read as "seven eighths" or as "seven divided by eight," since the line between the numerator and the denominator signifies division. Either a diagonal line / or a horizontal line — can be used to separate the numerator and the denominator.

Example 2

$\frac{72}{146}$ can be read as "seventy-two one hundred forty-sixths" or as "seventy-two divided by one hundred forty-six."

A third way to read a fraction is to describe how it looks (i.e., "seventy-two over one hundred forty-six"). This method is also acceptable.

Read the following fractions to yourself.

$\frac{9}{16}$	$\frac{41}{90}$	$\frac{5}{8}$	$\frac{4}{7}$	$\frac{16}{21}$	$\frac{193}{216}$
$\frac{17}{20}$	$\frac{1}{4}$	$\frac{10}{17}$	$\frac{1}{3}$	$\frac{3}{4}$	$\frac{7}{16}$
$\frac{17}{651}$	$\frac{5}{16}$	$\frac{5}{55}$	$\frac{3}{8}$	$\frac{21}{109}$	$\frac{7}{21}$

common fraction *improper fraction*

There are three types of fractions. The type that we have been working with is called a **common fraction.** This type has a denominator (bottom number) larger than the numerator (top number). An **improper fraction** has a numerator larger than or equal to the denominator, as shown in example 3.

Example 3

$\frac{19}{6}$ $\frac{40}{10}$ $\frac{93}{16}$ $\frac{3}{3}$ $\frac{19}{9}$ $\frac{8}{4}$ $\frac{6}{5}$

mixed numbers

Mixed numbers contain a whole number as well as a common fraction, as shown in example 4.

Example 4

$4\frac{1}{3}$ $7\frac{1}{2}$ $14\frac{63}{91}$ $5\frac{1}{7}$ $1\frac{9}{16}$

In most instances, fractions will be presented as common fractions or as mixed numbers. Improper fractions are found in arithmetic operations and as temporary answers. They are not acceptable as final answers unless specifically asked for in a problem.

Exercise A

Identify the following by type of fraction: common, mixed, or improper.

1. $\frac{9}{13}$ common
2. $\frac{46}{17}$ *improper*
3. $\frac{1{,}107}{1{,}000}$ *improper*
4. $\frac{41}{35}$ *improper*
5. $16\frac{2}{9}$ *mixed*
6. $\frac{4}{7}$ *common*
7. $\frac{11}{9}$ *improper*
8. $\frac{76}{81}$ *common*
9. $\frac{1}{4}$ *common*
10. $\frac{4}{5}$ *common*
11. $4\frac{1}{10}$ *mixed*
12. $16\frac{2}{3}$ *mixed*

Interchanging Fraction Types

Objective 2

It is sometimes necessary to interchange fraction types in order to perform the fundamental processes or to find the correct answer. All answers in the improper fraction form should be changed to a mixed number or a whole number.

Example 5

$$\begin{array}{r} 2 \\ 4\overline{)9} \\ \underline{8} \\ 1 \end{array}$$

The improper fraction $\frac{9}{4}$ can be changed to a mixed number by dividing nine by four, just as the fraction is read. The result is that there are two whole number integers (two 4s) in $\frac{9}{4}$ and one-fourth left over. The whole number is written. The remainder is then written over the divisor. $2\frac{1}{4}$

(9 ÷ 4 =)

Example 6

$$\frac{197}{9} = \begin{array}{r} 21 \\ 9\overline{)197} \\ \underline{18} \\ 17 \\ \underline{9} \\ 8 \end{array} = 21\frac{8}{9}$$

(197 ÷ 9 =)

Exercise B

Change the following improper fractions to whole or mixed numbers.

1. $\frac{13}{8} = 1\frac{5}{8}$
2. $\frac{67}{12} =$ *$5\frac{7}{12}$*
3. $\frac{14}{5} =$ *$2\frac{4}{5}$*
4. $\frac{47}{14} =$ *$3\frac{5}{14}$*
5. $\frac{19}{6} =$ *$3\frac{1}{6}$*
6. $\frac{5}{2} =$ *$2\frac{1}{2}$*
7. $\frac{19}{9} =$ *$2\frac{1}{9}$*
8. $\frac{93}{10} =$ *$9\frac{3}{10}$*
9. $\frac{43}{10} =$ *$4\frac{3}{10}$*
10. $\frac{9}{8} =$ *$1\frac{1}{8}$*
11. $\frac{16}{4} =$ *4*
12. $\frac{57}{8} =$ *$7\frac{1}{8}$*

If an improper fraction can be changed to a mixed number, then we should be able to do the reverse process (i.e., change a mixed number to an improper fraction).

Example 7

The mixed number $4\frac{3}{5}$ can be changed to an improper fraction by multiplying the whole number (4) by the number of equal parts in each unit (the denominator 5) and adding the numerator to the product.

$$4\frac{3}{5} = \frac{(5 \times 4) + 3}{5} = \frac{23}{5}$$

(4 × 5 = + 3 =)

Example 8

$$7\frac{4}{10} = \frac{(7 \times 10) + 4}{10} = \frac{74}{10}$$

(7 × 10 = + 4 =)

Exercise C

Change the following mixed numbers to improper fractions.

1. $6\frac{3}{4} =$ $\frac{27}{4}$
2. $9\frac{5}{6} =$ $\frac{59}{6}$
3. $7\frac{4}{5} =$ $\frac{39}{5}$
4. $1\frac{7}{8} =$ $\frac{15}{8}$
5. $110\frac{1}{5} =$ $\frac{551}{5}$
6. $3\frac{1}{2} =$ $\frac{7}{2}$
7. $18\frac{3}{8} =$ $\frac{147}{8}$
8. $6\frac{2}{3} =$ $\frac{20}{3}$
9. $4\frac{3}{4} =$ $\frac{19}{4}$
10. $1\frac{1}{2} =$ $\frac{3}{2}$
11. $5\frac{7}{16} =$ $\frac{87}{16}$
12. $10\frac{1}{2} =$ $\frac{21}{2}$

Reducing Fractions

Objective 3

Answers to problems in fractions are correct only if they are reduced to lowest terms. The part of a whole that is described as $\frac{3}{4}$ can also be described as $\frac{6}{8}$, $\frac{12}{16}$, $\frac{24}{32}$, etc. In each of these examples, the numerator has the same relationship to the denominator as it does in the fraction $\frac{3}{4}$. The fraction $\frac{3}{4}$ can be changed to $\frac{6}{8}$ by doubling both the numerator and the denominator. The fraction $\frac{6}{8}$ can be changed to $\frac{12}{16}$ by following the same process again.

In each case, the numerator and the denominator are being multiplied by an equal amount (2). Any fraction can be changed in this manner. The fraction $\frac{3}{7}$ can be changed to $\frac{15}{35}$ by multiplying both the numerator and denominator by 5.

You should also be able to do the opposite operation (i.e., divide both the numerator and denominator by the same number). When both the numerator and denominator are divided by 5, $\frac{15}{35}$ equals $\frac{3}{7}$. When the numerator and denominator no longer can be divided by a whole number other than 1, the fraction has been reduced to lowest terms.

prime numbers

How do you know when the fraction has been reduced to lowest terms? The best approach is to use **prime numbers** as divisors. When the prime numbers cannot be evenly divided into the numerator and denominator, the fraction is fully reduced. A prime number is divisible only by itself and by one (1). The smaller prime numbers are 2, 3, 5, 7, 11, and 13.

Trial-and-Error Approach

A trial-and-error approach can be used to reduce fractions. This method states that the prime numbers are used to find a number that will divide into both the numerator and denominator. It is helpful to use a system that will make this approach easy to follow. Start with the smallest prime number and divide it into both the numerator and the denominator. The first prime number that divides evenly is a common divisor. If you are unsuccessful with the first prime number, go to the next highest prime number until you find one that works. Repeat this process until no prime number will divide evenly. The fraction will then be reduced to its lowest terms.

Example 9

The fraction $\frac{12}{45}$ must be reduced.

Try prime number 2. $12 \div 2 = 6 \qquad 45 \div 2 = 22 \text{ r } 1$

Two does not divide evenly and therefore is not a common divisor. Try prime number 3. $12 \div 3 = 4 \qquad 45 \div 3 = 15$

Three is a prime number that evenly divides into 12 and 45, and therefore, it is a common divisor. The fraction is reduced to $\frac{4}{15}$.

(12 ÷ 3 = 45 ÷ 3 =)

Exercise D

Reduce the following fractions to their lowest terms.

1. $\frac{5}{10} = \frac{1}{2}$
2. $\frac{9}{12} = \frac{3}{4}$
3. $\frac{14}{20} = \frac{7}{10}$
4. $\frac{40}{100} = \frac{2}{5}$
5. $\frac{18}{36} = \frac{1}{2}$
6. $\frac{14}{16} = \frac{7}{8}$
7. $\frac{22}{50} = \frac{11}{25}$
8. $\frac{3}{6} = \frac{1}{2}$
9. $\frac{6}{8} = \frac{3}{4}$
10. $\frac{9}{15} = \frac{3}{5}$
11. $\frac{16}{22} = \frac{8}{11}$
12. $\frac{6}{16} = \frac{3}{8}$

Exercise E

Change the following fractions to fractions of equal value using the denominator indicated.

1. $\frac{3}{4} = \frac{15}{20}$
2. $\frac{3}{5} = \frac{33}{55}$
3. $\frac{9}{11} = \frac{90}{110}$
4. $\frac{7}{20} = \frac{28}{80}$
5. $\frac{3}{7} = \frac{15}{35}$
6. $\frac{9}{10} = \frac{27}{30}$
7. $\frac{4}{5} = \frac{16}{20}$
8. $\frac{1}{2} = \frac{5}{10}$
9. $\frac{41}{100} = \frac{410}{1{,}000}$
10. $\frac{6}{7} = \frac{12}{14}$
11. $\frac{5}{12} = \frac{15}{36}$
12. $\frac{1}{4} = \frac{5}{20}$

Addition and Subtraction

Objective 4

Fractions, like whole numbers, can be added in order to find the sum, or total.

Example 10

$$\frac{5}{13} + \frac{3}{13} = \frac{8}{13}$$

(5 + 3 =)

The denominator must be the same in each fraction in order to add them. Only the numerators are added.

Common Denominator

In the previous example the addends had **common denominators** (both denominators were 13s in the example given). When the denominators of a fraction to be added are not the same, they must be made alike before the addition can begin.

Product Approach

A common denominator can always be found by multiplying the denominators. The product, or common denominator, becomes the denominator of both fractions. The next step is to change the numerator to give an equivalent fraction.

Example 11

$\frac{3}{4} + \frac{1}{5} =$ (4 × 5 = 20, the common denominator)

To find the numerator, divide the old denominator into the common denominator and multiply the quotient by the numerator for each fraction.

$$\left[\frac{3 \times 5}{4 \times 5}\right] + \left[\frac{1 \times 4}{5 \times 4}\right] = \frac{15}{20} + \frac{4}{20} = \frac{19}{20}$$

Note: Remember to reduce the answer to lowest terms.

(20 ÷ 4 = 20 ÷ 5 =)

lowest common denominator (LCD)

Another way to determine a common denominator is to find the **lowest common denominator (LCD).** This will reduce the need for dealing with unnecessarily large fractions when performing additional functions. To use this approach, select the largest denominator and test multiples of it against the other denominators until the other denominators divide equally into the multiple of the first denominator.

Example 12

Find the lowest common denominator for the following fractions.

$$\frac{5}{6} + \frac{3}{8} =$$

The largest denominator is 8. Six will not divide equally into 8; therefore 8 is added to the original 8 to make 16. Six will not divide equally into 16; therefore 8 is added to 16 to make 24. Six will evenly divide into 24; therefore, 24 is the lowest common denominator.

$$\frac{20}{24} + \frac{9}{24} = \frac{29}{24} = 1\frac{5}{24}$$

(24 ÷ 6 = 24 ÷ 8 =)

The same approach is used when more than two fractions are to be added.

Example 13

$\frac{2}{3} + \frac{1}{2} + \frac{3}{5}$ 30 is the common denominator.

$$\left[\frac{2 \times 10}{3 \times 10}\right] + \left[\frac{1 \times 15}{2 \times 15}\right] + \left[\frac{3 \times 6}{5 \times 6}\right] =$$

$$\frac{20}{30} + \frac{15}{30} + \frac{18}{30} = \frac{53}{30}$$

$\frac{53}{30}$ reduces to $1\frac{23}{30}$

that is, $30\overline{)53}$ = 1
$\underline{30}$
23 or $\frac{23}{30}$

(30 ÷ 3 = 30 ÷ 2 = 30 ÷ 5 =)

Exercise F

Find the sum in the following problems.

1. $\frac{1}{6} + \frac{2}{3} + \frac{1}{4} =$

   $\frac{2}{12} + \frac{8}{12} + \frac{3}{12} =$

   $\frac{13}{12} = 1\frac{1}{12}$

2. $\frac{7}{8} + \frac{1}{2} + \frac{3}{4} =$

   $\frac{7}{8} + \frac{4}{8} + \frac{6}{8} = \frac{17}{8} = 2\frac{1}{8}$

3. $\frac{2}{7} + \frac{1}{5} + \frac{2}{3}$

   $\frac{30}{105} + \frac{21}{105} + \frac{70}{105}$

   $= \frac{121}{105} = 1\frac{16}{105}$

4. $\frac{7}{8} + \frac{3}{4} + \frac{3}{4} =$

   $\frac{7}{8} + \frac{6}{8} + \frac{6}{8} = \frac{19}{8} = 2\frac{3}{8}$

5. $\frac{1}{10} + \frac{1}{2} + \frac{3}{7} =$

   $\frac{7}{70} + \frac{35}{70} + \frac{30}{70} = \frac{72}{70} = 1\frac{2}{70} = 1\frac{1}{35}$

6. $\frac{9}{10} + \frac{3}{8} =$

   $\frac{36}{40} + \frac{15}{40} = \frac{51}{40} = 1\frac{11}{40}$

7. $\frac{7}{20} + \frac{1}{3} + \frac{1}{2} =$

   $\frac{21}{60} + \frac{20}{60} + \frac{30}{60} = \frac{71}{60} = 1\frac{11}{60}$

8. $\frac{2}{5} + \frac{5}{16} + \frac{3}{4}$

   $\frac{32}{80} + \frac{25}{80} + \frac{60}{80} = \frac{117}{80} = 1\frac{37}{80}$

9 $\frac{7}{16} + \frac{3}{8} + \frac{1}{4} =$

   $\frac{7}{16} + \frac{6}{16} + \frac{4}{16} = \frac{17}{16} = 1\frac{1}{16}$

10. $\frac{3}{7} + \frac{5}{6} + \frac{2}{3} =$

   $\frac{18}{42} + \frac{35}{42} + \frac{28}{42} = \frac{81}{42} = 1\frac{39}{42} = 1\frac{13}{14}$

The subtraction of common fractions is similar to addition, because both require common denominators. Subtraction requires finding the difference of the numerators.

Example 14 $\frac{3}{4} - \frac{1}{4} = \frac{2}{4}$, which reduces to $\frac{1}{2}$

(4 ÷ 2 =)

Example 15

$\frac{1}{3} - \frac{2}{9} =$

$\left[\frac{1 \times 3}{3 \times 3}\right] - \frac{2}{9} =$

$\frac{3}{9} - \frac{2}{9} = \frac{1}{9}$

(1 × 3 =)

Example 16

$\frac{3}{4} - \frac{1}{5} =$

$\left[\frac{3 \times 5}{4 \times 5}\right] - \left[\frac{1 \times 4}{5 \times 4}\right] =$

$\frac{15}{20} - \frac{4}{20} = \frac{11}{20}$

(3 × 5 = 1 × 4 =)

Exercise G

Find the differences in the following problems.

1. $\frac{3}{4} - \frac{1}{8} =$
   $\frac{6}{8} - \frac{1}{8} = \frac{5}{8}$
2. $\frac{5}{8} - \frac{1}{4} =$
   $\frac{5}{8} - \frac{2}{8} = \frac{3}{8}$
3. $\frac{9}{10} - \frac{2}{3} =$
   $\frac{27}{30} - \frac{20}{30} = \frac{7}{30}$
4. $\frac{1}{4} - \frac{1}{5} =$
   $\frac{5}{20} - \frac{4}{20} = \frac{1}{20}$
5. $\frac{17}{20} - \frac{2}{9} =$
   $\frac{153}{180} - \frac{40}{180} = \frac{113}{180}$
6. $\frac{7}{8} - \frac{5}{12} =$
   $\frac{21}{24} - \frac{10}{24} = \frac{11}{24}$
7. $\frac{4}{9} - \frac{1}{4} =$
   $\frac{16}{36} - \frac{9}{36} = \frac{7}{36}$
8. $\frac{2}{3} - \frac{1}{4} =$
   $\frac{8}{12} - \frac{3}{12} = \frac{5}{12}$
9. $\frac{7}{10} - \frac{1}{16} =$
   $\frac{56}{80} - \frac{5}{80} = \frac{51}{80}$
10. $\frac{1}{2} - \frac{3}{16} =$
    $\frac{8}{16} - \frac{3}{16} = \frac{5}{16}$

Mixed Numbers

Mixed numbers, like common fractions, can be added to find the sum. This process consists of three steps.

Example 17

In the mixed numbers $8\frac{3}{8} + 2\frac{1}{8}$, (1) the whole numbers are added; (2) then the fractions are added; and, as a last step, (3) the mixed-number answer is reduced.

$$\begin{array}{r} 8\frac{3}{8} \\ +2\frac{1}{8} \\ \hline 10\frac{4}{8} \end{array} = 10\frac{1}{2} \text{ reduced}$$

(8 + 2 = 3 + 1 =)

This three-step process is lengthened when the fractions of the mixed numbers have different denominators.

Example 18

$4\frac{2}{3} + 6\frac{5}{8} =$

A common denominator must be found, as it is when adding common fractions.

$$\begin{array}{rcr} 4\frac{2}{3} & = & 4\frac{16}{24} \\ +\,6\frac{5}{8} & = & +6\frac{15}{24} \\ \hline & & 10\frac{31}{24} \end{array} = 11\frac{7}{24} \text{ reduced}$$

(4 + 6 = 16 + 15 =)

Exercise H

Find the sum in the following problems. Remember to reduce your answers to lowest terms.

1. $6\frac{3}{4} + 8\frac{1}{2} =$

   $6\frac{3}{4} + 8\frac{2}{4} = 14\frac{5}{4}$

   $= 15\frac{1}{4}$

2. $9\frac{2}{3} + 4\frac{5}{8} =$

   $9\frac{16}{24} + 4\frac{15}{24} = 13\frac{31}{24}$

   $= 14\frac{7}{24}$

3. $3\frac{5}{16} + 1\frac{3}{4} =$

   $3\frac{5}{16} + 1\frac{12}{16} = 4\frac{17}{16}$

   $= 5\frac{1}{16}$

4. $8\frac{1}{3} + 2\frac{5}{8} =$

   $8\frac{8}{24} + 2\frac{15}{24} = 10\frac{23}{24}$

5. $7\frac{1}{2} + 4\frac{3}{10} =$

   $7\frac{5}{10} + 4\frac{3}{10} = 11\frac{8}{10} = 11\frac{4}{5}$

6. $5\frac{3}{8} + 3\frac{5}{9} =$

   $5\frac{27}{72} + 3\frac{40}{72} = 8\frac{67}{72}$

7. $17\frac{3}{50} + 7\frac{3}{4} =$

   $17\frac{6}{100} + 7\frac{75}{100} = 24\frac{81}{100}$

8. $1\frac{7}{8} + 6\frac{5}{6} =$

   $1\frac{21}{24} + 6\frac{20}{24} = 7\frac{41}{24} = 8\frac{17}{24}$

9. $2\frac{3}{4} + 8\frac{1}{2} + 4\frac{2}{3} =$

   $2\frac{9}{12} + 8\frac{6}{12} + 4\frac{8}{12} = 14\frac{23}{12} = 15\frac{11}{12}$

10. $9\frac{1}{6} + 5\frac{3}{8} + 3\frac{1}{4} = 9\frac{4}{24} + 5\frac{9}{24} + 3\frac{6}{24} = 17\frac{19}{24}$

The subtraction of mixed numbers uses a similar process.

Example 19

$$\begin{array}{rcr} 9\frac{5}{6} & = & 9\frac{5}{6} \\ -1\frac{1}{3} & = & -1\frac{2}{6} \\ \hline & & 8\frac{3}{6} = 8\frac{1}{2} \text{ reduced} \end{array}$$

(9 − 1 = 5 − 2 =)

Example 20

$$\begin{array}{r} 19\frac{1}{10} \\ -16\frac{7}{10} \\ \hline \end{array}$$

In this case, a 1, or ten tenths, must be borrowed from the 19 in order to complete the subtraction.

$$\begin{array}{rl} 18\frac{11}{10} & \left(18 + \frac{10}{10} + \frac{1}{10}\right) \\ -16\frac{7}{10} & \\ \hline 2\frac{4}{10} & = 2\frac{2}{5} \text{ reduced} \end{array}$$

(18 − 16 = 11 − 7 =)

Exercise I

Complete the following subtraction problems. Reduce your answers to lowest terms.

1. $14\frac{2}{3} - 7\frac{1}{3} = 7\frac{1}{3}$

2. $3\frac{3}{8} - 1\frac{1}{8} =$

   $3\frac{3}{8} - 1\frac{1}{8} = 2\frac{2}{8} = 2\frac{1}{4}$

3. $6\frac{3}{8} - 1\frac{1}{10} =$

   $6\frac{15}{40} - 1\frac{4}{40} = 5\frac{11}{40}$

4. $59\frac{1}{3} - 6\frac{3}{8} =$

   $59\frac{8}{24} - 6\frac{9}{24} =$

   $58\frac{32}{24} - 6\frac{9}{24} = 52\frac{23}{24}$

5. $5\frac{7}{10} - 3\frac{1}{3} =$

   $5\frac{21}{30} - 3\frac{10}{30} = 2\frac{11}{30}$

6. $6\frac{1}{8} - 1\frac{1}{10} =$

   $6\frac{5}{40} - 1\frac{4}{40} = 5\frac{1}{40}$

7. $7\frac{3}{4} - 1\frac{1}{6} =$

   $7\frac{9}{12} - 1\frac{2}{12} = 6\frac{7}{12}$

8. $173\frac{1}{2} - \frac{1}{3} =$

   $173\frac{3}{6} - \frac{2}{6} = 173\frac{1}{6}$

9. $9\frac{3}{4} - 6\frac{2}{3} =$

   $9\frac{9}{12} - 6\frac{8}{12} = 3\frac{1}{12}$

10. $8\frac{1}{9} - 1\frac{1}{2} = 8\frac{2}{18} - 1\frac{9}{18} = 7\frac{20}{18} - 1\frac{9}{18} = 6\frac{11}{18}$

Multiplication and Division

The multiplication of common fractions is simpler than addition or subtraction. There is no need for a common denominator.

Common Fractions

The denominators are multiplied, and the numerators are multiplied. The product should be reduced as in the previous fraction problems.

Example 21

$$\frac{5}{6} \times \frac{4}{7} = \frac{20}{42} = \frac{10}{21} \text{ reduced}$$

(5 × 4 = 6 × 7 =)

Example 22

Hal has decided to set aside one-fifth of the Sport-A-Merica store floor space for baseball equipment. Three-fourths of that space will be used for baseball leagues. How much of the 1,000-square-foot store will be used for baseball leagues?

$$\frac{1}{5} \times \frac{1{,}000}{1} \times \frac{3}{4} = \frac{3{,}000}{20} = 150 \text{ square feet}$$

(1 × 3 × 1,000 = 5 × 4 × 1 =)

❍ *A whole number can be changed into a fraction by placing a denominator of one under the whole number. After all, the whole number is only an expression of so many units or ones.*

A shortcut or work-saving step can be used to reduce the amount of multiplication. By reducing the fractions before the multiplication begins, the factors are made smaller and the product is also much smaller.

Example 23

$$\frac{9}{\cancel{10}_{5}} \times \frac{\cancel{2}^{1}}{5} = \frac{9}{25} \quad \text{(Both 10 and 2 are divisible by 2.)}$$

(9 × 1 = 5 × 5 =)

Reducing can take place any time you can divide any numerator and any denominator by the same number.

Example 24 $\frac{3}{4} \times \frac{4}{9} = \frac{\overset{1}{\cancel{3}}}{4} \times \frac{4}{\underset{3}{\cancel{9}}} = \frac{1}{\underset{1}{\cancel{4}}} \times \frac{\overset{1}{\cancel{4}}}{3} = \frac{1}{3}$

(1 × 1 =   1 × 3 =)

Example 25 $\frac{6}{11} \times \frac{1}{42} = \frac{\overset{1}{\cancel{6}}}{11} \times \frac{1}{\underset{7}{\cancel{42}}} = \frac{1}{77}$

(1 × 1 =   11 × 7 =)

A series of fractions may be multiplied just as in the two previous examples. Reducing before multiplication is still advisable.

Example 26 $\frac{3}{4} \times \frac{7}{8} \times \frac{1}{9} = \frac{\overset{1}{\cancel{3}}}{4} \times \frac{7}{8} \times \frac{1}{\underset{3}{\cancel{9}}} = \frac{7}{96}$

(1 × 7 × 1 =   4 × 8 × 3 =)

Example 27 $\frac{9}{10} \times \frac{3}{4} \times \frac{8}{9} = \frac{\overset{1}{\cancel{9}}}{\underset{5}{\cancel{10}}} \times \frac{3}{\underset{1}{\cancel{4}}} \times \frac{\overset{\overset{1}{\cancel{2}}}{\cancel{8}}}{\underset{1}{\cancel{9}}} = \frac{3}{5}$

(1 × 3 × 1 =   5 × 1 × 1 =)

❍ *Fractions will be used to compute the solutions of interest problems and depreciation problems.*

Exercise J

Find the product in the following problems. Reduce your answers.

1. $\frac{3}{16} \times \frac{8}{9} =$

   $\frac{\overset{1}{\cancel{3}}}{\underset{2}{\cancel{16}}} \times \frac{\overset{1}{\cancel{8}}}{\underset{3}{\cancel{9}}} = \frac{1}{6}$

2. $\frac{7}{16} \times 2\frac{1}{5} =$

   $\frac{7}{16} \times \frac{11}{5} = \frac{77}{80}$

3. $\frac{7}{8} \times \frac{5}{7} =$

   $\frac{\overset{1}{\cancel{7}}}{8} \times \frac{5}{\underset{1}{\cancel{7}}} = \frac{5}{8}$

4. $\frac{5}{7} \times \frac{1}{6} =$

   $\frac{5}{7} \times \frac{1}{6} = \frac{5}{42}$

5. $\frac{2}{3} \times \frac{3}{4} =$

   $\frac{\overset{1}{\cancel{2}}}{\underset{1}{\cancel{3}}} \times \frac{\overset{1}{\cancel{3}}}{\underset{2}{\cancel{4}}} = \frac{1}{2}$

6. $\frac{5}{9} \times \frac{7}{10} =$

   $\frac{\overset{1}{\cancel{5}}}{9} \times \frac{7}{\underset{2}{\cancel{10}}} = \frac{7}{18}$

(Continued on next page)

Exercise J Continued

7. $\frac{7}{16} \times 8 \times \frac{13}{14} =$

$\frac{\overset{1}{\cancel{7}}}{\underset{2}{\cancel{16}}} \times \frac{\overset{1}{\cancel{8}}}{1} \times \frac{13}{\underset{2}{\cancel{14}}} = \frac{13}{4} = 3\frac{1}{4}$

8. $4\frac{3}{4} \times 5 \times \frac{5}{6} =$

$\frac{19}{4} \times \frac{5}{1} \times \frac{5}{6} = \frac{475}{24} = 19\frac{19}{24}$

9. $\frac{5}{8} \times \frac{5}{9} \times \frac{6}{7} =$

$\frac{5}{\underset{4}{\cancel{8}}} \times \frac{5}{\underset{3}{\cancel{9}}} \times \frac{\overset{\overset{1}{\cancel{2}}}{\cancel{6}}}{7} = \frac{25}{84}$

10. $\frac{13}{21} \times \frac{3}{10} \times \frac{5}{16} =$ $\frac{13}{\underset{7}{\cancel{21}}} \times \frac{\overset{1}{\cancel{3}}}{\underset{2}{\cancel{10}}} \times \frac{\overset{1}{\cancel{5}}}{16} = \frac{13}{224}$

Division

- *No common denominator is necessary to multiply **or** to divide fractions.*
- *To divide, turn the divisor upside down and then multiply!*

Division of common fractions requires a few additional instructions. The first step is to review the way in which a division problem is read. In chapter 1 on whole numbers, you found that a problem such as 18 ÷ 6 is read "eighteen divided by six" or "six divided into eighteen." Eighteen is the dividend and six is the divisor. In the problem $\frac{3}{5} \div \frac{1}{2}$, $\frac{3}{5}$ is the dividend and $\frac{1}{2}$ is the divisor. You must be able to identify the divisor in order to be able to do the additional steps required in the division of fractions.

In the example $\frac{3}{5} \div \frac{1}{2}$, the divisor $\frac{1}{2}$ must be inverted (i.e., tipped upside down) and the (÷) sign changed to a (×) sign. You then continue as you did in the multiplication problems.

Example 28

$\frac{3}{5} \div \frac{1}{2} =$

$\frac{3}{5} \times \frac{2}{1} = \frac{6}{5} = 1\frac{1}{5}$

(3 × 2 = 5 × 1 =)

Example 29

$\frac{9}{10} \div \frac{2}{5} =$

$\frac{9}{\underset{2}{\cancel{10}}} \times \frac{\overset{1}{\cancel{5}}}{2} = \frac{9}{4} = 2\frac{1}{4}$

(9 × 1 = 2 × 2 =)

Exercise K

Find the quotient in the following problems. Reduce your answers.

1. $\frac{6}{7} \div \frac{1}{2} =$

$\frac{6}{7} \times \frac{2}{1} = \frac{12}{7} = 1\frac{5}{7}$

2. $\frac{9}{10} \div \frac{3}{5} =$

$\frac{\overset{3}{\cancel{9}}}{\underset{2}{\cancel{10}}} \times \frac{\overset{1}{\cancel{5}}}{\underset{1}{\cancel{3}}} = \frac{3}{2} = 1\frac{1}{2}$

3. $\frac{3}{4} \div \frac{7}{16} =$

$\frac{3}{\underset{1}{\cancel{4}}} \times \frac{\overset{4}{\cancel{16}}}{7} = \frac{12}{7} = 1\frac{5}{7}$

4. $\frac{5}{8} \div \frac{1}{4} =$

$\frac{5}{\underset{2}{\cancel{8}}} \times \frac{\overset{1}{\cancel{4}}}{1} = \frac{5}{2} = 2\frac{1}{2}$

5. $\frac{3}{8} \div \frac{2}{9} =$

$\frac{3}{8} \times \frac{9}{2} = \frac{27}{16} = 1\frac{11}{16}$

6. $\frac{6}{7} \div \frac{2}{3}$

$\frac{\overset{3}{\cancel{6}}}{7} \times \frac{3}{\underset{1}{\cancel{2}}} = \frac{9}{7} = 1\frac{2}{7}$

Exercise K Continued

7. $\frac{7}{10} \div \frac{1}{5} =$

$\frac{7}{\cancel{10}_{2}} \times \frac{\cancel{5}^{1}}{1} = \frac{7}{2} = 3\frac{1}{2}$

8. $\frac{3}{4} \div \frac{1}{8} =$

$\frac{3}{\cancel{4}_{1}} \times \frac{\cancel{8}^{2}}{1} = \frac{6}{1} = 6$

9. $\frac{5}{7} \div \frac{1}{6} =$

$\frac{5}{7} \times \frac{6}{1} = \frac{30}{7} = 4\frac{2}{7}$

10. $\frac{5}{6} \div \frac{2}{3} = \frac{5}{\cancel{6}_{2}} \times \frac{\cancel{3}^{1}}{2} = \frac{5}{4} = 1\frac{1}{4}$

Mixed Numbers

The multiplication of mixed numbers requires an additional step (i.e., changing the mixed numbers to improper fractions).

Example 30

$1\frac{1}{6} \times 3\frac{1}{2} =$

$\frac{7}{6} \times \frac{7}{2} = \frac{49}{12} = 4\frac{1}{12}$

$(7 \times 7 = \quad 6 \times 2 =)$

Example 31

$2\frac{9}{10} \times 5\frac{1}{6} =$

$\frac{29}{10} \times \frac{31}{6} = \frac{899}{60} = 14\frac{59}{60}$

$(29 \times 31 = \quad 10 \times 6 =)$

The division of mixed numbers requires the same step (i.e., changing the mixed numbers to improper fractions).

Example 32

$4\frac{1}{2} \div 3\frac{3}{8} =$

$\frac{9}{2} \div \frac{27}{8} =$

$\frac{\cancel{9}^{1}}{\cancel{2}_{1}} \times \frac{\cancel{8}^{4}}{\cancel{27}_{3}} = \frac{4}{3} = 1\frac{1}{3}$

$(1 \times 4 = \quad 1 \times 3 =)$

Exercise L

Find the product or quotient in the following problems. Reduce your answers to lowest terms.

1. $6\frac{1}{2} \times 3\frac{3}{16} \times 5\frac{1}{8} =$

$\frac{13}{2} \times \frac{51}{16} \times \frac{41}{8} =$

$\frac{27{,}183}{256} = 106\frac{47}{256}$

2. $8\frac{1}{4} \div 1\frac{3}{4} =$

$\frac{33}{\cancel{4}_{1}} \times \frac{\cancel{4}^{1}}{7} = \frac{33}{7} = 4\frac{5}{7}$

3. $6\frac{3}{4} \div 1\frac{5}{8} =$

$\frac{27}{\cancel{4}_{1}} \times \frac{\cancel{8}^{2}}{13} = \frac{54}{13} = 4\frac{2}{13}$

(Continued on next page)

Exercise L Continued

4. $4\frac{2}{3} \div 1\frac{1}{6} =$

$\frac{\overset{2}{\cancel{14}}}{\underset{1}{\cancel{3}}} \times \frac{\overset{2}{\cancel{6}}}{\underset{1}{\cancel{7}}} = 4$

5. $5\frac{1}{6} \div 2\frac{4}{5} =$

$\frac{31}{6} \times \frac{5}{14} = \frac{155}{84} = 1\frac{71}{84}$

6. $10\frac{1}{4} \div 3\frac{4}{5} =$

$\frac{41}{4} \times \frac{5}{19} = \frac{205}{76} = 2\frac{53}{76}$

7. $5\frac{6}{7} \div \frac{5}{9} =$

$\frac{41}{7} \times \frac{9}{5} = \frac{369}{35} = 10\frac{19}{35}$

8. $5\frac{4}{9} \div 3\frac{3}{5} =$

$\frac{49}{9} \times \frac{5}{18} = \frac{245}{162} = 1\frac{83}{162}$

Increases and Decreases

❍ *Watch the wording on these examples!*

Business and government leaders frequently make comparisons of sales, revenues, employees, etc. The way in which the comparison is made is very important. The wording is the key to understanding the change, or difference.

Example 33

The sales of Sport-A-Merica were \$400,000 in 1999. In 2000 the sales were \$500,000, a \$100,000 difference. This could be expressed as "an increase of $\frac{1}{4}$" or "$1\frac{1}{4}$ of last year's sales." Both mean that sales increased by \$100,000. Be careful to watch the wording in problems that mention increase or decrease.

Example 34

If the sales of Sport-A-Merica were to go from \$500,000 in 2000 to \$400,000 in 2001, the information could be communicated by stating that sales decreased "by $\frac{1}{5}$" or "to $\frac{4}{5}$ of last year's sales." Again, in either method of stating the information, the sales decreased by \$100,000 compared to the earlier period's sales of \$500,000.

Ratios
Objective 5

● ***Use quotations from current local business publications to show how these comparisons can be expressed.***

A **ratio** is the comparison of one value to another. This may be expressed as a fraction such as $\frac{5}{1}$. In this case, 5 is being compared to 1. The number on the bottom is the value that is being compared to; it is the basis of comparison. The ratio may also be written as 5:1. The colon replaces the — sign. The ratio is still read as "5 to 1."

This ability to compare one value to another is useful in many business situations. A comparison can be made of the number of male to female employees in a firm, the amount of profits to sales over a period, and the number of hours of machine time to the number of hours of labor to perform a task.

Example 35

Sport-A-Merica had the following initial allocation of space in their first store layout.

Soft goods display = 340 square feet
Equipment display = 560 square feet
Office space = 100 square feet

What is the ratio of office space to the sales display area?

Total display area = 900 square feet
Total office area = 100 square feet

The ratio is 100:900, or 1:9 (reduced)

(100 ÷ 100 = 900 ÷ 100 =)

Example 36

There are eighty employees in the Accounting Department. Ten of the employees are in managerial positions. What is the ratio of nonmanagerial employees to managerial employees?

80 − 10 = 70

Ratio of 7 to 1

(70 ÷ 10 = 10 ÷ 10 =)

Exercise M

Use the following data to develop the ratios in this exercise.
Number of employees in firm = 400
Number of employees 40 years old or older = 60
Number of males = 240
Number of employees with 10 or more years of experience = 320
Number of employees that are in clerical positions = 80
Number of nonclerical employees in the firm = 320
Number of employees 50 years old or older = 20

1. Ratio of males to females = 3:2. (240:160)
2. Ratio of those between 40 and 50 years old to the total = *1:10*. *(60 − 20 = 40 40:400)*
3. Ratio of nonclerical to clerical employees = *4:1*. *(320:80)*
4. Ratio of all employees to male employees = *5:3*. *(400:240)*
5. Ratio of all employees to female employees = *5:2*. *(400:160)*
6. Ratio of employees who have fewer than 10 years to those who have 10 or more years of experience = *1:4*. *(80:320)*

ASSIGNMENT Chapter 2

Name ______________________ Date ______________ Score ____________

Skill Problems

Identify the following type of fraction: common, mixed, or improper.

1. 8/5 improper **2.** 1/4 common **3.** 2 1/2 mixed **4.** 101/52 improper

Change the improper fraction to a common or mixed fraction.

5. 34/4 8 1/2 **6.** 12/5 2 2/5 **7.** 50/4 12 1/2 **8.** 146/144 1 1/72

Change the mixed numbers to improper fractions.

9. 5 1/3 16/3 **10.** 16 2/9 146/9 **11.** 6 6/7 48 **12.** 15 3/4 63/4

Reduce the fractions to the lowest terms.

13. 18/24 3/4 **14.** 25/50 1/2 **15.** 40/100 2/5 **16.** 30/90 1/3
17. 5/4 1 1/4 **18.** 84/8 10 1/2 **19.** 24/10 2 2/5 **20.** 220/25 8 4/5

Change the following fractions to fractions of equal value using the denominator indicated.

21. 3/4 = 12/16 **22.** 3/7 9/21 **23.** 6/7 18/21 **24.** 2/3 22/33

Match the letter with the definition.

A. Denominator B. Numerator C. Common Fraction D. Improper Fraction
E. Mixed Numbers F. Prime Numbers G. Common Denominator
H. Ratio I. Lowest Common Denominator (LCD)

25. A The bottom portion of the fraction that describes the number of equal parts.
26. C Fraction with a denominator larger than the numerator.
27. H The comparison of one value to another.

Complete the following time card. (Hint: Use number of minutes over 60 for fraction.)

Day	Time In	Time Out	Time In	Time Out	Hours for Day
Monday	8:00 am	11:30 am	12:00 pm	5:00 pm	8 1/2 hours
Tuesday	8:30 am	11:30 am	12:30 pm	5:00 pm	**28.** 7 1/2 hours
Wednesday	8:00 am	12:00 pm	12:30 pm	5:45 pm	**29.** 9 1/4 hours
Thursday	8:15 am	12:00 pm	1:00 pm	5:00 pm	**30.** 7 3/4 hours
Friday	8:00 am	1:00 pm	2:00 pm	7:15 pm	**31.** 10 1/4 hours
Total Week	*********	*********	*********	*********	**32.** 43 1/4 hours

Answer the following questions using the time card data and the following information.

The employee is supposed to punch in work at 8:00 am, receive an hour for lunch, and punch out at 5:00.

33. What is the ratio of days worked in excess of 8 hours to less than 8 hours? 3 : 2
34. What is the ratio of full lunch hours to partial lunch hours? 3 : 2
35. What is the ratio of timely departures to late departures? 3 : 2

Complete the following problems. Reduce your answer to lowest terms.

36. 1/6 + 2/9 = *3/18 + 4/18 = 7/18*

37. 8/11 − 7/13 = *104/143 − 77/143 = 27/143*

38. 5 3/4 + 8 2/3 = *5 + 8 = 13, 3/4 + 2/3 = 9/12 + 8/12 = 17/12 = 1 5/12, 13 + 1 5/12 = 14 5/12*

39. 6/7 × 4/5 = *24/35*

40. 9/5 ÷ 3/10 = *9/5 × 10/3 = 90/15 = 6*

41. $\frac{3}{4} + \frac{2}{3} =$

$\frac{9}{12} + \frac{8}{12} = \frac{17}{12} = 1\frac{5}{12}$

42. $\frac{9}{10} \times \frac{7}{8} =$

$\frac{9}{10} \times \frac{7}{8} = \frac{63}{80}$

43. $\frac{5}{8} + \frac{5}{6} =$

$\frac{15}{24} + \frac{20}{24} = \frac{35}{24} = 1\frac{11}{24}$

44. $\frac{3}{4} + \frac{7}{8} + \frac{1}{3} =$

$\frac{18}{24} + \frac{21}{24} + \frac{8}{24} = \frac{47}{24} = 1\frac{23}{24}$

45. $\frac{7}{10} - \frac{2}{3} =$

$\frac{21}{30} - \frac{20}{30} = \frac{1}{30}$

46. $1\frac{1}{4} \times \frac{7}{8} =$

$\frac{5}{4} \times \frac{7}{8} = \frac{35}{32} = 1\frac{3}{32}$

47. $2\frac{5}{8} - \frac{2}{5} =$

$2\frac{25}{40} - \frac{16}{40} = 2\frac{9}{40}$

48. $\frac{7}{8} + \frac{1}{4} + 1\frac{3}{5} =$

$\frac{35}{40} + \frac{10}{40} + \frac{64}{40} = \frac{109}{40} = 2\frac{29}{40}$

49. $5\frac{3}{10} - 1\frac{5}{8} =$

$4\frac{52}{40} - 1\frac{25}{40} = 3\frac{27}{40}$

50. $6\frac{1}{3} + 5 + 3\frac{3}{8} =$

$6\frac{8}{24} + 5 + 3\frac{9}{24} = 14\frac{17}{24}$

51. $2\frac{1}{4} - \frac{7}{8} =$

$1\frac{10}{8} - \frac{7}{8} = 1\frac{3}{8}$

52. $\frac{4}{5} + \frac{3}{8} + \frac{1}{2} + \frac{3}{5} =$

$\frac{32}{40} + \frac{15}{40} + \frac{20}{40} + \frac{24}{40} = \frac{91}{40} = 2\frac{11}{40}$

53. $\frac{3}{4} \times \frac{7}{8} =$

$\frac{3}{4} \times \frac{7}{8} = \frac{21}{32}$

54. $3\frac{1}{6} \div 2\frac{3}{5} =$

$\frac{19}{6} \times \frac{5}{13} = \frac{95}{78} = 1\frac{17}{78}$

55. $2\frac{1}{6} - \frac{4}{5} =$

$1\frac{35}{30} - \frac{24}{30} = 1\frac{11}{30}$

56. $\frac{4}{5} \times \frac{3}{8} \times \frac{1}{2} =$

$\frac{4}{5} \times \frac{3}{8} \times \frac{1}{2} = \frac{3}{20}$

57. $\frac{5}{8} \times \frac{4}{5} =$

$\frac{5}{8} \times \frac{4}{5} = \frac{20}{40} = \frac{1}{2}$

58. $3\frac{2}{5} \times \frac{3}{4} =$

$\frac{17}{5} \times \frac{3}{4} = \frac{51}{20} = 2\frac{11}{20}$

59. $\frac{3}{4} \div \frac{5}{8} =$

$\frac{3}{4} \times \frac{8}{5} = \frac{24}{20} = \frac{6}{5} = 1\frac{1}{5}$

60. $\frac{3}{8} \times \frac{5}{6} \times \frac{3}{4} =$

$\frac{3}{8} \times \frac{5}{6} \times \frac{3}{4} = \frac{15}{64}$

61. $\frac{1}{5} \times \frac{3}{4} \times \frac{8}{9} \times \frac{4}{5} =$

$\frac{1}{5} \times \frac{\overset{1}{\cancel{3}}}{\underset{1}{\cancel{4}}} \times \frac{\overset{2}{\cancel{8}}}{\underset{3}{\cancel{9}}} \times \frac{4}{5} = \frac{8}{75}$

62. $4\frac{1}{2} + 3\frac{2}{5} + \frac{7}{8} =$

$4\frac{20}{40} + 3\frac{16}{40} + \frac{35}{40} = 7\frac{71}{40} = 8\frac{31}{40}$

63. $2\frac{5}{6} - \frac{3}{4} =$

$2\frac{10}{12} - \frac{9}{12} = 2\frac{1}{12}$

64. $6 + \frac{5}{8} + \frac{2}{3} =$

$6 + \frac{15}{24} + \frac{16}{24} = 6\frac{31}{24} = 7\frac{7}{24}$

Business Application Problems

1. The fleet of trucks averaged 12 1/4 miles a gallon. The fleet used 316 gallons of fuel for the week. How many miles could be driven by the fleet? *12 1/4 × 316 = 3,871 miles*

2. The value of an item in stock increased by 1/5 over one year. The value of the item at the beginning of the year was $42,800. What is the amount of the increase in value? *1/5 × $42,800 = $8,560 increase*

3. Don's Upholstery used 4 3/4 yards of material to cover a chair. Another 1 2/3 yards were used to cover a foot stool. How much material was used in total? *4 3/4 + 1 2/3 = 5 17/12 = 6 5/12 used*

4. There are 44 men in the sales division in addition to 64 women. What is the ratio of men to women? *44:64 = 11:16 men to women*

5. Of the 560 runners entered in a race, 3/7 are women. How many women are entered? *560 × 3/7 = 240 women*

6. Sonny will need to purchase 38 packages of potato salad to be used at a luncheon for a group of seniors. Each package of potato salad will weigh 2/3 of a pound. How much potato salad will he need to purchase for the event? *38 × 2/3 = 25 1/3 pounds*

7. A wedding gown will require 7 yards of material at $36 a yard plus 4 1/3 yards of lace at $81 a yard. What is the cost of the material for the gown? *7 × $36 + 4 1/3 × $81 = $603 for material*

8. Returned products totaled 52 units of the 3,796 units that were initially sold last year. What is the ratio of returns to initial sales for the year? *52:3,796 = 1:73 returns to sales*

9. If there were approximately two million "millionaires" out of approximately 272,000,000 people in the United States, what is the ratio of millionaires to the general population? *2,000,000:272,000,000 = 2:272 = 1:136*

10. Allison earns an overtime rate of 1 1/2 times her regular rate of pay. Her regular rate of pay is $9.50. What is her overtime rate? *1 1/2 × $9.50 = $14.25*

11. A "1/4 off" sale has been offered by a storage company for the month of July. If the regular price is $28 for the 4′ × 8′ space, what is the sale price? *$28 × 3/4 = $21*

12. Pat Hough increased her sales by 1/10 over last month. Sales last month were $14,630. How much was the increase? *$14,630 × 1/10 = $1,463*

13. Lillian McTole found she should devote 1/4 of her available time to studying her business courses. She has three business courses that she feels deserve equal attention. Lillian has twenty-four hours of free time over the weekend. How much time should she spend on each business course?
24/1 × 1/4 = 6
6 × 1/3 = 2 hours per course

14. Amy Coulter worked the following hours during last week: Monday, 6 3/10 hours; Tuesday, 8 3/10 hours; Wednesday, 9 hours; Thursday, 8 1/10 hours; and Friday, 9 1/10 hours. How many total hours did she work?
6 3/10 + 8 3/10 + 9 + 8 1/10 + 9 1/10 = 40 8/10 = 40 4/5 hours

15. A desk and accessories sold for $300 five years ago. A replacement would cost $2,100 today. What is the ratio of today's price to the price of five years ago? ***7:1***

16. A man's coat was priced at $188. The proprietor of Jule and Justin's Men's Store decided to have a 1/4-off sale of all the merchandise in the store for the month. What would be the sale price on the coat?
$188 × 3/4 = $141

17. Big Time Movers increased their revenue by 2 1/2 times over the revenue of last year. Last year's revenue was $560,000. What is the increase in the revenue for this year in dollars?
$560,000 × 5/2 = $1,400,000

18. Jim worked 2 3/10 hours in the morning and later returned to work for 5 1/2 hours. How many total hours did he work for the day?
2 3/10 + 5 5/10 = 7 8/10 = 7 4/5 hours

19. There are 488 employees at Filmer Corporation. Of the total, 61 are more than 50 years old. What is the ratio of all employees to those who are over 50? ***8:1***

20. An item increased in value to five times its price in one year. The current price is $465. To the nearest ten dollars, what was the price one year ago?
$465 × 1/5 = $93 = $90

21. Bert earns $12 per hour. When she works overtime, she is paid at 1 1/2 times her regular rate. What is her overtime rate?
$12 × 1 1/2 = $12 × 3/2 = $18 overtime rate

22. The delivery truck for Teller's Tuxedo & Tailoring ran the following route Tuesday morning: to the cleaners, 4/10 mile; to the post office, 7 1/10 miles; to the trucking company, 5 3/10 miles; to the store, 6 9/10 miles; to the gas station, 1/10 mile; to the coffee shop, 2 miles; back to the shop, 1 1/10 miles. What was the total mileage for the delivery truck that morning?
4/10 + 7 1/10 + 5 3/10 + 6 9/10 + 1/10 + 2 + 1 1/10 = 21 19/10 = 22 9/10 miles

23. The electronics department uses 1/5 of the space of a store that pays $2,500 rent per month. What portion of the rent should be charged to the electronics department?
$2,500 × 1/5 = $500

24. A piece of yard goods must be cut into lengths of 2 1/4 feet each. The bolt of material is 500 feet long. How many whole pieces can be cut from the bolt?
500 ÷ 2 1/4 = 500 × 4/9 = 222 2/9 = 222 whole pieces

25. One-fifth of the college's enrollment is made up of juniors and seniors. How many are freshmen and sophomores if the total enrollment is 38,110?
38,110 × 4/5 = 30,488 freshmen and sophomores

26. According to recent U.S. Census data, a city's population has grown to 1 2/7 of its previous size. At the previous census, the population was 187,684. What is the present size?
187,684 × 1 2/7 = 241,308 citizens

27. The hours worked by employees in the shipping department were as follows:

A. Rimshaw	39 1/10
P. Borgstun	36 3/10
W. Billings	38 3/10
F. Stalworth	37 7/10
B. Feller	39 9/10
H. Tolunmar	36 4/10

Total the hours for the department.
39 1/10 + 36 3/10 + 38 3/10 + 37 7/10 + 39 9/10 + 36 4/10 = 225 27/10 = 227 7/10 hours

28. The cost of manufacturing a flavoring used in the food industry is made up of 1/4 labor cost and 3/4 material cost. The flavoring in its finished form sells for $180 per ton. What is the dollar amount of the labor factor in the production of 81 tons of the substance?

$ 180
×81
$14,580 × 1/4 = <u>$3,645 for labor</u>

29. Write a complete sentence in answer to this question. Of the 650 units of a housing development, four out of five were pre-sold prior to the start of construction. How many remain to be sold?

650 × 1/5 = 130
<u>There are 130 units that are unsold.</u>

30. Write a complete sentence in answer to this question. The turnover rate of employees at C.D.C. is 9 out of 10 over a period of one year. If the year began with 450 employees, how many will still be employed at the end of the year?

450 × 1/10 = 45
<u>There will be 45 still employed at the end of the year.</u>

★ Challenge Problem

The sales of Bilkton Marketing Corporation are three times the sales of a year ago. The increase in sales, one-half of which is due to Internet advertising, was welcome. The Board of Directors indicated that 1/4 of the increase will be paid out in dividends. The dividend would amount to $2.20 a share for each of the 1,630,000 shares. What was the amount of sales last year?

1,630,000 × $2.20 = $3,586,000 × 4
= $14,344,000 increase
$14,344,000 ÷ 2 = <u>$7,172,000 last year's sales</u>

MASTERY TEST Chapter 2

Name ______________________ **Date** ______________ **Score** ____________

1. Label the fraction parts below.

whole number → $8\frac{4}{7}$ ← *numerator* (4), *denominator* (7)

2. Identify the following types of numbers.

A. Common Fraction
B. Mixed Number
C. Improper Fraction

B 1. $6\frac{1}{3}$

A 2. $\frac{3}{4}$

C 3. $\frac{5}{4}$

A 4. $\frac{7}{8}$

B 5. $4\frac{2}{3}$

C 6. $\frac{15}{10}$

3. Add $\frac{3}{4} + \frac{4}{5}$.

$\frac{15}{20} + \frac{16}{20} =$

$\frac{31}{20} = 1\frac{11}{20}$

4. Subtract $\frac{2}{3}$ from $1\frac{1}{2}$.

$1\frac{1}{2} - \frac{2}{3} =$

$1\frac{3}{6} - \frac{4}{6} =$

$\frac{9}{6} - \frac{4}{6} = \frac{5}{6}$

5. Multiply $15 \times 3\frac{1}{3}$.

$15 \times \frac{10}{3} =$

$\frac{150}{3} = 50$

6. Divide $\frac{7}{8}$ by $\frac{1}{5}$.

$\frac{7}{8} \div \frac{1}{5} =$

$\frac{7}{8} \times \frac{5}{1} =$

$\frac{35}{8} = 4\frac{3}{8}$

7. Reduce $\frac{16}{3}$.

$\frac{16}{3} = 5\frac{1}{3}$

8. Sales for Campbell & Associates have increased by $\frac{1}{2}$ over last quarter's sales of $98,480. What are the sales for this quarter?

\$98,480 × $\frac{3}{2}$ = \$147,720 sales this quarter

9. Nine hundred eighty-six citizens of a small city are more than sixty-two years old. The city's population is 8,874. What is the ratio of all of the citizens to those over sixty-two?

9:1

10. James Roberts worked $8\frac{1}{2}$ hours on Wednesday. During the day he spent $3\frac{1}{10}$ hours on the phone following up on orders that he had placed earlier in the month. How many hours did he spend off of the phone?

$8\frac{1}{2} - 3\frac{1}{10} =$

$8\frac{5}{10} - 3\frac{1}{10} = 5\frac{4}{10} = 5\frac{2}{5}$ *off of the phone*

11. One-fifth of the inventory of the Cason Company is more than one month old. The total inventory value is $57,960. What is the value of the inventory that is dated more than one month old?

\$57,960 × $\frac{1}{5}$ = \$11,592 one month old

12. Ted and Kay each worked $6\frac{1}{2}$ hours on Thursday. They were paid $8 per hour. What was their combined total earnings?

\$8 × $6\frac{1}{2}$ =

\$8 × $\frac{13}{2}$ = \$52; \$52 × 2 = \$104 combined earnings

13. There are 800 employees in the accounting department of a software company. One-half of these people will be retrained due to new technology. Of the remainder, $\frac{1}{4}$ are women, and $\frac{1}{10}$ of the men are to retire within the next year. How many men will there be in the accounting department a year from now?

$$800 \times \frac{1}{2} = 400 \text{ not retrained}$$

$$400 \times \frac{3}{4} = 300 \text{ men}$$

$$300 \times \frac{9}{10} = \underline{270 \text{ men left}}$$

(If you missed problems 1 or 2, you need to review Description and Types; problems 3, 4, or 10, you need to review Addition and Subtraction; problems 5, 6, 8, 11, 12, or 13, you need to review Multiplication and Division; problem 7, you need to review Reducing Fractions; and problem 9, you need to review Ratios.) Record your score on the Record of Performance on Mastery Tests in the back of the book.

1.01	1.28	35.1	151.7	0.4
4.58	2.24	248.5	1,116.9	1.6
1.26	2.80	574.8	2,384.6	1.3
1.12	2.85	1,685.0	6,816.0	1.7
2.02	3.51	405.1	1,723.9	1.2
0.93	1.57	874.0	3,319.7	0.8
1.24	1.30	192.3	736.7	0.3
d0.58	1.10	708.0	2,488.1	Nil
2.44	1.48	41.7	171.5	Nil
1.37	1.43	70.5	258.0	Nil
0.67	1.69	196.6	811.0	0.8
1.46	1.64	50.5	201.0	2.1
0.14	1.36	590.1	2,016.8	0.0
0.47	0.17	19.4	103.2	Nil
1.53	1.72	196.1	812.5	0.2

Working with Decimal Numbers 3

Suggested Classroom Time for This Chapter

hours 1 2 3 4 5

Perceived Student Difficulty with Topical Material

easy moderate difficult

Objectives

After mastering the material in this chapter, you will be able to:

1. Read and write decimal numbers. p. 46
2. Round off decimal numbers accurately. p. 48
3. Perform all the fundamental processes on decimal numbers. p. 49
4. Work word problems that require the use of decimal numbers. p. 52
5. Change fractions to decimal numbers. p. 53
6. Understand and use the following terms:
   Place Values
   Rounding Rule
   Decimal Point Placement Rule

Sport-A-Merica

Hal notes that nearly every transaction with a customer, the bank, or a supplier involves the use of decimals. Dollars and cents, percents, decimal expressions of fractions and various sales, economic, trade reports, and price quotations all are in use every day. Yesterday, Kerry asked, "What were our sales for last Tuesday?" Hal answered, "$1,083.70 is the amount I deposited in the bank from the cash register." Kerry then asked, "Are there any blue shirts with the tennis racquet logo on the sleeve left?" "We now have them on sale for $29.95, which is a 30% reduction," Hal responded. "Then we must have sold over half of the stock," stated Kerry. "That's the level at which the Retail Sportswear Association says we should begin marking the remaining stock for sale," said Hal, and he smiled. Hal knew that the shirt was a real success. It had only been in the store for a month and it was nearly sold out as of today. He was happy the shirt would make a substantial contribution to the profits of the month.

Selecting the right price and then the right sale price was a function of dealing with decimal numbers. He had Brenda (their sales clerk) mark the sale price on the stock after hours on Saturday. It cost Sport-A-Merica $12.60 per hour to have Brenda mark the sale price. Hal would need to determine how much her 6.5 hours of work would cost. He was sure that the cost would be worth it. The sale the next day was a real success.

Reading and Writing Decimal Numbers
Objective 1

A decimal number is another way of expressing a fraction or portion of a whole unit. The figure below shows the representation of a decimal number as part of a whole. Note that this decimal number can be shown in a way similar to the fraction $\frac{3}{4}$ in the previous chapter. The decimal value .3 can be shown as follows:

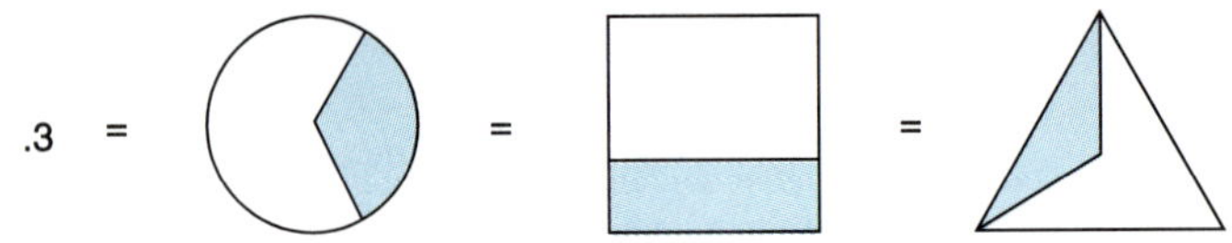

Additional information will be covered in this chapter that will allow you to adapt your knowledge of the fundamental processes to decimal numbers.

place values

To be able to read and write decimal numbers, you must be able to use the **place values** to the right of the decimal point. The following figure will help you identify the place values of decimal numbers.

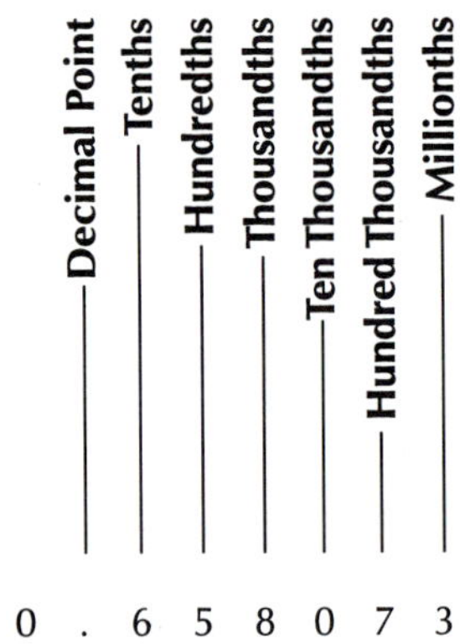

Notice that the spelling of the place values to the right of the decimal point has "ths" on the end of the place value name.

The digit in the thousandths place is an 8.

The digit in the tenths place is a 6.

The digit in the ten thousandths place is a 0.

The digit in the millionths place is a 3.

Exercise A

In the number 0.879325

1. The digit in the hundredths place is 7.
2. The digit in the tenths place is 8.
3. The digit in the millionths place is 5.
4. The digit in the ten thousandths place is 3.
5. The digit in the hundred thousandths place is 2.
6. The digit in the thousandths place is 9.

Exercise B

In the number .548061

1. The digit in the hundred thousandths place is a 6.
2. The digit in the *hundredths* place is a 4.
3. The digit in the *millionths* place is a 1.
4. The digit in the *ten thousandths* place is a 0.
5. The digit in the *tenths* place is a 5.
6. The digit in the *thousandths* place is an 8.

Check your progress by turning to the answers at the back of the book.

Now that you know the place values on both sides of the decimal point, you are able to read and write decimal numbers. The numbers to the right of the decimal place are read somewhat differently than those on the left.

Example 1 The number .507 is read five hundred seven thousandths. (Notice that the number is read, then the label is added on at the end.)

Example 2 The number .9637 is read nine thousand, six hundred thirty-seven ten thousandths.

Exercise C

Read the following decimal numbers to yourself.

1. .6057 2. .9871 3. .500 4. .936 5. .8710 6. .9057

Reading a number that has digits on both sides of the decimal point requires that you use the rules presented in chapter 1 along with those that you have just learned. One additional rule is necessary to make the reading complete. When reading the number, the decimal point is read as "and" or "point."

Example 3 The number 487.903 is read four hundred eighty-seven and nine hundred three thousandths *or* four hundred eighty-seven point-nine-zero-three.

Exercise D

Read the following numbers to yourself.

1. 1.9673 2. 400.5091 3. 87,365.851 4. 987.53 5. 896.57321

Reading values that include dollars and cents requires that you read all of the dollar values followed by the word "dollars" and then read "and ______ cents."

Example 4

Kerry must give a price quotation to a customer over the phone. The Sport-A-Merica price is $581.39. The price is read, "five hundred eighty-one dollars and thirty-nine cents." (Notice that the hyphens and the comma are still used in the writing of the values just as they were in chapter 1.)

Exercise E

Write the following values in words. Place commas and hyphens in the appropriate places.

1. **$823.67** Eight hundred twenty-three dollars and sixty-seven cents
2. **$2,105.46** Two thousand, one hundred five dollars and forty-six cents
3. **$867.28** Eight hundred sixty-seven dollars and twenty-eigth cents
4. **$234.87** Two hundred thirty four dollars and eighty-seven cents
5. **$3,869.61** Three thousand, eight hundred sixty-nine dollars and sixty-one cents

Exercise F

Write the following values using numbers. Insert commas and dollar signs where necessary.

1. Nine hundred thirty-five dollars and forty-nine cents.
   $935.49
2. Sixty-three dollars and twenty-six cents.
   $63.26
3. Five hundred seventy-two dollars and twenty-one cents.
   $572.21
4. Three thousand, eighteen dollars and seven cents.
   $3,018.07
5. Fifty-one thousand, three hundred seven dollars and thirty-six cents.
   $51,307.36

Rounding Off

Objective 2

rounding rule

● *Show students examples of interest rates, bond rates, and sales figures from the local newspaper to illustrate the variety of values used in everyday business reports.*

The rules that you learned in chapter 1 on rounding off will also hold true for decimal numbers.

The **rounding rule** is the same for decimal numbers as it is for whole numbers. If the instructions for rounding are to "round off your answer to the nearest thousandths," then you must be concerned with the digit in the thousandths place and the digit immediately to the right of it (i.e., the ten thousandths place). In the number .07583, the digit 5 is in the thousandths place. The 5 will be changed to a 6 if the digit immediately to the right of it is a 5 or larger. In the example given, the digit is an 8; therefore, the answer is .076. All of the digits to the right of the number to be rounded are dropped.

Example 5 Round off .603791 to the nearest hundred thousandths.

Digit in question (to be rounded)
Decision-making digit (5 or larger?)

.603791 = .60379

Exercise G

Round off the following numbers to the place value indicated.

1. .60578 to the nearest tenth .6
2. .4653 to the nearest hundredth *.47*
3. .0037 to the nearest tenth *.0*
4. .54621 to the nearest thousandth *.546*
5. .5467 to the nearest unit *1*

Exercise H

Round off the following values to the nearest cent.

1. $605.837 $605.84
2. $57.653 *$57.65*
3. $486.938 *$486.94*
4. $68,640.273 *$68,640.27*
5. $1,754.114 *$1,754.11*

Exercise I

Round off the following numbers to the place value indicated.

1. 607.8035 to the nearest hundredth 607.80
2. 758.9208 to the nearest hundred *800*
3. 759.92749 to the nearest tenth *759.9*
4. 547,647.8492 to the nearest ten thousand *550,000*
5. 14.79849 to the nearest hundredth *14.80*

Addition

Objective 3

The addition of decimal numbers requires the decimal points to be aligned before addition begins.

Example 6 Add the decimal numbers .03; 6.878; 5.67; and 181.006.

.03	.03
6.878	6.878
5.67	5.67
181.006	+181.006
	193.584

(.03 + 6.878 + 5.67 + 181.006 =)

Zeros may be added to the numbers if they will aid in the addition process. They should be added to the right of the last digit, thereby not changing the value of the number. The zeros fill in all the columns and, therefore, reduce errors.

Example 7

Find the sum of the following numbers: 6.53; 80.751; .7; 9.543; .00658; and .801.
With zeros added:

```
  6.53000
 80.75100
   .70000
  9.54300
   .00658
  +.80100
 98.33158
```

Put those decimal points in line; or make sure you don't forget where the decimal point belongs!

(*Note:* Any whole number has a decimal point to the right of the digit in the units or ones place.)

(6.53 + 80.751 + .7 + 9.543 + .00658 + .801 =)

Exercise J

Find the sum in the following problems.

1. 506.375; 9,831.6; 19.6025; 181; and 5,037.62 = 15,576.1975
2. 56.0983; 86.14; .9573; 658.9401; and 87 ***= 889.1357***
3. .456; 657.98; 76.98; .86946; and 8.941 ***= 745.22646***
4. 65.9201; 958; 65.937; and 75.937 ***= 1,165.7941***
5. 15.3; 8.0981; 817.04; and .05178 ***= 840.48988***

Exercise K

Find the sum in the following problems.

1. Eight and forty-three ten thousandths; six hundred twelve thousand; eighteen and three hundredths; and fifty-seven thousandths.

```
      8.0043
612,000.0000
     18.0300
      +.0570
612,026.0913
```

2. Three hundred four and six tenths; forty and twenty-seven thousandths; sixty-seven hundredths; and one and four ten thousandths.

```
304.6
 40.027
   .67
 +1.0004
346.2974
```

3. Fifty-three and six hundred two ten thousandths; four hundred sixteen thousandths; ten and four tenths; and sixteen.

```
 53.0602
   .416
 10.4
+16.
 79.8762
```

4. Thirty-two and twenty-six thousandths; fifty and ninety-five thousandths; four hundred twenty; and seventy-nine thousandths.

```
 32.026
 50.095
420.
  +.079
502.2
```

5. One hundred fifty-eight thousand; sixty and twenty-two ten thousandths; four hundred and six thousandths; and three and two hundred four ten thousandths.

```
158,000.
     60.0022
    400.006
     +3.0204
158,463.0286
```

Subtraction

Subtraction of decimal numbers also requires the decimal points to be aligned.

Example 8

Subtract 9.853 from 10.1

$$\begin{array}{r} 10.100 \\ -\ 9.853 \\ \hline .247 \end{array}$$

Note that zeros may be added to the minuend if it will aid you in the subtraction process.

(10.1 − 9.853 =)

Exercise L

Find the difference in the following problems.

1. 8.067 − .537 =
   $\begin{array}{r} -\ .537 \\ \hline 7.530 \end{array}$
2. 19.6 − 3.167 =
   16.433
3. .9853 − .547 =
   .4383
4. 65 − 8.957 =
   56.043
5. 86.958 − 45.095 =
   41.863

Exercise M

1. \$863.47 − \$27.98 =
   $\begin{array}{r} -27.98 \\ \hline \$835.49 \end{array}$
2. \$875.90 − \$345.85 =
   \$530.05
3. \$65.86 − \$17.69 =
   \$48.17
4. \$649.09 − \$52.10 =
   \$596.99
5. \$756.92 − \$478.00 =
   \$278.92

Multiplication

The multiplication of decimal numbers requires the additional step of the placement of the decimal point. When multiplying the whole numbers 37 and 8, we find the answer to be

$$\begin{array}{rl} 37 & \text{multiplicand} \\ \times\ 8 & \text{multiplier} \\ \hline 296 & \text{product} \end{array}$$

decimal point placement rule

The **decimal point placement rule** states that the answer should have as many decimal places to the right of the decimal point as the multiplier and the multiplicand have in total.

Example 9

37 × .8 =

$$\begin{array}{rr} 37 & \\ \times\ .8 & (+1) \\ \hline 29.6 & (1) \end{array}$$

(37 × .8 =)

Example 10

Hal and Kerry looked at their cost of labor for their sale at Sport-A-Merica. Brenda had worked 6.5 hours at $12.60 per hour. How much did she earn?

```
6.5 × $12.60 =        6.5   (+1)
                 ×$12.60   (+2)
                    6300
                   7560
                $ 81.900    (3)
```

(6.5 × 12.60 =)

Example 11

```
.0037 × .00008 =        .0037   (+4)
                  ×    .00008   (+5)
                  .000000296     (9)
```

Note: Zeros must be added to fill decimal place positions if there are blanks after stepping off the decimal point.

(.0037 × .00008 =)

Exercise N

Find the product in the following problems. Pay particular attention to the placement of the decimal point.

1.	2.	3.	4.	5.
982	3.4	76.2	5.63	459
× 8.6	×.67	× .650	× .405	× .006
5892	***2.278***	***49.53000***	***2.28015***	***2.754***
7856				
8,445.2				

Division

Objective 4

The division of decimal numbers uses the same rules as the division of whole numbers. The placement of the decimal point is the only additional problem in the division of decimal numbers. As shown in example 12, the division of 42 by 6 requires no decimal point placement, but 42 ÷.6 does require the decimal point placement.

Example 12

```
   7
6)42
```

❍ *Don't forget to put the decimal point in your answer. It could make a big difference!*

If the problem would have been to divide 42 by .6, the decimal place would need to be moved in the answer the same number of places as the divisor, as shown, in order to correctly place the decimal point.

```
                70.
.6)42.0    .6.)42.0.
            +1    +1
```

(43 ÷ .6 =)

Zeros must be added to the end of the dividend in order to fill in the spaces created by moving the decimal point. Additional zeros may be added to the right of the decimal place to aid in division. The number 420 has the same value as 420.0 or 420.000. The decimal point in the answer is placed directly above the decimal point in the dividend after the decimal point has been moved.

Example 13

530.1 ÷ .57 =

$$.57.\overline{)530.10.} = 930.$$

513
171
171
0

(530.1 ÷ .57 =)

Exercise O

Round off your answers to the nearest tenth in the following problems.

1. 25.0146 ÷ .7 =
   $.7.\overline{)25.0.146}$ = 35.7 (quotient 35.735)
2. .24 ÷ .6 = **.4**
3. 4.63 ÷ .015 =
   308.66666 = 308.7
4. 9.872 ÷ 5.9 =
   1.6732203 = 1.7
5. .045 ÷ 8.2 =
   .0054878 = 0

Changing Fractions to Decimal Numbers

Objective 5

Any fraction can be changed to a decimal number or its equivalent by dividing the numerator by the denominator.

Example 14

A quarter of a dollar is 25¢ or $.25. We interchange fractions and decimals all of the time!

Common Fraction

$$\frac{3}{4} = 4\overline{)3.000} = .75$$

2 8
20
20

Improper Fraction

$$\frac{13}{8} = 8\overline{)13.000} = 1.625$$

8
50
48
20
16
40
40

Mixed Number

$$4\frac{1}{8} = 4 \text{ plus } 8\overline{)1.000} = .125$$

8
20
16
40
40

Answer = 4.125

(3 ÷ 4 =)

Exercise P

Change the following fractions into decimal numbers.

1. $\frac{3}{8}$ = .375
2. $\frac{9}{16}$ = ***.5625***
3. $1\frac{3}{5}$ = ***1.6***
4. $\frac{12}{5}$ = ***2.4***
5. $2\frac{3}{16}$ = ***2.1875***

ASSIGNMENT Chapter 3

Name ______________________ **Date** ______________ **Score** ____________

Skill Problems

In the number 0.471803219

1. The digit in the hundredths place is ___*7*___.

2. The digit in the tenths place is ___*4*___.

3. The digit in the millionths place is ___*3*___.

4. The digit in the ten thousandths place is ___*8*___.

5. The digit in the hundred thousandths place is ___*0*___.

6. The digit in the thousandths place is ___*1*___.

Write the following value in words.

7. $856.45 ___*Eight hundred fifty-six dollars and forty-five cents*___

8. $3,271.99 ___*Three thousand, two hundred seventy-one dollars and ninety-nine cents*___

Round off the following numbers to the place value indicated.

9. .70568 to the nearest tenth ___*.7*___

10. $705.927 to the nearest cent ___*$705.93*___

11. 548,658.4587 to the nearest thousandth ___*548,658.459*___

12. 64.65738 to the nearest hundredth ___*64.66*___

Write the following using numbers.

13. Four hundred sixty-one thousandths ___*.461*___

14. Six and three hundredths ___*6.03*___

15. Fifty-two and twenty-seven ten thousandths ___*52.0027*___

16. Ten thousand, fifty-one ___*10,051*___

Complete the following Inventory List.

Item Code	Cost	Units in Stock	Value of Inventory
354068	$15.95	647	**17.** *$10,319.65*
654083	10.65	82	**18.** *873.30*
654082	11.65	1,264	**19.** *14,725.60*
654103	18.23	981	**20.** *17,883.63*

Change the following fractions into decimal numbers. Round off to the nearest thousandth.

21. 3/4 ___*.75*___ **22.** 5/10 ___*.5*___ **23.** 13/5 ___*2.6*___ **24.** 3/9 ___*.333*___

Select a letter for the definition.

A. Place Values B. Rounding Rule C. Decimal Point Placement Rule

25. ___*B*___ The number to the right must be 5 or larger to round up.

26. ___*C*___ The answer in multiplication problems should contain as many decimal places as the two values in the problem.

27. $42 \div .05 =$ ___*840*___

28. $5\frac{1}{4} \times .62 =$ ___*3.255*___

29. Add the decimal numbers .006; 67.953; 5.005; and 20.

$$\begin{array}{r} .006 \\ 67.953 \\ 5.005 \\ +\ 20. \\ \hline 92.964 \end{array}$$

30. Find the sum of the decimal numbers .7; .654; 65.47; and .047.

$$\begin{array}{r} .7 \\ .654 \\ 65.47 \\ +\ .047 \\ \hline 66.871 \end{array}$$

31. Subtract 9.3 from 12.634.

```
 12.634
−  9.3
  3.334
```

32. Find the difference between 20.6 and 2.654.

```
 20.600
− 2.654
 17.946
```

Find the product.

33. 657 ×9.4 = ***6,175.8***

34. 5.4 ×.24 = ***1.296***

35. 48.5 ×.460 = ***22.31***

36. .0058 ×2.0 = ***.0116***

Round off your answers to the nearest hundredth.

37. .032 ÷ 1.6 = ***.02***

38. 65.4183 ÷ .5 = ***130.84***

38. .56 ÷ .8 = ***.7***

40. $5.85 ÷ .5 = ***$11.70***

Complete the following problems.

41. 3.6 × .042 = ***.1512***

42. 14 ÷ .02 = ***700***

43.

```
     4.872
    12.87
 1,095.1037
 +345.81
 1,458.6557
```

44.

```
 876.5
−45.860
 830.64
```

45. $7\frac{1}{4}$ × .6 = ***4.35***

46. .042 ÷ 5 = ***.0084***

47. Round off your answer to the nearest tenth.

6.032
×8.087
48.780784 = 48.8

48. Round off your answer to the nearest cent.

$846.13 ÷ 5.6 =
$151.09464 = $151.09

49.

```
  63.08
   9.
    .075
 +  .0039
  72.1589
```

50. .3 × $\frac{7}{8}$ = ***.3 × .875 = .2625***

51. Round off your answer to the nearest hundredth.

.193 × 6.073 =
1.172089 = 1.17

52. $2\frac{1}{4}$ × .06 = ***2.25 × .06 = .135***

53. 1.29 ×.036 = ***.04644***

54. $19 ÷ 2.5 = ***$7.60***

55. 1,469. −6.3807 = ***1,462.6193***

Business Application Problems

Solve the following problems. Round off to the nearest cent.

1. The manager ordered two cases of copier paper. The price listed was $15.95 per case. Shipping cost is $5.25. What is the total for the two cases of paper? ***$37.15***

2. To create a square foot of finished concrete, we require $0.85 for labor. Our customer has $3,000.00 in their budget for outdoor improvements. If the materials cost $1,400, what is the largest amount of square footage we can complete for our customer (rounding off to the whole number)? ***1,882 sq. ft***

3. The employees in the production department increased their output by 1/5 over the standard amount of 420 for the week. Each piece produced is valued at $63.78. What is the value of the output for the week? ***420 × 1/5 = 84 420 + 84 = 504 504 × $63.78 = $32,145.12 value***

4. The Quality Assurance Manager reported to her superior that 1.3 units were defective for each hour worked in the Finish Department. The Finish Department worked 680 hours over a five-day period. How many defective parts were produced? ***1.3 × 680 = 884 defective parts***

5. The Procurement Department assigned each employee 58 items to oversee. The average value of each item was $386.45. There are five people in the Procurement Department. What is the value of the items that the department procures? ***58 × $386.45 × 5 = $112,070.50 value***

6. A piece of material that costs $6.83 plus $14 of labor goes into each of the 3,980 light fixtures produced by Jamison, Inc. each month. What is the total cost of the light fixtures produced by the company for the month? ***$6.83 + $14 = $20.83 × 3,980 = $82,903.40 total cost***

7. A jeweler made a bracelet of gold that weighed 10.7 ounces. The gold is priced at $280 per ounce. The jeweler charged $28 per hour for each of the 8.5 hours on the pr… What is the cost of the bracelet?
10.7 × $… = $238 $3,234 cost

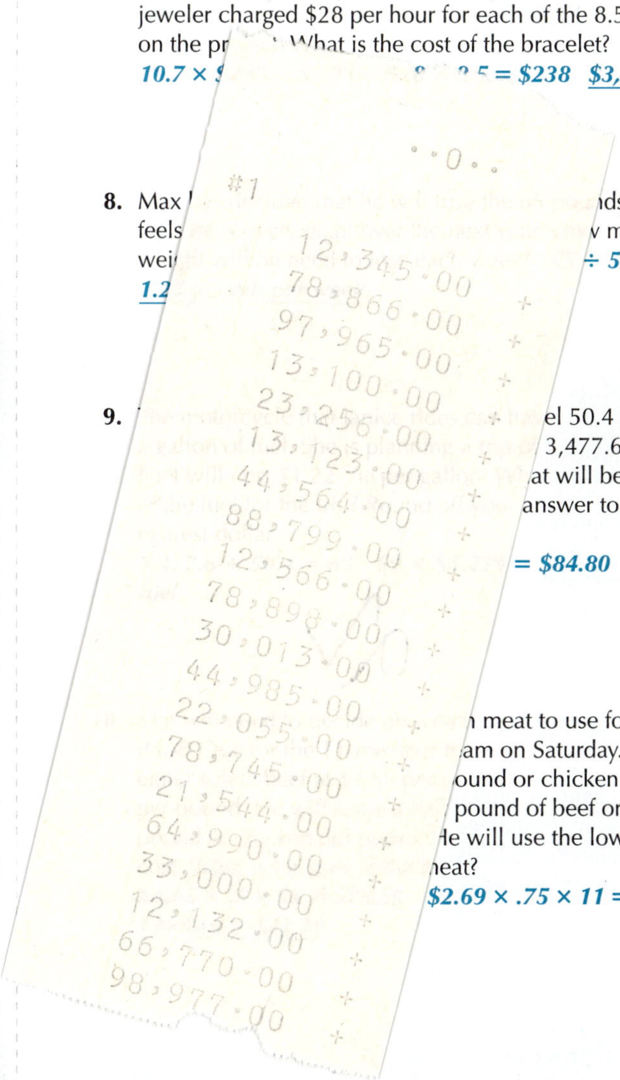

8. Max … nds that he feels … v much wei… ***÷ 52 =***
1.2…

9. … el 50.4 miles on … 3,477.6 miles. … at will be the cost … answer to the …
= $84.80 $85 for

… meat to use for the … am on Saturday. He can … ound or chicken at $2.69 … pound of beef or 3/4 … He will use the lowest total … neat?
$2.69 × .75 × 11 = $22.19

11. Dee is the quality assurance manager of a banquet firm. She has been asked to determine the cost of serving a group of twenty-five guests. The price will be $21.90 per pound. Each guest will be served .4 of a pound. What is the cost for the group of twenty-five guests?

21.90	***$8.76***
× .4	***× 25***
$8.76	***219.00 total cost***

12. Fynee is required to check all incoming invoices to make sure that the supplier has extended the totals correctly for each item and totaled the amount. Review her work to be sure that it is correct for the Jules Joyner Designs invoice.
$3,342.60 total

Jules Joyner Designs
Interior & Couture Fashion Design
104 W. Fourth Street Suite 302
ROYAL OAK, MICHIGAN 48067
(313) 545-9944

CUSTOMER'S ORDER NO.	PHONE	DATE
1981	*987-3765*	*January 8th*

NAME *Bevier Associates*
ADDRESS *1980 Dekca Place*
Kalamazoo, Michigan 48093

SOLD BY	CASH	C.O.D.	CHARGE	ON ACCT.	MDSE. RET'D.	PAID OUT
✓✓	✓					

QTY.	DESCRIPTION	PRICE	AMOUNT
1	*Custom Design Bridal Gown w/Silk and Satin French clear sequins and Diamond white tulle*	*3,126*	*3,126 00*
3	*Bridal headpieces*	*72.20*	*216 60*
		TAX	
RECEIVED BY		TOTAL	*3,342 60*

1600 All claims and returned goods MUST be accompanied by this bill. *Thank You*

Courtesy of Jules Joyner Designs, Inc.

13. Patricia Barrett worked 7.1 hours at the rate of $15.80 per hour. How much did she earn?
15.80
× 7.1
$112.18 earnings

14. The sales staff met to determine the total distance on the route through the city. The distance from the office to the warehouse was 4.6 miles; warehouse to the first stop, 3.2 miles; and sales route to the office, 54.1 miles. What is the total mileage?

4.6
3.2
+ 54.1
61.9 total miles

15. Allison is paid 8.3 ($.083) cents for each part that she produces. Yesterday she produced 893 parts. To the nearest cent, how much did she earn?

$.083
× 893
$74.119 = $74.12 earnings

16. Roberto Desin found that five-eighths of the 17.35 tons of chloride imported each year were used for agricultural purposes. To the nearest tenth of a ton, how much was used for this purpose?

17.35
× .625
10.84375 = 10.8 tons

17. Al LaFrance paid $10,500 for rent last year for his lock shop. The cost of the rent is $5 per month for each square foot of space. How many square feet of space did he rent?

$10,500 ÷ 12 = $875
$875 ÷ $5 = 175 square feet

18. Casoni Manufacturing, Inc. found that the price of needed raw materials had increased by 1.9 times over the past four years. The old price of four years ago was $14.50 per unit. What is the price increase?

$14.50
× 1.9
$27.55 price increase

19. Marie Kreger sold her second-hand clothing store for $87,300. The balance on the mortgage was $18,546.87. She paid the real estate agent a commission on the sale. The commission was $7,800. She also paid the bank a penalty of $567.12 for early payment on the mortgage. She decided to use the money from the sale to purchase a home that cost $93,900. How much money did she have to borrow to purchase the home?

$87,300
− 18,546.87
− 7,800
− 567.12
$60,386.01

$93,900.00
− 60,386.01
$33,513.99 need to borrow

20. The manager of the shipping department at Belin Fluids discovered that over the past month, 46.8 hours were lost because of accidents, 85 hours because of sickness, and 46 hours because of vacations. The department had scheduled 820 hours for the month. How many hours were actually worked?

820.
−46.8
−85.
−46.
642.2 hours actually worked

21. Joanne Bacher sold pennants at football games. The pennants cost thirty-seven cents each. They were sold for seventy-five cents. Joanne and her sales team sold 167,759 pennants during the season. What were the total sales for the season?

$.75
×167,759
$125,819.25 total sales

22. Bill Wilson said that he would like to expand his shop by 1.8 times its current area of 1,020 square feet. He expects the addition would cost about $150,000. How many square feet would the new portion of the building have?

1,020
×1.8
1,836 square feet

23. Julie Smith was offered a job paying $65 a day for eight hours' work. She would like to take the job, but she is not sure if the job pays more than the $8.20 an hour she is earning at her present job. Which is the higher hourly rate that Julie has as an option?

$65 ÷ 8 = $8.125 *$8.20 per hour*

24. The Lennolda Shoppe is taking inventory of the small pieces in the storage area of the shop. The inventory revealed the following: 45 pieces priced at $.54 each, 144 pieces priced at $.19 each, 20 units priced at $1.95 each, a box of 50 pieces priced at $.39 each, and 4 units priced at $.49 each. What is the total value of these small pieces of merchandise?

45 × $ .54 = $24.30
144 × $ .19 = 27.36
20 × $1.95 = 39.00
50 × $ .39 = 19.50
4 × $ .49 = +1.96
$112.12 total value

25. Ernest Seldon bought a bin of bird seed. The bird seed is made up of seven bushels of oats at $5.95 a bushel, ten bushels of cracked corn at $7.95 a bushel, and five bushels of cracked sunflower seeds at $14.95 a bushel. The mixture was sold for $7.50 a half bushel, which included the 10 cents for the bag in which it was packed. How much profit did he make?

7 × $5.95 = $41.65
10 × $7.95 = 79.50
5 × $14.95 = +74.75
22 bushels $195.90
×2
44 half bushels

$.10 × 44 = $4.40 bags
+195.90 seed
$200.30 costs

$7.50 × 44 = $330.00 sales
−200.30 costs
$129.70 profits

26. The lounge chairs of the Marina Bay Condominium are in need of repair. A review of the grounds indicated that there were 27 chairs and 27 loungers that needed to have new webbing installed. The manager received price quotes from area repair shops. The lowest price was $.71 per replaced web. There are 13 webs on each of the chairs and 22 webs on each lounger. How much will it cost to have the 54 units repaired?

27 × $.71 × 13 = $249.21
27 × $.71 × 22 = +421.74
$670.95

27. Mac McLeod projected sales figures for Barcelona, Inc. show the volume of "firm" and "possible" sales by representative. What is the difference between the firm and possible sales figures if each unit sells for $8.45?

Sales Rep.	Firm Sales	Possible Sales
A. Janor	410 units	650 units
B. K. Moore	301 units	800 units
N. Manoto	285 units	1,000 units
	996	*2,450*

2,450 − 996 = 1,454
1,454 × $8.45 = $12,286.30 difference

28. Write a complete sentence in answer to this question: Fuel for the fleet of trucks operated by Wiise and Company cost $5,689.40 last month. The company is able to purchase fuel at a cost of $1.17 per gallon. To the nearest tenth of a gallon, how many gallons of fuel did the firm purchase last month?

$5,689.40 ÷ $1.17 = 4,862.735 = 4,862.7
Wiise and Company purchased 4,862.7 gallons of fuel last month.

29. Write a complete sentence in answer to this question: Dr. Monroe's office will begin to deliver a small flower arrangement and card to each of the surgery patients as of the first of next month. There are 162 patients scheduled for surgery next month. The flowers will cost $8.50 and the cards will cost $1.25 each. The delivery charge will be $4 on the average. What will be the total cost of this service for the month?

$8.50
1.25
+4.00
$13.75 × 162 = $2,227.50
The total cost of the small flower arrangements and cards will be $2,227.50 for the month.

30. Write a complete sentence in answer to this question: The Boxon Corp. has offered to deduct one-eighth of the price of every unit of their product sold after the first 500,000 units are sold, per month. The product sells for $627.50 each. What will be the price of units sold over the 500,000 volume level? Round off your answer to the nearest cent.

$627.50 × 7/8 = $549.06
The price of the units sold after the first 500,000 will be $549.06.

★ Challenge Problem

The Gayle Company bought a roll of metal 4 feet wide with a length of 500 feet. The metal is to be cut into pieces 4 feet wide by 5.7 feet long. There were 14 pieces that had to be scrapped owing to errors made in cutting. The roll cost $4,678.95. If the firm makes a profit of $14.86 on each piece produced, what will be the total profit made from the sale of the full cut pieces?

500 ÷ 5.7 = 87.719298 pieces
−14 scrap pieces
73 full pieces
×$14.86
$1,084.78 total profit

MASTERY TEST Chapter 3

Name ______________________ Date ______________ Score ____________

1. Round off the following values as indicated.
   a. 7.9803 to the nearest ten *10*
   b. .08712 to the nearest tenth *.1*
   c. 647.096 to the nearest hundred *600*
   d. 64.0968 to the nearest thousandth *64.097*
   e. 43.916 to the nearest tenth *43.9*

2. Write the following value in words: \$385.70.
   Three hundred eighty-five dollars and seventy cents

3. Add 733.042; 8.367; and 13.758.
   755.167

4. Subtract 9.405 from 57.079.
   47.674

5. Multiply 1.2 × .56 and round off your answer to the nearest unit.
   .672 = 1

6. Divide \$3.72 by 8 and round off your answer to the nearest cent.
   \$.465 = \$.47

7. Change the following to decimal form.

   a. $\frac{5}{8} =$ *.625* **b.** $\frac{3}{10} =$ *.3* **c.** $4\frac{1}{5} =$ *4.2*

8. If a certain industrial product is made from plastic, it costs 6.7 cents (\$.067) per unit over a volume of 6,000 pieces. If the same product is made of rubber, it will cost \$450 for the same volume. The company has a contract that assures a revenue of \$1,560 for the 6,000 pieces made of either material. What is the maximum profit per piece rounded to the nearest cent?

   6,000)\$1,560 = \$.26 revenue per piece
   − .067 cost per piece (lowest)
   \$.193 = \$.19 per piece profit
   6,000)\$450 = \$0.075

9. AM Corp. has expanded its business 1.6 times over last year's volume of \$324,952. What is this year's volume?

   \$324,952
   × 2.6
   \$844,875.20 volume this year

10. Anthony James Ureel, the manager of an agribusiness firm, noted that seven-tenths of the 15,479 barrels of herbicide he bought last year were used for soybean production. To the nearest whole barrel, how many were used for soybeans?

    15,479
    × .7
    10,835.3 = 10,835 barrels

(If you missed problem 1, you need to review Rounding Off; problem 2, you need to review Reading and Writing Decimal Numbers; problem 3, you need to review Addition; problem 4, you need to review Subtraction; problems 5, 8, or 9, you need to review Multiplication; problem 6, you need to review Division; problem 7, you need to review Changing Fractions to Decimal Numbers; and problem 10, you need to review Multiplication, Changing Fractions to Decimal Numbers, and Rounding Off.) Record your score on the Record of Performance on Mastery Tests in the back of the book.

Banking Records 4

Suggested Classroom Time for This Chapter

hours 1 2 3 4 5

Perceived Student Difficulty with Topical Material

easy moderate difficult

Objectives

After mastering the material in this chapter, you will be able to:

1. Identify different types of endorsements. p. 66
2. Reconcile a bank statement. p. 67
3. Understand and use the following terms:
   Depositor
   Imprinting Checks
   Automatic Teller Machine (ATM
   Deposit Slip
   Check Register
   Check Stub
   Check
   Payee
   Blank Endorsement
   Special Endorsement
   Restrictive Endorsement
   Electronic Funds Transfer
   Bank Statement
   Deposits in Transit
   Outstanding Checks
   Service Charges
   Adjusted Balance

Sport-A-Merica

Hal receives the bank statement for Sport-A-Merica in the mail on October 3. Recognizing the fact that the company has bills to pay this week, Hal decides to review the bank statement and the company records. He notices that the bank included a $63 service charge for the month. That seems too high.

Kerry and Hal discuss the service charge. Kerry suggests moving the Sport-A-Merica accounts to First National Bank, where she has her personal account. Hal reasons that they would then have to sign new signature cards, transfer the balance, and order new checks. Would all of this be worth the bother? Kerry says she enjoys the ATM after-hours convenience, and in addition, First National Bank has a low minimum balance requirement of $1,000 before service charges are assessed.

Hal doesn't feel ATMs would be of much use to the firm, but understands why Kerry uses them. Furthermore, the Sport-A-Merica account balance is only $690.40 according to the check register, and Hal must pay bills this week. He is not sure of the *actual* balance in the account. Other charges, such as those for overdrafts and collection charges, seem more important from a business point of view.

On the other hand, Kerry knows that $600 will be deposited in *her* account automatically each month from her annuity investment. Her account balance must be adjusted each month. Both Kerry and Hal recognize the need to investigate banking costs to decide what they want to do with the money belonging to Sport-A-Merica.

Most debt payments made by business and by private individuals are made by check. The checking account provides a convenient as well as a safe payment method—a definite improvement over paying cash in person.

When a new checking account is opened, the parties who are authorized to sign checks are required to give a sample signature to the bank. A card containing the signature(s) is kept on file and is available to bank personnel at all times. The procedure of verifying the handwritten signature is used to provide security.

depositor

The new **depositor** is given a set of deposit slips and blank checks. A personalized set of checks is ordered at this time, also. The new, personalized set will contain pertinent information, such as the account holder's name, the account number, the address, the telephone number, and possibly the driver's license number. There is usually a charge for the **imprinting** process. Imprinted checks are easier for the bank to identify and process. The imprinting will also limit use to only the authorized parties whose names are imprinted on them or those authorized to sign for a company or organization.

imprinting checks

automatic teller machine (ATM)

An individual depositor may also request an **automatic teller machine (ATM)** card. This card allows the depositor to make deposits and withdrawals at many locations 24 hours a day by using the card along with a secret code number issued by the bank.

Deposits and Payments

When a depositor wants to deposit money in an account, a **deposit slip** or **deposit ticket** must be completed. The date, the amount of currency, the amount of coin, as well as each check to be deposited, are listed separately on the deposit slip. The deposit slip, along with the monies specified, is given to the bank teller or placed in the ATM machine by the depositor. A duplicate **deposit slip** or some acknowledgment is then given to the depositor. This is retained as proof of deposit, and the deposit is recorded in the check register.

deposit slip

● *Most checks deposited take a few days to clear the bank. A deposit does not necessarily assure that a check can be written against the deposited amount.*

DEPOSIT TICKET

Slow-But-Steady Company
1984 Walkers Cove
Springer, NJ 07459

June 2 20__

	DOLLARS	CENTS
CASH	483	10
CHECKS LIST SINGLY		
90-1313	248	65
90-3123	56	94
Subtotal	788	69
TOTAL ITEMS / TOTAL	788	69

55-138/212

BE SURE EACH ITEM IS PROPERLY ENDORSED

NO CARBON REQUIRED FOR CLEAR COPY, PRESS FIRMLY WITH BALL POINT PEN

Valley National Bank
325 PALISADE AVENUE
NEWARK, NEW JERSEY 07459

Christina Steady

⑆021201383⑆ 075⑈80936⑈5⑈ 1

DELUXE MHD-80 CHECKS AND OTHER ITEMS ARE RECEIVED FOR DEPOSIT SUBJECT TO THE PROVISIONS OF THE UNIFORM COMMERCIAL CODE OR ANY APPLICABLE COLLECTION AGREEMENT.

check register

The **check register** or **check stub** provides the form on which the depositor can determine the amount in the account. Deposits are added to the account balance, and issued checks are subtracted from the account balance. It is important to record the deposits and checks *and* to do the necessary addition or subtraction at the time the transaction occurs. This procedure will ensure that sufficient funds are in the account prior to issuing the next check.

Check Register

19___		BE SURE TO DEDUCT ANY PER PER ITEM CHARGES, SERVICE CHARGES OR FEES THAT MAY APPLY IN "OTHER" COLUMN.					
DATE		CHECK NUMBER	TRANSACTION DESCRIPTION	✓ T	(−) AMOUNT OF PAYMENT OR DEBIT	(+OR +) OTHER	(+) AMOUNT OF CREDIT	BALANCE FORWARD 2,611 80
10	17		TO:					320 00
			FOR:					2,931 80
10	17		TO: Travel Court Lodge		156 20			156 20
			FOR: meeting room					2,775 60
			TO:					
			FOR:					
			TO:					
			FOR:					
			TO:					
			FOR:					
			TO:					
			FOR:					

check
payee

When writing a **check,** the depositor must fill in the date; the party to whom the check is payable (the **payee**), which could be a person or an organization; the amount of the check in numbers as well as words; and lastly, authorized signature(s). The check register or check stub should also be filled out at this time. This provides a record that duplicates the information on the check itself.

Note that a deposit of $320 is added to the $2,611.80 balance brought forward; then check #47 for $156.20 is subtracted to determine the new balance of $2,775.60 for Abol Corporation.

When the bank receives the completed check, it handles it as a demand by the depositor to pay the payee the specified amount.

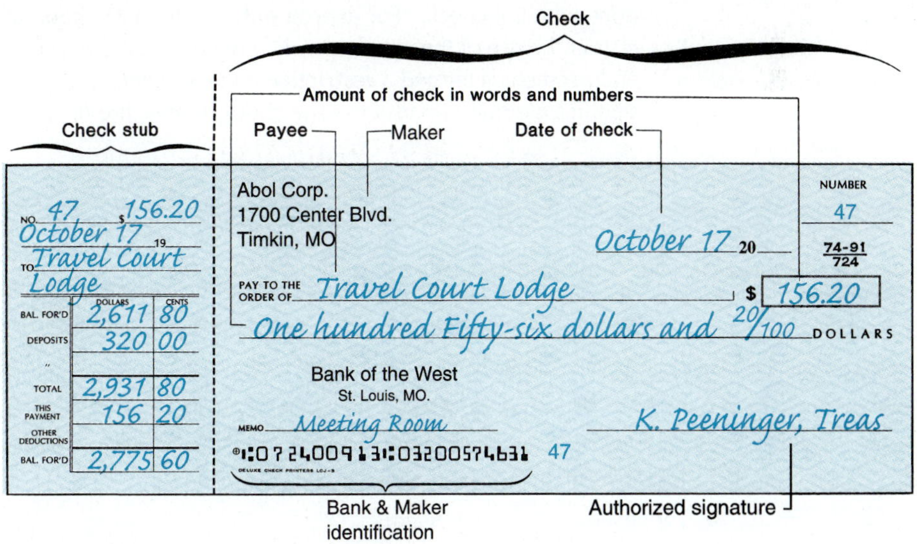

Types of Endorsements

Objective 1

Before the bank will exchange cash for the completed check, the payee must sign the check. This must be done on the back, on the same end that states "Pay to the order of." This signature must be the same name as that specified on the front of the check, such as "John M. Student." The federal government has requested that the endorsement be written in the top $1\frac{1}{2}$ inches of the check. This will allow the bank to use the rest of the area for processing. If the payee simply signs his or her name, the check can be cashed by anyone. The check is, therefore, transferable. This type of endorsement is termed a

blank endorsement

blank endorsement. Should a check with a blank endorsement be lost, it could possibly be cashed by any finder. The finder would simply have to sign his or her name below the endorsement of the payee to receive the specified amount of cash, unless the bank requests identification.

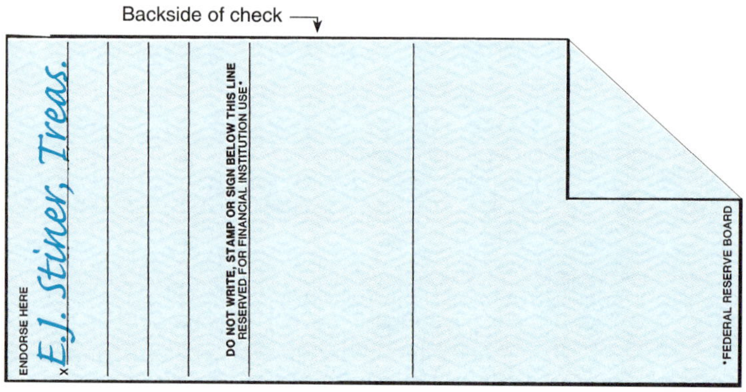

To provide protection to the payee, the depositor could add to the blank endorsement a limitation such as "Payable to J. Hawkins John M. Student." With J. Hawkin's signature, only J. Hawkins would be allowed to cash the check. This type of endorsement

special endorsement

is termed a **special endorsement.** This endorsement indicates a specific person or organization to which the check is payable.

Another restriction that can be added to the signature specifies exactly what is to be done with the check. "For deposit only John M. Student" and the payee's signature will restrict the use of the check to only a deposit in the payee's account. This type of

restrictive endorsement

endorsement is termed a **restrictive endorsement.** A restrictive endorsement usually signifies a terminal transfer of the check, that is, the final use of the check.

Exercise A

Identify the endorsements below as blank, special, or restrictive.

restrictive 1. For deposit to the account of S.S.X. Co. Abe Solen

special 2. Payable to the Green Co. Alice Madison

blank 3. James Amos

blank 4. P. D. Pierson

restrictive 5. For payment on promissory note #37802 P. Wolver

restrictive 6. For deposit only J. J. Adams

blank 7. The Big Horn Corp.

special 8. Payable to J. Swage Alp Smith

blank 9. P. & K. Turin Co.

blank 10. Iverson Co.

Checking services are available not only from banks, but also from savings and loan associations, credit unions, and investment firms. Not all services are available from all types of firms. In addition, many of these services are not available to corporations. It is worthwhile for the individual consumer and the corporate officer to investigate and compare available services.

Many of the "extra" services are available because of computers. An account holder can transfer funds from one account to another; receive interest on checking account balances; make deposits and withdrawals on a 24-hour basis from ATMs; have amounts such as a paycheck or retirement check automatically deposited; have installment payments or utility payments automatically withdrawn; and have purchases deducted through a debit card. The competition will change the services over time.

The account holder may also transfer funds from one account to another and make payments electronically through the use of a personal computer without actually writing a check by hand. Transfers and transactions made electronically, without a paper check, are termed **electronic funds transfer.**

electronic funds transfer

Reconciliation of a Bank Statement

Objective 2

Once a month (or less often by special arrangements) the bank sends a **bank statement** to the depositor. This statement is a record of all of the changes in the account that the bank is aware of since the last statement. It shows the checks that have been paid by the bank; the deposits received by the bank; service charges, if any; as well as other changes in the account balance.

bank statement

- *Ask how many students have their own checking accounts. Do they accept the balance that the bank provides?*

The balance on the statement will probably not agree with the balance in the check register or on the last check stub. This disagreement may not be due to inaccuracy on the part of either the bank or the depositor, because both parties have arrived at their balances by acting on the information available to them at the time. The bank statement and the checkbook records must be *reconciled*. This reconciliation procedure will determine the actual funds available in the account.

When the bank statement is received, the depositor must

1. Organize by check number the cancelled checks returned with the statement. These checks have been received by the bank, and their amounts have been subtracted from the account balance.
2. Identify what checks have *not* been received by the bank by comparing the returned checks with the check register or check stubs. These checks are termed **outstanding checks.** The total of their amounts must be subtracted from the bank statement balance. They represent a future reduction in the bank's account balance when the check is received by the bank.

   outstanding checks
3. Review the deposit slips returned with the bank statement or the indication of deposit received by the bank. This will identify the deposits made by the depositor but not received by the bank at the time the statement was prepared. These deposits are termed **deposits in transit.** The total deposits in transit must be added to the bank statement balance.

   deposits in transit
4. Review the bank statement to determine if any **service charges** have been specified. The total of these charges must be subtracted from the depositor's balance on the check register or check stub. The bank has already reduced the account balance for these charges as shown on the statement.

   service charges

   Service charges may be levied by the bank for many reasons, such as
   - for each check received
   - for each deposit made
   - for a monthly account use charge
   - for imprinting of checks
   - for an overdraft on the account balance
   - for a collection charge
   - for an ATM transaction

   All these charges are subtracted from the depositor's record (checkbook) as service charges.

5. Review the bank statement for other changes in the account balance. For example,
   An authorized automatic payment made from the account without a check written must be subtracted from the checkbook.
   A collection made by the bank and deposited to the account without a deposit slip must be added to the checkbook.
   A deposit made to the account without a deposit slip (i.e., payroll checks directly deposited to the account through wire transfer by the employer, Social Security checks directly deposited into the account by the federal government, or pension checks deposited by the employer) must be added to the checkbook.
   Checks returned for insufficient funds must be subtracted from the checkbook. Interest earned on the account balance must be added to the checkbook.

adjusted balance

The result of all of these changes in the bank statement balance is the determination of the actual funds available, known as the **adjusted balance.**

❍ *The adjusted balance means the amount of money that is available to spend.*

Reconciliation Model

Bank		**Checkbook**	
Balance (ending)	$ ______	Balance (ending)	$ ______
+ Deposits in Transit	+ ______	+ Credits	+ ______
− Outstanding Checks	− ______	− Service Charges	− ______
		− Debits	− ______
Adjusted Balance	$ ______	Adjusted Balance	$ ______

Example 1

Hal decided that first things should come first. He needed to find the adjusted balance in the current checking account so that he could pay the bills that are due. He reviewed the Sport-A-Merica, Inc. bank statement that showed an $863.70 balance. The check register showed a $690.40 balance. After organizing the cancelled checks in numerical order, he found that check number 1806 for $423.30 was still outstanding. A deposit for $267, made on the last day of the month, was not shown on the bank statement. Service charges totaled $63. A note for $80 was collected by the bank. The bank statement reconciliation appears here.

Sport-A-Merica, Inc.
Bank Statement Reconciliation
September 30, 20XX

Bank Statement Balance	$863.70	Check register balance	$690.40
Plus: Deposit in transit	+267.00	Less: Service charges	−63.00
	$1130.70		$627.40
Less: Outstanding check no. 1806	−423.30	Plus: Note collected	+80.00
Adjusted balance	$707.40	Adjusted balance	$707.40

(863.70 + 267 − 423.30 =   690.40 − 63.00 + 80.00 =)

Exercise B

Reconcile the information below to find the adjusted balance.

1. Statement balance = $877.98
Checkbook balance = $787.44
Service charge = $12.00

Checks outstanding =	No.	Amount
	471	$32.00
	477	25.17
	478	13.00
	479	32.37
		$102.54

Bank bal.	Chbk. bal.
$877.98	$787.44
−102.54	−12.00
$775.44	$775.44

2. Statement balance = $197.40
Checkbook balance = $168.80
Service charge = $9.00

Checks outstanding =	No.	Amount
	213	$14.80
	220	75.95
	221	21.85

Deposit in transit = $75.00

Bank bal.	*Ckbk. bal.*
$197.40	*$168.80*
+75.00	*−9.00*
$272.40	*$159.80*
−112.60	
$159.80	

3. Statement balance = $4,315.17
Checkbook balance = $3,901.07
Service charge = $6.00

Checks outstanding =	No.	Amount
	502	$312.00
	503	79.50
	504	131.26

Deposits in transit = $62.50, $40.16

Bank bal.	*Ckbk. bal.*
$4,315.17	*$3,901.07*
+102.66	*−6.00*
$4,417.83	*$3,895.07*
−522.76	
$3,895.07	

4. Statement balance = $1,430.00
Checkbook balance = $1,576.00
Service charge = $13.00

Checks outstanding =	No.	Amount
	32	$429.00
	37	362.70
	41	336.87
	42	133.43

Deposits in transit = $397.12, $65.36, $932.52

Bank bal.	*Ckbk. bal.*
$1,430.00	*$1,576.00*
+1,395.00	*−13.00*
$2,825.00	*$1,563.00*
−1,262.00	
$1,563.00	

Example 2

Kerry showed Hal her bank statement from August. Hal noted the service charge was only $31, which included $10 for new check imprinting and $1 each for the use of the ATM six times. She had a balance of $830 in her checkbook and $1,132 on the bank statement. The month's direct deposit of $600 from her annuity was shown on the bank statement, but not in her checkbook. Check number 107 for $83 was outstanding. A $350 deposit from her paycheck last week was not shown on the statement. Reconcile the balance.

BANK STATEMENT		CHECKBOOK	
Balance	$1,132	Balance	$830
Plus: Deposit in transit	350	Plus: Direct deposit	600
	$1,482		$1,430
Less: Check outstanding	83	Less: Service charges	31
Adjusted balance	$1,399	Adjusted balance	$1,399

(1,132 + 350 − 83 = *830 + 600 − 31 =)*

Exercise C

Reconcile the following information to find the adjusted balance. Note that these exercises include ATM, direct payment, and deposit activities.

1. Statement balance = $480
   Checkbook balance = $400
   Service charges = $1 for each ATM use (11 times this month)
   Direct deposit (by employer) = $400
   Deposits in transit = $309

Bank bal.	Ckbk. bal.
$480	$400
+309	−11
$789	+400
	$789

2. Statement balance = $683
   Checkbook balance = $718
   Note collected by bank = $500 (for the depositor)
   Note collection charge = $10
   Service charges = $18 monthly, $1 for each ATM use (12 times this month)
   Checks outstanding = No. 403 $350
   411 120
   Deposits in transit = $965

Bank bal.	*Ckbk. bal.*
$683	***$718***
+965	***+500***
−470	***−10***
$1,178	***−30***
	$1,178

3. Statement balance = $1,300
   Checkbook balance = $1,020
   Direct deposit (pension fund) = $380
   Direct deposit (social security check) = $176
   Checks outstanding = No. 142 $65.75
   145 50
   148 250
   149 10.45
   Deposits in transit = $652.20
   Service charges = $0 (due to high balance)

Bank bal.	*Ckbk. bal.*
$1,300.00	***$1,020***
+652.20	***+380***
−376.20	***+176***
$1,576.00	***$1,576***

4. Statement balance = $468
   Checkbook balance = $512
   Service charges = $22
   Direct deposit (by employer) = $238.88
   Checks outstanding = No. 178 $91.50
   179 50
   185 91.20
   Deposit in transit = $493.58

Bank bal.	*Ckbk. bal.*
$468.00	***$512.00***
+493.58	***−22.00***
−232.70	***+238.88***
$728.88	***$728.88***

5. Statement balance = $1,980
   Checkbook balance = $2,940
   ATM use = 7 times at $1.50 per use
   Service charge = $0 (high balance)
   Direct deposit (by pension fund) = $920
   Deposit in transit = $1,600
   Automatic payment = $309.50
   Checks outstanding = No. 981 $40

Bank bal.	*Ckbk. bal.*
$1,980	***$2,940.00***
+1,600	***−10.50***
−40	***+920.00***
$3,540	***−309.50***
	$3,540.00

ASSIGNMENT Chapter 4

Name ______________________ **Date** ______________ **Score** __________

Skill Problems

Identify the different types of endorsements.

1. *blank* Kim Funside
2. *special* Payable to Aree Singleton Kerri Donald
3. *restrictive* For deposit to account #22-807-5337 Sandra Belson
4. *restrictive* For deposit only Dennis Urie
5. *blank* Tulsa, Inc.
6. *special* Payable to the Reformed Church of Nicaragua Juan Cortez

Select a letter for each definition.

A. Depositor	B. Imprinting Checks	C. ATM	D. Deposit Slip
E. Check Register	F. Check Stub	G. Check	H. Payee
I. Bank Statement	J. Deposit in Transit	K. Outstanding Checks	
L. Service Charges	M. Adjusted Balance		

7. *E* The checking account balance from the records of the depositor.
8. *J* Coins, currency, and checks that have not been added to the records of the bank.
9. *C* A machine available 24 hours to accept deposits and provide cash withdrawals from accounts.
10. *K* Issued checks that have not cleared the bank.

Finish the following sentence with the proper term.

11. The *depositor* was given a set of deposit slips and blank checks.
12. Other charges against your checking account besides checks written may be costs for *imprinting checks* and bank *service charges*.
13. A record device to verify amount of money put into an account is a *deposit slip*.
14. Usually once a month, the bank sends a *bank statement* to the depositor.

Fill in the blanks with the above terms to complete the Reconciliation Model.

Reconciliation Model

Bank		Checkbook	
Balance (ending)	$ 1,282	Balance (ending)	$ 815
15. + *Deposits in Transit*	+ 200	+ Direct Deposit	+ 600
16. − *Outstanding Checks*	− 83	17. − *Service Charges*	− 16
18. *Adjusted Balance*	$ 1,399	*Adjusted Balance*	$ 1,399

Reconcile the information below to find the adjusted balance.

19. Statement balance = $987.26
    Checkbook balance = $1,182.53
    Service Charge = $16.50
    Deposits in transit = $500.00

Cks. Out:		*Bank bal.*	*Ckbk. bal.*
601	$ 10.99	*$ 987.26*	*$1,182.53*
605	6.78	*− 321.23*	*−16.50*
606	295.01	*+ 500.00*	*$1,166.03*
609	8.45	*$1,166.03*	
	$321.23		

Reconcile the information below to find the adjusted balance.

20. Statement balance = $54.36
Checkbook balance = $42.20
Service Charge = $7.50 monthly
6 ATM uses, $1 each
1 Overdraft, $27.50
Deposits in transit = $482.04

Cks. Out:	
822	$ 60.50
823	54.80
824	399.95
830	19.95
	$533.20

Bank bal.	Ckbk. bal.
$ 54.36	*$ 42.20*
− 535.20	*−41.00*
+ 482.04	*$ 1.20*
$ 1.20	

21. Statement balance = $4,631.40
Checkbook balance = $1,740
Deposit in transit = $1,287.60
Checks outstanding = $4,220
Service charge = $41

Bank bal.	Ckbk. bal.
$4,631.40	*$1,740.00*
+1,287.60	*− 41.00*
−4,220.00	*$1,699.00*
$1,699.00	

22. Statement balance = $4,057.50
Checkbook balance = $2,418.74
Service charge = $7.50 monthly; 255 checks, $.25 each; 20 deposits, $.35 each; Earnings credit on high balance, $78.25
Note collection fee, $6.00
Notes collected by bank for our company from our customer = $957.64

Cks. Out:	
14051	$687.12

Bank bal.	Ckbk. bal.
$4,057.50	*$2,418.74*
−687.12	*− 6.00*
$3,370.38	*+ 957.64*
	$3,370.38

23. Statement balance = $2,467.15
Checkbook balance = $2,607.12
Health ins. deduction = $917.00
Deposits in transit = $99.45, $41.05, $65.74, 124.35
Service charge = $14.35
Customer's check non-collectable = $12.44

Cks. Out:	
1406	$ 375.40
1440	214.00
1402	545.01
	$1,134.41

Bank bal.	Ckbk. bal.
$2,467.15	*$2,607.12*
−1,134.41	*− 917.00*
+ 330.59	*− 14.35*
$1,663.33	*− 12.44*
	$1,663.33

Using the information from question 23, complete the following check register.

Number	Date	Description	Amount of Check	✓ T	Amount of Deposit	Balance
	6/8/xx	Balance Forward				$ 3,296.09
1439	6/9/xx	Thuets Products-May	$245.05			**24.** *3,051.04*
1440	6/9/xx	PHBS-May	**25.** *214.00*			**26.** *2,837.04*
	6/9/xx	Monday's Receipts			$99.45	**27.** *2,936.49*
1441	6/10/xx	Gas & Go	15.50			**28.** *2,920.99*
	6/10/xx	Tuesday's Receipts			41.05	**29.** *2,962.04*
	6/11/xx	Wednesday's Receipts			65.74	**30.** *3,027.78*
1402	6/12/xx	Roger M Thayer-wages	**31.** *545.01*			**32.** *2,482.77*
	6/12/xx	Thursday's Receipts			124.35	**33.** *2,607.12*
	6/12/xx	Reconciliation				*2,607.12*
		34. *Service charge*	*14.35*			**35.** *2,592.77*
		36. *Returned Check*	*12.44*			**37.** *2,580.33*
		38. *Health Insurance*	*917.00*			**39.** *$1,663.33*

40. The answer in number 39 is termed the *adjusted balance* of the check register, which is the actual funds available.

Business Application Problems

Find the adjusted balance in the following problems unless otherwise directed.

1. Jamesson Company has a balance in its checking account of $3,587.50 at the end of the month. The statement indicated a balance of $3,662.50. The statement indicated a service charge of $28. Two checks for a total of $564 were not shown on the statement. A deposit of $461 was not shown on the statement. Reconcile the account.

Bank	*Checkbook*
$3,662.50	*$3,587.50*
+ 461.00	*− 28.00*
− 564.00	*$3,559.50*
$3,559.50	

2. Barnes, Thorner and Company's check register shows a balance of $586. Its accountant found the following by comparing the company records to the statement.

Checks Outstanding:		Deposits in Transit:
No. 492	$650	$912
495	497.50	261

The bank statement balance was $545.50. If the records of both the bank and the company are correct, what is the amount of the bank service charge?

Bank	*Checkbook*
$545.50	*$586*
+1,173.00	*−571*
−1,147.50	*$15 Service charge*
$571.00	

3. Bosditch, Inc. received their bank statement for last month showing a balance of $1,687. The checkbook shows a balance of $1,557.50. The accountant found a check was still outstanding for $226.90 and a deposit for $67.40 was not shown on the bank statement. A service charge of $30 was shown on the bank statement.

Bank	*Checkbook*
$1,687.00	*$1,557.50*
+67.40	*−30.00*
−226.90	*$1,527.50*
$1,527.50	

4. The check register for the Biggs Corporation is shown below. Except for checks 339 and 341, all the checks were shown on the bank statement. The bank statement balance was $5,044.70.

10 BE SURE TO DEDUCT ANY PER ITEM CHARGES, SERVICE CHARGES OR FEES THAT MAY APPLY IN OTHER COLUMN

DATE		CHECK NUMBER	TRANSACTION DESCRIPTION	✓ T	(−) AMOUNT OF PAYMENT OR DEBIT		(• OR •) OTHER		(+) AMOUNT OF CREDIT		BALANCE FORWARD	
10	12	335	TO: Jackson, Inc. FOR:		345	60					12,556	20
10	15	336	TO: Marked Charge Co. FOR:		569	-					11,987	20
10	15	337	TO: Al DeRow FOR:		45	-					11,942	20
10	22		TO: FOR:						1,982	50	13,924	70
10	25	338	TO: First Federal Bank FOR:		855	-					13,069	70
10	28	339	TO: Ace Properties FOR:		1,500	-					11,569	70
10	28	340	TO: Top Investments FOR:		8,000	-					3,569	70
10	31	341	TO: I.R.S. FOR:		2,250	-					1,319	70

From the incomplete information available, determine the adjusted bank balance.

Bank
$5,044.70
−3,750.00
$1,294.70 adjusted bank balance

5. From the incomplete information in problem 4, determine the service charge for the Biggs Corporation.
$1,319.70 − $1,294.70 = $25.00 service charge

6. Jay Marks reviewed The Mathews Inn bank statement for the month which showed a balance of $907.50. The checkbook shows a balance of $869.92. The business manager found that a check was still outstanding for $85.38 and a deposit for $24.90 was not shown on the bank statement. A service charge of $22.90 was shown on the bank statement.

Bank	***Checkbook***
$907.50	***$869.92***
+24.90	***−22.90***
−85.38	***$847.02***
$847.02	

7. Bands Corporation's check register is shown below. Check 511 is the only outstanding check. The deposit made on June 30 is not shown on the bank statement. The bank statement shows a balance of $3,674.95.

10 BE SURE TO DEDUCT ANY PER ITEM CHARGES, SERVICE CHARGES OR FEES THAT MAY APPLY IN OTHER COLUMN

DATE		CHECK NUMBER	TRANSACTION DESCRIPTION	✓ T	(−) AMOUNT OF PAYMENT OR DEBIT	(• OR •) OTHER	(+) AMOUNT OF CREDIT	BALANCE FORWARD
6	10	506	TO: Barker, Inc. FOR:		45 -			7,556 20
6	15	507	TO: Charter Bank FOR:		950 -			6,606 20
6	17		TO: FOR:				900 -	7,506 20
6	22	508	TO: Hill Electronics FOR:		1,459 50			6,046 70
6	25	509	TO: First National Bank FOR:		823 75			5,222 95
6	26	510	TO: Ace Properties FOR:		1,500 -			3,722 95
6	27	511	TO: Partical Quant. FOR:		700 -			3,022 95
6	30		TO: FOR:				1,750 -	4,772 95

From the incomplete information available, determine the adjusted bank balance.

Bank

$3,674.95
+1,750.00
−700.00
$4,724.95 adjusted bank balance

8. The bank statement for Toddler Toys, Inc. indicated the following information:

1. The checkbook balance was $3,975.80.
2. There was a service charge on the bank statement for $33.
3. The total of the checks outstanding was $197.50.
4. A deposit made last week for $2,550 was not shown on the bank statement.
5. The bank statement showed a current balance of $1,590.30.

Bank ***Checkbook***

$1,590.30 ***$3,975.80***
+2,550.00 ***−33.00***
−197.50 ***$3,942.80***
$3,942.80

9. Margin, Inc. received their bank statement for the month of July showing a balance of $6,063.50. The review of the cash account showed a balance of $7,341. A service charge for $83 was shown on the bank statement. There was al[illegible] a check outstanding in the amount of $255.50 and two recent deposits totaling $1,450 were not shown on the bank statement.

Bank ***Checkbook***

$6,063.50 ***$7,341***
+1,450.00 ***−83***
−255.50 ***$7,258***
$7,258.00

10. The checkbook for Melger and Associates indicates a balance of $2,528.60. However, the July bank statement showed a balance of $2,972.10. The accounting clerk found that a check was still outstanding for $1,452. The deposit made last week for $952.50 was not shown on the bank statement. A service charge of $56 was also listed on the bank statement.

Bank ***Checkbook***

$2,972.10 ***2,528.60***
+952.50 ***−56.00***
−1,452.00 ***$2,472.60***
$2,472.60

11. The Sperry Company check register shows a balance of $1,946.80. The bank statement shows a balance of $2,650.47. After a review of the bank statement, the following information is determined.

Checks Outstanding = No. 406 $79.80
471 20.00
472 720.50
481 327.33

Service Charges = $27.00
Deposits in Transit = $67.33, $109.50, $240.13

Bank bal. ***Ckbk. bal.***

$2,650.47 ***$1,946.80***
+416.96 ***−27.00***
$3,067.43 ***$1,919.80***
−1,147.63
$1,919.80

12. The Bengrin Company received the bank statement for the month of August on the thirty-first. The statement balance was $3,218.40, which compared to $3,176.41 in the cash account of the balance sheet. The $3,176.41 also agreed with the balance in the check register. The accountant reviewed the statement and found that the following points required reconciliation: check No. 493 for $357.18 had not been cashed, the deposits for $195 and $82.19 made on the thirtieth were not shown on the bank statement, and the service charge of ????? was not recorded on the check register.

Bank bal.

$3,218.40 ***$3,176.41***
+277.19 ***−3,138.41***
$3,495.59 ***$38.00 service charge***
−357.18
$3,138.41

13. The White Water Corporation bank statement revealed these items:

A balance of $147.30

A check for $31 deposited by the firm that was returned because there were insufficient funds in B. Bennett's account

Check 141 for $101 that was not cashed by the payee

A deposit for $78 that was not received by the bank when the statement was prepared

A service charge of $20 that has already been subtracted from the bank account balance

The firm's check register shows a $175.30 balance.

Bank bal. ***Ckbk. bal.***

$147.30 ***$175.30***
+78.00 ***−31.00***
$225.30 ***$144.30***
−101.00 ***−20.00***
$124.30 ***$124.30***

14. The Valley Inn has a balance of $5,698.03 on its check register. The bank statement shows a balance of $5,995.90. The following information is determined after a close review of the bank statement and the check register.

Checks Outstanding:		Deposits in Transit:
No. 13	$420.00	$165.00
23	270.50	203.60
31	100.00	103.03

Service charge (monthly account use): $21.00
Reconcile the bank statement.

Bank bal.	*Ckbk. bal.*
$5,995.90	*$5,698.03*
+471.63	*−21.00*
$6,467.53	*$5,677.03*
−790.50	
$5,677.03	

15. Cyndi found that the bank statement balance and the check register had the same balance of $89. After a review of the statement she found there was a $13 deposit in transit as well as a $29 outstanding check. She also found that she had failed to subtract a $4.50 check in her check register. The service charge amount on the statement was unreadable. What was the amount of the service charge?

Bank bal.	*Ckbk. bal.*	
$89.00	*$89.00*	*$84.50*
+13.00	*−4.50*	*−73.00*
$102.00	*$84.50*	*$11.50 service charge*
−29.00		
$73.00		

16. Scanloni Imports, Inc. received a bank statement on October 10 that revealed a balance of $1,680. The manager, Sally Scanloni, noted that the statement did not include the $150 deposit she made earlier in the week or the two checks she wrote yesterday. The checks totaled $483.75. There was also a service charge for $13. What should be the balance in the account?

Bank bal.
$1,680.00
+150.00
1,830.00
−483.75
$1,346.25 current balance

17. The Benji Company's check register balance is $73.60. The bank statement balance is $1,002. The following information is determined after a review of the bank statement and the check register:

Checks Outstanding:		Service Charges:
No. 140	$163.80	$18.50 for account use
181	980.00	8.00 for note collection
185	47.23	

Deposits in Transit: $18, $482, and $67.41
What was the value of the note collected by the bank?

Bank bal.	*Ckbk. bal.*	
$1,002.00	*$73.60*	*$378.38*
+567.41	*−26.50*	*−47.10*
$1,569.41	*$47.10*	*$331.28 value of note*
−1,191.03		
$378.38		

18. The check register for the Support Service Company is shown below. All the checks through No. 841 were included in the month's statement. The deposit of October 29 is not shown on the bank statement. The service charge is $24.

PLEASE BE SURE TO **DEDUCT** ANY PER CHECK CHARGES OR SERVICE CHARGES THAT MAY APPLY TO YOUR ACCOUNT

NUMBER	DATE	CHECKS ISSUED TO OR DESCRIPTION OF DEPOSIT	(−) AMOUNT OF CHECK	✓ T	(−) CHECK FEE (IF ANY)	(+) AMOUNT OF DEPOSIT	BALANCE
							1680.48
839	10/1	TO/FOR ABERESON C	16.00	✓			−16.00 BAL 1664.48
840	10/6	TO/FOR SMITH CORP.	822.20				−822.20 BAL 842.28
841	10/9	TO/FOR O. PETERSON	327.00				−327.00 BAL 515.28
842	10/10	TO/FOR SCHMIDT & SONS	5.00				−5.00 BAL 510.28
843	10/16	TO/FOR SMITH CORP.	200.00				−200.00 BAL 310.28
844	10/23	TO/FOR H.D. HEBNER	29.50				−29.50 BAL 280.78
		TO/FOR DEPOSIT			10/29	583.00	+583.00 BAL 863.78

The balance on the bank statement is $491.28. What is the adjusted balance?

Bank bal.	**Ckbk. bal.**
$491.28	**$863.78**
+583.00	**−24.00**
$1,074.28	**$839.78**
−234.50	
$839.78	

19. Susan Bigsby noted that her credit union had charged her account $1 for each of the six times she had used the automatic teller machine in the month of August. There were $10 in other service charges. Her statement balance was $265. She had made one deposit since the statement date for $160. The statement indicated an automatic payment from the account in the amount of $32. Her records showed a balance of $473. She had no outstanding checks. What is the adjusted account balance?

Bank bal.	*Ckbk. bal.*
$265	*$473*
+160	*−16*
$425	*−32*
	$425

20. The November bank statement for the Pondi Company is shown here. The statement does not show a deposit made on November 30 for $587.29, or two checks outstanding—No. 862 for $27 and No. 867 for $368.50. What is the adjusted balance for the month?

Bank bal.
$1,620.52
+587.29
$2,207.81
−395.50
$1,812.31

DESERT NATIONAL BANK
200 Knight Street
Woodman, AZ 85257
CHECKING ACCOUNT STATEMENT

ACCOUNT NUMBER: 307-4531-03

Pondi Company
1802 Centerline Road
Debonet, Illinois 06061

PERIOD
FROM Nov. 1
TO Nov. 30

DATE	TRANSACTION CODE	TRANSACTION AMOUNT	TRANSACTION CODE	TRANSACTION AMOUNT	TRANSACTION CODE	TRANSACTION AMOUNT	BALANCE
Nov. 3	1	223 40					1 085 50
Nov. 8	1	750 00	8	1 000 00			835 50
Nov. 10	8	321 80					513 70
Nov. 20	8	40 00	8	125 00	1	300 00	648 70
Nov. 21	1	1 000 00	8	87 50	8	500 00	1 061 20
Nov. 27	1	567 82					1 629 02
Nov. 30	SC	8 50					1 620 52

INCLUDED IN STATEMENT

BEGINNING BALANCE	LOW BALANCE DATE	LOW BALANCE AMOUNT	MO. AVE. AVAILABLE BALANCE
$862.10	Nov. 20	348.70	$537.20

NO. CREDITS	NO. DEPOSITED ITEMS	TOTAL ADDED FOR CREDITS
0	5	$2,841.22
NO. DEBITS	**NO. DR. CHARGED FOR**	**TOTAL DEDUCTED FOR DEBITS**
1	6	$2,082.80

TRANSACTION CODES

CREDITS
1–DEPOSIT
3–CREDIT MEMO
6–CREDIT CASH CONSOLIDATION
7–LIST CORRECTION
10–CHECK CORRECTION
12–CHARGE CORRECTION

DEBITS
BLANK-CHECK
5–DEPOSIT CORRECTION
8–LIST
29–CERTIFIED CHECK MEMO
40–MISC. BANK CHARGE
46–DEBIT CASH CONSOLIDATION
60–DEBIT MEMO
SC–SERVICE CHARGE
30–SERVICE CHARGE MEMO
– –OVERDRAWN BALANCE

PLEASE RECONCILE YOUR STATEMENT PROMPTLY

21. ProJet, Inc. check register shows a $450.40 balance. A computer printer skip on the bank statement conceals the statement balance, but a review of it does reveal the following information.

a. Service charges: $21.00

b. Deposits in transit: $87, $600, and $273

c. Checks outstanding: No. 87 $76.23

No. 87	$76.23
88	$185
98	$632.50

Ckbk. bal.
$450.40
−21.00
$429.40

22. Mary Travel, Inc. has a balance of $2,724.63 in its check register. The bank statement shows a balance of $3,027.40. A review of the check register and bank statement shows the following information:

Checks Outstanding:		Deposits in Transit:
No. 328	$421.90	$368.60
345	370.50	103.03

Service charge (monthly account use): $18
Reconcile the bank statement.

Bank bal.	*Ckbk. bal.*
$3,027.40	*$2,724.63*
+471.63	*−18.00*
$3,499.03	*$2,706.63*
−792.40	
$2,706.63	

23. Write a complete sentence in answer to this question. Able Movers found that their bank statement was not complete with regard to the deposits that they had made yesterday for $3,690. The statement balance was $1,856. There were four checks outstanding that totaled $290. A service charge of $56 for monthly service, transaction charges, and collections was also noted on the statement. The bookkeeper was not available to adjust the balance. What should the balance be according to the available information?

Bank bal.
$1,856
+3,690
−290
$5,256 balance
The bank information indicates a $5,256 balance.

24. Write a complete sentence in answer to this question. John's bank statement showed a charge for use of his automatic teller privileges in the amount of $1.50 per use. John used the auto teller seven times over the course of the month. On his account was also a flat service charge of $12.00. If his checkbook balance is correct at $61.30, what is his adjusted balance?

$61.30 Balance
−10.50 ($1.50 × 7) } *service charges*
−12.00
$38.80 adjusted balance
The adjusted balance is $38.80.

25. Write a complete sentence in answer to this question. Karen Wallace was notified by her bank that she had not properly endorsed a check for deposit. The check amount was $4,690. The bank indicated over the phone that her balance would be affected by this problem until she could come to the bank and straighten it out. Karen had also transferred $500 from her savings account to the checking account just yesterday. The balance shown in the checkbook is $3,852 before the transfer is recorded. What is her real balance at this point?

$3,852 − $4,690 + $500 = $338 overdrawn
The real balance is an overdraw of $338.

★ Challenge Problem

The Polk & Stern Company found that a $63.50 check had not been recorded in the check register. The register's uncorrected balance was $3,684.72. The bank statement balance was $1,990.32. A note had been collected by the bank for the depositor for $750 and a charge of $5 was levied against the account for this collection service. A $423.50 check deposited by the company was returned because it was improperly endorsed. New checks were imprinted by the bank for a charge of $25.80. The company had deposited $1,476.60 as the week's receipts since the bank prepared the monthly statement. A $450 monthly interest payment was shown on the bank statement as having been automatically withdrawn from the account. There was no monthly use service charge on the account because of the amount of the balance. What is the adjusted balance?

Bank bal.	*Ckbk. bal.*
$1,990.32	*$3,684.72*
+1,476.60	*+750.00*
$3,466.92	*$4,434.72*
	−5.00
	$4,429.72
	−423.50
	$4,006.22
	−25.80
	$3,980.42
	−450.00
	$3,530.42
	−63.50
	$3,466.92

MASTERY TEST Chapter 4

Name ______________________ Date ______________ Score ____________

1. Using your name as payee, give an example of:
   a. A restrictive endorsement *For deposit only G.S. Student*
   b. A blank endorsement *Gerry S. Student*
   c. A special endorsement *Payable to College, G.S. Student*

2. The Baird Corp. check register shows a $1,063.08 balance. The bank statement shows a balance of $1,088. The accountant reviews the bank statement and the check register and finds the following information:

Checks Outstanding:	Deposits in Transit:
No. 184—$161.92	$165.00
187—$180.00	$ 80.00
188—$105.50	$311.97
No service charges.	
Automatic payment from account from bank: $100	Deposit to account without deposit slip: $234.47

Reconcile the bank statement and find the adjusted balance.

Bank Bal.	*Ckbk. Bal.*
$1,088.00	*$1,063.08*
+ 556.97	*− 100.00*
$1,644.97	*$963.08*
− 447.42	*+ 234.47*
$1,197.55	*$1,197.55*

3. Calvin York has a balance of $92.61 in his check register as of April 30. The bank statement received on May 2 indicates a $47 balance. After reviewing both the check register and the bank statement, he finds that the deposit that he made on April 30 for $25 was omitted. He also finds that a check that he had deposited on April 25 was returned because the account holder did not have enough funds to cover the check. The amount of the check is $47.43. There are also two checks outstanding: No. 103 for $12.80 and No. 107 for $25. What is the service charge?

Bank Bal.	*Ckbk. Bal.*	
$47.00	*$92.61*	*$45.18*
+ 25.00	*− 47.43*	*− 34.20*
$72.00	*$45.18*	*$10.98 service charge*
− 37.80		
$34.20		

4. Jon Sasinowski has received a message from his bank that his account has been overdrawn. The last statement of balance was $89.70. Since the date of the statement there has only been one check written for $97.10. The statement had included a service charge of $16. There has not been a deposit made to the account since the statement. By how much has he overdrawn his account?

Bank Bal.
$89.70
− 97.10
$ 7.40 overdrawn

5. From the information given, reconcile the balance. There is a notation that the deposit on June 13 of a check for \$1,962.80 was returned to the account holder due to insufficient funds. A service charge of \$9 was also noted. Checks No. 474 and 476 were not shown on the statement. There were no outstanding checks from previous months.

PLEASE BE SURE TO **DEDUCT** ANY PER CHECK CHARGES OR SERVICE CHARGES THAT MAY APPLY TO YOUR ACCOUNT

NUMBER	DATE	CHECKS ISSUED TO OR DESCRIPTION OF DEPOSIT	(−) AMOUNT OF CHECK	✓ T	(−) CHECK FEE (IF ANY)	(+) AMOUNT OF DEPOSIT		BALANCE
								940 20
470	6/7	TO/FOR CAMPBELL ASSOC.	50 00	✓				-50 00
							BAL	890 20
471	6/10	TO/FOR BEARDSON CO.	682 40					-682 40
							BAL	207 80
		TO/FOR DEPOSIT			6/13	1962 80		+1962 80
							BAL	2170 60
472	6/20	TO/FOR LITTLE & SONS	423 16					-423 16
							BAL	1747 44
473	6/27	TO/FOR PEARSON CORP.	800 00					-800 00
							BAL	947 44
474	6/27	TO/FOR McTAGGART & CO	12 50					-12 50
							BAL	934 94
475	6/29	TO/FOR R.E. EMERSON	25 60					-25 60
							BAL	909 34
		TO/FOR DEPOSIT			6/29	1827 00		+1827 00
							BAL	2736 34
476	6/30	TO/FOR J. OLSONSITE CORP.	10 00					-10 00
							BAL	2726 34
		TO/FOR						
							BAL	
		TO/FOR						
							BAL	
		TO/FOR						
							BAL	
		TO/FOR						
							BAL	

REMEMBER TO RECORD AUTOMATIC PAYMENTS / DEPOSITS ON DATE AUTHORIZED.

Ckbk. Bal.

	\$2,726.34
−	**1,962.80**
	\$ 763.54
−	**9.00**
	\$ 754.54 balance

Payroll

5

Suggested Classroom Time for This Chapter

hours 1 2 3 4 5

Perceived Student Difficulty with Topical Material

easy moderate difficult

Objectives

After mastering the material in this chapter, you will be able to:

1. Determine gross earnings for an employee. p. 84
2. Determine net pay for an employee. p. 90
3. Determine the employer's share of payroll taxes. p. 95
4. Understand and use the following terms:
   Pay System
   Gross Earnings
   Time Card
   Payroll Register
   Net Pay
   Gross Sales
   Sales Returns
   Sales Allowances
   Net Sales
   Straight Commission
   Commission Formula
   Deductions
   Exemption
   Form W-4
   Taxable Base
   Tax Rate
   FUTA

Sport-A-Merica

The volume of business has been very good the past four months. Kerry is beginning to feel the extra burden of spending so many hours at the store. With the Christmas season about to begin, she is thinking about adding one or two more sales staff to ease the work load. Hiring the extra salespeople now would allow a little time to train them to use the cash register, improve their sales techniques, and inform them about the products.

New people added to the staff would increase the coverage on the sales floor but would also mean higher payroll costs. Kerry would be able to do some additional outside sales and public relations work while the new people handle the floor sales. After all, Kerry's success as an athlete was one of the reasons the business was able to gain momentum so quickly.

Kerry knows additional people will also mean more paperwork, government forms, taxes, and expenses above the simple rate of pay to the employee. On the other hand, Brenda, the current sales clerk, has complained about the overtime hours she has had to work in recent months. Additional employees would help reduce Brenda's overtime hours, reduce Sport-A-Merica's overtime expense, and allow more coverage of the floor as the holiday season begins. Kerry is sure that it costs Sport-A-Merica over $450 a week to have Brenda work all those hours. Hal agrees hiring one or two more employees is a good idea while business is booming.

Kerry would like to hire a sales representative to call on schools if he or she would work on a commission basis. She estimates that Sport-A-Merica is missing over $2,000,000 in business by not employing an outside sales representative. How much could Sport-A-Merica pay such a qualified employee?

Kerry and Hal will need to determine the cost of paying the employees as part of their planning.

Determining Gross Earnings

Objective 1

pay system

All organizations that have paid employees must maintain payroll records to determine the amount the employee has earned. The company and the employee enter into an agreement at the time of employment. This agreement determines the type of **pay system** on which the employee will be compensated. There are four basic types of pay systems:

1. A *salary* is paid to employees for a pay period during which the hours or days to be worked may not be limited. This type of pay system is usually part of an open-ended agreement that requires the employee to complete specified tasks. This type of pay system is often provided to managerial employees.

● ***Ask students to indicate examples of pay systems that they have experienced.***

2. An *hourly rate* pay system is used in a variety of situations and is designed to compensate an employee for the amount of time spent on the job. Specified work standards are established to indicate the expected level of output. Under this system, the more hours worked, the more earned. This is the most frequently used pay system. It is almost exclusively found in entry-level jobs and often in jobs involving repetitious work.

❍ *The piece-rate system is used in garment manufacturing, some other manufacturing businesses, and agriculture.*

3. A *piece-rate* pay system is used when an operation is repetitious. This system compensates an employee with more pay for more output. The employee is paid a specified amount for each operation, pieces produced, or cycles completed. The more an employee produces, the more he or she earns. Occasionally the employee is actually paid more per piece once a minimum has been produced. This is termed differential piece rate.
4. A *commission* pay system is used in conjunction with the selling of goods or services. This system compensates the salesperson for the amount of sales made, not the number of hours worked. A specified percent of the value of the items sold is paid to the salesperson as a commission. The more items sold, the more earned.

These four basic systems are often combined and are used in conjunction with each other to form a more flexible pay system.

Example 1

1. A combination salary and commission can be used to provide a basic income source for a salesperson in difficult times, as well as to motivate the employee to sell more.
2. A combination hourly and piece-rate system can be used for the same purpose.
3. A combination salary and hourly system can be used to compensate managerial employees for excessive work hours.

These are examples of how pay system combinations can be developed to meet the needs of the organization and its employees.

❍ *What the employee earns with each paycheck is determined in part by the number of paychecks issued each year.*

The frequency of payment is also determined at the time of employment. This is known as the *pay period.*

An employee on an annual salary of $20,800 could be paid as follows:

Pay Period	Number of Pay Periods per Year	Amount Paid per Period
Weekly	52	$ 400
Biweekly	26	800
Semimonthly	24	866.67
Monthly	12	1,733.33

Overtime

For employees paid on a hourly rate system, the Fair Labor Standards Act of 1938 requires payment of one and one-half times the regular wage rate on hours in excess of forty worked by an employee in one week. The effect of this is shown in example 2.

Example 2

Brenda, a Sport-A-Merica employee, works 46 hours each week. Her regular wage is $9.00 per hour. Do her gross earnings exceed $450 a week?

40 hours × $9.00/hour =	$360.00
6 hours × $13.50/hour $\left(\$9.00 \times 1\frac{1}{2}\right)$ =	+81.00
	$441.00

○ *In an hourly system, the more hours worked, the greater the gross earnings.*

Brenda's gross earnings do not exceed $450. *(40 × 9.00 6 × 13.50 =)*

gross earnings

The total, $441.00, is her **gross earnings** for the pay period. This is not the amount that she will receive on her paycheck, but it is the amount that she has earned. Brenda's hours may be recorded by a time clock. Each time she enters or leaves her place of employment she selects her **time card** from a rack and places it in the time clock. The time clock records the time on the card. The record on the card shows the hours she works and is used to compute her gross pay. Hourly employees are usually paid weekly.

time card

This information—hours worked, hourly rate of pay, and gross earnings—is entered into a **payroll record** (sometimes termed a *payroll register*) by a payroll clerk. This computerized data is used to determine **net pay.**

payroll record
net pay

Exercise A

Redi Transport Company pays its employees weekly. The hours worked in excess of forty per week are paid at the rate of one and one-half times the regular wages as prescribed by law. Determine the gross earnings for the following employees.

Name	Title	Hourly Wage	Hours Worked	Regular Earnings	Overtime Earnings	Total Gross Earnings
1. A. Bennett	Associate	$10.00 (10 × 40 =)	43 (15 × 3 =)	$400.00	$45.00	$445.00
2. J. Dixon	Associate	9.50 *(9.50 × 40 =)*	42 *(14.25 × 2 =)*	*380.00*	*28.50*	*408.50*
3. L. Pierce	Associate	8.00 *(8 × 40 =)*	41 *(12 × 1 =)*	*320.00*	*12.00*	*332.00*
4. O. Russell	Sr. Associate	22.00 *(22 × 40 =)*	42 *(33 × 2 =)*	*880.00*	*66.00*	*946.00*
5. B. Smith	Trainee	8.00 *(8 × 40 =)*	40	*320.00*	*0*	*320.00*
6. K. Williams	Supervisor	18.00 *(18 × 38 =)*	38	*684.00*	*0*	*684.00*
7. J. Wilson	Associate	9.00 *(9 × 40 =)*	41 *(13.50 × 1 =)*	*360.00*	*13.50*	*373.50*

Exercise B

Williamson Manufacturing pays its employees weekly. The company compensates its employees one and one-half times for all hours worked in excess of eight each day (Monday through Friday), whether or not they work in excess of forty each week. All hours worked on Saturday are one and one-half times regular pay. All Sunday hours are triple time. Determine the gross earnings for the following employees.

Name	M	T	W	Th	F	Sa	Su	Hourly Rate	Regular Earnings	Overtime Earnings	Gross Earnings
1. L. Becker	8	9	7	9	9	8	5	$ 8.00	$312.00	$252.00	$564.00
(8.00 × 39 =)	(12.00 × 11 =)			(24.00 × 5 =)							
2. J. Crimmins	7	6	7	8	8			8.20	*295.20*	*0*	*295.20*
(8.20 × 36 =)											
3. B. Fortos	6	8	8	9	9	9	8	8.10	*307.80*	*328.05*	*635.85*
(8.10 × 38 =)	*(12.15 × 11 =)*			*(24.30 × 8 =)*							
4. T. Jones	9	9	8	7	8			12.50	*487.50*	*37.50*	*525.00*
(12.50 × 39 =)	*(18.75 × 2 =)*										
5. B. Lemis	5	7	8	10	8			16.00	*576.00*	*48.00*	*624.00*
(16.00 × 36 =)	*(24.00 × 2 =)*										
6. K. Mason	10	9	6	8	8	8	5	9.50	*361.00*	*299.25*	*660.25*
(9.50 × 38 =)	*(14.25 × 11 =)*			*(28.50 × 5 =)*							
7. A. Rust	8	10	7	8	9	9	3	8.50	*331.50*	*229.50*	*561.00*
(8.50 × 39 =)	*(12.75 × 12 =)*			*(25.50 × 3 =)*							
8. P. Zist	8	9	7	9	9	8		10.00	*390.00*	*165.00*	*555.00*
(10.00 × 39 =)	*(15.00 × 11 =)*										

❍ The more pieces made on a piece-rate system, the greater the gross earnings.

If an employee is paid on a piece-rate basis, the weekly wage is determined by multiplying the number of pieces produced by the amount paid to produce each piece. If the employee produces 1,463 pieces in a week and will be paid 32¢ for each piece produced, the employee's gross earnings for that week will be $468.16 (1,463 × .32). If 1,289 pieces are produced the next week, the gross earnings will be $412.48 (1,289 × .32).

gross sales

For an employee who is paid on a commission system, the pay is based on the value of the item sold. The total value of the sales is termed **gross sales.** The product of price per unit and the number of units sold will determine gross sales. Occasionally the amount of the sale is changed because of occurrences after the sale is completed. Should a customer find that the product is not what was expected—the wrong color, size, or quantity, or other such circumstances—the merchandise may be returned to the seller. The seller gives the buyer credit for the return by reducing the total invoice amount. This action also reduces the sale amount for the salesperson. This reduction in sales is termed **sales returns.**

sales returns

sales allowances

Another factor that may reduce sales is termed **sales allowances.** Instead of asking the buyer to return the merchandise for credit, the seller may choose to offer a price reduction to the buyer. If the buyer keeps (and sells) merchandise that is damaged or is the wrong color, size, etc., the buyer is given a sales allowance.

❍ The more items sold on a commission pay system, the greater the gross earnings.

net sales

Sales returns and sales allowances reduce the amount of actual merchandise sold and, therefore, the amount of commission earned. By subtracting the value of sales returns and sales allowances from gross sales, **net sales** can be determined. This is the amount on which a commission is calculated.

gross sales
−sales returns
−sales allowances
net sales

Example 2

Sharon Pierson sold 4,072 units of her firm's product in one week. The selling price per item was $4.95. The selling firm had twenty-eight units returned and made a 20 percent (.20) price allowance on another forty units. What were the net sales for Sharon for the week?

4,072	= Units sold
×$4.95	= Price per unit
$20,156.40	= Gross sales
−138.60	= Less sales returns (28 units × $4.95 per unit)
−39.60	= Less sales allowances (40 units × $4.95 × 20%)
$19,978.20	= Net sales for the week

(4,072 × 4.95 = −138.60 − 39.60 =)

Exercise C

Determine the net sales for each employee.

Employee	Units Sold	Price/Unit	Units Returned	Sales Allowance Units	Sales Allowance Percent	Net Sales
1. L. Potter (963 − 12 = 951 951 × 60 =)	963	$60.00	12	2	10% (2 × 60 × .10 =)	$57,048.00
2. M. Wierick ***(900 − 420 = 480 480 × 2.20 =)***	900	2.20	420	10	18 ***(10 × 2.20 × .18 =)***	***1,052.04***
3. T. Bowden ***(2,500 × 8 =)***	2,500	8.00	0	9	7 ***(9 × 8 × .07 =)***	***19,994.96***
4. M. Maxwell ***(9,500 − 2 = 9,498 9,498 × 41.75 =)***	9,500	41.75	2	0	—	***396,541.50***
5. N. Hildredth ***(700 − 35 = 665 665 × 23.80 =)***	700	23.80	35	80	50 ***(80 × 23.80 × .50 =)***	***14,875.00***

straight commission

A salesperson compensated by a specified percent of net sales is paid on a **straight commission** basis. Net sales are multiplied by a percent agreed upon by the firm and the salesperson. The product of net sales and the percent rate of commission determine gross earnings.

Example 3

Dale Kirkby is a manufacturer's representative for Patio Playtime Furniture. He earns a commission of 3 percent on all net sales. Net sales for the month ended June 30 were $230,000. What is his commission for the period?

$230,000	= Net sales
×.03	= Straight commission rate
$6,900	= Gross earnings

(230,000 × .03 =)

Exercise D

Determine the gross earnings in the following exercise.

Gross Sales		Sales Returns		Sales Allowances		Net Sales		Straight Commission Rate		Gross Earnings
1. $84,600	−	$400	−	$ 710	=	$83,490	×	3% (.03)	=	$2,504.70
2. 37,000	−	700	−	1,950	=	*34,350*	×	6 (.06)	=	*2,061.00*
3. 38,000	−	260	−	0	=	*37,740*	×	7 (.07)	=	*2,641.80*
4. 97,000	−	300	−	1,250	=	*95,450*	×	4 (.04)	=	*3,818.00*
5. 98,520	−	850	−	13,570	=	*84,100*	×	3.5 (.035)	=	*2,943.50*

Commission Formula

Extra sales beyond a certain level may be more difficult to achieve. A salesperson may not feel that the extra effort is really worth the extra compensation. Additionally, the selling firm may enjoy lower cost of selling as sales increase for a sales representative. In these cases, an additional incentive may be necessary in order to gain the extra sales. With a **commission formula,** the additional incentive is a salary or a commission rate that increases as sales increase. This makes every extra sale worth even more to the salesperson.

Example 4

A commission formula can motivate an employee to sell even more items.

Kerry found a person she felt could do a good job representing Sport-A-Merica to colleges and other schools in the area. She explained that the position would be a commissioned job, and therefore, gross pay would be determined by performance. The job would pay 1 percent commission on all net sales, and an additional 3 percent on net sales above $140,000 per month. Kerry felt that the additional 3 percent would provide a real incentive. She told the new employee that once some groundwork had been completed, she expected net sales to be about $185,000 per month. She explained how the commission would be determined (as shown here).

$185,000	= Net sales per month
×.01	= Commission rate on all net sales
$1,850	= Commission
$185,000	
−140,000	
$45,000	= Net sales above $140,000
×.03	= Commission rate on net sales above $140,000
$1,350	= Commission
$1,850	= (1% commission)
+1,350	= (3% commission)
$3,200	= Gross earnings per month

(185,000 × .01 =   45,000 × .03 =)

Example 5

Lori Driscol has been the top sales associate for Piedmont & Co. for the past six years. She is paid a commission of 1 percent on net sales up to $100,000, 2.5 percent on net sales between $100,000 and $150,000, and 4 percent on net sales above $150,000. Lori's net sales for the month totaled $164,000. What are her gross earnings for this month?

$100,000	= (First $100,000 of net sales)
×.01	= (1% commission rate)
$1,000	= Commission
$50,000	= (Net sales between $100,000 and $150,000)
×.025	= (2.5% commission rate)
$1,250	= Commission
$14,000	= (Net sales above $150,000) ($164,000 − 150,000 =)
×.04	= (4% commission rate)
$560	= Commission
$1,000	= (1% commission)
1,250	= (2.5% commission)
+560	= (4% commission)
$2,810	= Gross earnings

(100,000 × .01 = 50,000 × .025 = 14,000 × .04 =)

Commission Model for Example 5

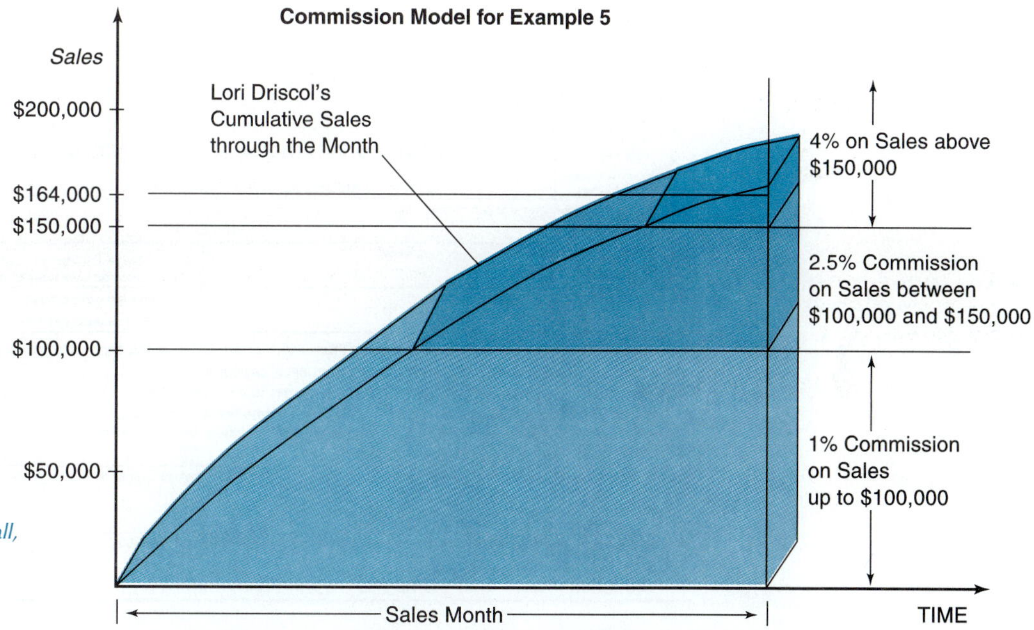

Watch for the key words all, above, and between in commission formula problems.

Exercise E

Determine the gross earnings in these exercises.

Net Sales	Commission Formula	Gross Earnings
1. $86,000	1% (.01) on net sales and 2% (.02) on sales over $60,000 (86,000 × .01 = 26,000 × .02 =)	$1,380.00
2. 127,000	2% (.02) on net sales above $35,000 ***(92,000 × .02 =)***	***1,840.00***
3. 134,900	2% (.02) on net sales and 3.5% (.035) on sales over $120,000 ***(134,900 × .02 = 14,900 × .035 =)***	***3,219.50***
4. 208,500	2% (.02) on net sales between $150,000 and $200,000, 3% (.03) on net sales over $200,000 ***(50,000 × .02 = 8,500 × .03 =)***	***1,255.00***
5. 168,300	1.5% (.015) on net sales above $150,000 and 2% (.02) on net sales above $200,000 ***(18,300 × .015 =)***	***274.50***

Determining Net Pay

Objective 2

deductions
net pay

The amount of money that an employee earns through any of the pay systems is not the amount received on payday. From the total earnings, or gross pay, a variety of **deductions** must be subtracted. After these have been subtracted, the remainder, known as **net pay,** is payable to the employee.

Deductions are of two types: mandatory and voluntary. Payroll taxes—local and state (where applicable) and federal—are mandatory. They are mandatory for the employee to pay, as well as for the employer to collect as an agent for the taxing authority. The second type is voluntary. Voluntary deductions are discussed on page 93.

Mandatory Deductions

Federal Income Tax

The laws of the United States require that the **federal income tax** be paid as wages are earned. The employer is required to collect or withhold from the employee's paycheck a specified amount each pay period. This amount and the amounts withheld from other employees' paychecks are then periodically sent to the federal government's Internal Revenue Service (IRS). A similar procedure is used for state and local income taxes.

To determine the amount of income tax to be withheld, most employers use a set of charts provided by the IRS. The charts are for weekly, monthly, and biweekly pay periods. Income tax is based on an employee's ability to pay (i.e., the more money earned, the greater the amount of tax). The tax is computed on the amount of gross earnings and the number of exemptions an employee claims. The employee is entitled to claim one **exemption** for himself or herself and one for each dependent. The employee makes this claim on **Form W-4,** which is shown below.

exemption
Form W-4

Form W-4 has been completed for employee Jacqulyn S. Retzloff. She has claimed herself, her spouse, and two children—a total of four. This information with the gross earnings is used to determine the amount of income tax from the charts shown on page 91.

Form **W-4** | **Employee's Withholding Allowance Certificate** | OMB No. 1545-0010 **2000**

Department of the Treasury Internal Revenue Service | ▶ For Privacy Act and Paperwork Reduction Act Notice, see page 2.

1 Type or print your first name and middle initial *Jacqulyn Sue*	Last name *Retzloff*	**2** Your social security number *378 : 58 : 7122*
Home address (number and street or rural route) *300 Agler*	**3** ☐ Single ☒ Married ☐ Married, but withhold at higher Single rate. Note: *If married, but legally separated, or spouse is a nonresident alien, check the Single box.*	
City or town, state, and ZIP code *Your City, State 01010*	**4** If your last name differs from that on your social security card, check here. **You must call 1-800-772-1213 for a new card** . . . ▶ ☐	

5 Total number of allowances you are claiming (from line **H** above **OR** from the applicable worksheet on page 2) **5** *4*

6 Additional amount, if any, you want withheld from each paycheck . . . . **6** $ *0*

7 I claim exemption from withholding for 2000, and I certify that I meet **BOTH** of the following conditions for exemption:
- Last year I had a right to a refund of **ALL** Federal income tax withheld because I had **NO** tax liability **AND**
- This year I expect a refund of **ALL** Federal income tax withheld because I expect to have **NO** tax liability.

If you meet both conditions, write "EXEMPT" here . . . . ▶ **7**

Under penalties of perjury, I certify that I am entitled to the number of withholding allowances claimed on this certificate, or I am entitled to claim exempt status.
Employee's signature
(Form is not valid unless you sign it) ▶ *Jacqulyn Sue Retzloff* **Date ▶** *November 16, 2000*

8 Employer's name and address (Employer: Complete lines 8 and 10 only if sending to the IRS.) *BEST Wages, Inc.* *100 Main Street Big City, MO 010101*	**9** Office code (optional)	**10** Employer identification number *42 : 064 1173*

Example 6

If Ms. Retzloff's earnings for the week are $508, the amount of the income tax would be $25. The amount of tax is found in the "Married Persons—Weekly" chart. By determining the correct range in which her $508 income for the week would fall and by moving across until the column for the number of exemptions claimed is a 4, to correspond with the W-4 certificate, the value of $25 is then found.

MARRIED Persons—**WEEKLY** Payroll Period

(For Wages Paid in 2000)

If the wages are–		And the number of withholding allowances claimed is—										
At least	But less than	0	1	2	3	4	5	6	7	8	9	10
		The amount of income tax to be withheld is—										
400	410	42	34	26	18	10	2	0	0	0	0	0
410	420	44	36	27	19	11	3	0	0	0	0	0
420	430	45	37	29	21	13	5	0	0	0	0	0
430	440	47	39	30	22	14	6	0	0	0	0	0
440	450	48	40	32	24	16	8	0	0	0	0	0
450	460	50	42	33	25	17	9	1	0	0	0	0
460	470	51	43	35	27	19	11	3	0	0	0	0
470	480	53	45	36	28	20	12	4	0	0	0	0
480	490	54	46	38	30	22	14	6	0	0	0	0
490	500	56	48	39	31	23	15	7	0	0	0	0
500	510	57	49	41	33	25	17	9	1	0	0	0
510	520	59	51	42	34	26	18	10	2	0	0	0
520	530	60	52	44	36	28	20	12	4	0	0	0
530	540	62	54	45	37	29	21	13	5	0	0	0
540	550	63	55	47	39	31	23	15	7	0	0	0
550	560	65	57	48	40	32	24	16	8	0	0	0
560	570	66	58	50	42	34	26	18	10	2	0	0
570	580	68	60	51	43	35	27	19	11	3	0	0
580	590	69	61	53	45	37	29	21	13	5	0	0
590	600	71	63	54	46	38	30	22	14	6	0	0
600	610	72	64	56	48	40	32	24	16	8	0	0
610	620	74	66	57	49	41	33	25	17	9	1	0

SINGLE Persons—**WEEKLY** Payroll Period

(For Wages Paid in 2000)

If the wages are–		And the number of withholding allowances claimed is—										
At least	But less than	0	1	2	3	4	5	6	7	8	9	10
		The amount of income tax to be withheld is—										
300	310	38	30	22	14	6	0	0	0	0	0	0
310	320	40	32	23	15	7	0	0	0	0	0	0
320	330	41	33	25	17	9	1	0	0	0	0	0
330	340	43	35	26	18	10	2	0	0	0	0	0
340	350	44	36	28	20	12	4	0	0	0	0	0
350	360	46	38	29	21	13	5	0	0	0	0	0
360	370	47	39	31	23	15	7	0	0	0	0	0
370	380	49	41	32	24	16	8	0	0	0	0	0
380	390	50	42	34	26	18	10	2	0	0	0	0
390	400	52	44	35	27	19	11	3	0	0	0	0
400	410	53	45	37	29	21	13	5	0	0	0	0
410	420	55	47	38	30	22	14	6	0	0	0	0
420	430	56	48	40	32	24	16	8	0	0	0	0
430	440	58	50	41	33	25	17	9	1	0	0	0
440	450	59	51	43	35	27	19	11	3	0	0	0
450	460	61	53	44	36	28	20	12	4	0	0	0
460	470	62	54	46	38	30	22	14	6	0	0	0
470	480	64	56	47	39	31	23	15	7	0	0	0
480	490	65	57	49	41	33	25	17	9	0	0	0
490	500	67	59	50	42	34	26	18	10	2	0	0
500	510	68	60	52	44	36	28	20	12	3	0	0
510	520	70	62	53	45	37	29	21	13	5	0	0
520	530	71	63	55	47	39	31	23	15	6	0	0

SINGLE Persons—**MONTHLY** Payroll Period

(For Wages Paid in 2000)

If the wages are–		And the number of withholding allowances claimed is—										
At least	But less than	0	1	2	3	4	5	6	7	8	9	10
		The amount of income tax to be withheld is—										
3,240	3,280	578	513	447	382	317	281	246	211	176	141	106
3,280	3,320	589	524	459	393	328	287	252	217	182	147	112
3,320	3,360	600	535	470	404	339	293	258	223	188	153	118
3,360	3,400	612	546	481	416	350	299	264	229	194	159	124
3,400	3,440	623	557	492	427	361	305	270	235	200	165	130
3,440	3,480	634	569	503	438	373	311	276	241	206	171	136
3,480	3,520	645	580	515	449	384	319	282	247	212	177	142
3,520	3,560	656	591	526	460	395	330	288	253	218	183	148
3,560	3,600	668	602	537	472	406	341	294	259	224	189	154
3,600	3,640	679	613	548	483	417	352	300	265	230	195	160
3,640	3,680	690	625	559	494	429	363	306	271	236	201	166
3,680	3,720	701	636	571	505	440	375	312	277	242	207	172
3,720	3,760	712	647	582	516	451	386	320	283	248	213	178
3,760	3,800	724	658	593	528	462	397	332	289	254	219	184
3,800	3,840	735	669	604	539	473	408	343	295	260	225	190
3,840	3,880	746	681	615	550	485	419	354	301	266	231	196
3,880	3,920	757	692	627	561	496	431	365	307	272	237	202
3,920	3,960	768	703	638	572	507	442	376	313	278	243	208

Social Security Tax

> ● ***Prior to 1991, the Social Security and Medicare taxes were combined and were known as the F.I.C.A. tax. They used the same taxable base until 1991.***

The Social Security tax is the second major mandatory deduction. The Social Security tax is levied on the gross earnings of the employee. Again, the employer (with a few exceptions) is obligated to withhold a specified amount from the employee's gross earnings. Payment is later made to the federal government by the employer. The Social Security tax differs from the federal income tax in that it is not based on an ability to pay, and there are no exemptions to consider. Deductions are made based on the first $76,200 that the employee earns from each employer each year.

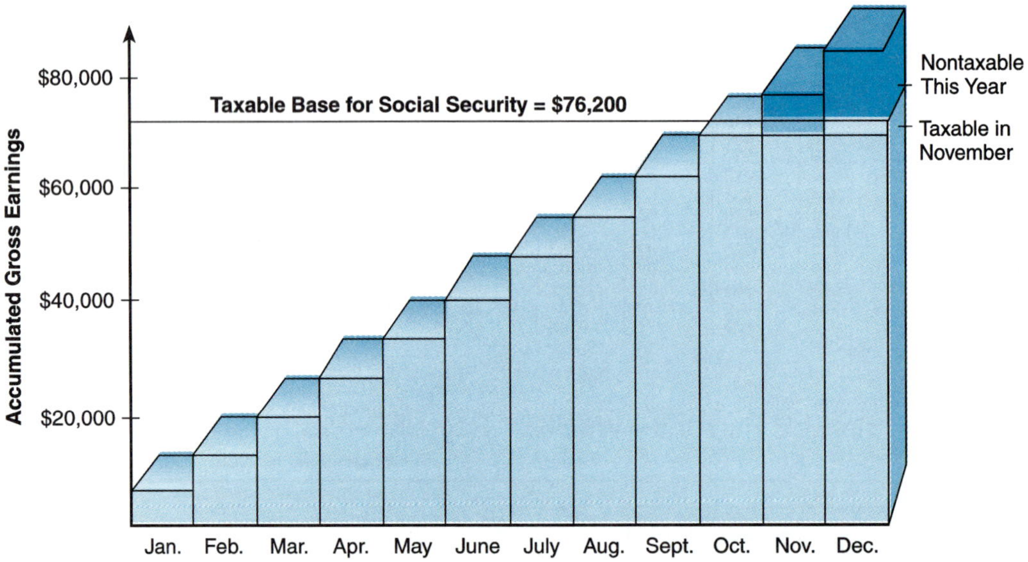

taxable base

At year's end, the employee pays tax on only $76,200, even if several employers withheld Social Security tax totaling more than the tax based on an income of $76,200. The $76,200 represents the **taxable base.** Earnings beyond this amount are not subject to Social Security for that employer in that year.

Example 7

If Ms. Retzloff's year-to-date earnings before this pay period were $75,900 from this employer, then only the amount necessary to bring the accumulated taxable gross earnings to the $76,200 maximum will be taxed this pay period. Once again using $508 as her gross earnings, the taxable amount must first be found.

	$76,200	
	−75,900	
$	300	taxable this pay period

Under these circumstances, the Social Security tax for this period would be $18.60 ($300 × .062). For the next pay period, Social Security will not be deducted, because the first $76,200 of earnings have already been taxed for the year.

Ms. Retzloff's Medicare tax deduction would be unchanged. This deduction would amount to $7.37 ($508 × .0145).

(300 × .062 =)

tax rate

The $76,200 is taxable at a **tax rate** of 6.2 percent as it is earned. Charts are also available for determining the amount of this deduction. The taxable base ($76,200) and the tax rate (6.2 percent) are subject to change by Congress.

Medicare Tax

The Medicare tax rate is 1.45 percent on all earnings. There is not a taxable base maximum and there are no deductions or exemptions considered. The Medicare tax is levied on the gross earnings of the employee.

Example 8

Using the previous example, if Ms. Retzloff's gross earnings for the week were $508, her Social Security tax would be $31.50. This amount is found by multiplying the gross earnings by the tax rate, that is, $508 × .062. Her Medicare tax would be $7.37 ($508 × .0145).

(508 × .062 =) *(508 × .0145 =)*

State and Local Income Taxes

Most states and many cities have an income tax. The structure of the tax specifications differs widely, but is similar to that of the federal income tax, Social Security, or Medicare tax. The tax may be a graduated tax (*progressive,* in the sense that the tax rate increases as the gross income increases), a flat rate taxed on all gross income, or a rate taxed on wages up to a maximum taxable base (similar to Social Security).

Example 9

Sarah has weekly earnings of $830. She lives in a city that has an income tax of 1 percent of all earnings. The city is in a state that has a progressive income tax. The state tax is 2 percent on the first $750 of earnings a week, $2\frac{1}{2}$ percent on the next $200 of earnings a week, and 3 percent on earnings above $1,000 a week. What is the amount of city and state income tax that Sarah will have deducted from her paycheck? (*Note:* 1% = .01, 2% = .02, $2\frac{1}{2}$% = .025, and 3% = .03)

$830 × .01 = $8.30 city income tax

$830
−750 × .02 = 15.00
$80 × .025 = +2.00
$17.00 state income tax

(830 × .01 = 750 × .02 = 80 × .025 =)

Example 10

Veronica Smathers lives in a state that has an income tax that is a flat 3.7 percent of the first $15,000 earned per year. Veronica has earned $14,849 so far this year. This week's paycheck will be based on gross earnings of $650. What will be the deduction for state income tax for this week? (*Note:* 3.7% = .037)

$15,000
−14,849
$151 taxable this pay period
×.037
$5.59 state income tax

(151 × .037 =)

Voluntary Deductions

The second type of deduction is voluntary deductions authorized by the employee. A payroll deduction authorization slip is signed by the employee and given to the payroll clerk. This okays deductions from gross pay for any one of a variety of purposes specified by the employee.

Some examples of voluntary deductions are union dues, uniform rental, credit union payments, savings contributions, savings bonds, retirement programs, profit-sharing plans, Individual Retirement Act (IRA) plans, and direct deposits to bank accounts. Some of these "voluntary" deductions may be required by union contract, by a court action or settlement, the result of a financial contract (credit union payment, etc.), but they are not necessarily mandated *by law.*

Once all of the deductions have been calculated, they are subtracted from gross earnings to determine net pay. This amount is made payable to the employee on the paycheck and is the actual amount that the employee has available to spend.

Example 11

Tom has been a Sport-A-Merica employee for a short time. He is paid a salary of $500 a week. Tom has a wife and three children, and claims all of them as exemptions. His earnings prior to this paycheck were $3,500 for the year. He has deductions of $40 per paycheck for a credit union payment, and $10 for the purchase of U.S. savings bonds. What is his net pay?

$500.00 = Gross earnings
− 17.00 = Federal income tax (five exemptions)
−31.00 = Social Security tax
− 7.25 = Medicare tax
−40.00 = Credit union payment
−10.00 = U.S. savings bonds
$394.75 = Net pay

(500 − 17.00 − 31.00 − 7.25 − 40.00 − 10.00 =)

Exercise F

Determine the net pay of each employee at Burson Techniques, Inc. for the first week of the year from the information below.

Employee	Gross Earnings	Federal Income Tax	State Income Tax	Medicare Tax	Social Security Tax	Union Dues	IRA Deduction	Net Pay
1. Johnny Biddy	$ 600.00	$ 53	$28.00	8.70	$37.20	$6	$ 0	$467.10
2. James Carson	1,263.40	196	61.73	*18.32*	*$78.33*	0	40	*869.02*
3. J. Grason	898.00	96	42.88	*13.02*	*55.68*	6	120	*564.42*
4. Kay Lasher	756.83	75	37.16	*10.97*	*46.92*	6	0	*580.78*
5. Sharron Orr	694.20	66	34.72	*10.07*	*43.04*	6	25	*509.37*

Exercise G

Determine net pay for the week from the incomplete payroll register below. All the employees are married.

Payroll Register

Employee	Gross Earnings	Exemptions Claimed	Federal Income Tax	Gross Earnings Before Pay Period	Medicare Tax	Social Security Tax	Net Pay
1. Sarah Blair	$435.00	2	$*30.00*	$31,000	$*6.31*	$*26.97*	$*371.72*
2. Jock Lieber	521.00	2	*44.00*	29,795	*7.55*	*32.30*	*437.15*
3. Fran Smith	487.00	1	*46.00*	47,900	*7.06*	*30.19*	*403.75*
4. John Ulrick	507.00	5	*17.00*	24,958	*7.35*	*31.43*	*451.22*
5. Kay Wilson	496.00	0	*56.00*	41,800	*7.19*	*30.75*	*402.06*
6. Bill Puhl	462.00	1	*43.00*	42,000	*6.70*	*28.64*	*383.66*
7. A. Lingle	595.00	2	*54.00*	77,850	*8.63*	*0*	*532.37*
8. Jim Seal	460.00	1	*43.00*	47,940	*6.67*	*28.52*	*381.80*
9. Jean King	427.00	2	*29.00*	38,900	*6.19*	*26.47*	*365.34*
10. Ed Ely	479.00	4	*20.00*	76,891	*6.95*	*0*	*452.05*

Employer's Payroll Taxes

Objective 3

● ***Remind students that gross earnings are not a firm's only cost of labor. Other costs include employer's payroll taxes, fringe benefits, and workers' compensation insurance.***

The employer, like the employee, is obligated to pay taxes based on the employees' gross earnings. The taxes are an extra expense to the employer, increasing the cost of labor.

The most significant of these taxes is the Social Security tax. The employer is required to pay an amount matching the employee's contribution. Six point two percent (6.2 percent or .062) of the first $76,200 each employee earns for each year must be paid to the federal government. When the employer makes the payment, the employer is really forwarding an amount equal to 12.4 percent of the employee's earnings. The 12.4 percent results from a 6.2 percent deduction from the employee plus the 6.2 percent additional tax from the employer.

The employer will also pay Medicare tax at 1.45 percent (2.9 percent with the employee share). The Medicare tax is based on all earnings for each employee. That is, there is not a taxable base maximum as with Social Security. As a result, very high-income individuals will pay more in Medicare tax than in Social Security tax in a year.

A third employer payroll tax is the two-part unemployment tax. The two parts are the Federal Unemployment Tax (FUTA) and the state unemployment tax (may have different names by state). Both taxes are based on the first $7,000 of gross earnings of each employee. This is an employer-only tax.

The FUTA rate is 6.2 percent (.062). A credit of 5.4 percent may be granted against the FUTA rate for the unemployment tax levied by the state. This means that the FUTA tax rate is really .8 percent (6.2 percent − 5.4 percent). This tax is remitted to the federal government quarterly for amounts due of $100 or more and annually for amounts due of less than $100.

The state may vary the 5.4 percent tax rate based on the unemployment history within the state and/or the unemployment history of the individual employer. New companies will initially pay the highest rate. This tax is remitted to the state government.

These payroll taxes levied against the employer are, in effect, increases in the cost of labor, even though these taxes do not appear as part of the employee's gross earnings.

Example 12

The accountant has provided Hal with the payroll information for the current period. The amount of gross earnings for Sport-A-Merica employees for the period equals $4,620. The company must remit the Social Security tax, Medicare tax, and state and federal unemployment taxes for the period to the federal and state governments. The state unemployment tax rate is 5.4% (.054). The federal unemployment tax is .8% (.008). All wages are to be taxed. How much must be remitted?

$4,620 taxable earnings this pay period
× .124 (6.2% × 2)
$572.88 (for Social Security tax)

$4,620
× .029 (1.45% × 2)
$133.98 for Medicare tax

$4,620
× .054
$249.48 for state unemployment tax

$4,620
× .008
$ 36.96 for federal unemployment tax

$572.88
133.98
+36.96
$743.82 = amount to be remitted to federal government
$249.48 = amount to be remitted to state government

(4,620 × .124 =) *(4,620 × .029 =)* *(4,620 × .008 =)* *(4,620 × .054 =)*

Exercise H

Determine the amount due the federal government for the Social Security tax, Medicare tax, and F.U.T.A. and to the state government for state unemployment tax. The amount should include both the employee and employer share when appropriate.

Gross Earnings of Employees for the Pay Period	Social Security Tax	Medicare Tax Due	F.U.T.A. Due (.8%)	State Unemployment Tax Due (5.4%)
$4,890	1. $ 606.36	6. $141.81	11. $ 39.12	16. $ 264.06
12,850	2. *1,593.40*	7. *372.65*	12. *102.80*	17. *693.90*
3,478	3. *431.27*	8. *100.86*	13. *27.82*	18. *187.81*
23,600	4. *2,926.40*	9. *684.40*	14. *188.80*	19. *1,274.40*
8,920	5. *1,106.08*	10. *258.68*	15. *71.36*	20. *481.68*

ASSIGNMENT Chapter 5

Name ______________________ **Date** ______________ **Score** ___________

Skill Problems

Select the best term/phrase for the statements below.

A. Pay System	B. Gross Earnings	C. Time Card	D. Payroll Register
E. Net Pay	F. Gross Sales	G. Sales Returns	H. Sales Allowances
I. Net Sales	J. Deductions	K. Exemptions	L. Form W-4
M. Tax Rate	N. Taxable Base	O. Straight Commission	P. Commission Formula

1. ___*C*___ An item of control that allows a payroll department to determine the number of hours worked by employees.

2. ___*E*___ The amount due a person after deductions have been subtracted from gross pay.

3. ___*I*___ The dollar value of sales after sales returns and sales allowances have been subtracted from gross sales.

4. ___*B*___ The amount earned by the employee in a pay period.

5. ___*D*___ A form showing gross earnings, deductions, and net pay for all employees.

6. ___*K*___ Legal dependents and the employee; can be claimed by the employee in order to reduce the amount of income tax withheld.

7. ___*J*___ Amounts subtracted from gross pay for such items as taxes, contributions, and savings.

8. ___*F*___ All the sales of a firm for a time period.

9. ___*P*___ A combination of ever-increasing rates of pay for sales designed to motivate an employee to sell more.

10. ___*G*___ Merchandise returned to the seller because of damage, errors in ordering, etc.

11. ___*H*___ Adjustments made in the price of the merchandise in order to compensate the buyer for keeping damaged or incorrectly shipped merchandise.

12. ___*N*___ The maximum amount of earnings that can be taxed for Social Security and Medicare tax purposes for that year.

Johson Haber, Associates pays its employees weekly. The hours worked in excess of forty per week are paid at the rate of one and one-half times the regular wages as prescribed by law. Determine the gross earnings for the following employees.

Name	Hourly Wage	Hours Worked	Regular Earnings	Overtime Earnings	Total Gross Earnings
13. J. Yale	11.50	46	*$460.00* *(11.50 × 40 =)*	*$103.50* *(11.5 × 1.5 × 6 =)*	*$563.50* *(460 + 103.50 =)*
14. R. Ott	9.75	44.25	*390.00* *(9.75 × 40 =)*	*62.16* *(9.75 × 1.5 × 4.25 =)*	*452.16* *(390 + 62.16 =)*
15. P. Birkson	10.25	28	*287.00* *(10.25 × 28 =)*	*0*	*287.00*
16. L. Parker	8.85	39	*345.15* *(8.85 × 39 =)*	*0*	*345.15*
17. K. Mathews	9.40	46.5	*376.00* *(9.4 × 40 =)*	*91.65* *(9.4 × 1.5 × 6.5 =)*	*476.65* *(376 + 91.65 =)*

Determine the net sales in the following exercise.

Units Sold	Price Per Unit	Units Returned	Sales Allowance Units	Sales Allowance Percent	Net Sales
18. 845 *(845 × 65.20 = 55,094)*	$65.20	5 *(5 × 65.20 = 326)*	3 *(3 × 65.20 × .2 = 39.12)*	20%	***$54,728.88 (55,094 − 326 − 39.12 =)***
19. 201 *(201 × 65.20 = 13.105.20)*	65.20	9 *(9 × 65.20 = 586.80)*	8 *(8 × 65.20 × .3 = 156.48)*	30%	***12,361.92 (13,105.20 − 586.80 − 156.48 =)***
20. 540 *(540 × 69.50 = 37,530)*	69.50	0	0	20%	***37,530***
21. 942 *(942 × 60.10 = 56,614.20)*	60.10	9 *(9 × 60.10 = 540.90)*	7 *(7 × 60.10 × .15 = 63.11)*	15%	***56,010.19 (56,614.20 − 540.90 − 63.11 =)***
22. 150 *(150 × 75.00 = 11,250)*	75.00	2 *(2 × 75.00 = 150.00)*	0	20%	***11,100.00 (11,250 − 150.00 =)***

Determine the net sales and gross earnings in the following exercise.

Gross Sales	Sales Returns	Sales Allowances	Net Sales	Straight Commission Rate	Gross Earnings
23. $64,500	$ 540	$ 780	***$63,180.00 (64,500 − 540 − 780 =)***	4%	***$2,527.20 (63,180 × .04 =)***
24. 21,450	280	85	***21,085.00 (21,450 − 280 − 85 =)***	2	***421.70 (21,085 × .02 =)***
25. 5,410	1,450	0	***3,960.00 (5,410 − 1,450 =)***	25	***990 (3,960 × .25 =)***
26. 200,460	4,870	2,401	***193,189.00 (200,460 − 4,870 − 2,401 =)***	.5	***965.95 (193,189 × .005 =)***

Determine the gross earnings in these exercises.

Net Sales	Commission Formula	Gross Earnings
27. $910,140	2% on net sales up to $800,000, 3% on sales over $800,000 ***(800,000 × .02 = 16,000, 110,140 × .03 = 3,304.20)***	***$19,304.20 (16,000 + 3,304.20 =)***
28. 4,845	10% on first $4,000 of net sales, 25% on sales over $4,000 ***(4,000 × .1 = 400, 845 × .25 = 211.25)***	***611.25 (400 + 211.25 =)***
29. 245,670	2% on net sales above $95,000 ***(245,670 − 95,000 = 150,670, 150,670 × .02 = 3,013.40)***	***3,013.40***
30. 2,780	10% on net sales and additional 2% on sales over $3,000 ***(2,780 × .1 = 278)***	***278.00***
31. 5,641	15% on net sales and additional 10% on sales over $4,500 ***(5,641 × .15 = 846.15, 5,641 − 4,500 = 1,141, 1,141 × .10 = 114.10)***	***960.25 (845.15 + 114.10 =)***

Determine the net pay for each employee for the first week of the year from the information below. All employees are married and claim two exemptions.

Employee	Gross Earnings	Federal Income Tax	Social Security Tax	Medicare Tax	Credit Union	Union Dues	Net Pay
32. H. Barker	$431	*$30.00*	*$26.72*	*$6.25*	$ 0	$3.50	*$364.53*
33. L. Lowery	587	*53.00*	*36.39*	*8.51*	40.00	4.20	*444.90*
34. P. Norris	412	*27.00*	*25.54*	*5.97*	17.00	3.50	*332.99*
35. F. Tunks	490	*39.00*	*30.38*	*7.11*	78.00	3.50	*332.01*
36. B. Willard	525	*44.00*	*32.55*	*7.61*	25.00	4.00	*411.84*

Determine the net pay for each salaried employee for the month of March from the information below. All employees are single.

Employee	Gross Earnings	Exemptions Claimed	Federal Income Tax	Medicare Tax	Social Security Tax	Net Pay
37. J. Cousins	$3,868.00	5	*$419.00*	*$56.09*	*$239.82*	*$3,153.09*
38. K. Dale	3,751.00	1	*647.00*	*54.39*	*232.56*	*2,817.05*
39. J. Kerns	3,460.00	0	*634.00*	*50.17*	*214.52*	*2,561.31*
40. M. Kurt	3,421.00	3	*427.00*	*49.60*	*212.10*	*2,732.30*
41. P. Wisfal	3,672.00	3	*494.00*	*53.24*	*227.66*	*2,897.10*

Complete the following payroll ledger. All of the employees are single.

Employee	Gross Earnings (Week)	Exemptions Claimed	Federal Income Tax	Gross Earnings Before Pay Period	Medicare Tax	Social Security Tax	Net Pay
Tait, J.	$356.00	4	*$13.00*	$ 5,612.00	*$ 5.16*	*$22.07*	*$ 315.77*
Willis, C	391.00	2	*35.00*	65,035.00	*5.67*	*22.63*	*327.70*
Woods, D	357.50	5	*5.00*	4,652.50	*5.18*	*22.17*	*325.15*
Yale, M.	384.25	0	*50.00*	85,421.80	*5.57*	*0*	*328.68*
Totals	**42.** *$1,488.75*	****	**43.** *$103.00*	****	**44.** *$21.58*	**45.** *$66.87*	**46.** *$1,297.30*

47. Using the ledger above, what is the total amount due to the federal government for Social Security and Medicare tax only? The amount should include employee and employer share.

Social Security Tax	*$66.87*
Medicare Tax	*+21.58*
subtotal	*88.45*
Double for employer	*×2*
Total due	*$176.90*

48. Using the ledger above, what is the total amount due to the federal government for federal unemployment? The rate for this company is .8% of the first $7,000.00.
$356.00 + 357.50 = 713.50 × .008 = $5.71

49. Using the ledger above, what is the total amount due to the state government for unemployment? The rate for this company is 2.7% of the first $7,000.
$356.00 + 357.50 = 713.50 × .027 = $19.26

50. Using the ledger above, what is the total amount due to the federal government?
$176.90 + 5.71 + 68.00 = $250.61

Business Application Problems

1. Susie earns a salary of $380 a week on her job as a data entry clerk. She is single and claims no withholding allowances. What is the deduction for federal withholding tax?
***$50** (from table)*

2. An employee mentioned to a friend that she had $34.32 deducted from her paycheck for Social Security tax in the first week of January. What were her gross earnings?
$34.32 ÷ .062 = $553.55 gross

3. Gerrard has a salary of $2,000 per month as the manager of a used car lot. He also earns a commission of 5 percent (.05) on the sales made by the lot. Last month the lot sold 12 vehicles totaling $56,470. What was Gerrard's gross earnings for the month?
$56,470 × .05 = $2,823.50 + $2,000 = $4,823.50 for the month

4. Karen has earned a salary of $2,488 a week for the first 32 weeks of this year. What is the Social Security deduction for this week?
$2,488 × 32 = $79,616 0 deduction

5. A commission of 4.5 percent on all sales as well as a salary of $200 per week is paid to all sales associates at World Tickets. Last week Manuel sold $17,600 in travel packages. What is his gross income for the week?
$17,600 × .045 = $792 + $200 = $992 gross income

6. Alex worked 52 hours last week. He worked 58 hours this week. His employer pays time and a half for hours in excess of 40 per week. Alex earns $9.80 per hour. How much more did Alex earn this week?
58 − 52 = 6 × 1.5 × $9.80 = $88.20 more

7. The employees at Bedford Inc. had earnings of $14,860 for the week. The Social Security and Medicare tax amount on the earnings must be noted in the accounting records. What is the total amount for these taxes combined for the employer and employee?
$14,860 × .062 × 2 = $1,842.64
$14,860 × .0145 × 2 = $430.94
$1,842.64 + $430.94 = $2,273.58 total

8. A combination piece rate and hourly rate is earned by Beth. She worked 38 hours last week and produced 648 pieces. Her hourly rate is $9. She earns $.32 for each finished piece. What are her gross earnings for the week?
38 × $9 = $342. $.32 × 648 = $207.36
$342 + $207.36 = $549.36 gross earnings

9. First National employs Suzanne as a mortgage broker. She had a record month in June with sales of $365,000 in new mortgages. She earns a commission of 1.3 percent on all sales above $50,000. What is her gross earnings amount?
$365,000 − $50,000 = $315,000 × .013 = $4,095 gross earnings

10. Dale earns his income from Daytime Productions through a commission formula of 1 percent on all sales up to $60,000, 2.5 percent on all sales between $60,000 and $100,000, and 4 percent on all sales above $100,000. What are his gross earnings for the month if his sales were $120,800 for the period?
.01 × $60,000 = $600 .025 × $40,000 = $1,000
.04 × $20,800 = $832 $600 + $1,000 + $832 = $2,432 gross earnings

11. Brad is reviewing Gwen's paycheck authorization. She earned $385 last week. Her W-4 form indicates that she claims two exemptions and that she is single. She has also authorized a deduction for $40 toward retirement. Her gross earnings to date are $35,670. What is the net pay amount for this check?
$385 × .062 = $23.87 $385 × .0145 = $5.58
− 23.87 Social Security tax
− 5.58 Medicare tax
− 34.00 federal income tax
− 40.00 retirement
$281.55 net pay

12. Determine the amount of gross earnings for an employee from the time card information shown below.

Days	M	T	W	Th	F
	10	8	9	7	10

Notes:
1. Pay rate = $9 per hour
2. Overtime paid at one and one-half times the regular pay rate for all hours in excess of 40 per week.

40 × $9 = $360
4 × 1½ × $9 = 54
$414 gross

13. The Bridgeton Manufacturing Company pays a commission of 3 percent on all sales, and an extra $1\frac{1}{2}$ percent on sales above $250,000. The top sales associate sold $267,970 last month. There were thirty-five units returned during the month. Each unit is priced at $659.00. What was the commission earned by the top salesperson this month?

$267,970
−23,065 = 35 × $659 returns
$244,905 × .03 = $7,347.15 commission

14. Kay Smite was paid a salary of $1,490 a week. She worked each week of the year. What was the deduction for Social Security on the last (52nd) week of the year?

$1,490	*$76,200*
×51	*−75,990*
$75,990	*$210 × .062*
	= $13.02 for the 52nd week

15. What was the total Social Security tax deduction from Ms. Smite's earnings during the year?

$76,200
×.062
$4,724.40 for the year

16. Kolan Industries, Inc., employs Kerry as a telemarketer. Her gross earnings this week are made up of a salary of $320, plus $138 in commissions she earned on sales. She is married, has one child, and claims one exemption. What is her net pay if her gross earnings to date are $25,837?

$458.00	*gross earnings*
−42.00	*federal income tax*
−28.40	*Social Security tax*
−6.64	*Medicare tax*
$380.96	*net pay*

17. Rob is paid a salary of $600 a month, plus a 2 percent commission on net sales over $40,000 and an additional 1 percent on net sales over $100,000. Rob sold $164,800 of merchandise this month, with only $3,200 of returns. What were Rob's gross earnings?

$164,800	*$161,600*	*$161,600*
−3,200	*−40,000*	*−100,000*
$161,600	*$121,600*	*$61,600*
	×.02	*×.01*
	$2,432	*$616*

$600	*salary*
2,432	*commission*
+616	*commission*
$3,648	*gross earnings*

18. Sun Labs, Inc. employs Sam on a piece-rate basis at $.50 per piece. His W-4 form shows that he claims a total of two exemptions and that he is single. His earnings to date are $82,412. What is his net pay?

Output Record

Monday	Tuesday	Wednesday	Thursday	Friday
193	189	109	128	141

760	*pieces*
×$ .50	
$380.	*gross earnings*
−34.	*federal income tax*
−5.51	*Medicare tax*
$340.49	*net pay*

19. Darcy had a gross income of $549.35 for the past week. He lives in a city that has an income tax of 1 percent. The state also has an income tax of 2 percent on the first $30,000. Darcy has an accumulated earnings of $29,288 prior to this week. What is Darcy's net income for the week if the deduction for federal income tax is $87.00?

$549.35	*gross earnings*
−87.00	*federal income tax*
−34.06	*Social Security tax*
−7.97	*Medicare tax*
−5.49	*city income tax*
−10.99	*state income tax*
$403.84	*net income*

20. What would be the net income for Darcy for the following week if his gross income was the same as problem 19?

$29,288	*$30,000.00*
+549.35	*−29,837.35*
$29,837.35	*$162.65 × .02 = $3.25*

$549.35	*gross earnings*
−87.00	*federal income tax*
−34.06	*Social Security tax*
−7.97	*Medicare tax*
−5.49	*city income tax*
−3.25	*state income tax*
$411.58	*net income*

21. Vee has determined that the company must pay the Social Security tax due the federal government. The amount withheld from employees totals $4,680. What is the amount of the check payable to the federal government for Social Security tax (employee plus employer)?

$4,680
×2
$9,360 (employer and employee share)

22. Deductions by the Payroll Office of $18 for union dues, $27.50 for uniform rental, $50 for a savings program, and $10 for tool rental are authorized by Peter Wilson. The payroll clerk also notes that he claims one exemption even though he is married. He has earned $14,900 prior to this pay period. His hourly rate is $13. He worked forty-two hours this week. The company pays double the regular rate for hours in excess of forty. What is his net pay?

40 × $13 = $520
2 × 2 × $13 = +52
$572

$572.00	*gross earnings*
−60.00	*federal income tax*
−35.46	*Social Security tax*
−8.29	*Medicare tax*
−18.00	*union dues*
−27.50	*uniform rental*
−50.00	*savings*
−10.00	*tool rental*
$362.75	*net pay*

23. The earnings for Dixie Brown total $9,867 for the year to date. Dixie is married but claims no exemptions. Her gross earnings for this week are $410. She has $50 deducted from her check each week for savings. She also has a $15 deduction for parking each week. The city in which Dixie works has a 1 percent tax for residents and a $\frac{1}{2}$ percent tax for nonresidents on the first $10,000 of gross earnings. Dixie is not a resident of the city. What is her net income for the week?

$10,000
−9,867
$133
×.005
$.67 city income tax

$410.00	*gross earnings*
−44.00	*federal income tax*
−25.42	*Social Security tax*
−5.95	*Medicare tax*
−50.00	*savings*
−15.00	*parking*
−.67	*city income tax*
$268.96	*net income*

24. Perie has a full-time job that pays $8 per hour at Adain Products, Inc. He worked 43 hours last week. The firm pays time and one-half for all hours in excess of 40 per week. Perie is single. He had worked double shifts and a great deal of overtime earlier in the year, bringing his year-to-date gross earnings to $41,800. He has also agreed to have $50 deducted from his check each week for savings. What is his net pay?

40 × $8 =	*$320*	*$356.*	*gross earnings*
3 × $1\frac{1}{2}$ × $8 =	*36*	*−22.07*	*Social Security tax*
	$356	*−5.16*	*Medicare tax*
		−38.00	*federal income tax*
		−50.00	*savings*
		$240.77	*net pay*

25. Write a complete sentence in answer to this question. Janice Noteri is considering changing the number of exemptions she claims in order to increase her take-home pay. Janice is a single mother and has a salary of $350 per week. She now only claims herself, but she is considering claiming three exemptions. What will be the difference in her take-home pay if she makes the change?

$38 with one exemption
−21 with three exemptions
$17
The difference will be $17.00

26. Write a complete sentence in answer to this question. Al Derton receives a salary of $2,365 per month. He claims one exemption and is single. This paycheck included a bonus of $1,500 in addition to his salary. How much will Al receive in this monthly paycheck if his earnings to date total $23,000?

$2,365
+1,500
$3,865 gross earnings

$3,865.00	*gross earnings*
−681.00	*federal income tax*
−239.63	*Social Security tax*
−56.04	*Medicare tax*
$2,888.33	

Al will receive $2,888.33 in his monthly paycheck.

27. Write a complete sentence in answer to this question. If an employee is now earning an annual salary of $41,800 and receives a raise of $5,000, what will be the additional take-home pay for the year? The W-4 form indicates that the employee claims three exemptions and is single. She receives a monthly paycheck.

at	*$41,800.00*	*gross earnings*
	−2,591.60	*Social Security tax*
	−606.10	*Medicare tax*
	−5,388.	*federal income tax (449 × 12 =)*
	$33,214.30	*net income*
at	*$46,800.00*	*gross earnings*
	−2,901.60	*Social Security tax*
	−678.60	*Medicare tax*
	−6,732.	*federal income tax (561 × 12 =)*
	$36,487.80	*net income*

$36,487.80
−33,214.30
$3,273.50

The additional take-home pay will be $3,273.50

★ Challenge Problem

Catty is employed with the Marks Corporation. She earns a salary of $200 a week, as well as a commission of $1\frac{1}{2}$ percent of sales. Her commission last month was $840. She will receive one-fourth of that amount this week in her gross earnings. She is married, has five children, but only claims three exemptions. She also earns $40 a week by doing artwork on the side out of her own home. She collects this income in cash. Catty has earned $14,900 to date from her sales for the Marks Corporation and $3,880 for her artwork. She has $47 taken out of her paycheck each week for savings and $100 is directly deposited into her bank account. What is her paycheck amount for the week?

$200.00	*salary*
+210.00	*commission*
$410.00	*gross income*
−19.00	*federal income tax*
−25.42	*Social Security tax*
−5.95	*Medicare tax*
−47.00	*savings*
−100.00	*direct deposit*
$212.63	*net pay*

MASTERY TEST Chapter 5

Name __________ Date __________ Score __________

True or False

T (F) 1. Gross earnings + deductions = net pay.

(T) F 2. A W-4 form is used to determine the amount of withholding allowances for an employee.

T (F) 3. Overtime pay is not subject to F.I.C.A. tax because it is $1\frac{1}{2}$ times the usual rate of pay.

T (F) 4. Net pay is determined from reading the numbers on the time card.

T (F) 5. FUTA is deducted from each employee's wages.

(T) F 6. Gross earnings is the amount that is used to determine federal income tax.

7. Compute the employee's gross earning's from the piece-rate information below.

Output Record

Days	M	T	W	Th	F
Units of output	74	83	91	59	87
Hours worked	8	6	8	7.5	8.1

Piece-rate = $1.15 per unit

394 × $1.15 = $453.10 gross earnings

8. Determine the amount of federal income tax to be withheld if the gross earnings are $400 for the week, the number of exemptions claimed is one, and the employee is single.

$45 for federal income tax

9. The payroll clerk noted that $911.57 was withheld from the employees of the company for Social Security tax for the week. What is the total amount that the employer must send the federal goverment for Social Security tax for the week?

$911.57
× 2
$1,823.14 must be sent

10. Peter earns an hourly rate of $9.50. He is paid weekly and worked forty-seven hours last week. He is paid overtime for hours in excess of forty at the rate of one and one-half times. His earnings to date are $26,280. He is married and claims two exemptions. What is his net pay?

40 × $9.50 = $380.00
$7 \times 1\frac{1}{2} \times \$9.50$ = + 99.75
= $479.75
− 36.00 federal income tax
− 29.74 Social Security tax
− 6.96 Medicare tax
$407.05 net pay

11. Jeff Hazel earns $9 per hour. He worked 42 hours last week. He is paid double time for all hours worked in excess of 40 each week. His earnings to date are $7,947. He is single and claims one exemption. What is his gross pay for the week?

40 × $9 = $360
+ 2 × $18 = 36
$396 gross pay

12. Compute the gross earnings for an employee who is paid on a combination hourly and piece-rate system.

Day	M	T	W	T	F
Hours worked	9	7.6	8	7.2	8
Units of output	93	112	141	86	152

Hourly rate = $7.00
Piece rate = 7.8¢ per unit

39.8 × $7 = $278.60
584 × $.078 = + 45.55
$324.15 gross earnings

Business Measurements 6

Suggested Classroom Time for This Chapter					
hours	1	2	3	4	5

Perceived Student Difficulty with Topical Material		
easy	moderate	difficult

Objectives

After mastering the material in this chapter, you will be able to:

1. Perform all of the fundamental processes in problems dealing with measurements. p. 108
2. Find the solution to problems dealing with linear, surface, and volume measure (using the traditional measurement system). p. 108
3. Find the solution to problems dealing with linear, surface, and volume measure (using the metric system). p. 112
4. Convert measurements to the metric system or to the traditional measurement system. p. 113
5. Use currency exchange rates in business transactions. p. 115
6. Understand and use the following terms:
   Linear Measure
   Surface Measure
   Volume Measure
   Currency Exchange

Sport-A-Merica

The addition of two new sales clerks helps relieve some of the pressure for Kerry, but the relief doesn't last long. As the holiday buying season begins to build, it is apparent that Sport-A-Merica has established itself in the retail community. The customer count is growing at such a rate that it is clear the store will soon be bulging at the seams. The stock on the floor is interfering with the traffic flow of customers throughout the store. The sales clerks, Kerry, and Hal are almost on top of one another. The merchandise processing room needs to be at least 20 ft. by 20 ft. Hal would also like to bring in a line of Italian bicycles. He has no room to show them in the present store. With the favorable exchange rate he feels that Sport-A-Merica is losing potential income.

Hal realizes that as soon as the holiday season is over, a new store location must be found. Sport-A-Merica must expand before competition comes along and takes business away. Though the need to move is apparent, a new and bigger store in a good location will be difficult to find. The right location will mean the possibility of even greater growth in the future. This decision is not only important for the immediate future, but will possibly have an effect on the next growth move.

Realizing that the moving decision cannot wait until after the holidays, Kerry and Hal scour the classified ads, speak with realtors, and spend weekends looking for a suitable, affordable store. They did not realize the full amount of the considerations that come into play on a business expansion decision. The square footage of floor space is a prime issue. The length of exposure the store has to the street, the volume of cubic feet of air to be heated or cooled (Hal has budgeted for $7,000 a year), the cost of improvements, the existence of nearby parking—all are items that must be given serious attention. Though time consuming, the prospect of enlarging the operation, as well as the profits, keeps the search under way. Persistence will pay off!

Understanding dimensional measurements is useful to a business student because this can be applied in the areas of building decoration and landscaping (carpeting, painting, sodding, and seeding), construction (determining the amount of earth removal, insulation required, and the area to be heated or cooled), and establishing retail and wholesale space allocations (determining overhead expenses by area, shelf allocations, and sales by floor space).

Objective 1

The first part of this chapter includes using some basic measuring principles. The second part introduces the metric measurement system and problems relating to metric conversion as well as to metric measurement. You will be able to make use of your knowledge in working with whole numbers, fractions, decimals, and the fundamental processes in this chapter. Do your work carefully to be sure to:

1. perform accurately the basic fundamentals on measurement information
2. select the correct conversion factor from the measurement table
3. reduce or borrow units from larger or smaller units

The following is a table of measurements. This table will be useful to you in working the problems in the first part of the chapter.

Measurements Table

Linear or Straight Measure

12 inches	=	1 foot
3 feet	=	1 yard
$5\frac{1}{2}$ yards	=	1 rod
$16\frac{1}{2}$ feet	=	1 rod
66 feet	=	1 chain
320 rods	=	1 mile
5,280 feet	=	1 mile
1,760 yards	=	1 mile

Cubic or Volume Measure

1,728 cubic inches	=	1 cubic foot
27 cubic feet	=	1 cubic yard
128 cubic feet	=	1 cord of wood

Liquid Measure

4 gills	=	1 pint
2 pints	=	1 quart
4 quarts	=	1 gallon
$31\frac{1}{2}$ gallons	=	1 barrel
$7\frac{1}{2}$ gallons	=	1 cubic foot
32 ounces	=	1 quart

Dry Measure

2 pints	=	1 quart
8 quarts	=	1 peck
4 pecks	=	1 bushel

Surface or Square Measure

144 square inches	=	1 square foot
9 square feet	=	1 square yard
$30\frac{1}{4}$ square yards	=	1 square rod
160 square rods	=	1 acre
640 acres	=	1 square mile or section

Counting

12 units	=	1 dozen
12 dozen	=	1 gross

Avoirdupois Weight

16 ounces	=	1 pound
100 pounds	=	hundredweight
20 hundredweights	=	1 ton
2,000 pounds	=	1 ton
2,240 pounds	=	1 long ton

Time

60 seconds	=	1 minute	100 years	=	1 century
60 minutes	=	1 hour	$4\frac{1}{3}$ weeks	=	1 month
24 hours	=	1 day	30 days	=	1 month
7 days	=	1 week	52 weeks	=	1 year
365 days	=	1 year	13 weeks	=	1 quarter (year)
366 days	=	1 leap year	12 months	=	1 year

Linear Measure

Objective 2

linear measure

Linear measure considers only one dimension: length.

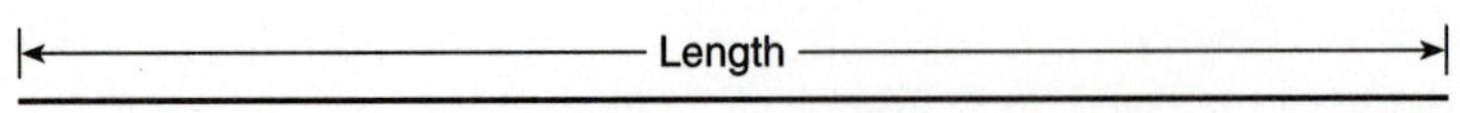

● *Use a yardstick or tape measure to demonstrate linear measure.*

To perform the basic fundamentals on measurement information you must begin in the same manner as you would any other whole-number problem. The first step is to set up the problem; then perform the necessary function.

Example 1

4 yd, 1 ft, 5 in. × 2 = 4 yd, 1 ft, 5 in.
×2
8 yd, 2 ft, 10 in.

(4 × 2 = 1 × 2 = 5 × 2 =)

Example 2

4 hr, 37 min, 14 sec + 1 hr, 40 sec = 4 hr, 37 min, 14 sec
+1 hr 40 sec
5 hr, 37 min, 54 sec

(4 + 1 = 14 + 40 =)

○ *How much time in a day remains after subtracting your time in class and studying?*

The examples so far have not required borrowing or reducing in order to complete the problem. In the following examples, you must either borrow or reduce in order to find the correct answer.

When reducing measurement problems, always begin with the smallest measurement (in size) and work up to the largest.

Example 3

17 yd, 2 ft, 9 in. × 3 = 17 yd, 2 ft, 9 in.
×3
51 yd, 6 ft, 27 in.

This answer must be reduced to lowest terms.

2 ft, 3 in.
12)27 in.

51 yd, 6 ft
+2 ft, 3 in.
51 yd, 8 ft, 3 in.

2 yd, 2 ft
3)8 ft

51 yd 3 in.
+2 yd, 2 ft
53 yd, 2 ft, 3 in.

(17 × 3 = 2 × 3 = 9 × 3 =)

Example 4

7 yd, 1 ft, 3 in. − 5 yd, 2 ft, 10 in.

In this example, ten inches cannot be subtracted from three inches; so, a foot's worth of inches must be borrowed and added to the three inches. The sum, fifteen inches, is enough to allow the subtraction. The next step is to subtract the feet. A yard's worth of feet must be borrowed and added to the zero feet. The sum, three feet, allows the subtraction of two feet. The last step is to subtract the yards.

○ *How far is it from your desk to the chalkboard? How wide is your desk?*

7 yd, 1 ft, 3 in.	7 yd, 0 ft, 15 in.	6 yd, 3 ft, 15 in.
−5 yd, 2 ft, 10 in.	−5 yd, 2 ft, 10 in.	−5 yd, 2 ft, 10 in.
		1 yd, 1 ft, 5 in.

(6 − 5 = 3 − 2 = 15 − 10 =)

Exercise A

Solve the following problems. Remember to reduce your answers.

1. 13 yd, 2 ft, 7 in. × 4 = 52 yd, 8 ft, 28 in.
   = <u>55</u> yd, <u>1</u> ft, <u>4</u> in.
2. 8 lb, 6 oz ÷ 6 = <u>*1*</u> lb <u>*$6\frac{1}{3}$*</u> oz *(8 × 16 = 128   128 + 6 = 134   134 ÷ 6 =)*
3. 1 hr, 6 min, 10 sec ÷ 2 = <u>*33*</u> min, <u>*5*</u> sec *(66 ÷ 2 = 33   10 ÷ 2 = 5)*
4. 60 lb, 11 oz × 7 = <u>*424*</u> lb, <u>*13*</u> oz *(60 × 7 = 420   11 × 7 = 77)*
   4 76 60
5. 5 hr, 17 min − 3 hr, 50 min, 13 sec = <u>*1*</u> hr, <u>*26*</u> min, <u>*47*</u> sec
   −3 50 13

Surface Measure

surface measure

Surface measure (or square measure) considers two dimensions: length and width. Answers are labeled as square feet, square yards, etc. Surface measure is the result of finding the product of the length and width (i.e., L × W).

● *Use 12″ linoleum floor tile laid side-by-side to demonstrate how "squares" fit together. One tile can be marked with 1″ squares.*

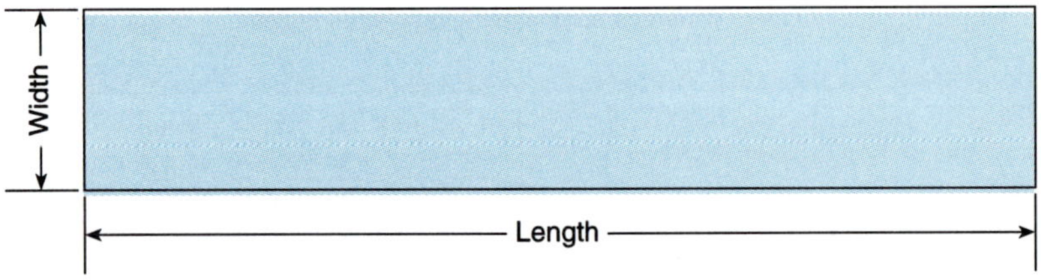

Example 5

The area Sport-A-Merica currently has includes a merchandise processing room that measures 19 ft long and 11 ft wide. How many square feet are there in the room?

```
  19 ft
 ×11 ft
  19
 19
 209 sq. ft
```

(19 × 11 =)

Example 6

Surface measure problems also need to be reduced. In the previous example, the answer should be reduced to square yards and square feet. The measurements table indicates there are nine square feet in a square yard.

○ *Can you estimate the size of a room by the number of floor or ceiling tiles?*

```
   23 sq. yd, 2 sq. ft
9)209 sq. ft
  18
   29
   27
    2
```

(209 ÷ 9 =)

Exercise B

Complete the problems below and reduce your answers.

1. 4 sq. yd, 2 sq. ft + 22 sq. yd,
   8 sq. ft, 7 sq. in. = 26 sq. yd, 10 sq. ft,
   7 sq. in. = 27 sq. yd, 1 sq. ft., 7 sq. in.
2. 71 sq. yd ÷ 3 = ___23___ sq. yd, ___6___ sq. ft *(71 ÷ 3 = 23 Remainder: 2 × 3 = 6)*
3. $9\frac{1}{2}$ ft × $6\frac{1}{2}$ ft = ___6___ sq. yd, ___$7\frac{3}{4}$___ sq. ft *(9.5 × 6.5 = 61.75)*
4. 3.2 mi × 6.8 mi = ___21.76___ sq. mi *(3.2 × 6.8 =)*
   4 9
5. 5 sq. yd − 3 sq. yd, 2 sq. ft = ___1___ sq. yd, ___7___ sq. ft
   −3 2

Volume Measure

volume measure

Volume measure (or cubic measure) considers the three dimensions of length, width, and depth. Answers are labeled as cubic feet, cubic yards, etc. Cubic measure is the result of finding the product of the length, width, and depth of an object. In some problems, height may be substituted for the depth factor (i.e., L × W × D).

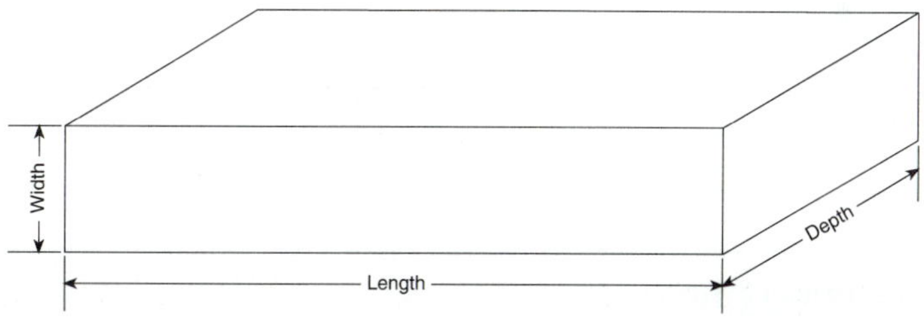

Example 7

Hal needs to estimate the cost of the heating/cooling bill for the new space Sport-A-Merica is considering. How many cubic feet of air are there in a one-story building measuring 80 feet long, 32 feet wide, with a ceiling 8 feet high?

L	×	W	×	D	=
80 ft	×	32 ft	×	8 ft	= 20,480 cu. ft

Hal has budgeted $7,000 for heating/cooling the new space for one year. The utility company estimates the annual cost of heating/cooling this type of building at $.32 per cubic foot. What is the estimated cost of heating/cooling for the year?

20,480 cubic ft × $.32 = $6,553.60

Volume measure problems, as well as surface measure problems, may need to be reduced; in the previous problem, the answer could be reduced to cubic yards and cubic feet. After consulting the measurements table, you will find that there are 27 cubic feet in a cubic yard.

27)20,480 = 758 cu. yd 14 cu. ft

(80 × 32 × 8 = 20,480 × .32 =)

Exercise C

Complete the problems below and reduce your answers.

1. 6 yd × 3 ft × 2.5 ft = 135 cu. ft = __5__ cu. yd
2. 8 ft × 10 ft × 3 ft = __8__ cu. yd, __24__ cu. ft *(8 × 10 × 3 = 240 240 ÷ 27 =)*
3. 42 sq. ft by 7 ft high = __10__ cu. yd, __24__ cu. ft *(42 × 7 = 294 294 ÷ 27 =)*
4. 16 cu. yd, 19 cu. ft + 3 cu. yd, 12 cu. ft = __20__ cu. yd, __4__ cu. ft *(16 + 3 = 19 19 + 12 = 31)*
5. A pool is 40 feet long, 20 feet wide, and averages 5 feet deep. Find the total volume to the nearest cubic foot.
   __4000__ cu. ft *(40 × 20 × 5 =)*

The Metric System
Objective 3

● *Use a two-liter bottle to demonstrate the approximate volume of a liter. Nearly all packaged goods use metric measure descriptions.*

Most packaged goods sold in the United States are currently labeled in our traditional measurement system, as well as in the metric system. All other industrialized nations use the metric system. Our government has made a commitment to integrate the metric system into our business activities. Since exports and imports are ever greater factors in our economy, it is imperative that we fully learn this international language of measure.

First we will look at the metric system and see how it is structured. Then we will discuss how to convert traditional measurements to the metric system. Finally, we include some practice problems in measurement conversion.

The liter, the meter, and the gram are the basic units of measure. For a rough comparison:

1 liter = approximately 1 quart

1 meter = approximately 1 yard

1 kilogram = approximately 2 pounds

The Metric Measurement System

Linear Measure

10 millimeters (mm)	=	1 centimeter (cm)
10 centimeters	=	1 decimeter (dm)
10 decimeters	=	1 meter (m)
10 meters	=	1 decameter (dcm)
10 decameters	=	1 hectometer (hm)
10 hectometers	=	1 kilometer (km)
10 kilometers	=	1 myriameter (mym)

Square Measure

100 square millimeters (sq. mm)	=	1 square centimeter
100 square centimeters (sq. cm)	=	1 square decimeter
100 square decimeters (sq. dm)	=	1 square meter
100 square meters (sq. m)	=	1 square decameter
100 square decameters (sq. dcm)	=	1 square hectometer
100 square hectometers (sq. hm)	=	1 square kilometer
100 square kilometers (sq. km)	=	1 square myriameter (sq. mym)

Cubic Measure

1,000 cubic millimeters (cu. mm)	=	1 cubic centimeter
1,000 cubic centimeters (cu. cm, cc)	=	1 cubic decimeter
1,000 cubic decimeters (cu. dm)	=	1 cubic meter
1,000 cubic meters (cu. m)	=	1 cubic decameter
1,000 cubic decameters (cu. dcm)	=	1 cubic hectometer
1,000 cubic hectometers (cu. hm)	=	1 cubic kilometer
1,000 cubic kilometers (cu. km)	=	1 cubic myriameter (cu. mym)

Liquid and Dry Measure

10 milliliters (ml)	=	1 centiliter (cl)
10 centiliters	=	1 deciliter (dl)
10 deciliters	=	1 liter (L)
10 liters	=	1 decaliter (dcl)
10 decaliters	=	1 hectoliter (hl)
10 hectoliters	=	1 kiloliter (kl)
10 kiloliters	=	1 myrialiter (myl)

Weight Measure

10 milligrams (mg)	=	1 centigram (cg)
10 centigrams	=	1 decigram (dg)
10 decigrams	=	1 gram (g)
10 grams	=	1 decagram (dcg)
10 decagrams	=	1 hectogram (hg)
10 hectograms	=	1 kilogram (kg)
10 kilograms	=	1 myriagram (myg)
10 myriagrams	=	1 quintal (q)
10 quintals	=	1 tonneau (T)

Notice that in all the metric units of measure, the prefix is the same (i.e., milli, centi, deci, etc.).

kilo- (larger)	hecto-	deca-	basic unit of measure	deci-	centi-	milli- (smaller)

Factors

Notice that all the metric system measurements have a base of 10 (i.e., by dividing 1 meter by 10, the result is 1 decimeter, 10 hectometers equals 1 kilometer, and so forth). Once you become familiar with the prefixes of this system, it is much easier than the 12 inches to a foot, 3 feet to a yard, 1,760 yards to a mile, etc., that we have been using.

The square measure unit is found by the same method as it is in the traditional measurement system (i.e., length times (×) width). If 1 meter is equal to 10 decimeters, then 1 square meter must be equal to 100 square decimeters (10 × 10 = 100).

The same holds true for volume or cubic measure. Length × width × depth is the method used to determine cubic measure. This is the reason that 1 cubic meter is equal to 1,000 cubic decimeters (10 × 10 × 10 = 1,000).

The previous explanations should give you a basic understanding of the manner in which the measurement units are found.

The base of 10 allows for easy conversion from meters to, say, decameters; 53.6 meters equals 5.36 decameters, since there are 10 meters to a decameter ($10\overline{)53.6}$ = 5.36).

7.6 square kilometers = 760 square hectometers (100 square hectometers equals one square kilometer (i.e., 100 × 7.6 = 760)).

53,065.2 cubic millimeters = 53.0652 cubic centimeters (53,065.2 ÷ 1,000 = 53.0652).

Exercise D

Convert the following measurements to the base indicated.

1. 963.2 centimeters to meters 9.632
2. 856 square hectometers to square kilometers *8.56*
3. 806,329 cubic centimeters to cubic meters *.806329*
4. .76 liters to centiliters *76*
5. 47,357 grams to decigrams *473,570*

Objective 4

It may be necessary for you to convert from the traditional measurement system to the metric system, or vice versa. The conversion table will aid in that process.

Example 8

A hole 47 ft. long, 18 in. wide, and 42 in. deep equals ____________ cubic meters.

47 ft. × 1.5 ft. × 3.5 ft. = 246.75 cu. ft.

246.75 cu. ft.
×.02832 (1 cu. ft. = .02832 cu. m) (from conversion table)
6.98796 cu. m

(47 × 1.5 × 3.5 = 246.75 × 0.2832 =)

Example 9 437 kilometers equals ________________ miles.

```
    437  kilometers
  ×.6214 (1 km = .6214 mi) (from table)
271.5518 miles
```

(437 × .6214 =)

Conversion Table

Linear Measure

1 in.	= 2.54 cm	1 mm	= .03937 in.
1 ft	= .3048 m	1 cm	= .3937 in.
1 yd	= .9144 m	1 dm	= .3281 ft
1 rd	= 5.029 m	1 m	= 39.37 in.
1 mi	= 1.6093 km	1 m	= 3.281 ft
		1 m	= 1.0936 yd
		1 dcm	= 1.9884 rd
		1 km	= .6214 mi

Liquid and Dry Measure

1 dry qt	= 1.101 L	1 L	= .908 dry qt
1 liquid qt	= .9463 L	1 L	= 1.0567 liquid qt
1 liquid gal	= .3785 dcl or 3.785 L	1 dcl	= 2.6417 liquid gal
1 pk	= .881 dcl or 8.81 L	1 dcl	= 1.135 pk
1 bu	= .3524 hl	1 hl	= 2.8377 bu

Weight Measure

1 gr troy	= .0648 g	1 g	= 15.432 gr troy
1 oz troy	= 31.104 g	1 g	= .03215 oz troy
1 oz avoir.	= 28.35 g	1 g	= .03527 oz avoir.
1 lb troy	= .3732 kg	1 kg	= 2.679 lb troy
1 lb avoir.	= .4536 kg	1 kg	= 2.2046 lb avoir.
1 T (short)	= .9072 met. t	1 met. t	= 1.1023 T (short)

Square Measure

1 sq. in.	= 6.452 sq. cm	1 sq. mm	= .00155 sq. in.
1 sq. ft	= .0929 sq. m	1 sq. dm	= .1076 sq. ft
1 sq. yd	= .8361 sq. m	1 sq. km	= .3861 sq. mi
1 sq. mi	= 259 ha or 2.589 sq. km	1 sq. cm	= .155 sq. in.
1 sq. rd	= 25.293 sq. m	1 sq. m	= 1.196 sq. yd
		1 ha	= 2.471 acre

Cubic or Volume Measure

1 cu. in.	= 16.3872 cu. cm (cc)	1 cu. cm (cc)	= .06102 cu. in.
1 cu. ft	= 28.317 cu. dm or .02832 cu. m	1 cu. dm	= .0353 cu. ft
1 cu. yd	= .7646 cu. m	1 cu. m	= 1.308 cu. yd
1 cd	= 3.624 steres (st)	1 st	= .2759 cd

Exercise E

Fill in the missing values. *Note:* Your answers may vary from the answers in the back of the book because of your choice of the rounded conversion factors from the table.

1. 4 yd, 7 in. equals __3.8354__ m (meters). (4 × .9144 = 7 × 2.54 =)
2. 4 yd, 2 ft, 6 in. equals __*4.4196*__ m (meters). ***(4 × .9144 = 2 × .3048 = 6 × 2.54 =)***
3. 36.1 m equals __*39.47896*__ yd (yards). ***(36.1 × 1.0936 =)***
4. 16 liquid qt equals __*15.1408*__ L (liters). ***(16 × .9463 =)***
5. 91 sq. ft equals __*8.4539*__ sq. m (square meters). ***(91 × .0929 =)***
6. 7 sq. mi equals __*18.123*__ sq. km (square kilometers). ***(7 × 2.589 =)***
7. 2.6 sq. m equals __*3.1096*__ sq. yd (square yards). ***(2.6 × 1.196 =)***
8. 4 cu. ft, 18 cu. in. equals __*.1135749*__ cu. m (cubic meters). ***(4 × .02832 = 18 × 16.3872 =)***
9. 3 lb, 14 oz avoir. equals __*1.7577*__ kg (kilograms). ***(3 × .4536 = 14 × 28.35 =)***
10. 7.6 metric tons equals __*8.37748*__ tons. ***(7.6 × 1.1023 =)***

Exchange Rates
Objective 5

The United States is one of the largest economic markets in the world in terms of its population and its spending power. Recent increases in international trade have made it very important for us to provide product information in the metric system. The exchange of currency has also become an important aspect of trade to understand as international transactions increase. Economies of the various nations of the world are constantly changing, and thus the demand for their currencies is also changing. The rate of exchange

EXCHANGE RATES

Country	U.S. $ equiv.	Currency per U.S. $
Argentina (Peso)	1.0005	.9995
Australia (Dollar)	.6524	1.5328
Austria (Schilling)	.07573	13.204
Bahrain (Dinar)	2.6525	.3770
Belgium (Franc)	.0258	38.7100
Brazil (Real)	.5811	1.7210
Britain (Pound)	1.6013	.6245
Canada (Dollar)	.6784	1.4740
Chile (Peso)	.002025	493.85
China (Renminbi)	.1208	8.2780
Columbia (Peso)	.0005934	1685.33
Czech. Rep. (Koruna)		
Commercial rate	.02768	36.130
Denmark (Krone)	.1401	7.1390
Ecuador (Sucre)		
Floating rate	.00010753	9300.00
Finland (Markka)	.1753	5.7055
France (Franc)	.1589	6.2946
Germany (Mark)	.5328	1.8768
Greece (Drachma)	.003209	311.59
Hong Kong (Dollar)	.1290	7.7549
Hungary (Forint)	.004185	238.94
India (Rupee)	.02330	42.915
Indonesia (Rupiah)	.0001231	8125.00
Ireland (Punt)	1.3233	.7557
Israel (Shekel)	.2427	4.1195
Italy (Lira)	.0005382	1858.05
Japan (Yen)	.008229	121.52
Jordan (Dinar)	1.4104	.7090
Kuwait (Dinar)	3.2616	.3066
Lebanon (Pound)	.0006631	1508.00
Malaysia (Ringgit)	.2632	3.8000
Malta (Lira)	2.4777	.4036
Mexico (Peso)		
Floating rate	.1030	9.7100
Netherland (Guilder)	.4729	2.1147
New Zealand (Dollar)	.5356	1.8671
Norway (Krone)	.1261	7.9273
Pakistan (Rupee)	.01935	51.670
Peru (new Sol)	.3007	3.3253
Philippines (Peso)	.02630	38.020
Poland (Zloty)	.2504	3.9942
Portugal (Escudo)	.005198	192.38
Russia (Ruble) (a)	.04092	24.440
Saudi Arabia (Riyal)	.2666	3.7505
Singapore (Dollar)	.5796	1.7252
Slovak Rep. (Koruna)	.02291	43.643
South Africa (Rand)	.1600	6.2510
South Korea (Won)	.0008432	1186.00
Spain (Peseta)	.006263	159.66
Sweden (Krona)	.1161	8.6118
Switzerland (Franc)	.6547	1.5275
Taiwan (Dollar)	.03054	32.743
Thailand (Baht)	.02696	37.090
Turkey (Lira)	.00000248	403522.00
United Arab (Dirham)	.2723	3.6728
Uruguay (New Peso)		
Financial	.08938	11.118
Venezuela (Bolivar)	.001670	598.75

● *The reciprocal exchange can always be found by making a fraction of the exchange rate with the rate as the denominator and one as the numerator; i.e., 1 U.S. dollar =* $\frac{1}{3.1626}$ *= .3066 Kuwait dinar*

of one currency for another currency is of daily concern to business firms. The rate of currency exchange is printed in the business section of most newspapers and business periodicals daily.

The member nations of the European Union (E.U.) have agreed to use a single currency, the Euro. Beginning in the year 2001, the Euro will be used for common exchange. Currently, all economic data, major transactions, and economic forecasts use Euros as the basis of valuation. This is an important issue as the E.U. is the largest economic unit in the world in terms of population and a very sizable force in economic buying power. The Euro, like other currencies, changes value. The Euro is valued at approximately one U.S. dollar.

The table of exchange rates shows the value of the monetary units of other countries compared to that of the U.S. dollar. Notice that some currencies are worth more than the U.S. dollar (Kuwait's, for example) and others are worth less (Japan's).

Example 10

Mota International, Inc., sold 400 units of their product to a firm in Kuwait at a price of $58 per unit. How many Kuwait dinars would it take to pay the invoice?

400 × $58 = $23,200 total invoice price in U.S. dollars
$23,200 × .3066 = 7,113.12 Kuwait dinars

(23,200 × .3066 =)

Example 11

If the same order were sold to a customer in Japan, how many yen would it take to pay the invoice?
$23,200 × 121.52 = 2,819,264 Japanese yen

(23,000 × 121.52 =)

Exercise F

Determine the amount of foreign currency needed to make the following purchases.

Customer Country	Invoice Amount (U.S.)	Currency Needed
1. Australia	$4,600 (4,600 × 1.5328 =)	7,050.88 dollars (Australian)
2. Norway	15,900 ***(15,900 × 7.9273 =)***	***126,044.07*** krone
3. Chile	7,000 ***(7,000 × 493.85 =)***	***3,456,950*** pesos (Chile)
4. Canada	230,000 ***(230,000 × 1.4740 =)***	***339,020*** dollars (Canadian)
5. Mexico	450 ***(450 × 9.71 =)***	***4,369.5*** pesos (Mexico)

The currency exchange rate is equally important for U.S. firms that buy from producers in other countries. The U.S. firm will need to make a similar type of calculation to determine the amount of U.S. dollars needed to pay the invoice.

Example 12

Hal is quoted a price of 835,000 lira for each of the 20 bicycles that Sport-A-Merica will purchase from a firm in Italy. What is the amount of the invoice in U.S. dollars?

20 × 835,000 lira = 16,700,000 lira
16,700,000 lira × .0005382 = 8,987.94 U.S. dollars

(16,700,000 × .0005382 =)

Exercise G

Determine the number of U.S. dollars necessary to make the following purchases.

Purchase from	Invoice Amount	U.S. Dollars Needed
1. South Africa	600 rands (600 × .16 =)	$96.00
2. India	7,000 rupees *(7,000 × .0233 =)*	***$163.10***
3. Finland	500 markkas *(500 × .1753 =)*	***$87.65***
4. South Korea	9,200,000 wons *(9,200,000 × .0008432 =)*	***$7,757.44***
5. Philippines	50,000 pesos *(50,000 × .02630 =)*	***$1,315.00***

The exchange tables are printed daily because the exchange rates may vary from one day to the next. When a firm agrees to a purchase, they must think about the possibility of exchange rates changing between the date of the purchase and the date that the invoice is to be paid.

Example 13

Wiedmont Associates had agreed to purchase an item for 161,000 Peruvian sols. The exchange rate on the date of purchase was .3007. The exchange rate on the date that the invoice was due is .3019. How much more in U.S. dollars did it take to pay the invoice when it was due than it would have taken when the purchase was made?

$$\begin{array}{r} .3019 \\ -.3007 \\ \hline .0012 \end{array} \quad .0012 \times 161{,}000 \text{ sols} = \$193.20$$

(.3019 − .3007 = .0012 × 161,000 =)

ASSIGNMENT Chapter 6

Name ______________________ Date ______________ Score __________

Skill Problems

Solve the following problems. Remember to reduce your answers.

1. 4 yd, 2 ft, 10 in. − 1 yd, 11 in. =
−1 11
3 yd, 1 ft, 11 in.

2. 14 lb, 8 oz ÷ 4 = *3 lb, 10 oz*
3 lb: 4)14, 12, 2
10 oz: 4)32 + 8 = 40

3. 4 hr, 30 min, 25 sec ÷ 5 = *54 min, 5 sec*
(4 × 60 × 60 = 14,400, 30 × 60 = 1,800, 14,400 + 1,800 + 25 = 16,225 ÷ 5 = 3,245 ÷ 60 = 54 r 5)

4. 20 lb, 8 oz × 3 = *61 lb, 8 oz*
(60 lb, 24 oz =)

2 82 60
5. 3 hr, 23 min − 1 hr, 40 min, 24 sec
−1 40 24
= 1 hr, 42 min, 36 sec

6. 8 sq. yd, 2 sq. ft + 15 sq. yd, 7 sq. ft, 4 sq. in. =
24 sq. yd, 4 sq. in. (23 sq. yd, 9 sq. ft, 4 sq. in.)

7. 46 sq. yd ÷ 3 = *15 sq. yd, 3 sq. ft*
(46 ÷ 3 = 15, remainder: 1 × 3 = 3)

8. 5 1/2 ft × 8 1/2 ft = *46 3/4 sq. ft or 46 sq. ft, 108 sq. in. (5.5 × 8.5 =)*

9. 4.8 mi × 8.3 mi = *39.84 sq. mi (4.8 × 8.3 =)*

10. 7 hr, 40 min, 30 sec ÷ 6 = *1 hr, 16 min, 45 sec*
1 hr: 6)7, 6, 1
16 min: 6)60 + 40 = 100, 6, 40, 36, 4
45 sec: 6)240 + 30 = 270

11. 18 yd × 2 ft × 1.5 ft = *6 cu. yd*
(54 × 2 × 1.5 = 162 cu. ft, 162/27 =)

12. 9 ft × 11 ft × 5 ft = *18 cu. yd, 9 cu. ft*
(9 × 11 × 5 = 495 cu. ft, 495/27 =)

13. 14 cu. yd, 21 cu. ft + 4 cu. yd, 12 cu. ft =
19 cu. yd, 6 cu. ft (14 + 4 = 18, 21 + 12 = 33)

14. 50 ft long, 12 ft wide, and 6 in. deep equals how many cubic feet (nearest cubic foot)?
300 cu. ft (50 × 12 × .5 =)

15. 3 yd, 2 ft, 8 in. × 2 = ___7___ yd, ___2___ ft., ___4___ in. *(3 × 2 = 2 × 2 = 8 × 2 =)*

16. 36 lb, 8 oz ÷ 4 = ___9___ lb, ___2___ oz *(36 ÷ 4 = 8 ÷ 4 =)*

17. 7 min, 20 sec × 83 = ___10___ hr, ___8___ min, ___40___ sec *(7 × 83 = 20 × 83 =)*

18. 2 sq. yd, 7 sq. ft × 7 = ___19___ sq. yd, ___4___ sq. ft *(2 × 7 = 7 × 7 =)*

19. 5 yd, 2 ft × 4 ft × 7 yd, 1 ft = ___55___ cu. yd, ___11___ cu. ft *(17 ft. × 4 ft. × 22 ft. =)*

20. 2.3 sq. ft × 5 ft = ___11.5___ cu. ft *(2.3 × 5 =)*

Convert the following measurements to the base indicated

21. 4,362 dcm = ___43,620___ meters. *(4,362 × 10 =)*

22. 18 L equals ___18,000___ cl *(18 × 100 =)*

23. 786.5 cm to meters: *7.865*

24. 642 cubic hectometers to cubic decameters: *642,000*

25. 807,604 cu. cm to cubic meters: *.807604*

26. .46 L to centileters: *46*

27. 37,480 g to decigrams: *374,800*

Fill in the missing values.

28. 870 tons = ___789.264___ metric tons.
(870 × .9072 =)

29. 8 yd, 9 in. = ___7.5438___ m (meters).
(8 × .9144 = 7.3152) + (9 × 2.54/100 = .2286)

30. 7 lb, 12 oz avoir. = ___3.5154___ kg (kilograms).
(7 × .4536 = 3.1752) + (12 × 28.35/1000 = .3402)

31. 5 yd, 2 ft = ___5.1816___ meters.
(5 × .9144 = 4.572) + (2 × .3048 = .6096)

32. 9 cu. ft, 16 cu. in. = ___.255142___ cu. m (cubic meters). *(9 × .02832 = .25488) + (16 × 16.3872/1000000 = .0002621952)*

33. 47 m = ___51.3992___ yd (yards).
(47 × 1.0936 =)

34. 8.4 sq. m = ___10.0464___ sq. yd (square yards).
(8.4 × 1.196 =)

35. 47 liquid qt = ___44.4761___ L (liters).
(47 × .9463 =)

36. 42 sq. mi = ___108.738___ sq. km (square kilometers).
(42 × 2.589 =)

37. 95 sq. ft = ___8.8255___ sq. m (square meters).
(95 × .0929 =)

38. 781 cm = ___307.4797___ in. (inches).
(781 × .3937 =)

Customer Country	Invoice Amount (in customers' currency)		Currency Needed	
39. United States	$5,971	*(5,971 × 1.8768 =)*	*11,206.372*	marks (Germany)
40. Canada	10,943	*(10,943 × .6784 =)*	*7,423.73*	U.S. dollars
41. Greece	98,740	*(98,740 × .003209 =)*	*316.86*	U.S. dollars
42. Kuwait	345,158	*(345,158 × 3.2616 =)*	*1,125,767.30*	U.S. dollars
43. Taiwan	324,482	*(324,482 × .03054 =)*	*9,909.68*	U.S. dollars
44. South Africa	654,871	*(654,871 × .16 =)*	*104,779.36*	U.S. dollars
45. United States	421	*(421 × 1.7252)*	*726.31*	dollars (Singapore)
46. Mexico	25,480	*(25,480 × .1030 =)*	*2,624.44*	U.S. dollars
47. Brazil	7,804	*(7,804 × .5811 =)*	*4,534.90*	U.S. dollars
48. Israel	16,300	*(16,300 × .2427 =)*	*3,956.01*	U.S. dollars
49. United States	5,000	*(5,000 × 121.52 =)*	*607,600*	yen (Japan)
50. Germany	6,850	*(6,850 × .5328 =)*	*3,649.68*	U.S. dollars

Business Application Problems

1. A piece of land sold as a building site for $27,000. The piece of land was 1.89 acres. What was the cost per acre? ***$27,000/1.89 = $14,285.71 per acre***

2. The owner of the land in problem 1 will build a single-story house with 3,500 square feet and a shop with 2,000 square feet on the site. How many square feet of land will not be covered by buildings?
3,500 + 2,000 = 5,500 sq. ft
1.89 × 160 × 30.25 × 9 = 82,328.4 sq. ft
82,328.4 − 5,500 = 76,828.4 sq. ft

3. The owner of the land in problem 1 will spray the land not covered with buildings with an insecticide. The spray costs $13 per liter. Each liter will cover 5,000 square feet. What is the cost of the spray material? Only full liters are available for purchase.
76,828.4/5,000 = 15.36568 16 × $13 = $208 cost

4. A contract valued at 42,000,000 Indian rupees has an expected delivery of one year. At the end of the completed contract, the Indian rupee had dropped in value by .00011. What is the amount of the savings to the buyer in dollars, if the old rate was .02386 Indian rupees to one U.S. dollar?
42,000,000 × .02386 = $1,002,120
−.00011
42,000,000 × .02375 = $ 997,500
$4,620 savings

5. An Idaho firm contracted to sell goods to a firm in Argentina. The contract is valued at 50,000 Argentine pesos. What is the value of the deal in U.S. dollars?
50,000 × 1.0005 = $50,025 value

6. Kariva will paint the four walls of her store, which measures 28 feet by 40 feet. The walls are 8 feet high. The paint she is considering will cover 500 square feet per gallon. Each gallon costs $19.95. Ignoring the area of windows and doors, what will be the cost of the paint? (*Note:* She must purchase full gallons.)
(2) × 28 × 8 = 448 sq. ft
+(2) × 40 × 8 = 640 sq. ft
1,088 sq. ft ÷ 500 = 3 gallons

7. Taxson, Inc. purchased electronic parts from a firm in Japan last year. The price of each piece was 81.4 yen. Taxson purchased 48,000 pieces at intervals of 4,000 each month over the year. The contract called for payment at the end of each month. The average exchange rate of the yen was .0094 U.S. dollars. What was the total cost of the purchase in U.S. dollars?
48,000 × 81.4 × .0094 = $36,727.68 value

8. A company will purchase 5,000 pounds of a chemical material known as "Simcon-2" from one of two suppliers. The American supplier's price is \$3.95 per pound. The Venezuelan supplier's price is 4,840 bolivars per kilogram (avoir). Both prices include delivery. What is the lowest total price for the 5,000 pounds?
5,000 × \$3.95 = \$19,750 5,000 × .4536 =
2,268 kg × 4,840 = 10,977,120 bolivars × .00167 =
\$18,331.79 Venezuelan supplier

9. Kem-Kleene, Inc. purchases soap in a 55-gallon container. The soap is packaged into 6-ounce packages for use in dispensers for their commercial customers. How many 6-ounce packages can be developed from one 55-gallon container? *55 × 4 × 32 = 7,040 ounces*
7,040/6 = 1,173.3 1,173 packages

10. If Kem-Kleene, Inc. in problem 9 were to enter the European market, the packages would be .5 liter in size and would sell for \$1.95 each. What is the value of a 55-gallon container of soap packaged for the European market?
55 × 4 = 220 quarts 220 × .9463 = 208.186 liters
208.186/.5 = 416.372
416 × \$1.95 = \$811.20 value

11. Carol Huntington was assigned a new task by her supervisor. The task is expected to take her 1 hour and 53 minutes. She must complete the task two times a day. How much time will be left out of an 8-hour day for other activities?
1 hr, 53 min
×2
2 hr, 106 min = 3 hr, 46 min
8 hr
−3 hr, 46 min
4 hr, 14 min left

12. Moore and Associates imported raw materials from Thailand at a quoted price of 57,000 baht. How many U.S. dollars will be required to pay the invoice?
57,000 Thailand baht
×.02696
\$1,536.72

13. The shipping department must deliver 500 boxes that are $1\frac{1}{2}$ cubic feet in size. The boxes will be packed into a container that is able to accommodate 75 cubic feet. How many containers will be necessary to ship the 500 boxes?
500 boxes
$\times 1\frac{1}{2}$ cu. ft $\frac{750 \text{ cu. ft}}{75 \text{ cu. ft}}$ = 10 containers
750 cu. ft

14. Bayburn & Jackson purchased machinery priced at 500,000 South Korean won. When the invoice was due, the exchange rate had changed from .000841 to .000832. How much money did Bayburn & Jackson make or lose by delaying payment?
.000841
−.000832
.000009 .000009 × 500,000 = \$4.50 made

15. The Shaw family is considering a new pool for their summer home. They contact the Cool Pool Company, a local distributor. The pool company representative stated that the first job would be to remove the sod from an area 30 ft × 60 ft. Next, a hole would be dug 25 ft × 55 ft × an average of 6 ft deep. The pool would then be built in the hole. The finished pool would measure 20 ft × 50 ft × 5 ft deep (average). How many square yards of sod must be removed from the lawn before the digging can begin?
30 ft
×60 ft
1,800 sq. ft
1,800 sq. ft ÷ 9 = 200 sq. yd to be removed

16. How much would it cost the Shaw family for earth removal at \$46 a cubic yard (nearest cubic yard)?
25 ft
×55 ft
1,375 sq. ft
×6 ft
8,250 cu. ft
8,250 cu. ft ÷ 27 =
305.55555 = 306 cu. yd
306 cu. yd.
× \$46
\$14,076 cost

17. The Shaws are concerned about the amount of time it would take to fill the pool if their well will only pump 225 gallons per hour. How many hours will it take to fill the pool to within one foot of the top?
50 ft
×20 ft
1,000 sq. ft × 4 ft = 4,000 cu. ft
×7.5
30,000 gal ÷ 225
$= 133\frac{1}{3}$ hr

18. George and Delores will mix one cubic meter of chocolate, 400 cubic decimeters of nuts, and 700 cubic decimeters of coconut together in their candy shop. From this mixture they will make Mother's Day hearts that will be 50 cubic centimeters in size. How many hearts will they be able to make?

1 cu. m chocolate = 1,000,000 cu. cm
400 cu. dm nuts = 400,000 cu. cm
700 cu. dm coconut = 700,000 cu. cm
2,100,000 ÷ 50
= 42,000 hearts

19. Write a complete sentence in answer to this question. The Dive Shop will have a new sign on the top of their building measuring 10 feet high and 27 feet long. The material that will be used costs $41 per square meter. What will be the cost of the material?

10 ft × 27 ft = 270 sq. ft
×.0929
25.083 sq. m × $41 = $1,028.40
The cost of the material will be $1,028.40.

20. Write a complete sentence in answer to this question. The mayor's office must be recarpeted. The room is 30 ft wide and 32 ft long. An 8-ft by 4-ft area near the desk in the corner of the room is to be left uncarpeted. The carpet store has suggested a nylon beige tweed at $37 a square yard. The carpet can only be sold in full square yards. How much will it cost to carpet the portion of the room as described?

30 ft × 32 ft = 960 sq. ft
− 8 ft × 4 ft = − 32 sq. ft
928 sq. ft ÷ 9
= 103.11 sq. yd
= 104 sq. yd
×$37
$3,848
It will cost $3,848 to carpet the room as described.

★ Challenge Problem

The J. B. Colson Company will build a new building in July. In order to determine the potential savings in utilities expense gained by installing insulation, an estimation of costs and savings must be developed. The building will be 75 ft wide by 260 ft long. The side walls will be 10 ft high in the single-story structure. All of the ceilings will be insulated to a depth of 7 in., except that of the Shipping and Receiving Department, which has an area of 40 ft × 65 ft. The local utility company has stated there will be a $60 per month savings on heating and cooling expenses if the building is insulated to the 7-in. depth in the areas described. The insulating material will cost $2.78 for each cubic-meter bag. How many months of fuel savings will it take to repay the cost of the insulation?

75 ft × 260 ft = 19,500 sq. ft
−40 ft × 65 ft = −2,600 sq. ft

$$16{,}900 \text{ sq. ft} \times \frac{7}{12} \text{ ft}$$

= 9,858.3333 cu. ft
×.02832
279.18799 cu. m
×$2.78
$776.14 cost ÷ $60
= 12.94 months

MASTERY TEST Chapter 6

Name ______________________ Date ______________ Score ____________

1. The Harry Cornell Co. has just received a contract to construct a bicycle path across a part of the county. The company had three groups of men to work on the project, which was 4 miles, 895 yards long. The company has decided to divide the distance equally among the three groups. To the nearest foot, how long of a distance will each group be responsible?

   1 mi 885 yd
   3)4 mi 895 yd
   3 +1,760
   1 2,655
   2,655

2. If Clara's job operation requires 37 seconds to complete, how many complete operations can she do in a period of 5 hours?

   60 min × 5 hr = 300 min in 5 hr
   300 min × 60 = 18,000 sec in 5 hr
   37)18,000 = 486 operations

3. Bell Korten is about to remodel his insurance office. The room is 30 ft × 42 ft with 8-ft side walls. He will paint the walls and ceiling all one color. The paint that he picked out will cover 500 sq ft per gallon. How many gallons must be purchased to do the job if he does not allow for doors, windows, etc.? (Partially used gallons can not be returned.)

   (2) 30 ft × 8 ft = 480 sq. ft
   (2) 42 ft × 8 ft = 672 sq. ft
   1,152 sq. ft ÷ 500 = 2.304 = 3 gal.

4. A building must have an addition that measures a total of 400 feet (for the three sides). The purchasing agent must determine the amount of cement to be used to fill a footing trench $2\frac{1}{2}$ ft deep by 18 in. wide. Determine, to the nearest cubic yard, how many cubic yards of cement will be needed to fill the trench.

   400 ft × 2.5 ft × 1.5 ft =
   1,500 cu. ft; 1,500 ÷ 27 = 55.55 = 56 cu. yd

5. 53 sq. m ÷ 17 equals *33.55835* sq. ft.

   53 × 1.196 = 63.388 sq. yd × 9 =
   570.492 sq. ft ÷ 17 = 33.55835

6. A dump truck hauls 7.6 cubic yards of gravel per load. How many cubic meters would the truck haul?

   7.6 cu. yd
   × .7646
   5.81096 cu. m

7. Ninety pounds of candy is to be sold at \$.50 for each 70-gram box. How many boxes can be made from the 90 pounds?
   (Use avoir.)

   90
   × .4536
   40.824 kg = 40.824 g 40,824 g ÷ 70 = 583.2 boxes

8. 8 yd, 2 ft, 9 in. − 5 yd, 1 ft, 11 in. = *3 yd 10 in.* ?

   8 yd 2 ft 9 in.
   −5 1 11
   3 yd 10 in.

9. Convert 143 square decameters to squre meters: *14,300*.

10. What is the U.S. dollar value of an invoice from China for 42,300 renminbi?

    42,300 × .1208 = \$5,109.84

11. The owner of a store contacted on air conditioning firm for an estimate. The owner indicated that the dimensions of the store were 60 ft by 120 ft, with 12-ft-high side walls. The air conditioning firm stated that their Model X-1,000 could handle 75,000 cubic feet. Would the Model X-1,000 be large enough for the store?

    60 ft × 120 ft × 12 ft = 86,400 cu. ft No!

12. Bolton, Inc., sold a machine priced at \$3,900 to a firm in Ecuador. What amount of sucre will Bolton, Inc., expect in payment for the sale?

    \$3,900 × 9,300 = 36,270,000 sucre

PERCENT 7

Suggested Classroom Time for This Chapter

hours 1 2 3 4 5

Perceived Student Difficulty with Topical Material

easy moderate difficult

OBJECTIVES

After mastering the material in this chapter, you will be able to:

1. Define what *percent* means. p. 128
2. Change percent to a fraction. p. 128
   percent to a decimal. p. 128
   decimal to a fraction. p. 128
   decimal to a percent. p. 128
3. Write all the above with fractional percent. p. 129
4. Identify the base, rate, and amount in word problems. p. 130
5. State the formulas to be used to find the base, the rate, and the amount. p. 130
6. Solve for the unknown using one of three methods. p. 131
7. Work word problems that apply percent to business situations. p. 134
8. Understand and use the following terms and concepts:
   Percent
   Base
   Rate
   Amount

SPORT-A-MERICA

Kerry and Hal have found two possible new store locations for Sport-A-Merica. One store, on Washington Avenue, is a perfect location, but it is a little smaller than Hal had wanted. The space is only 3,000 square feet. Current store space is 1,000 square feet. The second location has 3,700 square feet and is in a good area of town, on Cedar Street, a busy highway area. Both stores have the space divided into several rooms, which would provide more display space on the walls.

Hal realizes that now is the time to take some serious measurements. Will the new space allow for more merchandise? Should the store add more lines of merchandise? The store currently handles 890 items. Hal wants to add 47 new items. What percent increase will that be? Should they advertise the percent increase? Will there be room to park the new delivery vehicle they purchased last month? Utilities will be a big consideration. Hal will need to investigate the cost of electricity for a space that is 200 to 270 percent larger than the present store. Should the cost of utilities be divided based on the area of floor space or based on the type of merchandise in a department? The additional utilities will add to operating costs. This would mean that the price of the merchandise would have to be increased, unless Sport-A-Merica can increase sales by at least as much as the increase in costs.

Making the move to a larger store is clearly necessary, as Hal sees it. There is a very good indication that more business would be available if they could expand. To make the right decision, Hal must compare the current space with the new possibilities and with new product opportunities. Hal and Kerry know that the amount of space, costs, and profits will not stay the same.

What *Percent* Means

Objective 1

Percent is a way of expressing the relationship between two numbers. It is shown in hundredths (i.e., 1 percent is the same as $\frac{1}{100}$). In business applications, percent is used to compute payroll deductions, cash and trade discounts, depreciation, interest, returns, markup, and taxes—to name just a few. The importance of percent cannot be emphasized too much.

percent

The sign % signifies percent and, therefore, hundredths. You can also express hundredths as a fraction by showing a denominator of 100 (i.e., $\frac{?}{100}$). In chapter 3, we saw that the second position to the right of the decimal point is the hundredths position. One hundredth is written .01, five hundredths is .05, and forty-three hundredths is .43, and so on.

Percent to a Fraction

Objective 2

To change a number expressed as a percent to a fraction with a denominator of 100: (1) drop the percent sign, and (2) add a denominator of 100.

Example 1

1. $47\% = \frac{?}{100}$
2. $47\% = \frac{47}{100}$

Percent to a Decimal

In chapter 3, we found that any fraction can be changed into a decimal number by dividing the numerator by the denominator. In the previous example, 47 percent was changed into a fraction of $\frac{47}{100}$. This fraction can be changed into a decimal number by dividing 47 by 100, which gives an answer of .47 (i.e., forty-seven hundredths).

Example 2

$7\% = .07$
$135\% = 1.35$

(7 ÷ 100 = 135 ÷ 100 =)

Decimal to a Fraction

If a fraction can be changed to a decimal number, we should be able to reverse the process and change a decimal to a fraction: (1) drop the decimal point, and (2) put the number over 100.

Example 3

1. $.76 = \frac{?}{100}$
2. .76 is read as seventy-six hundredths, which can be expressed as $\frac{76}{100}$

Example 4

$.08 = \frac{8}{100} \qquad 1.53 = \frac{153}{100}$

(.08 × 100 = 1.53 × 100 =)

Decimal to a Percent

A decimal can similarly be changed to a percent. Again, recalling that *percent* is hundredths, .67 can be changed to 67 percent; .67 is read "sixty-seven hundredths" or "sixty-seven percent." To complete the process: (1) drop the decimal point, and (2) add the percent sign.

Example 5

$.06 = 6\%$
$1.15 = 115\%$

(.06 × 100 = 1.15 × 100 =)

You should now be able to interchange percents, fractions, and decimals in any order. Use the previous page of this chapter as a guide to do the first few problems of exercises A and B.

Exercise A

Express each of the following as a decimal.

1. $17\% = \frac{17}{100} =$ .17
2. $56\% = \frac{56}{100} =$ *.56*
3. $11\% = \frac{11}{100} =$ *.11*
4. $9\% = \frac{9}{100} =$ *.09*
5. $156\% = \frac{156}{100} =$ *1.56*

Exercise B

Fill in the blanks in the following problems.

1. 41 $\% = \frac{41}{100} =$ .41
2. *62* $\% = \frac{62}{100} = .62$
3. *127* $\% = \frac{127}{100} = 1.27$
4. *10* $\% = \frac{10}{100} = .1$
5. $21\% = \frac{21}{100} =$ *.21*

Fractional Percent

Objective 3

❍ A fractional percent is very small. It is less than $\frac{1}{100}$.

Quite often, a percent is expressed as a mixed number, such as $8\frac{1}{2}$ percent or $20\frac{1}{4}$ percent. In order to use these percents in calculations, you must be able to change the mixed number, or *fractional percent,* to a decimal. The process requires these steps: (1) change the fraction to a decimal, (2) place the decimal behind the whole number, (3) drop the percent sign, and (4) divide by 100.

Example 6

$$9\frac{3}{4}\% = 9.75\% = \frac{9.75}{100} = .0975$$

$$\frac{1}{5}\% = .2\% = \frac{.2}{100} = .002$$

(9.75 ÷ 100 =   .2 ÷ 100 =)

When you read a fractional percent that does not include a whole number, such as $\frac{1}{4}$ percent, the percent can be read as "one-fourth percent" or, more clearly, as "one-fourth of one percent." One-fourth percent is realistically a fraction of 1 percent. It should not be confused with 25 percent.

Exercise C

Fill in the blanks in the following problems.

1. $8\frac{1}{5}\% = \frac{8.2}{100} =$ .082
2. $1\frac{3}{8}\% = \frac{1.375}{100} =$ *.01375*
3. *.6* $\% = \frac{.6}{100} =$ *.006*
4. $52.9\% = \frac{52.9}{100} =$ *.529*
5. *29.5* $\% = \frac{29.5}{100} = .295$

Percent Problems
Objective 4

In this portion of the chapter, you will learn how to put the percent skills you have learned to work for you. By using the percent skills and the relationships to other information, you will be able to solve business problems. Before we proceed, you must become very familiar with the descriptions of the following three terms. Understanding these descriptions will make your efforts in problem solving begin correctly.

base

The **base** is: always the whole.
always 100%.
always what something else is compared to.
the number we take some percent "of."
the number that follows the word "of" in a problem sentence.
defined as "the factor of which so many hundredths is taken."
not always the largest number.

● ***Encourage students to know the descriptions for base, rate, and amount.***

The base would be the number of students in your class, for example. In a business situation, it might be the total number of employees.

rate

The **rate** is: a percent.
usually expressed in a percent form, but can also be expressed as a fraction or a decimal (i.e., 25%, or $\frac{1}{4}$, or .25).
defined as "a certain number of hundredths."

For example, the rate would be the percent of students in your class over 5 feet 8 inches tall. In the business situation noted previously, it could be the percent of females in the firm.

amount

The **amount** is: related to the base because of the size of the rate.
not always the smallest number (may be larger than the base!).
sometimes termed *part, portion,* or *percentage.*
defined as "the result of taking a certain number of hundredths of the base."

❍ *Basic algebra substitution can also be used to solve for the unknown.*

The amount would be the *actual number* of students over 5 feet 8 inches tall. In the business situation noted earlier, it could be the 176 females in the company.

Take time now to develop an understanding of the three terms and their meanings.

There are three ways to begin solving a percent problem. You could memorize the formulas shown here, you could use the model as a memory aid, or you could learn to solve for the unknown. In order to provide you with as many options as possible, all three ways will now be explored.

The Formulas
Objective 5

In working percent problems, two of the factors will always be known or can be easily found. You must find the third unknown factor. The formulas are as follows:

$$\text{Amount} = \text{Base} \times \text{Rate} \qquad A = B \times R$$

$$\text{Base} = \frac{\text{Amount}}{\text{Rate}} \qquad B = A \div R$$

$$\text{Rate} = \frac{\text{Amount}}{\text{Base}} \qquad R = A \div B$$

The Model

By putting your finger on the factor that you wish to find, the remaining uncovered portion of the model reveals the math operations that are used to find the unknown factor. The model is as follows:

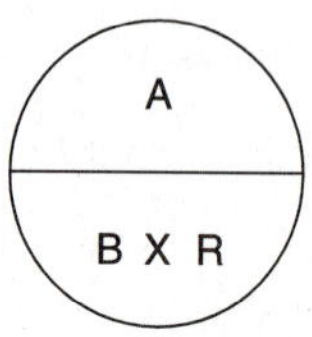

Example 7 If you know Amount and Rate in a problem but need to find the Base, put your finger over B.

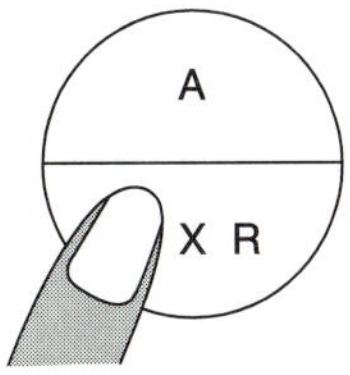

Base is found by dividing A by R.

Solving for the Unknown

Objective 6

● *The solving for the unknown approach is used in examples 8–20.*

We often make statements like, "You have \$15 and I have \$8. Together we have \$23." This could be expressed in another way by saying, "How much money do we have together if you have \$15 and I have \$8?" This is an expression. We could alter the expression a little more by changing the words to "M is equal to the amount of money you have (\$15) and I have (\$8)." By making the statement in mathematical terms only, we could write M = \$15 + \$8. This is known as a *mathematical expression*. What it says is that the values (letters or numbers or a combination of both) on one side of the equals sign are equal to the values on the other side. It is as simple as that!

It is also possible to restate the same facts in other ways to make other expressions. For example, "Between the two of us we have \$23. You have \$15. How much do I have?" That expression could be written as \$23 = \$15 + I. We know that if we were to try to find what value I is, we would need to determine what must be added to \$15 to make \$23. The answer is \$8. This could be solved in yet another way: by subtraction. You may do any fundamental operation (add, subtract, multiply, or divide) to an expression as long as you do the operation *on both sides of the expression*. Using the last expression, \$23 = \$15 + I, \$15 can be subtracted from both sides of the expression to determine the answer.

$$\begin{array}{rcr} \$23 & = & \$15 + 1 \\ -\$15 & & -\$15 \\ \hline \$8 & = & \$0 \;\; + 1 \end{array}$$

or 1 = \$8

Using the same information, another expression can be developed: "The total amount of money we have between us, less \$10, is equal to \$13. What is the amount of money we have between us?" The mathematical expression can be written as T − \$10 = \$13. The solution can be found by addition.

$$\begin{array}{rcr} T - \$10 & = & \$13 \\ + \$10 & & + \$10 \\ \hline T + \;\; \$0 & = & \$23 \end{array}$$

or T = \$23

A few ground rules should be covered at this point in order to create some consistency and understanding regarding this type of math. The letters that are used are termed *variables*. The variable is usually shown on the left side of the expression. The multiplication and division of variables and numbers can be expressed in several ways. These expressions mean the same thing.

4S, 4(S), $4 \times S$, and $4 \cdot S$

(Note that both the parentheses and the dot mean "multiply.")

$\frac{1}{5}R$, $\frac{1}{5} \times R$, $\frac{R}{5}$, and $R \div 5$

Expressions can include multiplication and division, as well as addition and subtraction. An expression can be written as 5A = 40. To find A, you need to divide both sides of the expression by 5.

$\frac{5A}{5} = \frac{40}{5}$ The answer is A = 8.

The multiplication function can be used in much the same fashion. $\frac{H}{2} = 90$ can be solved by multiplying both sides of the expression by 2.

$2 \times \frac{H}{2} = 90 \times 2$ The answer is H = 180.

Exercise D

In the following expressions, solve for the variable.

1. $J + 20 = 56$
   $\underline{-20 - 20}$
   $J = 36$

2. $E - 2 = 9$
   $\underline{+2 + 2}$
   $E = 11$

3. $M - 3 = 56$
   $\underline{+3 + 3}$
   $M = 59$

4. $\frac{P \times 7}{\mathit{7}} = \frac{42}{\mathit{7}}$ ***P = 6***

5. $\frac{L}{4} = 80$
   $4 \times \frac{L}{4} = 80 \times 4$ L = 320

6. $\frac{8K}{\mathit{8}} = \frac{10 + 6}{\mathit{8}}$ ***K = 2***

7. $\frac{7(A)}{\mathit{7}} = \frac{42}{\mathit{7}}$ ***A = 6***

8. $T \cdot 4 + 10 = 50$
   $4T + 10 = 50$
   $\underline{-10 - 10}$
   $4T = 40$ T = 10

9. $H + 5 = 30$
   $\underline{-5 - 5}$
   $H = 25$

10. $G \times \frac{1}{2} = 32$
    $2 \cdot \frac{G}{2} = 32 \cdot 2$ G = 64

When solving a problem with a variable, the first step is to put the problem into a statement that is an expression. This statement can, in turn, be made into a mathematical expression. You can then solve for the unknown in that expression.

Some additional hints in working toward an expression: The word "of" indicates that the multiplication function is to be used. As an example, one-fourth of the students is equal to 20. What is the total? The mathematical expression for the two previous sentences is $\frac{1}{4}S = 20$. This statement says that $\frac{1}{4}$ of S is equal to 20. The word "of" prior to the word "students" indicates that $\frac{1}{4}$ should be multiplied by S (for students) in order to find the total.

In the percent problems that follow, there will always be two factors known or easily found, and one unknown. You now know how to solve for the unknown using an expression. Some example problems will help show how the percent problems are solved. Use the formulas, the model, or solving for the unknown to follow along in the solutions.

Example 8

Base = 400	A = B × R
Rate = 20%	A = 400 × .20
Amount = Unknown	A = 80

(400 × .20 =)

Example 9

B = 60 $\qquad$ $A = B \times R$

R = 140% $\qquad$ $A = 60 \times 1.40$

A = Unknown $\qquad$ A = 84

(60 × 1.40 =)

Example 10

A = 918 $\qquad$ $B = \frac{A}{R}$

R = 153% $\qquad$ $B = \frac{918}{1.53}$

B = Unknown $\qquad$ B = 600

(918 ÷ 1.53 =)

Example 11

The current electric bill is about $450 per month. The new building for Sport-A-Merica will cost $90 more per month for electricity. What is the percent of increase?

B = $450 current bill $\qquad$ $R = \frac{A}{B}$

A = $90 increase $\qquad$ $R = \frac{\$90}{\$450}$

R = Unknown $\qquad$ R = 20%

(90 ÷ 450 =)

Four-Step Process

Notice that a four-step process can always be used:

1. Identify the factors (base, rate, and amount).
2. Select the formula for the unknown.
3. Substitute the values for the factors.
4. Complete the fundamental process called for in the problem.

In the following problems, find the unknown factor. Round off your answer to the nearest unit or nearest percent.

Exercise E

1. B = 93
   R = 107%
   A = 100 (93 × 1.07 =)
2. B = 116
   R = *86%* *(100 ÷ 116 =)*
   A = 100
3. B = 700
   R = $3\frac{3}{4}$% *(700 × .0375 =)*
   A = *26*
4. B = 492
   R = 14%
   A = *69* *(492 × .14 =)*
5. B = 92
   R = *11%* *(10 ÷ 92 =)*
   A = 10
6. B = *500* *(180 ÷ .36 =)*
   R = 36%
   A = 180

Exercise F

1. B = \$17
   R = 42%
   A = \$7.14 (17 × .42 =)
2. B = \$180 (54 ÷ .30 =)
   R = 30%
   A = \$54
3. B = \$860
   R = 512% (4,400 ÷ 860 =)
   A = \$4,400
4. B = \$23.20 (58 ÷ 2.50 =)
   R = 250%
   A = \$58
5. B = \$48
   R = $14\frac{3}{4}$%
   A = \$7.08 (48 × .1475 =)
6. B = \$47
   R = 43% (20 ÷ 47 =)
   A = \$20

Objective 7

You will find it much easier to use this four-step process when working the word problems at the end of this chapter. The next several examples include an expression that will aid you in solving for the unknown.

Example 12

Forty percent of the students at Bisvern College are in the Vocational Division. There are 6,800 students at the college. How many are in the Vocational Division?

40% of the students are in the Vocational Division.
(R) (B)
40% of 6,800 = A
40% × 6,800 = A
2,720 = A

(.40 × 6,800 =)

If the same information were used in the problem with a different unknown value, the approach would be the same regarding the use of the four-step process.

Example 13

Of the 6,800 students at Bisvern College, 2,720 are in the Vocational Division. What is the percent of Vocational Division students?

2,720 students of 6,800 students is ______________ percent?
(A) (B)
2,720 ÷ 6,800 is equal to ______________ percent?
2,720 ÷ 6,800 = R
40% = R

(2,720 ÷ 6,800 =)

Again, the same information can be phrased differently to ask for a different unknown value. The four-step process would still be used.

Example 14

The 2,720 Vocational Division students represent 40 percent of the total students at Bisvern College. What is the total number of students at Bisvern College?

2,720 students represent 40 percent of the total.
2,720 students represent 40 percent of B.
2,720 = 40% × B
6,800 = B

(2,720 ÷ .40 =)

Increases and Decreases

The same formula can be used to find increases and decreases in an amount. The way the problem is worded is very important to the solution. In most cases, you will begin by finding the difference in the previous figure and the current figure.

Example 15

Mark is the manager of the produce section of the local grocery store. On Monday he had 14 boxes of lettuce. On Wednesday he had only 5 boxes left. What percent was sold? Nine of the total (14) were sold.

$$\begin{array}{r} 14 \\ -5 \\ \hline 9 \end{array}$$ were sold (difference)

B = 14	R = A ÷ B
A = 9	R = 9 ÷ 14
R = ?	R = 64%

(9 ÷ 14 =)

Example 16

If the question in example 15 were to read "What percent was left?", the solution would be a little different. Five of the total (14) were left.

B = 14	R = A ÷ B
A = 5 (left)	R = 5 ÷ 14
R = ?	R = 36%

(5 ÷ 14 =)

Example 17

Because of increased competition, the sales of the Morten Company decreased from $459,098 to $398,013 over the period of one year. What is the percent of decrease? $61,085 is what percent of $459,098?

$459,098
−398,013
$61,085 (difference or decrease)

B = $459,098	R = A ÷ B
A = $61,085	R = $61,085 ÷ $459,098
R = ?	R = 13%

(459,098 − 398,013 = 61,085 ÷ 459,098 =)

The same approach is used to find increases.

Example 18

Ellen had 147 credit customers at the beginning of the quarter. At the quarter's end, she had 159. What is the percent of increase in credit customers? Twelve is what percent of 147?

$$\begin{array}{r} 159 \\ -147 \\ \hline 12 \end{array}$$ (increase or difference)

B = 147	R = A ÷ B
A = 12	R = 12 ÷ 147
R = ?	R = 8%

(12 ÷ 147 =)

Example 19

There were 890 items in stock at Sport-A-Merica at the beginning of the year. During the year the company added 47 new items. What is the percent of increase in inventory items? Forty-seven is what percent of 890?

B = 890	R = A ÷ B
A = 47	R = 47 ÷ 890
R = ?	R = 5%

(47 ÷ 890 =)

Example 20 Jason was earning an hourly rate of \$8.00. He asked his boss for a raise. In his next paycheck he noticed his new rate of \$9.60. What is the rate of the increase? \$1.60 is what percent of \$8?

\$9.60	B = \$8	R = A ÷ B
−8.00	A = \$1.60	R = \$1.60 ÷ \$8
\$1.60 (increase or raise)	R = ?	R = 20%

(1.60 ÷ 8 =)

Exercise G

In the following problems, find the unknown factor. Round off your answer to the nearest unit, cent, or percent.

1. Student enrollment grew from 8,000 to 10,000 or a __25__ percent increase. (2,000 ÷ 8,000 =)
2. Sales decreased from \$670,000 to \$600,000 or a __*10*__ percent decrease. ***(70,000 ÷ 670,000 =)***
3. Employment increased from 93 to 106 employees or a __*14*__ percent increase. ***(13 ÷ 93 =)***
4. The state budget increased by \$16 million from \$633 million or a __*3*__ percent increase. ***(16 ÷ 633 =)***
5. The cost of a computer printer decreased by \$21 from \$189 or a __*11*__ percent decrease. ***(21 ÷ 189 =)***
6. Serious accidents increased by 10 percent from the 850 from last year or to __*935*__ accidents this year. ***(850 × .10 =)***
7. Jim sold 20 percent less this month than the \$500,000 that he sold last month or __*\$400,000*__ sales for this month. ***(500,000 × .20 =)***
8. Advertising increased by 30 percent this month to a level of \$150,000. The advertising was __*\$115,384.61*__ last month. ***(150,000 ÷ 1.30 =)***
9. The number of clients grew to sixty from fifty last month or a __*20*__ percent increase. ***(10 ÷ 50 =)***
10. The price of a product decreased from \$5.50 to \$5 or a __*9*__ percent decrease. ***(.50 ÷ 5.50 =)***

ASSIGNMENT Chapter 7

Name ______________________ **Date** ______________ **Score** ____________

Skill Problems

Select the letter for the definition.

A. Amount B. Base C. Rate

1. ___*B*___ The basis of comparison; the number of which so many hundredths is taken.

2. ___*C*___ A means of expressing some part of 100.

Fill in the blanks in the following problems.

3. $18\% = \frac{\mathbf{18}}{100}$

4. $47.5\% =$ ___*.475*___

5. $.41 = \frac{\mathbf{41}}{100}$

6. $2.5 =$ ___*250*___ %

7. $\frac{7}{10} =$ ___*.7*___

8. $\frac{1}{5}\% =$ ___*.002*___

9. $1.759 =$ ___*175.9*___ %

10. $\frac{4.6}{100} =$ ___*4.6*___ %

11. $458\% =$ ___*4.58*___

12. $\frac{1}{4}\% =$ ___*.0025*___

13. ___*30*___ $\% = \frac{30}{100} = .30$

14. $12\% = \frac{\mathbf{12}}{100} = .12$

15. $47\% = \frac{47}{100} =$ ___*.47*___

16. $.5\% = \frac{.5}{100} =$ ___*.005*___

17. $17\% = \frac{17}{100} =$ ___*.17*___

18. ___*134*___ $\% = \frac{134}{100} = 1.34$

19. ___*26*___ $\% = \frac{26}{100} = .26$

20. $81\% = \frac{\mathbf{81}}{100} = .81$

21. $10\ 1/5\% = \frac{10.2}{100} =$ ___*.102*___

22. ___*1 3/4*___ $\% = \frac{1.75}{100} = .0175$

23. $3.6\% = \frac{\mathbf{3.6}}{100} = .036$

24. $30\ 1/2\% = \frac{\mathbf{30.5}}{100} = .305$

Solve for the unknown factor. Round off your answer to the nearest unit, cent, or percent.

25. B = 62
R = 410%
A = ___*254*___ *(62 × 4.1 =)*

26. B = ___*13,108*___ *(852 ÷ .065 =)*
R = $6\frac{1}{2}\%$
A = 852

27. B = $17,500
R = ___*91*___ % *(16,000 ÷ 17,500 =)*
A = $16,000

28. B = $56
R = ___*1,429*___ % *(800 ÷ 56 =)*
A = $800

29. B = ___*$64.80*___ *(81 ÷ 1.25 =)*
R = 125%
A = $81.00

30. B = $420
R = 140%
A = ___*$588*___ *(420 × 1.4 =)*

31. B = $16
R = 30%
A = ___*4.80*___ *(16 × .30 =)*

32. B = ___*$120*___ *(48 ÷ .40 =)*
R = 40%
A = $48

33. B = $8,000
R = ___*150%*___ *(12,000 ÷ 8,000 =)*
A = $12,000

34. B = ___*$20.29*___ *(71 ÷ 3.5 =)*
R = 350%
A = $71

35. B = $70
R = 15 3/4%
A = ___*$11.03*___ *(70 × .1575 =)*

36. B = $82
R = ___*18%*___ *(15 ÷ 82 =)*
A = $15

37. B = 80
R = 105%
A = ___*84*___ *(80 × 1.05 =)*

38. B = 600
R = ___*17*___ *(100 ÷ 600 =)*
A = 100

39. B = 440
R = 2 1/2% *(440 × .025 =)*
A = ___*11*___

40. B = 40
R = 18%
A = *7* *(40 × .18 =)*

41. B = 87
R = *14%* *(12 ÷ 87 =)*
A = 12

42. B = *771* *(185 ÷ .24 =)*
R = 24%
A = 185

In the following expressions, solve for the variable.

43. A + 21 = 56
− 21 = −21
A = 35

44. B − 8 = 54
+8 = +8
B = 62

45. $\frac{C \times 6}{6} = \frac{36}{6}$
C = 6

46. $\frac{D}{5} = 60$
5 × D/5 = 60 × 5
D = 300

47. $\frac{9E}{9} = \frac{10 + 8}{9}$
E = 2

48. $\frac{3(F)}{3} = \frac{21}{3}$
F = 7

49. G*8 + 6 = 22
8G + 6 = 22
− 6 = −6
8G = 16
G = 2

50. H × 1/3 = 45
*3*H/3 = 45*3*
H = 135

In the following problems, find the unknown factor. Round off your answer to the nearest unit, cent or percent.

51. Sales increased from $125,000 to $260,000 or a *108* percent increase. *(135,000 ÷ 125,000 =)*

52. The price of a product decreased from $8.50 to $6.80 or a *20* percent decrease. *(1.70 ÷ 8.50 =)*

53. Complaints increased by 10 percent from 20 from last year or to *22* complaints this year. *(20 × .10 =)*

54. The budget must be cut by 9% from last year's $3,000,000 or a *$270,000* cut. *(3,000,000 × .09 =)*

55. Joe received a 25% commission of $2,046. How much did Joe sell? *$8,184.* *(2,046 ÷ .25 =)*

Business Application Problems

Round your answer to the nearest unit, cent, or percent.

1. A U.S. manufacturer has 23 percent of their parts produced outside of the country. If their product costs $18,470, what is the value of the foreign-made parts? *R × B = A .23 × $18,470 = $4,248.10 foreign made*

2. Martin and Associates purchased $562,960 worth of products from Allied Corp. last year. Total purchases from all sources for the year were $3,628,950. What percent of total purchases were made from Allied Corp.? *A/B = R $562,960./$3,628,950. = .155 = 16% from Martin and Associates*

3. The average price of used homes went from $108,900 to $127,200 in one year. What is the percent of increase? *R = A ÷ B $18,300/$108,900 = .168 = 17% increase*

4. The average employee of Maxon Incorporated earns $644 per week. The average deduction is $136. What is the percent of take-home pay for the average employee? *A/B = R $136/$644 = .21 79% take-home pay*

5. Of the 6,000 shareholders of Tarmix Corporation, 420 attended the annual meeting and voted in person. Another 200 voted through proxy ballots. What percent voted? *A/B = R*
420 + 200 = 620/6,000 = .10 10% voted

6. A tourist site analyzed their customers to determine how many were from out of state. In the past three months that number totaled 5,890. The out-of-state customers represented 68 percent of the total. How many customers did the firm have in total? *A/R = B*
5,890/.68 = 8,661.7 8,662 customers

7. The Book Nook sales for the month were $82,645. Book sales accounted for 62 percent of the total sales. What is the dollar value of the book sales for the month? *B × R = A $82,645 × .62 = $51,239.90 book sales*

8. The number of default loan customers for the Grand Tower Bank dropped to 84, or 1.5 percent, for the month of March. How many loan customers does the bank have? *A/R = B 84/.015 = 5,600 customers*

9. Tomison Technical College enrollment includes 480 students who are being retrained as a result of downsizing in the area last year. The balance of the student population of 916 has recently graduated from high school. What is the percent of recent high school graduates? *A/B = R 480 + 916 = 1,396*
916/1,396 = .656 66% recent grads

10. The Needam Nuggets minor league baseball team attendance was 4,756 last night. The stadium was 82 percent sold out. What is the capacity of the stadium?
A/R = B   4,756/.82 = 5,800 capacity

11. The Eye Care Center reviewed their customer list and found that 20 percent of their 340 customers used a credit card to make their purchases. How many used a credit card?
B = 340   A = B × R   340
R = 20%   ×.20
A =   68 card users

12. There were 14,600 paid visitors to the theme park by noon on Saturday. The manager estimated that those arriving before noon represented 40 percent of those that would visit the park that day. How many will visit the park?
B =   B = A ÷ R   .40)14,600
R = 40%   = 36,500 will visit
A = 14,600

13. An ad indicated that a new computer priced at $600 could be purchased by putting $102 down. What percent is the down payment required?
B = $600   R = A ÷ B $600)$102 = 17%
R =
A = $102

14. The price of a product increased 17 percent. The increase brought the price up to $1,000. How much did the product originally sell for?
B =   B = A ÷ R
R = 117%   B = $1,000 ÷ 1.17 = $854.70 originally
A = $1,000

15. Mr. Albert E. Hodge found that 70 percent of his assets were in land and buildings. A real estate broker indicated that he could sell the land and buildings for $140,000. How much are Mr. Hodge's assets worth?
B =   B = A ÷ R
R = 70%   B = $140,000 ÷ 7 = $200,000 assets
A = $140,000

16. Bill Otis noted that his total mileage for the month was 2,400. He realized that only 600 miles were for family reasons. What percent were for family reasons?
B = 2,400   R = A ÷ B
R =   R = 600 ÷ 2,400 = 25% for family reasons
A = 600

17. A company wished to buy a building priced at $30,000. They had a down payment of 20 percent. How large of a loan will they have to make?
B = $30,000   A = B × R
R = 80%   A = $30,000 × .80 = $24,000 loan
A =

18. The Ace Security Company has decided to give Fred Jantz a 5 percent raise over his current rate of $4.80 per hour. What will be his new rate?
B = $4.80   A = B × R
R = 105%   A = $4.80 × 1.05
A =   = $5.04 per hour

19. Nancy Pryor noticed that her paycheck stated that she had earned $247 for last week's work. She also took note of the fact that $32.50 was deducted from her check for taxes. What percent was deducted?
B = $247   R = A ÷ B
R =   R = $32.50 ÷ $247
A = $32.50   = 13% for taxes

20. Spot Deliveries purchased an item by putting $5,000 down and financing the balance. The down payment was 20 percent of the cost of the item. What was the cost of the item?
B =   B = A ÷ R
R = 20%   B = $5,000 ÷ .2 = $25,000 cost
A = $5,000

21. Sammy is a salesperson at a men's store. Yesterday he sold $865 of merchandise. Sammy stated that this was 110 percent of the amount he sold on Tuesday. How much did Sammy sell on Tuesday?
B =   B = A ÷ R
R = 110%   B = $865 ÷ 1.10
A = $865   = $786.36 on Tuesday

22. The Deli section of the grocery store enjoyed sales of $890 last Tuesday. The store had sales of $39,740 on the same day. What percent of the sales came from the Deli?
B = $39,740   R = A ÷ B
R =   R = $890 ÷ $39,740
A = $890   = 2% of sales

23. The sales staff of Reilly and Associates made up 80 percent of the total employees. The sales staff and the legal staff together made up 90 percent of the total employees. There are 12 members of the legal staff. How many employees are there in total?

B =   *B* = *A* ÷ *R*
R = 10%   *B* = 12 ÷ .1 = 120 employees
A = 12

24. A corporate budget was increased by 2.4 percent over the previous year. The previous budget called for expenditures of $45,970,000. What is the new budget amount?

B = $45,970,000   *A* = *B* × *R*
R = 102.4%   *A* = $45,970,000 × 1.024
A =   = $47,073,280 for new budget

25. The average expenditure for food in the Malory family is $80 per week. The family spends 16% of gross income on food. What is the family gross income?

B =   *B* = *A* ÷ *R*
R = 16%   *B* = $80 ÷ .16
A = $80   = $500 gross income

26. The Home Mortgage Company found that 164 of their clients were in default for the month of July. Those who defaulted represented 2 percent of the total mortgages held by the company. How many mortgages does the company hold?

B =   *B* = *A* ÷ *R*
R = 2%   *B* = 164 ÷ .02 = 8,200 mortgages
A = 164

27. Carrie spends $480 for rent each month. She has totaled her expenditures on food and has found that she spends $260, or 16 percent of her gross income, on food. What percent of her gross income does she spend on rent?

B =   *B* = *A* ÷ *R*
R = 16%   *B* = $260 ÷ .16 = $1,625
A = $260

B = $1,625   *R* = *A* ÷ *B*
R =   *R* = $480 ÷ $1,625 = .295
A = $480   = 30%

28. Write a complete sentence in answer to this question. If the unemployment rate changed from 7.3 percent to 7.5 percent, how many additional workers would be unemployed in a work force of 90,000,000 people?

B = 90,000,000   *A* = *B* × *R*
R = .2%   *A* = 90,000,000 × .002
A =   = 180,000
There would be 180,000 additional workers unemployed.

29. Write a complete sentence in answer to this question. All of the sales staff of the Donaldson Corp. made the quota for the month. The quota was a sales volume of $260,000. Seven of the ten sales representatives went over the quota by 15 percent. One additional sales representative exceeded the quota by 20 percent. What was the excess of quota achievement using the figures given?

$260,000 × .15 = $39,000 × 7 = $273,000
B × *R* = *A*
$260,000 × .20 = $52,000 × 1 =   + 52,000
$325,000
The sales in excess of quota were $325,000 for the month.

30. Write a complete sentence in answer to this question. Grain and Company will increase their sales this year to $3,500,000. The increase will come from $200,000 received from sales of their "Carl" product, plus a 10 percent increase over last year in sales of their other product lines. What were sales last year?

$3,500,000
−200,000
$3,300,000

B =
R = 110%
A = $3,300,000

B = *A* ÷ *R*
B = $3,300,000 ÷ 1.1
B = $3,000,000
Sales last year were $3,000,000.

★ Challenge Problem

The Groome Corp. has determined to give 5 percent of their employees a 5 percent raise. The total of the salaries paid to employees is $3,000,000. The average salary is $25,000. What will be the total cost of salaries after the raise?

$25,000)$3,000,000 = 120 employees

B = 120   *A* = *B* × *R*
R = 5%   *A* = 120 × .05
A =   *A* = 6 employees

$25,000
×.05
$1,250 Raise
×6
$7,500
+3,000,000
$3,007,500 Total cost

MASTERY TEST Chapter 7

Name ______________________ **Date** ______________ **Score** ____________

Fill in the blanks in the following problems.

1. 9 % = ***.09***

2. $\frac{7}{8}$ = ***87.5*** %

3. .21 = ***21*** %

4. $9\frac{1}{2}\% = \frac{9.5}{100}$

5. (T) F A = R × B

6. T (F) R = B ÷ A

7. T (F) 250 = 50% of 50

8. B = $600 ***A = B × R***
R = 25% ***A = $600 × .25 = $150***
A =

9. The Sales Department of Jacklen Marketing Co. reported that gross sales were $642,800 for the first quarter of the year. Sales returns were $12,856 for the same period. What percent were the returns to gross sales (nearest percent)?
B = $642,800 R = A ÷ B
R = R = $12,856 ÷ $642,800 = 2% returns
A = $12,856

10. The cost of materials for the boat manufacturing industry increased by $8\frac{1}{2}$ percent over last year. An item that sells for $24 wholesale this year would have been priced lower last year by what amount?
B = B = A ÷ R
R = $108\frac{1}{2}$% B = $24 ÷ 1.085 = $22.12
A = $24

$24.00
− 22.12
$1.88 lower

11. Food is about 16 percent of the average family budget. What will be the expenditures for food by a family that has an income of $1,500 a month?
B = $1,500 A = B × R
R = 16% A = $1,500 × .16 = $240 for food
A =

12. The work force in Vitar County is approximately 85,000 people. The unemployment rate recently decreased .7 percent. How many additional people were put to work?
B = 85,000 A = B × R
R = .7% A = 85,000 × .007 = 595 put to work
A =

13. An apartment building increased in appraised value by $7,000 over a period of one year to $86,000. What was the percent increase in appraised value?
B = $79,000 R = A ÷ B
R = R = $7,000 ÷ $79,000 = 8.9% increase
A = $7,000

14. A bank had 40 percent of its holdings in stocks and bonds. The amount invested totaled $68 million in all investments. How much was invested in stocks and bonds?
B = $68,000,000 A = B × R
R = 40% A = $68,000,000 × .4 =
A = $27,200,000 in stocks and bonds

15. The bank in problem 14 added $816,000 to total investments the following year. What is the percent increase in investments?
B = $68,000,000 R = A ÷ B
R = R = $816,000 ÷ $68,000,000 =
A = $816,000 1.2% increase

INSURANCE

8

Suggested Classroom Time for This Chapter

hours 1 2 3 4 5

Perceived Student Difficulty with Topical Material

easy moderate difficult

OBJECTIVES

After mastering the material in this chapter, you will be able to:

1. State the purpose of insurance. p. 144
2. Calculate premiums in fire, life, and workers compensation insurance problems. p. 144
3. Calculate the amount of payment on a fire loss covered by coinsurance. p. 144
4. Identify different types of life insurance coverage. p. 146
5. Understand and use the following terms and concepts:
   Premiums
   Policy
   Coinsurance
   Beneficiary
   Term Life Policy
   Straight Life Policy
   Cash Surrender Value
   Limited Pay Life Policy
   Endowment Life Policy
   Universal Life Policy
   Variable Life Policy
   Workers' Compensation

SPORT-A-MERICA

Hal met Mark Barink, a State Mutual Insurance agent, yesterday at two o'clock. "Mark," Hal said, "We have made some rather important changes at Sport-A-Merica over the past few months. You have seen us grow since we started out as a small three-person store to what we are today. We expect to grow even bigger over the next few years. We now have a larger store we'll soon be moving into, a new delivery vehicle on the road, and a greater dependence on management, in addition to other insurance considerations we'd like to discuss with you."

Mark replied, "As you continue to grow, your insurance needs will grow with you."

"That is why I thought we should meet now," said Hal. As the conversation continued, the more questions Hal asked, the more questions he developed. Hal recognizes the need for insurance, which spreads the risk and reduces potential losses for Sport-A-Merica, and he also realizes he needs to learn some basic insurance information by relying on the expertise of an insurance representative to cover the details for the business. He especially needs to know the cost of the fire insurance for the new store. And, what would it cost for a life insurance policy on him? After all, he is now 47 years old! What about Sport-A-Merica employees getting injured on the job? When Mark mentioned that a discount would apply if Sport-A-Merica had two kinds of insurance, Hal began thinking about the future needs of the firm even more.

Objective 1

The purpose of insurance is to minimize the cost of unforeseen losses. If a firm or person foresees a chance of high risk, some types of insurance coverage may be warranted. There are many types of insurance—health, fire, casualty, life, crop, automobile, marine, workers' compensation, and even rain insurance. This financial protection can be purchased for groups of people, for individuals, and even for another person. Some insurance may be purchased electronically using a computer at home. In some instances, it can be purchased from the federal government or from privately owned firms. Private firms are either stock companies (owned by stockholders of that firm) or mutual companies (owned by policyholders of that firm). An insurance policy is usually sold through an insurance agent who is a local representative of an insurance company. Due to the vastness of this topic, only a brief coverage of fire insurance, life insurance, and workers' compensation insurance will be included.*

premiums

Insurance protection is purchased by making fixed payment amounts at periodic intervals. The payment amounts are termed **premiums.** The premium amount is established by the insurance company, the insurer, by first determining the chance of loss through statistical analysis, and then adding on the cost of doing business. Premiums are usually paid quarterly, semiannually, or annually. There is an economic savings to the insured who pays annually because this decreases the handling costs of the insurance company. A **policy,** or a written contract, is then given to the insurance owner; this policy identifies the areas and amounts of coverage.

policy

Property Insurance

Property damage or loss is a common type of insurance. There are many types of property insurance, including that which covers damage due to wind, fire, flood, vandalism, and breakage. Fire is the greatest damage threat and is included in most property coverage. Fire insurance rates vary by the amount of fire protection available nearby, the age of the dwelling, its construction materials, and the presence of fire alarm system or extinguishers on the premises, among other factors. The higher the risk, the higher the premium. Premium rates are stated in terms of $100 of coverage. Rates for coverage over $100 are multiplied by the number of $100s of coverage desired.

Objective 2

Example 1

Determine the fire insurance premium on the new space for Sport-A-Merica for $110,000 of coverage. The rate per $100 of coverage is $.325 on this type of building.

Number of 100s × Premium per 100 = Annual premium

$110,000 = 1,100 hundreds
×$.325 per $100 of coverage
$357.50 annual premium

(1,100 × .325 =)

Property insurance, including fire protection, is often purchased for periods longer than one year. Policies for three to five years offer a reduction in the amount of the premium because of the reduced cost of doing business.

Objective 3 *coinsurance*

Fire Insurance

A common clause in most fire insurance policies is the **coinsurance** clause. This clause states that the policyholder must provide a minimum percent of coverage on the property. A common coinsurance percent is 80 percent. A building that is valued at $41,000 must be insured for $32,800 ($41,000 × .80; insurance required) in order to receive maximum payment if a loss occurs. If the property was insured for only $24,600 (60 percent of $41,000; insurance carried) and was damaged by $14,000, the insurance

*Further information is available from your local insurance agency or by writing the Consumer Information Center, Pueblo, Colorado 81009, for these free booklets: Insurance for Renters and Homeowners 018D, Insurance for Your Health, Car, and Life 019D, Questions and Answers about No-Fault Auto Insurance 256D.

company would not pay for the full amount of the loss. A partial payoff of $10,500 would be made. This amount is a prorated payment. It is found by

$$\frac{\text{insurance carried}}{\text{insurance required}} \times \text{amount of loss}$$

$$\frac{\$24{,}600}{\$32{,}800} \times \$14{,}000 = \$10{,}500 \text{ (amount of payoff)}$$

Example 2

James Garison's warehouse is covered by an insurance policy that has an 85 percent coinsurance clause. The building is valued at $120,000. It is insured for $78,000. Fire did $40,000 worth of damage to the building. What is the amount of the settlement from the insurance company?

building value × coinsurance clause = insurance required
$120,000 × .85 = $102,000

$$\frac{\text{insurance carried}}{\text{insurance required}} \times \text{amount of loss} = \text{payoff}$$

$$\frac{\$78{,}000}{\$102{,}000} \times \$40{,}000 = \$30{,}588.24$$

(78,000 ÷ 102,000 = × 40,000 =)

Though the property owner may carry more insurance than the coinsurance clause specifies, the payoff will never exceed the face value of the policy. Also, the payoff will never exceed the value of the loss. The maximum payoff on property insured to the coinsurance limit will always be the face value of the policy or the amount of the loss, whichever is lower. Keep this in mind when completing the assigned problems.

Exercise A

Calculate the amount of the annual premium in the problems below.

	Value of Property	Amount of Coverage	Premiums Per $100 of Coverage/Yr.	Total Premium Per Year
1.	$46,000	$36,000	$.310 (360 × .310 =)	$111.60
2.	31,900	24,000	.396 *(240 × .396 =)*	*95.04*
3.	54,000	42,000	.462 *(420 × .462 =)*	*194.04*
4.	67,000	51,000	.332 *(510 × .332 =)*	*169.32*
5.	21,300	16,500	.405 *(165 × .405 =)*	*66.83*
6.	87,000	70,000	.358 *(700 × .358 =)*	*250.60*
7.	36,000	27,000	.490 *(270 × .490 =)*	*132.30*
8.	61,900	50,000	.318 *(500 × .318 =)*	*159.00*
9.	54,800	42,500	.325 *(425 × .325 =)*	*138.13*
10.	21,700	17,000	.431 *(170 × .431 =)*	*73.27*

Exercise B

Calculate the amount of payoff in these problems. Do not round off the insurance carried to the insurance required factor before multiplying the amount of loss. This will provide the most accurate answer.

	Value of Property	Amount of Coverage	Coinsurance Clause Specification	Amount of Loss	Amount of Payoff
1.	$46,000	$36,000	80% (36,000/36,800 ×	$12,000 =)	$11,739.13
2.	65,000	52,000	80 *(52,000/52,000 ×*	20,000 *=)*	*20,000.00*

(Continued on next page)

Exercise B Continued

Calculate the amount of payoff in these problems. Do not round off the insurance carried to the insurance required factor before multiplying the amount of loss. This will provide the most accurate answer.

	Value of Property	Amount of Coverage	Coinsurance Clause Specification	Amount of Loss	Amount of Payoff
3.	$41,500	$ 30,000	75 *(30,000/31,125 ×*	$ 6,000 *=)*	*$ 5,783.13*
4.	60,000	40,000	80 *(40,000/48,000 ×*	40,000 *=)*	*33,333.33*
5.	260,000	180,000	75 *(180,000/195,000 ×*	30,000 *=)*	*27,692.31*
6.	95,000	76,000	80 *(76,000/76,000 ×*	10,000 *=)*	*10,000.00*
7.	120,000	60,000	80 *(60,000/96,000 ×*	80,000 *=)*	*50,000.00*
8.	80,000	48,000	75 *(48,000/60,000 ×*	15,000 *=)*	*12,000.00*
9.	60,000	45,000	75 *(45,000/45,000 ×*	22,500 *=)*	*22,500.00*
10.	300,000	200,000	80 *(200,000/240,000 ×*	28,500 *=)*	*23,750.00*

Life Insurance

Life insurance has many purposes, the most common of which is to provide some financial security or protection for the survivors of the insured. The survivors that are to receive the financial benefits are named by the insured in the policy. These named persons or organizations are termed **beneficiaries.**

beneficiary

Life insurance can be purchased in many forms by making monthly, quarterly, semiannual, or annual premium payments. There is an economic savings earned by paying annually. Some of the many types of life insurance are discussed here.

Objective 4

Term Life

The **term life policy** is often offered by employers as part of a fringe benefit package. It provides protection only as long as the premiums are paid. So long as an employee remains with the employer, the term life insurance policy will remain in effect. Should the employee be terminated, the term life insurance is then terminated. It is important, therefore, to have separate insurance coverage not provided by the employer. This type of policy is the most inexpensive for the amount of protection.

Term life premium rates are determined by measuring life expectancy and then adding the cost of doing business and an amount of profit. The premium cost rises each year due to the reduced life expectancy. A review of the Annual Premium table on page 148 shows how rates change for each year of age.

Straight Life

The **straight life policy** also termed whole life, is similar to the term life policy. The insured is protected only as long as premiums are made. Straight life insurance premiums are fixed per year and are higher than term life insurance premiums because part of the premium is set aside by the company for the future. The surplus portion of the premium above the cost of a term life policy in the early years is retained to cover the cost of the insurance in later years. The straight life policy allows the insured to keep the same premium rate over the life of the policy.

Cash Surrender Value

The surplus premium over the cost of term insurance is termed **cash surrender value** or **loan value**. The surplus is paid to the insured by the insurer if the policy is cancelled. They can also be borrowed if the insured needs cash and is willing to pay an interest charge on the amount borrowed. The cash surrender value increases with the length of time that the insurance is in force. As you will note from the following table, the type of insurance purchased also determines the amount of cash surrender value that the insured's policy is worth. This table is predetermined in the policy.

Cash Surrender Value/Loan Value, Age of Insured 25, per $1,000 of Coverage

Years Policy in Force	Straight Life	Limited (20) Pay Life	Endowment Life (20)
3	$ 15	$ 45	$ 91
5	31	97	170
10	57	235	402
15	105	360	673
20	194	518	990

Example 3

If Guy, a 25-year-old, purchased a $50,000 Limited Pay Life policy 10 years ago, how much can he borrow on the policy?

50 = Number of $1,000 of coverage
×$235 = Loan value per $1,000 of coverage for 25-year-old
$11,750 = Amount Guy can borrow on policy

(50 × 235 =)

Limited Pay Life

The **limited pay life policy** is similar to the straight life policy. Part of the premium is invested for the insured, but the insured is required to make only a limited number of premium payments such as 20 years' worth. This reduces the financial pressure on the insured when earnings decline, such as after retirement. Because of the limited number of premium payments, the premium is higher than the premium of term or straight life insurance.

Endowment Life

The **endowment life policy** is similar to the limited pay life policy. It has a limited number of premium payments and has a savings factor. The savings factor is larger in this policy than in limited pay life. Endowment life is used to prepare financially for retirement, college, a trip, or a business venture. Because of the greater savings factor, the premium payments are the highest of all the types of life insurance discussed.

Premiums are stated per $1,000 of life insurance coverage. The premium amount varies from company to company due to the efficiency of the firm, the quality of investments made, or type of insurance company. A representative schedule of premiums follows.

Annual Premium per $1,000 of Insurance

Age of Insured*	Term Life	Straight Life	Limited Pay Life	Endowment Life
18	$ 4.42	$14.98	$21.35	$35.45
19	4.43	15.09	21.68	35.48
20	4.45	15.11	22.02	35.50
21	4.47	15.46	22.40	35.52
22	4.49	15.80	22.85	35.55
23	4.53	16.17	23.30	35.58
24	4.58	16.55	23.75	35.62
25	4.61	16.96	24.25	35.67
26	4.67	17.40	24.76	35.74
27	4.72	17.82	25.26	35.81
28	4.80	18.27	25.80	35.87
29	4.89	18.75	26.35	35.95
30	4.98	19.26	26.92	36.05
31	5.08	19.75	27.50	36.17
32	5.22	20.33	28.15	36.28
33	5.39	20.95	28.80	36.43
34	5.60	21.55	29.40	36.60
35	5.77	22.20	30.11	36.79
36	6.07	22.91	30.82	36.99
37	6.40	23.62	31.55	37.25
38	6.76	24.40	32.30	37.52
39	7.17	25.20	33.15	37.81
40	7.60	26.05	33.99	38.15
41	8.09	26.95	34.85	38.55
42	8.63	27.85	35.75	38.95
43	9.23	28.84	36.70	39.40
44	9.86	29.92	37.80	39.90
45	10.55	30.99	38.81	40.38
46	11.35	32.10	39.90	40.95
47	12.20	33.35	40.89	41.55
48	13.20	35.70	42.10	42.25
49	14.21	36.00	43.35	43.00
50	15.39	37.50	44.70	43.80

*Premiums for ages two weeks through eighty years old and beyond are also available.

The annual premium for a $15,000 limited pay life insurance policy would be $342.75 if the insured were twenty-two years old at the time the policy was purchased.

$15,000 = 15 = $1,000 of coverage
×$22.85 = Annual premium for a twenty-two-year-old's limited pay life insurance policy
$342.75 = Annual premium

Example 4

What will be the annual premium on Hal at Sport-A-Merica for a term life policy of $8,000 if he is forty-seven years old?

Number of 1,000s × Premium per 1,000 = Premium

8 = $1,000 of coverage
×$12.20 = Annual premium for forty-seven-year-old term life
$97.60 = Annual premium

(8 × 12.20 =)

Exercise C

Calculate the annual premium in the following problems.

	Amount of Insurance Coverage	Type of Insurance	Age of Insured	Annual Premium
1.	$ 20,000	Straight	18 (20 × 14.98 =)	$299.60
2.	55,000	Term	23 *(55 × 4.53 =)*	*249.15*
3.	100,000	Limited pay	20 *(100 × 22.02 =)*	*2,202.00*
4.	20,000	Term	28 *(20 × 4.80 =)*	*96.00*
5.	60,000	Straight	32 *(60 × 20.33 =)*	*1,219.80*
6.	5,000	Endowment	24 *(5 × 35.62 =)*	*178.10*
7.	20,000	Limited pay	30 *(20 × 26.92 =)*	*538.40*
8.	25,000	Straight	45 *(25 × 30.99 =)*	*774.75*
9.	20,000	Term	26 *(20 × 4.67 =)*	*93.40*
10.	120,000	Endowment	35 *(120 × 36.79 =)*	*4,414.80*

Exercise D

Calculate the cash surrender value for a 25-year-old in the following problems.

Amount of Coverage	Type of Insurance	Years in Force	Cash Surrender Value	
1. $64,000	Limited pay	5	$ 6,208	(64 × 97 =)
2. 125,000	Endowment	20	*123,750*	*(125 × 990 =)*
3. 300,000	Straight	3	*4,500*	*(300 × 15 =)*
4. 150,000	Term	10	*0*	*(150 × 0 =)*
5. 200,000	Endowment	15	*134,600*	*(200 × 673 =)*

The life insurance industry has developed additional insurance products that are very flexible. It is possible to design a life insurance coverage program that is tailored to one single person. Within a flexible insurance program would undoubtedly be found a universal life policy or a variable life policy, in addition to a personally funded term policy. While the term policy provides the best amount of protection in the form of the amount of the policy, universal or variable life will provide the best amount of flexibility.

Universal Life

Universal life policies allow for changes in premium amounts and, thereby, changes in the amount of insurance coverage, all under the same policy. As the income of the insured changes, the premium amount can be changed. This works especially well for those starting their career. As income and responsibility increases, the amount of insurance can be increased. The surplus portion of the premium for a universal life policy is usually invested in fixed-rate-return vehicles, such as bonds or U.S. Treasury Bills.

Variable Life

Variable life policies are very much like universal life policies in the sense that the premium and coverage amount can be changed over the life of the policy based upon affordability or need. The major difference between these two types of flexible policies is that variable life policy premium surpluses are invested in stocks and mutual funds. The return from these investments is not fixed. In periods of prosperity, the returns will be very high. When the economy slumps, the returns are apt to be poor.

In addition, it is possible to make only one large premium payment. The surplus is invested and then used for future premium payments. A policyholder can also indicate that no more premiums will be paid. A paid-up policy can then be issued. Or, the insured can simply withdraw the cash value of the policy. These possibilities—and many others—require that each person and each business stay in close contact with a qualified insurance agent.

Workers' Compensation Insurance

Employers in all fifty states are required to have some type of **workers' compensation** insurance. This type of insurance provides compensation to employees who have been injured on the job. It pays for medical costs for injuries and work-related disease. Workers are also compensated for lost time from work due to injuries or disease. This type of insurance is a major cost consideration for employers in the mechanized and technical part of our economy. The current total national premium for workers' compensation insurance is in excess of $40 billion per year. In some industries, the cost of workers' compensation insurance premiums may be as much as 60 percent of employee wages. In other industries, the cost could be less than 1 percent.

The amount of the premium is determined by the accident experience of the industry or the individual firm. A firm operating in an industry that is associated with a low level of employee injury or disease will pay a low premium. However, if the firm or industry has a record of high risk to employees, indicating that injury, disease, or death is likely to occur, the premium will be higher.

Example 5

What is the semiannual premium for workers' compensation insurance for a group of 15 employees at Gerrard Grate and Company? The insurance agent has quoted a premium of $400 per employee per year.

Premium per employee × Number of employees = Annual premium
$400 × 15 = $6,000
Annual premium ÷ 2 = Semiannual premium
$6,000 ÷ 2 = $3,000

(400 × 15 = 6,000 ÷ 2 =)

A business manager will need to consider the cost of workers' compensation insurance when determining the total cost of labor.

Example 6

LeFlure Corporation employs 73 workers at their Wisconsin installation. The insurance agent has quoted a premium of $7,300 per employee for workers' compensation insurance coverage. What is the amount of the insurance premium per hour for each employee, if each employee works 2,000 hours per year?

Annual premium ÷ Hours worked annually = Cost per hour
$7,300 ÷ 2,000 = $3.65 per hour

(7,300 ÷ 2,000 =)

Exercise E

Calculate the total annual premium in the following problems.

	Number of Employees	Annual Premium per Employee	Total Annual Premium for Firm
1.	43	$3,650 (43 × 3,650 =)	$ 156,950
2.	67	1,200 *(67 × 1,200 =)*	***$ 80,400***
3.	1,258	2,360 *(1,258 × 2,360 =)*	***$2,968,880***
4.	259	18,900 *(259 × 18,900 =)*	***$4,895,100***
5.	372	9,374 *(372 × 9,374 =)*	***$3,487,128***

Exercise F

Calculate the hourly cost of the premium for each employee in exercise E. Assume that each employee works 2,000 hours per year.

1. ___\$1.83___ (3,650 ÷ 2,000 =)
2. ___*\$.60*___ *(1,200 ÷ 2,000 =)*
3. ___*\$1.18*___ *(2,360 ÷ 2,000 =)*
4. ___*\$9.45*___ *(18,900 ÷ 2,000 =)*
5. ___*\$4.69*___ *(9,374 ÷ 2,000 =)*

ASSIGNMENT Chapter 8

Name ______________________ Date ______________ Score ______________

Skill Problems

Select the best answer for the statements below.

A. Premiums	B. Policy	C. Coinsurance
D. Beneficiary	E. Term Life Policy	F. Universal Life Policy
G. Limited Pay Life Policy	H. Endowment Life Policy	I. Workers' Compensation

1. ___*E*___ Provides protection only for as long as premiums are paid. No surplus or savings factor built into premium.
2. ___*I*___ Provides protection for injuries or illness because of occupational activities.
3. ___*C*___ The policy holder and the insurance company share in the financial loss of property damage.
4. ___*H*___ A large portion of this premium is directed toward a savings plan that may be used for college, retirement, etc.
5. ___*A*___ Payments for insurance protection.

Calculate the amount of the annual premium in the problems below. Round off to the nearest cent.

Fire Insurance

Value of Property	Amount of Coverage	Premium Per $100 of Coverage/Yr.		Total Premium Per Year	
$200,000	$170,000	$.412	**6.**	*$ 700.40*	*(1,700 × .412 =)*
130,000	100,000	.410	**7.**	*410*	*(1,000 × .410 =)*
80,000	60,000	.359	**8.**	*215.40*	*(600 × .359 =)*
350,000	280,000	.375	**9.**	*1,050*	*(2,800 × .375 =)*
62,500	53,000	.415	**10.**	*219.95*	*(530 × .415 =)*
65,000	54,000	.246	**11.**	*132.84*	*(540 × .246 =)*
45,000	40,000	.310	**12.**	*124.00*	*(400 × .31 =)*
52,500	60,000	.332	**13.**	*199.20*	*(600 × .332 =)*
7,650	5,000	.490	**14.**	*24.50*	*(50 × .49 =)*
101,650	100,000	.325	**15.**	*325.00*	*(1,000 × .325 =)*
34,050	30,000	.294	**16.**	*88.20*	*(300 × .294 =)*
25,470	24,000	.805	**17.**	*193.20*	*(240 × .805 =)*
90,650	70,000	.758	**18.**	*530.60*	*(700 × .758 =)*
250,860	200,000	.651	**19.**	*1,302.00*	*(2,000 × .651 =)*

Life Insurance

Amount of Coverage	Type of Insurance	Age of Insured	Annual Premium
20. $150,000	Straight	43	*$4,326* *(150 × 28.84 =)*
21. 50,000	Term	23	*226.50* *(50 × 4.53 =)*
22. 20,000	Endowment	22	*711.00* *(20 × 35.55 =)*
23. 250,000	Straight	50	*9,375.00* *(250 × 37.50 =)*

	Amount of Coverage	Type of Insurance	Age of Insured		Annual Premium
24.	$ 75,000	Term	23 *(75 × 4.53 =)*		*$339.75*
25.	40,000	Limited pay	27 *(40 × 25.26 =)*		*1,010.40*
				Years in Force	**Cash/Loan Value**
26.	150,000	Endowment	25 *(150 × 402 =)*	10	*60,300*
27.	20,000	Straight	25 *(20 × 15 =)*	3	*300*
28.	20,000	Term	25 *(20 × 0 =)*	20	*0*

Workers' Compensation Insurance

	Number of Employees	Annual Premium per Employee	Total Annual Premium for Firm
29.	357	$ 650 *(357 × 650 =)*	*$232,050*
30.	730	200 *(730 × 200 =)*	*146,000*
31.	258	5,600 *(258 × 5,600 =)*	*1,444,800*
32.	90	750 *(90 × 750 =)*	*67,500*
33.	25	980 *(25 × 980 =)*	*24,500*
34.	26	1,340 *(26 × 1,340 =)*	*$34,840*
35.	1,546	4,860 *(1,546 × 4,860 =)*	*7,513,560*
36.	544	16,470 *(544 × 16,470 =)*	*8,959,680*
37.	130	9,405 *(130 × 9,405 =)*	*1,222,650*

Calculate the amount of payoff in the problems below. For the most accurate answer, do not round off insurance carried to insurance required before multiplying.

Value of Property	Amount of Coverage	Coinsurance Clause Specification	Amount Required by Clause	Amount of Loss	Amount of Payoff
$ 65,000	$50,000	80%	**38.** $*52,000* *(65,000 × .8 =)*	$14,000	**39.** $*13,461.54* *(50,000/52,000 > 100%)*
45,000	30,000	80	**40.** *36,000* *(45,000 × .8 =)*	42,000	**41.** *35,000.00* *(30,000/36,000 × 42,000 =)*
52,500	30,000	70	**42.** *36,750* *(52,500 × .7 =)*	10,000	**43.** *8,163.27* *(30,000/36,750 > 100%)*
29,000	5,500	75	**44.** *21,750* *(29,000 × .75 =)*	4,000	**45.** *1,011.49* *(5,500/21,750 × 4,000 =)*
101,650	50,000	80	**46.** *81,320.00* *(101,650 × .8 =)*	80,000	**47.** *49,188.38* *(50,000/81,320 × 80,000 =)*
40,000	30,000	85	**48.** *34,000* *(40,000 × .85 =)*	23,000	**49.** *20,294.12* *(30,000/34,000 × 23,000 =)*
31,500	24,000	90	**50.** *28,350* *(31,500 × .9 =)*	12,000	**51.** *10,158.73* *(24,000/28,350 × 12,000 =)*
90,650	70,000	85	**52.** *77,052.50* *(90,650 × .85 =)*	4,500	**53.** *4,088.12* *(70,000/77,052.50 × 4,500 =)*
250,860	200,000	95	**54.** *238,317.00* *(250,860 × .95 =)*	16,000	**55.** *13,427.49* *(200,000/238,317 × 16,000 =)*

	Value of Property	Amount of Coverage	Coinsurance Clause	Amount of Loss	Amount of Payoff
56.	$24,000	$17,250	80% *(17,250/19,200 ×*	$10,000 *=)*	*$8,984.38*
57.	80,000	50,000	80 *(50,000/64,000 ×*	40,000 *=)*	*31,250*
58.	62,000	46,500	75 *(46,500/46,500 ×*	6,500 *=)*	*6,500*
59.	100,000	80,000	85 *(80,000/85,000 ×*	90,000 *=)*	*80,000*
60.	20,000	14,000	80 *(14,000/16,000 ×*	9,500 *=)*	*8,312.50*

Business Application Problems

1. Ken purchased a $300,000 straight life insurance policy at the age of 30. What is the amount of annual premium?
300 × $19.26 = $5,778 premium

2. What is the amount of premium for a 46-year-old male who will purchase a $350,000 endowment life policy?
350 × $40.95 = $14,332.50 premium

3. An office building was purchased for $260,000. The new owner will purchase $200,000 of fire insurance though the fire insurance company has an 80 percent coinsurance clause. The rate per $100 is $.395. What is the annual premium?
$200,000/$100 = 2,000 × $.395 = $790 premium

4. If a fire caused $148,500 of damage in problem 3, what is the amount of payoff? *$260,000 × .8 = $208,000*
$200,000/$208,000 × $148,500 = $142,788.45 payoff

5. If there had been $160,000 of fire insurance coverage in problem 3, what would have been the payoff if the building was a complete loss? *$160,000 policy amount*

6. Baxter Associates has a workers' compensation premium that is 73 percent of wages. Gene is paid $16 per hour as a skilled worker. What is the cost of the workers' compensation insurance per hour for Gene?
$16 × .73 = $11.68 cost

7. Adam paid the first premium on his $56,000 limited pay life insurance policy when he was 32 years old. What is the annual premium amount?
56 × $28.15 = $1,576.40 premium

8. Adam, in problem 7, is now 9 years older. How much has he paid in premiums to date?
56 × $28.15 × 9 = $14,187.60 in premiums

9. What would be the additional premium if Adam, in problem 7, were to purchase another $56,000 of limited pay life insurance at his current age (41)?
56 × $34.85 = $1,951.60 premium

10. A new firm in the warehouse business will be charged a rate of $6.10 per hour worked for each employee for workers' compensation insurance. The firm will begin operations with 23 employees. How much should the accountant plan to pay for workers' compensation insurance in the first year if employees average 42 hours a week?
42 × 52 × 23 × $6.10 = $306,415.20 premium

11. The Burton Corp. wishes to purchase a term life insurance policy on their president, Ann Pierson. The company officials feel that if she died, there would be a decline in earnings because of the absence of her leadership. The company will be named the beneficiary. What will be the cost of a $125,000 term life policy on Ann Pierson if her age is fifty?
125 × $15.39 = $1,923.75 premium

12. Phil Buckner has decided that he can afford $12 now each month for life insurance protection. He is 28 years old, married, and has two children. He is not sure which type of policy to purchase, but has narrowed the decision down to term life or straight life. Once selected, he will maintain that type of policy regardless of premium changes in the future. If he dies at age 35, how much more will his family receive if he decides to purchase term life?

$12
× 12 months
$144 annual premium

Term:
$4.80)$144 = $30,000

Straight Life:
$18.27)$144 = $7,881.77

$30,000.00
−7,881.77
$22,118.23 more benefits

13. The Rectan Corp. allows their employees to select from a group of insurance options (dental, optical, life, medical, etc.). The firm will pay up to $120 per month per employee for insurance. Eddie, age 47, has decided to put $80 of his allowance per month into term life insurance. How much insurance will the firm be able to purchase for Eddie?

$80
× 12 months
$960 per year

$960 ÷ $12.20 = $78,688.52 of term insurance

14. Personnel Plus has purchased an office building for $160,000. Their insurance agent indicated a rate of $.405 per $100 of fire insurance coverage. The insurer has a 75 percent coinsurance clause on fire protection. What will be the annual premium if the firm follows the 75 percent clause stipulation?

$160,000
× .75
$120,000

1,200 × $.405 = $486 premium

15. Polisar Corp. incurred a $35,000 fire loss on an empty warehouse. The building is valued at $100,000. The insurance company specified a 75 percent coinsurance clause in the fire protection policy. What amount of insurance coverage was necessary to receive the maximum payment on the loss?

$100,000 × .75 = $75,000 policy

16. Lambtree Corporation will hire 42 additional employees for summer work. The monthly premium for workers' compensation is $80 per employee. What is the amount of additional premium per month for this workers' compensation insurance for the new employees?

42 × $80 = $3,360 additional premium

17. The Lawson home was totally destroyed by fire at an estimated loss of $74,800. The insurance policy contained an 80 percent coinsurance clause. The face value of the policy was $38,000. What was the amount of the payoff to the Lawson family?

$$\frac{\$38{,}000}{\$59{,}840} \times \$74{,}800 = \$47{,}500$$

Payoff = $38,000 (value of policy)

18. Ben Wolfson will hire two employees for the men's sportswear department. The cost of workers' compensation insurance will be $2.70 per hour for each employee. What is the annual premium for both employees if each employee works 2,000 hours per year?

2,000 × $2.70 × 2 = $10,800 annual premium

19. A total loss due to fire was experienced by the Burkholdt Corp. The building had a market value of $450,000 and was insured for $360,000. The insurance policy stipulated a coinsurance clause of 80 percent. What is the payoff?

$450,000
× .8
$360,000

$$\frac{\$360{,}000}{\$360{,}000} \times \$450{,}000 = \$360{,}000 \text{ amount of coverage}$$

20. A company has added to the staff over the last month. Each new employee will receive a term life insurance policy in the amount of $10,000. The premiums are paid by the company. The employees are listed with their respective ages. What is the total cost of the insurance coverage to the company?

Bob Akers, 24 years old
Andy Watts, 37 years old
Carla White, 45 years old

10 × $ 4.58 = $45.80
10 × $ 6.40 = 64.00
10 × $10.55 = +105.50
$215.30 total premium

21. A new building was insured for $120,000 for fire protection. The structure was appraised at $160,000. A fire in the second story caused $42,000 in damage. What was the payoff to the owner if the insurance policy specified an 80 percent coinsurance clause?

$160,000 × .80 = $128,000 required coverage

$\frac{\$120,000}{\$128,000} \times \$42,000 = \underline{\$39,375 \text{ payoff}}$

22. An employee can be hired at $8.00 per hour plus fringe benefits that cost $3.00 per hour. An additional cost of an annual (2,000 work hours) workers' compensation insurance premium of $3,500 must be considered. A labor service offers to provide a worker for $12 per hour. Which is less expensive?

$3,500 ÷ 2,000 = $1.75 insurance
+3.00 benefits
+8.00 wage rate
$12.75 total

Labor service

23. A new commercial building for Beck and Company will be completed on the first of the month. The insurance agent has indicated that the property insurance policy will specify a 75 percent coinsurance clause. The building will have a market value of $580,000. What is the amount of insurance coverage that Beck and Company will need to carry? Round off your answer to the nearest ten thousand dollars.

$580,000 × .75 = $435,000 = $440,000 coverage

24. Write a complete sentence in answer to this question. Carol Reddy, a 43-year-old, has $200 per month to use for additional insurance coverage. She is considering term and straight life policies. She is interested in the difference in the amount of coverage that she can obtain from the same investment. What is the difference in coverage that she can purchase? Round off your answer to the nearest $1,000 of coverage.

$200/mo.
×12
$2,400/yr.

Term:	*$2,400 ÷ $ 9.23 =*	*$260,000*
Straight:	*$2,400 ÷ $28.84 =*	*−83,000*
		$177,000

The difference in coverage is $177,000.

25. Write a complete sentence in answer to this question. What is the hourly cost for workers' compensation insurance for a group of 15 employees? The annual premium per person is $800. All employees work 2,000 hours per year.

$800 × 15 = $12,000 total premium
$12,000 ÷ 2,000 = $6 per hour premium
The hourly cost is $6 per hour.

★ Challenge Problem

The Denberton Company pays an annual fire insurance premium of $185.50, at $.35 per $100 of coverage of fire insurance. The building is valued at $85,000. Last week they had a $50,000 fire. There is a 75 percent coinsurance clause. What is the amount of the loss to the Denberton Company not covered by insurance?

$185.50 ÷ $.35 = $53,000 of insurance coverage

$\frac{\$53,000}{\$63,750} \times \$50,000 = \$41,568.63 \text{ payoff}$

$50,000.00
−41,568.63
$8,431.37 loss

MASTERY TEST Chapter 8

Name ______________________ **Date** ______________ **Score** __________

True or False

(T)	F	**1.**	An insurance policy is a contract that sets forth the rights of the insured and the amount of coverage.
(T)	F	**2.**	Endowment life insurance has the highest cost per thousand.
T	(F)	**3.**	The money paid at regular intervals to an insurance company is termed coverage.
(T)	F	**4.**	The beneficiary will receive the face value of the life insurance policy.
(T)	F	**5.**	When an insurance company states that the insured must share in the risk of insurance by absorbing part of the losses, the firm is specifying a coinsurance clause.
(T)	F	**6.**	A term life insurance policy will buy more coverage for a stated amount of dollars than any other type of policy.
(T)	F	**7.**	The purpose of insurance is to reduce the risk of loss.
T	(F)	**8.**	If the homeowner has a total loss and the insurance coverage exceeds the value of the property, the insurance company will pay out the value of the policy to the homeowner.

9. The Wilson Manufacturing Company will purchase term life insurance for its employees up to the amount of their individual annual income. The insurance is a fringe benefit for employees after one year of employment. Beverly Seldon, a 40-year-old receptionist, earns $18,000 a year. What is the annual cost of this fringe benefit to the Wilson Manufacturing Company?

18 × $7.60 = <u>$136.80 premium</u>

10. How much term life insurance can be purchased by a 30-year-old male who has an extra $70 per month? Round off your answer to the nearest $1,000.

$ 70/mo.
× 12
$840/yr

$840 ÷ $4.98 = <u>$169,000 coverage</u>

11. The Shaw family home burned down at a total loss of $34,000. They had insurance on the property. The insurance company had a 75 percent coinsurance clause in the contract. What will the payout be if the family had only 60 percent coverage?

$34,000 × .75 = $25,500   *$34,000 × .60 = $20,400*

$$\frac{\$20,400}{\$25,500} \times \$34,000 = \underline{\$27,200 \text{ payoff}}$$

Simple Interest 9

Suggested Classroom Time for This Chapter

hours 1 2 3 4 5

Perceived Student Difficulty with Topical Material

easy moderate difficult

Objectives

After mastering the material in this chapter, you will be able to:

1. Explain what *interest* means. p. 162
2. Compute the time factor. p. 162
3. Compute both ordinary and exact interest, using the principal, rate, and time factors. p. 166
4. Compute any one unknown factor in the interest formula (I = P × R × T) when the other three factors are known. p. 168
5. Understand and use the following terms:
   Interest
   Principal
   Rate
   Time
   The Calendar
   Leap Year
   The Interest Formula (I = P × R × T)
   Ordinary Interest
   Exact Interest

Sport-A-Merica

As a result of the conversation with Mark Barink about the discount on insurance, Hal has decided to purchase an additional delivery vehicle for Sport-A-Merica. The cost of the vehicle is $26,000. Hal feels the money can be borrowed from a bank for a short period of time, and he calls Community National Bank for loan information after reading their ad in the business section of the newspaper. Marion McGregor, a loan officer with the bank, responds, "A ninety-day loan, using the vehicle as collateral, could be negotiated with a favorable rate." Hal feels that Sport-A-Merica will have the cash to pay off the loan prior to the end of the ninety-day period. He will need to determine the day the loan must be repaid and the cost of borrowing the money. After checking the rates at First National Bank, where Sport-A-Merica has its checking account, Hal makes an appointment to arrange the loan with Marion McGregor at Community National Bank.

What Interest Means
Objective 1

There are undoubtedly many answers to the question "What is interest?" The lender may say that it is a return to cover the risk on his investment, a borrower may state that it is a penalty for having to borrow, and the economist would probably state that it is related to the supply and demand of risk capital. **Interest** is a rental amount or a charge for the use of money.

interest

The total amount of interest is determined by the amount of money borrowed, the rate of interest or charge rate, and the length of time the money is borrowed.

When borrowing, or renting, money, the borrower is obligated to pay back the amount borrowed at the end of the period, as well as the interest or rent on the amount borrowed. It should be noted that both the amount borrowed and the interest are owed by the borrower.

In order to be able to compute interest, you must know three factors. The amount of money to be borrowed is termed the **principal.** The charge rate is termed the **rate,** and the length of time the money is to be borrowed is termed the **time.** To solve an interest problem, you must be able to determine the *number of days* that the money has been borrowed. You may in fact be given the date that the money was borrowed, termed the date of loan, and the date the money was paid back, termed the repayment date, or *maturity date*. You will need to compute the number of days that have elapsed.

principal, rate
time

The Calendar
Objective 2

In order to accomplish this task you must be familiar with the **calendar.** This can be accomplished in at least three ways. You can memorize the number of days per month from the table below.

Days per Month

Month	Days	Month	Days
January	31	July	31
February	28 or 29	August	31
March	31	September	30
April	30	October	31
May	31	November	30
June	30	December	31

● ***At this point in the chapter, students tend to discount the importance of accurately determining the correct time of a loan. You may want to introduce the interest formula briefly and then expand upon the importance of the time factor.***

A second way to become familiar with the calendar is to memorize the poem: "Thirty days has September, April, June, and November. All the rest have thirty-one, except for February which has twenty-eight and in leap year has twenty-nine." You probably remember the poem from your childhood. It is a good tool to recall for your work in this chapter.

A third way to become familiar with the calendar is to use the knuckles of your hands. By making a fist with both hands and starting the months on the first knuckle as shown below, you can determine the months with thirty-one days and those which have less. The months on top of a knuckle have 31 days. Those months (other than February) that fall

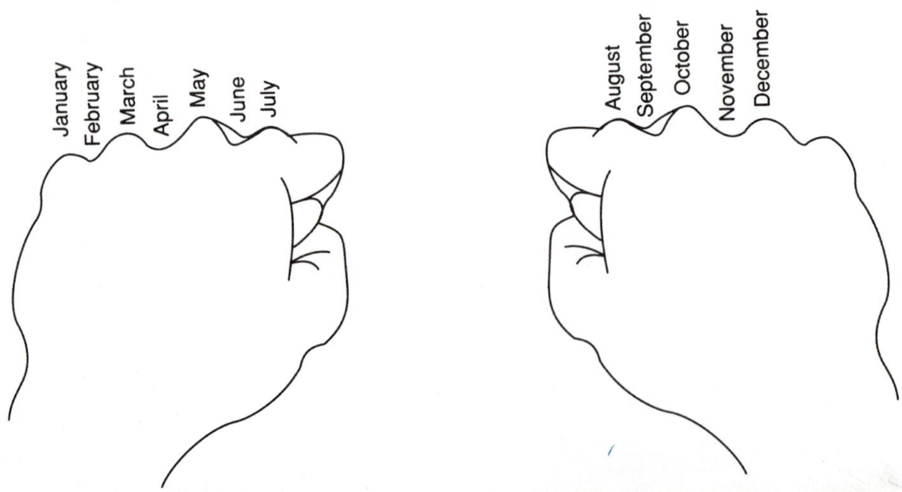

between a knuckle have 30 days. By remembering that February is a special exception depending on leap year, you should be able to determine the correct number of days in a month.

leap year

February has twenty-eight days, with the exception of **leap year,** when it has twenty-nine days. Leap year is any year that is evenly divisible by four. When computing the time factor, you may find that the dates given for the length of time the money was borrowed may not be at the beginning or the end of a month. You must then compute the elapsed time period in order to find the time factor. Below is a model to be used to compute the time.

Time Calculation Model

days in the month that borrowing took place
−date money is borrowed
days that money was borrowed in this month
+days in each succeeding month
+date of repayment
total time factor

Notice that you are actually not counting the day that the money was borrowed, but you are counting, as a full day, the day the money was returned.

Example 1

A loan made on July 16 was repaid on October 22. What is the time of the loan?

July has	31 days
Loan was made on July	−16
Borrowed money for	15 days in July
Borrowed money in August for	31 days
Borrowed money in September for	30 days
Borrowed money in October for	22 days (date of repayment)
Total time of loan =	98 days

(31 − 16 = +31 + 30 + 22 =)

Example 2

Julie borrows $7,000 at 8 percent on December 4 to be repaid on February 16 of the following year.

December has	31 days
Loan was made on December	−4
Borrowed money for	27 days in December
Borrowed money in January for	31 days
Borrowed money in February for	16 days (date of repayment)
Total time of loan =	74 days

(31 − 4 = +31 + 16 =)

The time factor can also be determined by using the time table shown on page 164. Each day of the year is assigned a number that begins with the first day of the year. By subtracting the number of the date of the loan from the date to repayment, the number of days in the loan can be determined.

Time Table

Day of Month	Jan.	Feb.	March	April	May	June	July	Aug.	Sept.	Oct.	Nov.	Dec.	Day of Month
1	1	32	60	91	121	152	182	213	244	274	305	335	1
2	2	33	61	92	122	153	183	214	245	275	306	336	2
3	3	34	62	93	123	154	184	215	246	276	307	337	3
4	4	35	63	94	124	155	185	216	247	277	308	338	4
5	5	36	64	95	125	156	186	217	248	278	309	339	5
6	6	37	65	96	126	157	187	218	249	279	310	340	6
7	7	38	66	97	127	158	188	219	250	280	311	341	7
8	8	39	67	98	128	159	189	220	251	281	312	342	8
9	9	40	68	99	129	160	190	221	252	282	313	343	9
10	10	41	69	100	130	161	191	222	253	283	314	344	10
11	11	42	70	101	131	162	192	223	254	284	315	345	11
12	12	43	71	102	132	163	193	224	255	285	316	346	12
13	13	44	72	103	133	164	194	225	256	286	317	347	13
14	14	45	73	104	134	165	195	226	257	287	318	348	14
15	15	46	74	105	135	166	196	227	258	288	319	349	15
16	16	47	75	106	136	167	197	228	259	289	320	350	16
17	17	48	76	107	137	168	198	229	260	290	321	351	17
18	18	49	77	108	138	169	199	230	261	291	322	352	18
19	19	50	78	109	139	170	200	231	262	292	323	353	19
20	20	51	79	110	140	171	201	232	263	293	324	354	20
21	21	52	80	111	141	172	202	233	264	294	325	355	21
22	22	53	81	112	142	173	203	234	265	295	326	356	22
23	23	54	82	113	143	174	204	235	266	296	327	357	23
24	24	55	83	114	144	175	205	236	267	297	328	358	24
25	25	56	84	115	145	176	206	237	268	298	329	359	25
26	26	57	85	116	146	177	207	238	269	299	330	360	26
27	27	58	86	117	147	178	208	239	270	300	331	361	27
28	28	59	87	118	148	179	209	240	271	301	332	362	28
29	29	. . .	88	119	149	180	210	241	272	302	333	363	29
30	30	. . .	89	120	150	181	211	242	273	303	334	364	30
31	31	. . .	90	. . .	151	. . .	212	243	. . .	304	. . .	365	31

Note: One day must be added to dates after February 28 on leap year problems.

Example 3

Using the information in example 1 and the time table, a loan is made on July 16 and repaid on October 22. What is the time of the loan?

October 22 is day	295
July 16 is day	−197
	98 days

(295 − 197 =)

Example 4

Use the time table to determine the time of a loan that was made on January 10, 2004 and repaid on May 23, 2004.

May 23 is day	144 (143 + 1 for leap year)
January 10 is day	−10
	134 days

(144 − 10 =)

Exercise A

In the following problems, find the number of days or the term of the loan.

Date of Loan	Date of Repayment	Days
1. July 17	August 21 (31 − 17 = 14 14 + 21 =)	35
2. May 17	June 30 *(31 − 17 = 14 14 + 30 =)*	*44*
3. March 10	June 1 *(31 − 10 = 21 21 + 30 + 31 + 1 =)*	*83*
4. September 16	December 3 *(30 − 16 = 14 14 + 31 + 30 + 3 =)*	*78*
5. November 22	January 3 *(30 − 22 = 8 8 + 31 + 3 =)*	*42*

When the date of the loan is known and the number of days, or the term, of the loan is known, you have the necessary information to compute the date of repayment. The determination of the repayment date is very similar to the approach used in determining the term of the loan.

Example 5

The $26,000 loan made to Sport-A-Merica for the new vehicle was made on January 16, 2001 for ninety days. Determine the date of repayment or the maturity date.

January has	31 days
Loan was made on January −	16
Borrowed money for	15 days in January
There are	28 days in February
There are	31 days in March
	74 days in total to date
→	16 more days required to make a total of
	90 days

Loan will come due on April 16.

(31 − 16 =   +28 + 31 + 16 =)

Example 6

Mark Watson borrowed $562 on June 25 for sixty days at 9 percent interest. What is the date of repayment, or the maturity date?

June has	30 days
Loan was made on June	−25
Borrowed money for	5 days in June
There are	31 days in July
	36 days in total to date
→	24 more days required to make total
	60 days

Loan will mature on August 24.

(30 − 25 =   +31 + 24 =)

The time table can be used to determine the due date on a loan. Using the date of the loan number from the table, add the term of the loan to the number, and find the new number (sum) in the table. The new number will be the maturity date.

Example 7

Using the information from example 6, find the maturity date of a loan made on June 25 for sixty days.

June 25 is day 176
+60 term of note
236 number to be found on the time table

Day 236 is August 24.

(176 + 60 =)

Exercise B

In the following problems, find the date of repayment or the maturity date. Watch for leap year!

Date of Loan	Term of Loan	Date of Repayment
1. December 8, 2001	90 days (31 − 8 = 23 23 + 31 + 28 + 8 =)	March 8, 2002
2. October 12, 2001	45 *(31 − 12 = 19 19 + 26 =)*	*November 26*
3. June 3, 2004	60 *(30 − 3 = 27 27 + 31 + 2 =)*	*August 2*
4. August 30, 2002	120 *(31 − 30 = 1 1 + 30 + 31 + 30 + 28 =)*	*December 28*
5. January 23, 2004	60 *(31 − 23 = 8 8 + 29 + 23 =)*	*March 23*

The Formula: I = P × R × T

Objective 3

time

Time is one of the three factors needed in the formula to compute interest (I = P × R × T). Time is usually expressed as a fraction of a year. The numerator is the length of time that the money is to be borrowed (term of loan). The denominator is either 360 or 365. A denominator of 360 is used to compute what is termed **ordinary interest.** This approach earns more interest for the lender and is used for commercial loans to businesses. When a 365-day denominator is used to compute interest, the answer is termed **exact interest.** Exact interest earns the lender less interest because the denominator for time is larger. By law, exact interest is used for loans to consumers. Time may also be described as the term or term of loan.

ordinary interest

exact interest

rate

❍ *The old adage of "time is money" is especially true in interest calculations. The more the time, the more the money in interest!*

The **rate** of interest is always expressed as a percent. The percent or interest rate is an annual rate, that is, for one year. When the loan is for less than one year, the time factor must be used to adjust the annual rate for the shorter time period. The rate factor multiplied by the time factor will result in an adjusted rate factor in the interest formula. (*Note:* If the loan is for a year, the time factor will be 1, or $\frac{1}{1}$; for four years it will be 4 or $\frac{4}{1}$; for six months, or $\frac{1}{2}$ of a year, it will be $\frac{6}{12}$ or $\frac{1}{2}$.)

principal

● ***Until the passage of the Truth in Lending Act, consumers could legally be charged ordinary interest.***

The amount of money borrowed is termed the **principal**. This is the amount that the borrower is "renting," and therefore, the borrower must pay interest, or "rent," on this amount. The more that is borrowed at a given rate and time factor, the higher the interest.

You should now be able to understand all three elements of the interest formula: principal, rate, and time. To compute interest, you will be multiplying; dividing; using fractions, decimals, and percent; computing time; and lastly, rounding off.

Refer to appendix B for additional information on shortcuts in this area.

Example 8

Principal = $15,000
Rate = 6%
Time = 90 days
Interest = ______? (using ordinary interest)

$I = P \times R \times T$

$I = \$15{,}000 \times .06 \times \frac{90}{360}$

You can cancel in the above formula as shown.

$I = \$\overset{7,500}{\cancel{15{,}000}} \times \overset{.03}{\cancel{.06}} \times \frac{\overset{1}{\cancel{90}}}{\underset{\underset{1}{\cancel{2}}}{\cancel{360}\ \cancel{4}}}$

$I = \frac{\$7{,}500}{1} \times \frac{.03}{1} \times \frac{1}{1}$

$I = \$225$

(7,500 × .03 =)

> ● ***Show an example of how one day of time can cost a firm more interest.***
> ***$I = P \times R \times T$***
> ***$I = \$50{,}000 \times .09 \times \frac{1}{360}$***
> ***$I = \$12.50$***

Example 9

Principal = $1,600
Rate = 7%
Time = June 23 to August 30 (using exact interest)

$I = P \times R \times T$

$I = \$\overset{320}{\cancel{1{,}600}} \times .07 \times \frac{68}{\underset{73}{\cancel{365}}}$

$I = 320 \times .07 \times \frac{68}{73}$

$I = \$20.87$

Time Calculation
June 30
–June 23
7 days in June
+31 days in July
+30 days in August
68 days in total

(320 × .07 × 68 ÷ 73 =)

Exercise C

Compute interest in the following problems (use ordinary interest).

1. Principal = $600
Rate = 8%
Time = 42 days
Interest = $5.60 $\left(600 \times .08 \times \frac{42}{360} =\right)$

2. P = $1,000
R = 14%
T = 25 days
I = ***$9.72*** $\left(1{,}000 \times .14 \times \frac{25}{360} =\right)$

3. P = $6,000
$R = 12\frac{1}{4}\%$
T = 120 days
I = ***$245.00*** $\left(6{,}000 \times .1225 \times \frac{120}{360} =\right)$

4. P = $408
$R = 9\frac{1}{2}\%$
T = From June 16 to October 20
I = ***$13.57*** $\left(408 \times .095 \times \frac{126}{360} =\right)$

5. P = $260
R = 8%
T = January 17, 2004 to June 3, 2004
I = ***$9.71*** $\left(260 \times .08 \times \frac{168}{360} =\right)$

6. P = $3,500
R = 12%
T = February 29, 2004 to May 19, 2004
I = ***$93.33*** $\left(3{,}500 \times .12 \times \frac{80}{360} =\right)$

Exercise D

In the following problems, compute interest using the exact interest method (365-day year).

1. P = \$600
   R = 8%
   T = 45 days
   I = 5.92 $\left(600 \times .08 \times \frac{45}{365} =\right)$

2. P = \$3,000
   R = $12\frac{1}{4}$%
   T = 180 days
   I = ***\$181.23*** $\left(3{,}000 \times .1225 \times \frac{180}{365} =\right)$

3. P = \$365
   R = 15%
   T = 52 days
   I = ***\$7.80*** $\left(365 \times .15 \times \frac{52}{365} =\right)$

4. P = \$850
   R = $13\frac{1}{2}$%
   T = 36 days
   I = ***\$11.32*** $\left(850 \times .135 \times \frac{36}{365} =\right)$

5. P = \$1,000
   R = 10%
   T = 36 days
   I = ***\$9.86*** $\left(1{,}000 \times .10 \times \frac{36}{365} =\right)$

6. P = \$450
   R = 14%
   T = 75 days
   I = ***\$12.95*** $\left(450 \times .14 \times \frac{75}{365} =\right)$

Other Interest Problems

Objective 4

It may be necessary to determine the amount of principal, rate, or time on a loan when the interest and two other factors of the I = P × R × T formula are known. The formula I = P × R × T can be manipulated to develop the following formulas.

$$P = \frac{I}{R \times T} \qquad R = \frac{I}{P \times T} \qquad T = \frac{I}{P \times R}$$

○ Algebra can be used with the I = P × R × T formula to find any one of the four unknowns.

You will need to know these formulas; use the model shown or use your skill at finding the unknown to solve these problems.

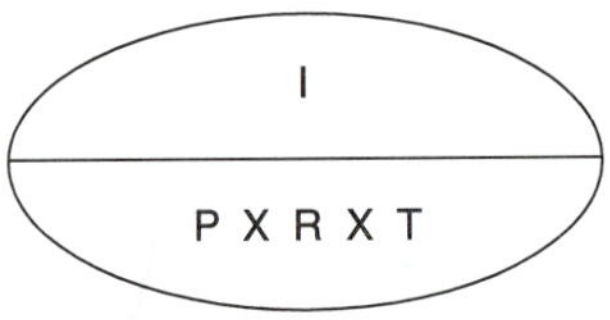

By putting your finger on the factor that you wish to find, the remaining portion of the model reveals the formula.

Example 10

If you know Interest, Principal, and Time, but need to find Rate, place your finger over R for Rate. Rate is found by dividing I by the product of P × T.

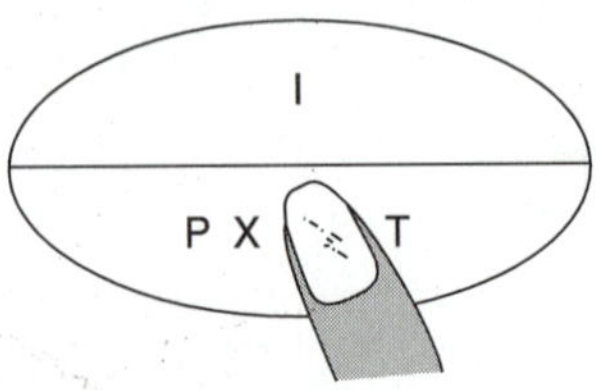

Computing Principal

The principal amount in a loan problem can be found when the other three factors are known.

Example 11

The interest on a loan to Sport-A-Merica was $520. The loan ran for ninety days at an 8 percent interest rate. What was the principal amount?

$$P = \frac{I}{R \times T} \qquad P = \frac{\$520}{.08 \times \frac{90}{360}} = \$26{,}000$$

(520 ÷ (.08 = × 90 = ÷ 360 =))

Computing Rate

When the rate on a loan is to be found, the formula $R = \frac{I}{P \times T}$ must be used.

Example 12

The same data from the earlier example will be used to illustrate this formula.

$$R = \frac{I}{P \times T} \qquad R = \frac{\$520}{\$26{,}000 \times \frac{90}{360}} = .08 = 8\%$$

(520 ÷ (26,000 = × 90 = ÷ 360 =))

Computing Time

The time factor can be determined by using the rate formula with one additional step. The decimal results equal a portion of a year. Multiply the decimal by 360 days to obtain the answer in days.

Example 13

$$T = \frac{I}{P \times R} \qquad T = \frac{\$520}{\$26{,}000 \times .08} = .25 \qquad .25 \times 360 \text{ days} = 90 \text{ days}$$

(520 ÷ 26,000 = ÷ .08 = × 360 =)

Exercise E

Fill in the missing information in these problems (use ordinary interest). Round off your answer to the nearest percent, day, or cent.

	Interest	Principal	Rate of Interest	Term of Loan
1.	$8.00	$720.00	8% (8 ÷ (.08 × 50 ÷ 360 =))	50 days
2.	*30.56*	5,500.00	10 *(5,500 × .10 × 20 ÷ 360 =)*	20
3.	4.50	400.00	*14* *(4.50 ÷ (400 × 30 ÷ 360 =))*	30
4.	14.63	1,300.00	15 *(14.63 ÷ 1,300 ÷ .15 × 360 =)*	*27*
5.	17.50	*750.00*	14 *(17.50 ÷ (.14 × 60 ÷ 360 =))*	60
6.	80.00	2,700.00	*12* *(80 ÷ (2,700 × 90 ÷ 360 =))*	90
7.	*212.50*	10,000.00	17 *(10,000 × .17 × 45 ÷ 360 =)*	45
8.	18.00	*1,080.00*	10 *18 ÷ (.10 × 60 ÷ 360 =))*	60
9.	8.98	4,200.00	11 *(8.98 ÷ 4,200 ÷ .11 × 360 =)*	*7*
10.	40.00	1,200.00	*13* *(40 ÷ (1,200 × 90 ÷ 360 =))*	90
11.	73.00	*4,132.09*	12 *(73 ÷ (.12 × 53 ÷ 360 =))*	53
12.	100.00	*6,000.02*	15 *(100 ÷ (.15 × 40 ÷ 360 =))*	40
13.	3.90	450.00	13 *(3.90 ÷ 450 ÷ .13 × 360 =)*	*24*
14.	70.00	750.00	*12* *(70 ÷ (750 × 270 ÷ 360 =))*	270
15.	83.67	1,000.00	12 *(83.67 ÷ 1,000 ÷ .12 × 360 =)*	*251*
16.	*1,155.00*	42,000.00	11 *(42,000 × .11 × 90 ÷ 360 =)*	90
17.	12.00	1,000.00	*10* *(12 ÷ (1,000 × 45 ÷ 360 =))*	45
18.	75.00	*4,615.38*	13 *(75 ÷ (.13 × 45 ÷ 360 =))*	45
19.	53.37	2,500.00	13 *(53.37 ÷ 2,500 ÷ .13 × 360 =)*	*59*
20.	41.00	*1,757.15*	14 *(41 ÷ (.14 × 60 ÷ 360 =))*	60

ASSIGNMENT Chapter 9

Name ______________________ **Date** ______________ **Score** ____________

Skill Problems

Select the best answer for the statements below.

A. Interest B. Principal C. Rate D. Time
E. Calendar F. The Interest Formula G. Ordinary Interest H. Exact Interest

1. ___*E*___ The number of days in each month of a year.
2. ___*A*___ A financial return to the lender for the risks undertaken during the term of a loan.
3. ___*C*___ The expression for finding the annual charge for the use of money.
4. ___*H*___ An interest rate that uses a 365-day year; used for personal loans only.
5. ___*G*___ An interest rate that uses a 360-day year; used for commercial loans only.
6. ___*B*___ The amount borrowed.

In the following problems, find the number of days or the term of the loan.

Date of Loan	Date of Repayment		Days
7. January 3, 2002	August 2, 2002	*(31 − 3 + 28 + 31 + 30 + 31 + 30 + 31 + 2 =)*	*211*
8. March 27, 2001	December 6, 2003	*(365 + 365 + 31 − 27 + 30 + 31 + 30 + 31 + 31 + 30 + 31 + 30 + 6 =)*	*984*
9. December 5, 1998	February 29, 2000	*(365 + 31 − 5 + 31 + 29 =)*	*451*
10. January 2, 2000	March 1, 2000	*(31 − 2 + 29 + 1 =)*	*59*
11. May 23, 2002	October 5, 2002	*(31 − 23 + 30 + 31 + 31 + 30 + 5 =)*	*135*

In the following problems, find the date of repayment or the maturity date.

Date of Loan	Term of Loan		Date of Repayment
12. February 3, 2001	120 days	*(28 − 3 + 31 + 30 + 31 + 3 =)*	*June 3, 2001*
13. November 24, 2002	60 days	*(30 − 24 + 31 + 23 =)*	*January 23, 2003*
14. April 14, 2003	90 days	*(30 − 14 + 31 + 30 + 13 =)*	*July 13, 2003*
15. June 22, 2004	45 days	*(30 − 22 + 31 + 6 =)*	*August 6, 2004*
16. December 18, 2003	120 days	*(31 − 18 + 31 + 29 + 31 + 16 =)*	*April 16, 2004*

Find the number of days or the term of the loan.

Date of Loan	Date of Repayment	Days
17. July 19 *(31 − 19 = 12 12 + 31 + 10 =)*	September 10	*53*
18. March 23 *(31 − 23 = 8 8 + 30 + 31 + 30 + 1 =)*	July 1	*100*
19. July 30 *(31 − 30 = 1 1 + 31 + 30 + 2 =)*	October 2	*64*
20. May 2 *(31 − 2 = 29 29 + 1 =)*	June 1	*30*
21. September 1 *(30 − 1 = 29 29 + 31 + 30 + 6 =)*	December 6	*96*

Find the date of repayment or maturity date.

Date of Loan	Term of Loan	Date of Repayment
22. January 10, 2000	65 days *(31 − 10 = 21 21 + 29 + 15 =)*	*March 15*
23. June 3, 2001	120 *(30 − 3 = 27 27 + 31 + 31 + 30 + 1 =)*	*October 1*
24. May 13, 2002	40 *(31 − 13 = 18 18 + 22 =)*	*June 22*
25. March 6, 2002	40 *(31 − 6 = 25 25 + 15 =)*	*April 15*
26. January 12, 2004	90 *(31 − 12 = 19 19 + 29 + 31 + 11 =)*	*April 11*

Compute interest in the following problems (use ordinary interest).

27. Principal = $4,000
Rate = 8%
Time = 45 days
Interest = *$40.00*
(4,000 × .08 × 45 ÷ 360 =)

28. P = $7,500
R = 10%
T = 90 days
I = *$187.50*
(7,500 × .10 × 90 ÷ 360 =)

29. P = $24,500
R = 7%
T = January 5, 2002 to May 23, 2003
I = *$2,396.24*
(365 + 31 − 5 + 28 + 31 + 30 + 23 =)
(24,500 × .07 × 503 ÷ 360 =)

30. P = $10,000
R = 15%
T = 25 days
I = *$104.17*
(10,000 × .15 × 25 ÷ 360 =)

31. P = $400
R = 7 1/4%
T = From July 1 to November 12
I = *$10.79*
(31 − 1 + 31 + 30 + 31 + 12 =)
(400 × .0725 × 134 ÷ 360 =)

32. P = $15,750
R = 12.5%
T = January 4, 2000 to March 29, 2000
I = *$464.84*
(31 − 4 + 29 + 29 =)
(15,750 × .125 × 85 ÷ 360 =)

Compute interest in the following problems (use exact interest).

33. Principal = $800
Rate = 8%
Time = 45 days
Interest = *$7.89*
(800 × .08 × 45 ÷ 365 =)

34. P = $10,740
R = 10%
T = 74 days
I = *$217.74*
(10,740 × .10 × 74 ÷ 365 =)

35. P = $1,000
R = 11%
T = 32 days
I = *$9.64*
(1,000 × .11 × 32 ÷ 365 =)

36. P = $8,760
R = 11 1/4%
T = From July 3 to August 29
I = *$153.90*
(31 − 3 + 29 =)
(8,670 × .1125 × 57 ÷ 365 =)

37. P = $806
R = 6%
T = 120 days
I = *$15.90*
(806 × .06 × 120 ÷ 365 =)

38. P = $105,500
R = 7 1/2%
T = 90 days
I = *$1,951.03*
(105,500 × .075 × 90 ÷ 365 =)

Compute interest in the following problems.

39. Principal = \$2,000

Rate = $12\frac{1}{2}\%$

Time = 30 days

Interest = \$ ***20.83*** (ordinary)

$\left(2{,}000 \times .125 \times \frac{30}{360} =\right)$

40. Principal = \$1,800

Rate = 9%

Time = 13 days

Interest = \$ ***5.85*** (ordinary)

$\left(1{,}800 \times .09 \times \frac{13}{360} =\right)$

41. Principal = \$30,000

Rate = 16%

Time = 62 days

Interest = \$ ***826.67*** (ordinary)

$\left(30{,}000 \times .16 \times \frac{62}{360} =\right)$

42. Principal = \$600

Rate = 9%

Time = 60 days

Interest = \$ ***8.88*** (exact)

$\left(600 \times .09 \times \frac{60}{365} =\right)$

43. Principal = \$9,500

Rate = $15\frac{3}{4}\%$

Time = 1 year

Interest = \$ ***1,496.25*** (exact) ***(9,500 × .1575 × 1 =)***

Fill in the missing information in these problems (use ordinary interest). Round off your answer to the nearest percent, day, or cent.

	Interest	Principal	Rate of Interest	Term of Loan	
44.	\$9.14	***\$940.12***	7%	50 days	***(9.14 ÷ (.07 × 50 ÷ 360 = .0097222))***
45.	***110.00***	4,800.00	11	75	***(4,800 × .11 × 75 ÷ 360 =)***
46.	143.00	1,300.00	***15***	264	***(143.00 ÷ (1,300 × 264 ÷ 360 = 953.33333))***
47.	323.00	48,450.00	12	***20***	***(323 ÷ 48,450 ÷ .12 × 360 =)***
48.	175.56	***7,900.21***	10	80	***(175.56 ÷ (.1 × 80 ÷ 360 = .0222222))***
49.	2.33	150.00	***8***	70	***(2.33 ÷ (150 × 70 ÷ 360 = 29.17))***
50.	***366.67***	55,000.00	6	40	***(55,000 × .06 × 40 ÷ 360 =)***
51.	27.30	***780.00***	14	90	***(27.30 ÷ (.14 × 90 ÷ 360 = .035))***
52.	54,078.75	940,500.00	11	***188***	***(54,078.75 ÷ 940,500 ÷ .11 × 360 =)***
53.	15.41	460.00	***7***	170	***(15.41 ÷ (460 × 170 ÷ 360 = 217.22222))***
54.	165.00	***11,000.00***	9	60	***(165.59 ÷ (.09 × 60 ÷ 360 = .015))***
55.	600.00	***60,000.00***	12	30	***(600 ÷ (.12 × 30 ÷ 360 = .01))***
56.	87.50	***7,000.00***	15	30	***(87.50 ÷ (.15 × 30 ÷ 360 =))***
57.	500.00	20,000.00	***10***	90	***(500 ÷ (20,000 × 90 ÷ 360 =))***
58.	36.00	900.00	16	***90***	***(36 ÷ (900 × .16) × 360 =)***
59.	14.70	760.00	***10***	68	***(14.70 ÷ (760 × 68 ÷ 360 =))***
60.	38.33	5,000.00	12	***23***	***(38.33 ÷ (5,000 × .12) × 360 =)***
61.	46.67	***3,500.26***	16	30	***(46.67 ÷ (.16 × 30 ÷ 360 =))***

Business Application Problems

Round off your answer to the nearest unit, percent, or cent.

1. Mari paid \$78.90 in interest to her credit union for a ninety-day, 8 percent loan that indicated exact interest. To the nearest dollar, what amount did she borrow?
\$78.90/(.08 × 90/365) = \$3,999.80 = \$4,000 borrowed

2. A loan for \$50,000 was repaid by Latire Company to their bank through the issuance of a check for \$50,489.58. The loan carried an ordinary interest rate of 11.75 percent. What was the term of the loan?
\$489.58/(\$50,000 × .1175) = .0833327; .0833327 × 360 = 30 days

3. A commercial loan in the amount of $14,500 was repaid, including $161.11 in interest at 10 percent, on June 20th. What was the date that the ordinary interest loan was made? ***$161.11/($14,500 × .1) = .1111103; .1111103 × 360 = 40*** ***May 11th***

4. Marti Cook lent a friend $500 for a period of 85 days. She charged 15 percent interest. What was the amount due at the end of the 85 days? (Use exact interest.)
$500 × .15 × 85/365 = $17.47
$500 + 17.47 = $517.47 due

5. The National Bank received $15,166.67 on a $15,000 note that was outstanding for 40 days. What was the rate of interest on the note? (Use ordinary interest.)
$166.67/($15,000 × 40/360) = 10% rate

6. What is the amount of a commercial loan that carried an interest rate of 15.5 percent for a term of sixty days if the total interest on the loan was $4,133.33? (Use ordinary interest.) Round off your answer to the nearest dollar.
$4,133.33/(.155 × 60/360) = $160,000.07 =
$160,000 loan amount

7. What is the payoff on the loan in problem 6?
$160,000 + $4,133.33 = $164,133.33 payoff

8. What would be the amount of payoff in problem 6 if the loan had been an exact interest loan?
$160,000 × .155 × 60/365 = $4,076.71 + $160,000 =
$164,076.71 payoff

9. Birkhut and Masters Company paid $1,250 on a $25,000, 120-day ordinary interest loan. What was the interest rate on the loan?
$1,250/($25,000 × 120/360) = .15 = 15 percent rate

10. Alisandra borrowed $400 for a period of six months. She paid off the exact interest loan, including $20 in interest, on time. What rate of interest did she pay?
$20/($400 × 6/12) = 10 percent rate

11. Sight and Sound Video borrowed $15,000 from a supplier for a period of sixty days at 14 percent ordinary interest. What is the amount of interest due at the end of the loan?

$I = P \times R \times T$

$I = \$15{,}000 \times .14 \times \frac{60}{360}$

$I = \$350$ ***interest***

12. Chad deposited his bonus check for $2,000 in his savings account on January 15, 2004. He hopes to have enough saved to take a vacation on July 5, 2004. The savings account pays $5\frac{1}{4}$ percent interest. What will his deposit be worth on July 5? (Use ordinary interest.)

$I = P \times R \times T$

$I = \$2{,}000 \times .0525 \times \frac{172}{360}$

$I = \$50.17$

$2,000.00 deposit
+50.17 interest
$2,050.17 balance

13. A ninety-day loan was secured from the First National Bank. The loan was made by Jackie Lee for $12,000. The bank is charging a 14 percent rate. What will be the amount of interest on the loan if it runs the full ninety-day period? (Use exact interest.)

$I = P \times R \times T$

$I = \$12{,}000 \times .14 \times \frac{90}{365}$

$I = \$414.25$

14. James B. Mackie borrowed $9,900 from a bank. The bank must charge 13 percent interest on loans to make a profit. Mr. Mackie's loan was for ninety-three days. How much interest did the bank earn? (Use ordinary interest.)

$I = P \times R \times T$

$I = \$9{,}900 \times .13 \times \frac{93}{360}$

$I = \$332.48$ ***interest***

15. The Severance National Bank of Adrain, Michigan, charged Harold Denlar $9 worth of interest on a ninety-day loan of $1,000 at $10\frac{1}{2}$ percent interest. Mr. Denlar went all the way to the bank president, Mr. Ronald Severance, to complain that he had been overcharged. After careful examination of the facts, Mr. Severance stated that Mr. Denlar was in fact *under*charged. Who was correct—Mr. Denlar or Mr. Severance? (Use exact interest.)

$I = P \times R \times T$

$I = \$1{,}000 \times .105 \times \frac{90}{365}$

$I = \$25.89$

Mr. Severance is correct.

16. A loan was repaid with a check in the amount of $5,758.81. The loan was made on January 31, 2002 and repaid on June 1, 2002. The loan was for $5,500. To the nearest percent, what was the rate of interest on the loan? (Use ordinary interest.)

$R = \frac{I}{P \times T}$

$R = \frac{\$258.81}{\$5{,}500 \times \frac{121}{360}}$

$R = .1400024 = \underline{14\%}$

17. Tony Cox borrowed $1,983 from the Pratt Union. The credit union charges a $12\frac{3}{4}$ percent interest rate on loans. What is the amount due on the loan if it runs for sixty days? (Use exact interest.)

$I = P \times R \times T$

$I = \$1{,}983 \times .1275 \times \frac{60}{365}$

$I = \$41.56$

$1,983.00
+41.56
$2,024.56 due

18. A loan for $850 dated July 27 was repaid on December 20. The rate of interest was 16 percent. What was the value of the check? (Use exact interest.)

$I = P \times R \times T$

$I = \$850 \times .16 \times \frac{146}{365}$

$I = \$54.40$

$ 850
+54.40
$904.40 value of check

19. Mike Bruen agreed to repay an education loan of $12,000 to his father plus 9 percent interest. The loan was from a period starting on August 31, 2002 to July 15, 2003. What was the amount due his father on July 15? (Use ordinary interest.)

$I = P \times R \times T$

$I = \$12{,}000 \times .09 \times \frac{318}{360}$

$I = \$954$

$12,000
+954
$12,954 due

20. A loan for $22,500 at 10 percent interest was paid back with a check in the amount of $22,950 on July 10, 2001. Using ordinary interest, determine the date the note was made.

$T = \frac{I}{P \times R}$

$T = \frac{\$450}{\$22{,}500 \times .10}$

$T = .2 \times 360 = 72 \text{ days} = \underline{\text{April 29, 2001}}$

21. Orson Ingrahm loaned a business associate $5,000 on December 20, 2000. The loan was for three months at $14\frac{3}{4}$ percent. What is the amount of the interest on the loan? (Use ordinary interest.)

$I = P \times R \times T$

$I = \$5{,}000 \times .1475 \times \frac{3}{12}$

$I = \underline{\$184.38}$

22. What was the principal of a commercial loan that carried a rate of 16 percent for a period of ninety days if the total interest paid was $14,000? (Use ordinary interest.)

$P = \frac{I}{R \times T}$

$P = \frac{\$14{,}000}{.16 \times \frac{90}{360}}$

$P = \underline{\$350{,}000 \text{ principal}}$

23. What is the difference in interest on a loan using ordinary versus exact interest if the loan value is $5,000 for a term of 120 days at $16\frac{1}{2}$ percent?

$I = P \times R \times T$

$I = \$5{,}000 \times .165 \times \frac{120}{360} \text{ or } \frac{120}{365}$

$I = \$275 \text{ or } \$271.23$

$275.00
−271.23
$ 3.77 difference

24. Write a complete sentence in answer to this question. A student loan for $6,000 must be refinanced. The lender had loaned the money to the student two years ago at 14 percent. The student has not made any payments on the principal or interest. How much will the student need to refinance?

$I = P \times R \times T$

$I = \$6{,}000 \times .14 \times \frac{2}{1}$

$I = \$1{,}680$

$$\begin{array}{r} \$6{,}000 \\ +1{,}680 \\ \hline \$7{,}680 \end{array}$$

The student will need to refinance $7,680.

25. Write a complete sentence in answer to this question. Sherry borrowed $1,500 from a loan company on June 23. She repaid the loan on August 30. The loan specified a 15 percent rate (ordinary interest). What is the amount of payoff on the loan?

$I = P \times R \times T$

$I = \$1{,}500 \times .15 \times \frac{68}{360}$

$I = \$42.50$

$$\begin{array}{r} \$1{,}500.00 \\ +42.50 \\ \hline \$1{,}542.50 \end{array}$$

The amount of payoff on the loan is $1,542.50.

★ Challenge Problem

The Malon Corporation borrowed $52,000 from the Manufacturers Commerce Bank. The bank charges $9\frac{3}{4}$ percent interest for loans, and pays $4\frac{1}{2}$ percent compounded semiannually on savings. The bank is forty-seven years old and will soon finance a new $4.3 million Science and Trade Center. How much interest will the Malon Corporation have to pay for their ninety-day loan? (Use ordinary interest.)

$I = P \times R \times T$

$I = \$52{,}000 \times .0975 \times \frac{90}{360}$

$I =$ *$1,267.50 interest*

MASTERY TEST Chapter 9

Name ______________________ **Date** ______________ **Score** ____________

Complete the following problems.

1. Interest can be defined as
   a. The amount of money borrowed.
   b. The amount of money that is to be returned to the lender.
   ⓒ The money that the borrower must pay the lender for the use of the principal.
   d. The results of time multiplied by rate.

2. Date of loan: March 6
Date of repayment: August 1
Days: *148*

3. Date of loan: February 20, 2002
Days of term of loan: Sixty days
Date of repayment: *April 21, 2002*

4. P = \$50,000
R = 16%
T = 30 days
I = \$ *666.67* (ordinary interest)

5. What is the answer to problem 4 using exact interest?

$I = P \times R \times T$

$I = \$50,000 \times .16 \times \frac{30}{365}$

I = *657.53*

6. Viola Austin borrowed \$1,000 from a friend at the rate of $10\frac{3}{4}$ percent interest. The loan ran for forty-five days. What is the repayment date for this loan made on May 12?

June 26

7. What is the amount of the interest due on the loan in problem 6? (Use ordinary interest.)

$I = P \times R \times T$

$I = \$1,000 \times .1075 \times \frac{45}{360}$

$I = \$13.44$

8. What is the total amount of money that Viola must repay when the loan comes due in problem 6?

\$1,000.00
+ 13.44
\$1,013.44 due

9. Bud borrowed \$1,800 from a friend for sixty days at $15\frac{1}{2}$ percent interest. What was the amount of interest on the loan at the repayment date? (Use ordinary interest.)

$I = P \times R \times T$

$I = \$1,800 \times .155 \times \frac{60}{360}$

$I = \$46.50$

10. Patty Emerson lent a business associate \$5,000 on May 2. The loan was to be repaid on October 12 along with 13 percent interest. What was the amount due on October 12? (Use exact interest.)

$I = P \times R \times T$

$I = \$5,000 \times .13 \times \frac{163}{365}$

$I = \$290.27$

\$5,000.00
+ 290.27
\$5,290.27 due

11. Julie Snider borrowed \$5,000 for sixteen days at 17 percent interest. How much must she pay back? (Use exact interest.)

$I = P \times R \times T$

$I = \$5,000 \times .17 \times \frac{16}{365}$

$I = \$37.26$

\$5,000.00
+ 37.26
\$5,037.26 due

12. A loan for \$900 at 12 percent interest runs for a period of seventy-one days. The borrower asks the bank to indicate the amount of interest owed when the loan comes due. How much interest will be due on the loan at the maturity date? (Use ordinary interest.)

$I = P \times R \times T$

$I = \$900 \times .12 \times \frac{71}{360}$

$I = \$21.30$ *interest due*

Promissory Notes 10

Suggested Classroom Time for This Chapter

hours 1 2 3 4 5

Perceived Student Difficulty with Topical Material

easy moderate difficult

Objectives

After mastering the material in this chapter, you will be able to:

1. Describe the legal implications of a promissory note. p. 180
2. Compute the maturity value of a promissory note using ordinary or exact interest. p. 180
3. Compute the proceeds of a discounted noninterest-bearing or discounted interest-bearing promissory note. p. 181
4. Understand and use the following terms:
   Promissory Note
   Maker
   Payee
   Face Value
   Term of Note
   Maturity Value
   Noninterest-bearing Note
   Interest-bearing Note
   Discount Rate
   Term of Discount
   Discount Amount
   Proceeds

Sport-A-Merica

Although the holiday season is over, the sales volume of Sport-A-Merica continues to increase. Hal has been working hard managing the business in the store, while Kerry has been concentrating her efforts on selling to commercial accounts. After several months of calling on Shape-Up, Inc., Kerry obtains a $7,000 wholesale order for sportswear and equipment. Kerry's persistence has paid off again! Shape-Up had been interested in Sport-A-Merica merchandise for some time, but was unable to purchase in a quantity large enough to qualify for a discount from Sport-A-Merica. This order will be profitable for both firms, but Sport-A-Merica will have to accept a promissory note from Shape-Up in order to make the deal. With the promissory note due soon, Kerry is considering the effect this will have on the cash situation for Sport-A-Merica.

Legal Environment

Objective 1

promissory note *maker* *payee*

A **promissory note** is used in business as an IOU. It is a legally enforceable written contract. The **maker** of the note is obligated to repay the amount of the note, along with interest, to the **payee.** The Uniform Commercial Code defines a promissory note as "an unconditional promise in writing made by one person to another, signed by the maker engaging to pay on demand or at a particular future time a certain sum of money to order or to the bearer."

face value *term of note*

The amount written on the note is known as the **face value** and is equal to the amount borrowed. The note is a negotiable instrument (i.e., it is transferable from one party to another) for cash or merchandise. The time period that the note is to run is known as the **term of note.** This will be the actual number of days from the time the note is made until it is repaid. Some notes run for years and are usually for larger sums of money.

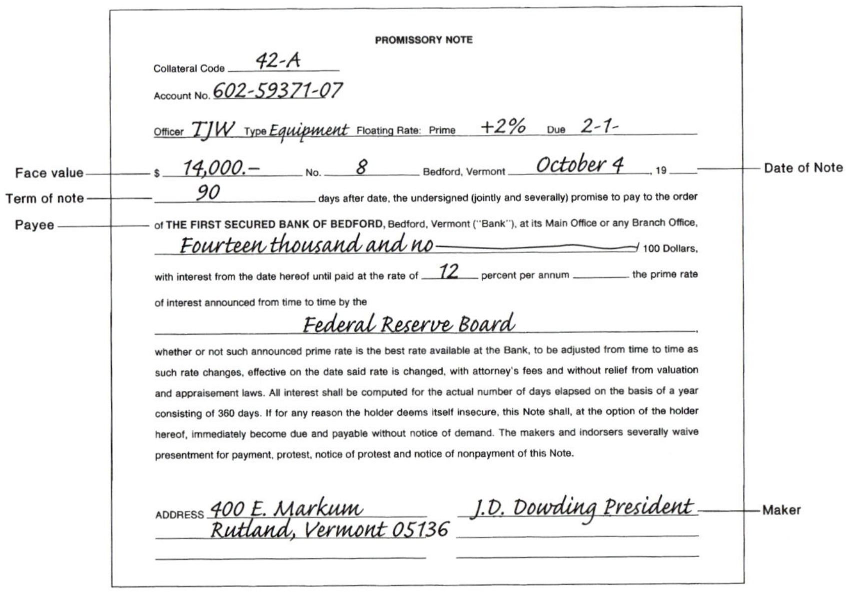

PROMISSORY NOTE

Collateral Code 42-A

Account No. 602-59371-07

Officer TJW Type Equipment Floating Rate: Prime +2% Due 2-1-

$ 14,000.– No. 8 Bedford, Vermont October 4, 19

90 days after date, the undersigned (jointly and severally) promise to pay to the order of THE FIRST SECURED BANK OF BEDFORD, Bedford, Vermont ("Bank"), at its Main Office or any Branch Office,

Fourteen thousand and no 100 Dollars,

with interest from the date hereof until paid at the rate of 12 percent per annum the prime rate of interest announced from time to time by the

Federal Reserve Board

whether or not such announced prime rate is the best rate available at the Bank, to be adjusted from time to time as such rate changes, effective on the date said rate is changed, with attorney's fees and without relief from valuation and appraisement laws. All interest shall be computed for the actual number of days elapsed on the basis of a year consisting of 360 days. If for any reason the holder deems itself insecure, this Note shall, at the option of the holder hereof, immediately become due and payable without notice of demand. The makers and indorsers severally waive presentment for payment, protest, notice of protest and notice of nonpayment of this Note.

ADDRESS 400 E. Markum Rutland, Vermont 05136 J.D. Dowding President

Objective 2

maturity value

A promissory note can be given, as an example, to the seller of merchandise by the buyer or as an extension of credit by the seller to the buyer. In this example, the maker, J. D. Dowding, promises to pay the payee, The First Secured Bank of Bedford, the sum of \$14,000 plus 12 percent interest for the 90-day term of the note. This amount to be repaid is due 90 days after October 4, the date the note was made. The value of the note on that date is known as the **maturity value,** and will be \$14,420 [\$14,000 + (\$14,000 $\times$ .12 $\times$ $\frac{90}{360}$ = \$420)]. The maker is obligated to pay \$14,420, the maturity value, to the payee or bearer on that date.

The numerical values can be shown graphically.

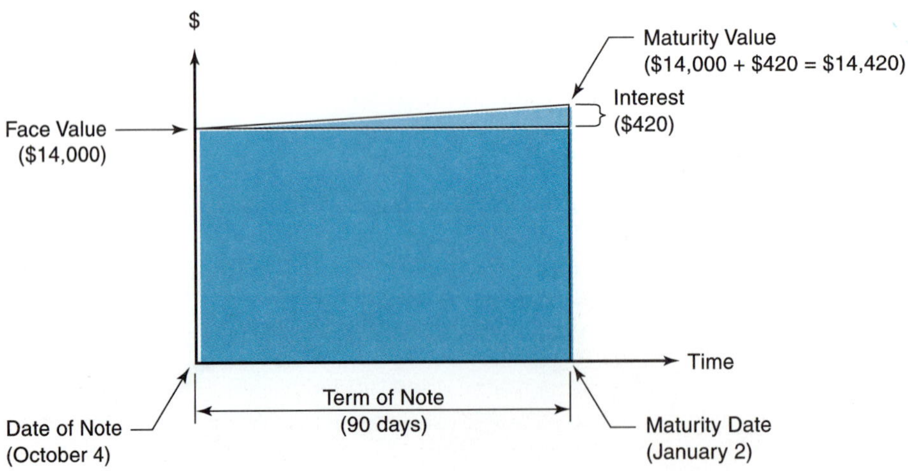

noninterest-bearing note

interest-bearing note

Occasionally, the payee will not charge the drawer interest on the note. This type of note is known as a **noninterest-bearing note.** In the promissory note shown on page 180, the face value of $14,000 would also be the maturity value ($14,000) as a noninterest-bearing note. Otherwise, and most frequently, the note will be an **interest-bearing note,** indicating that the maker must repay both face value ($14,000) and interest ($420) at maturity, a total of $14,420.

Exercise A

Determine the maturity value of the interest-bearing and noninterest-bearing notes described here. Use ordinary interest.

Face Value	Term of Note	Interest Rate	Maturity Value
1. $6,500	60 Days	15% (6,500 × .15 × 60 ÷ 360 =)	$6,662.50
2. 4,000	60	12 *(4,000 × .12 × 60 ÷ 360 =)*	*4,080.00*
3. 13,600	60	17 *(13,600 × .17 × 60 ÷ 360 =)*	*13,985.33*
4. 25,000	100	$10\frac{3}{4}$ *(25,000 × .1075 × 100 ÷ 360 =)*	*25,746.53*
5. 9,000	40	None	*9,000.00*
6. 7,000	120	13 *(7,000 × .13 × 120 ÷ 360 =)*	*7,303.33*
7. 8,500	30	$17\frac{1}{2}$ *(8,500 × .175 × 30 ÷ 360 =)*	*8,623.96*
8. 5,000	30	None	*5,000.00*
9. 60,000	73	14 *(60,000 × .14 × 73 ÷ 360 =)*	*61,703.33*
10. 3,000	1 Year	18 *(3,000 × .18 × 1 =)*	*3,540.00*

Discounting

Objective 3

● ***Remind students not to mix up the interest rate and the discount rate.***

discount rate

term of discount

When the payee of a promissory note (noninterest-bearing or interest-bearing) has a shortage of cash, the firm (or person) may decide to sell a promissory note. This will turn a semiliquid asset into spendable cash, allowing the payee to pay invoices, take advantage of cash discounts, and meet payroll obligations. The note is negotiable, and therefore, it can be sold to a bank or another firm that has ready cash. The payee must determine the rate that this other lending institution will charge the payee to gain cash for the note. This rate is called the **discount rate.** The length of time the note will be discounted at this rate is known as the **term of discount.**

Computing Discount

● ***Students tend to use the complement of the term of discount.***

If a 90-day note dated October 4 has a maturity value of $14,420, and is discounted for cash on December 13 of that year, then the term of discount will be twenty days.

October has	31 days	January	2—Maturity date
−October	4	December	+31
	27	−December	−13—Date of discount
November	30		20 days = Term of discount
December	31		
January	+2—Maturity date		
	90 days		

discount amount

If the discount rate was 10 percent, the **discount amount** would be computed as follows:

○ *The discount is always computed on the maturity value.*

discount amount = maturity value × discount rate × term of discount

$$\$80.11 = \$14,420 \times .10 \times \frac{20}{360}$$

Computing Proceeds

proceeds

The discount amount is charged by the lending institution for cash and will be subtracted from the maturity value in order to find the amount of cash that the payee will receive, the **proceeds.** The proceeds = $14,420 − $80.11 = $14,339.89. The payee receives $14,339.89 on December 13; the lending institution will receive the $14,420 maturity value of the note from the maker the following year on January 2 as the bearer of the note.

The numerical values to determine proceeds can also be shown graphically.

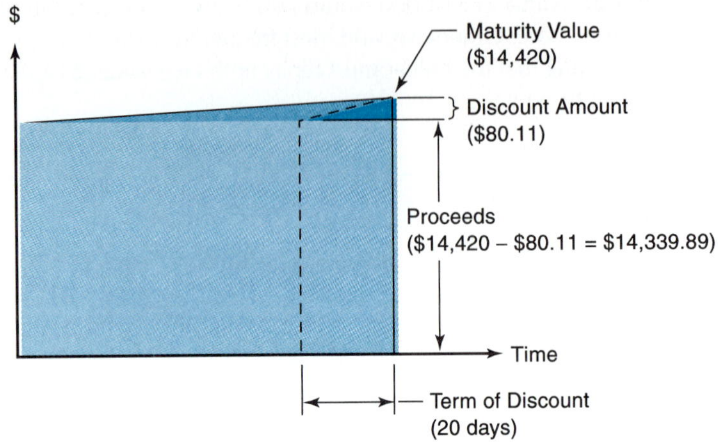

Another example should clarify any questions.

Example 1

Information

Face value of note	$780
Rate of interest	$9\frac{1}{2}\%$
Date of note	June 17, 2001
Term of note	60 days
Interest	$______
Maturity value	$______
Discount rate	$10\frac{1}{4}\%$
Date of discount	July 30, 2001
Term of discount	______
Discount amount	$______
Proceeds	$______

Solution

$I = P \times R \times T$

$I = \$780 \times .095 \times \frac{60}{360}$

$I = \$12.35$

$\$780.00 + \$12.35 = \$792.35 =$ maturity value

Discount = maturity value × discount rate × term of discount

$\text{Discount} = \$792.35 \times .1025 \times \frac{17}{360}$

Discount = $3.84

● *Have students look at their answers for reasonableness.*

June 30	July has 31
−June 17	−July 30
13	1
+July 31	+Aug. 16
44	17 days = Term of discount
+Aug. 16—Maturity date	
60	

Maturity value	=	$792.35
Amount of discount	=	−3.84
Proceeds	=	$788.51

(780 × .095 × 60 = ÷ 360 = 792.35 × .1025 × 17 = ÷ 360 =)

In the next example, the calculations are shown in an abbreviated form. This model should be useful for further assignments in this area.

Example 2

Kerry finds that the $7,000 note she accepted from Shape-Up must be discounted for cash in order to pay the current bills on May 17, 2002. The note will be discounted at 9 percent even though it has only earned 8 percent since it was given on March 11, 2002. What are the proceeds on this 90-day note?

Face value of note	$7,000
Rate of interest	8%
Date of note	March 11, 2002
Term of note	90 days
Discount rate	9%
Date of discount	May 17, 2002
Proceeds	$________

Solution

$$\text{Interest} = \$7{,}000 \times .08 \times \frac{90}{360} = \$140$$

$$\underline{+140} \text{ interest}$$

$$\text{Discount} = \$7{,}140 \times .09 \times \frac{23}{360} = \$41.06$$

$$\underline{-41.06} \text{ discount}$$

$$\text{Proceeds} = \$7{,}098.94$$

Note: The proceeds may be greater or less than the face value.

(7,000 × .08 × 90 = ÷ 360 = 7,140 × .09 × 23 = ÷ 360 =)

X

Exercise B

Determine the term of discount in each of the following exercises.

Date of Note	Term of Note	Date of Discount	Term of Discount
1. June 4, 2002	60 days (20 + 3 =)	July 11, 2002	23 days
2. Aug. 20, 2001	60 *(4 + 19 =)*	Sept. 26, 2001	*23*
3. Sept. 14, 2000	90 *(26 + 30 + 13 =)*	Oct. 5, 2000	*69*
4. May 12, 2002	60 *(10 + 11 =)*	June 20, 2002	*21*
5. Jan. 7, 2000	95 *(15 + 11 =)*	Mar. 16, 2000	*26*
6. Dec. 14, 2003	150 *(2 + 31 + 30 + 12 =)*	Feb. 27, 2004	*75*
7. April 12, 2002	30 *(10 + 12 =)*	April 20, 2002	*22*
8. Jan. 30, 2000	60 *(0 + 30 =)*	Feb. 29, 2000	*30*

Exercise C

Compute the answers to these problems. Use ordinary interest (360-day year).

	1.		2.
Face value	$450		$5,000
Rate of interest	10%		16%
Date of note	Oct. 3, 2002		Mar. 6, 2001
Term	45 days		60 days
Interest (450 × .10 × 45 ÷ 360 =)	$5.63	*(5,000 × .16 × 60 ÷ 360 =)*	*$133.33*
Maturity value	$455.63		*$5,133.33*
Discount rate	10%		11%
Date of discount	Nov. 1, 2002		Mar. 12, 2001
Term of discount	16 days		*54 days*
Discount am't. (455.63 × .10 × 16 ÷ 360 =)	$2.03	*(5,133.33 × .11 × 54 ÷ 360 =)*	*$84.70*
Proceeds	$453.60		*$5,048.63*

	3.		4.
Face value	$5,000		$6,000
Rate of interest	14%		12%
Date of note	May 3, 2001		July 16, 2002
Term	90 days		60 days
Interest *(5,000 × .14 × 90 ÷ 360 =)*	*$175.00*	*(6,000 × .12 × 60 ÷ 360 =)*	*$120.00*
Maturity value	*$5,175.00*		*$6,120.00*
Discount rate	13%		13%
Date of discount	July 3, 2001		Aug. 30, 2002
Term of discount	*29 days*		*15 days*
Discount am't. *(5,175 × .13 × 29 ÷ 360 =)*	*$54.19*	*(6,120 × .13 × 15 ÷ 360 =)*	*$33.15*
Proceeds	*$5,120.81*		*$6,086.85*

	5.		6.
Face value	$1,800		$170
Rate of interest	13%		$13\frac{3}{4}$%
Date of note	March 1, 2002		June 16, 2003
Term	120 days		19 days
Interest *(1,800 × .13 × 120 ÷ 360 =)*	*$78*	*(170 × .1375 × 19 ÷ 360 =)*	*$1.23*
Maturity value	*$1,878*		*$171.23*
Discount rate	12%		14%
Date of discount	May 14, 2002		June 30, 2003
Term of discount	*46 days*		*5 days*
Discount am't. *(1,878 × .12 × 46 ÷ 360 =)*	*$28.80*	*(171.23 × .14 × 5 ÷ 360 =)*	*$.33*
Proceeds	*$1,849.20*		*$170.90*

	7.		8.
Face value	$800		$1,100
Rate of interest	11%		12%
Date of note	January 3, 2004		April 1, 2001
Term	40 days	1100 x .12 x 75	75 days
Interest *(800 × .11 × 40 ÷ 360 =)*	*$9.78*	*(1,000 × .12 × 75 ÷ 360 =)*	*$25*
Maturity value	*$809.78*		*$1,025*
Discount rate	$9\frac{3}{4}$%		9%
Date of discount	Feb. 3, 2004		May 15, 2001
Term of discount	*9 days*		*31 days*
Discount am't. *(809.78 × .0975 × 9 ÷ 360 =)*	*$1.97*	*(1,025 × .09 × 31 ÷ 360 =)*	*$7.94*
Proceeds	*$807.81*		*$1,017.06*

(Continued on next page)

Exercise C Continued

	9.		10.
Face value	$1,000		$3,500
Rate of interest	13%		$12\frac{1}{2}\%$
Date of note	May 3, 2001		May 31, 2002
Term	60 days		45 days
Interest *(1,000 × .13 × 60 ÷ 360 =)*	***$21.67***	*(3,500 × .125 × 45 ÷ 360 =)*	***$54.69***
Maturity value	***$1,021.67***		***$3,554.69***
Discount rate	$13\frac{1}{2}\%$		15%
Date of discount	June 30, 2001		July 1, 2002
Term of discount	*2 days*		*14 days*
Discount am't. *(1,021.67 × .135 × 2 ÷ 360 =)*	***$.77***	*(3,554.69 × .15 × 14 ÷ 360 =)*	***$20.74***
Proceeds	***$1,020.90***		***$3,533.95***

ASSIGNMENT Chapter 10

Name ______________________ **Date** ______________ **Score** ____________

Skill Problems

Write T for true or F for false in blank for the following statements.

1. ___*F*___ A promissory note is not legally enforceable.

2. ___*T*___ A promissory note is used in business as an IOU.

3. ___*T*___ The actual number of days from the time the note is made until it is repaid is the term of the note.

4. ___*F*___ Noninterest-bearing notes are used the most frequently.

5. ___*F*___ The face value is the amount a business is able to pay, not necessarily the amount of the note.

Select the best term for the description below.

A. Face Value B. Maker C. Maturity Date D. Payee E. Term of Note

6. ___*A*___ The amount borrowed (written on the note).

7. ___*E*___ The length of time that the note is to run.

8. ___*D*___ The party lending the face value of the note.

Determine the maturity value of the interest-bearing and noninterest-bearing notes described here. Use ordinary interest. Round off to the nearest cent.

Face Value	Term of Note	Interest Rate	Maturity Value	
9. $20,000	60 days	14%	*$20,466.67*	*((20,000 × .14 × 60 ÷ 360 =) + 20,000 =)*
10. 6,000	90	11	*6,165.00*	*((6,000 × .11 × 90 ÷ 360 =) + 6,000 =)*
11. 460,800	120	None	*460,800*	
12. 1,250	60	7 1/2	*1,265.63*	*((1,250 × .075 × 60 ÷ 360 =) + 1,250 =)*
13. 8,500	100	13	*8,806.94*	*((8,500 × .13 × 100 ÷ 360 =) + 8,500 =)*
14. 60,750	1 Year	15	*69,862.50*	*((60,750 × .15 =) + 60,750 =)*
15. 3,050	30	None	*3,050.00*	

Compute the answers to these problems. Use ordinary interest. Round off to the nearest cent.

Face value of note	$940	**16.** Interest	*$24.68*	*(940 × .105 × 90/360 =)*
Rate of interest	10.5%	**17.** Maturity value	*$964.68*	*(940 + 24.68 =)*
Date of note	Jan. 2, 2002	**18.** Term of discount	*36 days*	*(90 − (31 − 2 + 25) =)*
Term of note	90 days	**19.** Discount amount	*$10.61*	*(964.68 × .11 × 36/360 =)*
Discount rate	11%	**20.** Proceeds	*$954.07*	*(964.68 − 10.61 =)*
Date of discount	Feb. 25, 2002			

Face value of note	$7,400	**21.** Interest	*$259.00*	*(7,400 × .07 × 180/360 =)*
Rate of interest	7%	**22.** Maturity value	*$7,659.00*	*(7,400 + 259 =)*
Date of note	May 12, 2001	**23.** Term of discount	*131 days*	*(180 − (31 − 12 + 30) =)*
Term of note	180 days	**24.** Discount amount	*$250.83*	*(7,659 × .09 × 131/360 =)*
Discount rate	9%	**25.** Proceeds	*$7,408.17*	*(7,659 − 250.83 =)*
Date of discount	June 30, 2001			

Face value of note	$10,000	**26.** Interest	*$100*	*(10,000 × .12 × 30/360 =)*
Rate of interest	12%	**27.** Maturity value	*$10,100*	*(10,000 + 100 =)*
Date of note	Oct. 31, 2000	**28.** Term of discount	*20 days*	*(30 − 10 =)*
Term of note	30 days	**29.** Discount amount	*$84.17*	*(10,100 × .15 × 20/360 =)*
Discount rate	15%	**30.** Proceeds	*$10,015.83*	*(10,100 − 84.17 =)*
Date of discount	Nov. 10, 2000			

Compute the answers to these problems. Use exact interest.

Face value of note	$5,000	**31.** Interest	*$263.01*	*(5,000 × .16 × 120/365 =)*
Rate of interest	16%	**32.** Maturity value	*$5,263.01*	*(5,000 + 263.01 =)*
Date of note	Nov. 24, 2003	**33.** Term of discount	*10 days*	*(120 − (30 − 24 + 31 + 31 + 29 + 13) =)*
Term of note	120 days	**34.** Discount amount	*$25.95*	*(5,263.01 × .18 × 10/365 =)*
Discount rate	18%	**35.** Proceeds	*$5,237.06*	*(5,263.01 − 25.95 =)*
Date of discount	March 13, 2004			

Face value of note	$1,140	**36.** Interest	*$0.00*	*(1,140 × .0 × 45/365 =)*
Rate of interest	0%	**37.** Maturity value	*$1,140.00*	*(1,140 + 0 =)*
Date of note	Feb. 3, 2001	**38.** Term of discount	*4 days*	*(45 − (28 − 3 + 16) =)*
Term of note	45 days	**39.** Discount amount	*$1.25*	*(1,140 × .1 × 4/365 =)*
Discount rate	10%	**40.** Proceeds	*$1,138.75*	*(1,140.00 − 1.25 =)*
Date of discount	March 16, 2001			

	41.	**42.**	**43.**	**44.**	**45.**
Face value	$65,000	$6,400	$4,500	$1,500	$40,000
Rate of interest	15%	11%	10%	15%	$14\frac{1}{2}$%
	(65,000 × .15 × 90 ÷ 360 =)	*(6,400 × .11 × 60 ÷ 360 =)*	*(4,500 × .10 × 45 ÷ 360 =)*	*(1,500 × .15 × 80 ÷ 360 =)*	*(40,000 × .145 × 30 ÷ 360 =)*
Date of note	May 3	March 8	Oct. 15	Jan. 13	June 10
Term of note	90 days	60 days	45 days	80 days	30 days
Date of discount	June 10	May 1	Nov. 2	Feb. 1	July 1
Rate of discount	16%	$10\frac{1}{2}$%	11%	12%	12%
	(67,437.50 × .16 × 52 ÷ 360 =)	*(6,517.33 × .105 × 6 ÷ 360 =)*	*(4,556.25 × .11 × 27 ÷ 360 =)*	*(1,550 × .12 × 61 ÷ 360 =)*	*(40,483.33 × .12 × 9 ÷ 360 =)*
Proceeds	*$65,878.94*	*$6,505.92*	*$4,518.66*	*$1,518.48*	*$40,361.88*

Compute the answers to the problems listed. Use exact interest. Round off to the nearest cent.

	46.	**47.**	**48.**	**49.**	**50.**
Face value	$3,000	$7,000	$6,500	$420	$900
Rate of interest	11%	13%	12%	None	10%
	(3,000 × .11 × 90 ÷ 365 =)	*(7,000 × .13 × 100 ÷ 365 =)*	*(6,500 × .12 × 90 ÷ 365 =)*	*(—)*	*(900 × .10 × 45 ÷ 365 =)*
Date of note	July 10	Sept. 30	Oct. 5	May 4	Aug. 19
Term of note	90 days	100 days	90 days	120 days	45 days
Date of discount	Aug. 30	Oct. 15	Dec. 16	July 30	Sept. 3
Rate of discount	12%	15%	15%	16%	11%
	(3,081.37 × .12 × 39 ÷ 365 =)	*(7,249.32 × .15 × 85 ÷ 365 =)*	*(6,692.33 × .15 × 18 ÷ 365 =)*	*(420 × .16 × 33 ÷ 365 =)*	*(911.10 × .11 × 30 ÷ 365 =)*
Proceeds	*$3,041.86*	*$6,996.09*	*$6,642.83*	*$413.92*	*$902.86*

Business Application Problems

1. On May 12, Jackson Associates agreed to accept a promissory note in exchange for the $16,000 on account owed by Balentine and Company. The note carried a 13 percent interest rate and a term of 90 days. On June 1, Jackson Associates discounted the note at 15 percent for cash at the First Trust Bank. What is the maturity value of the note? (Use ordinary interest.) ***$16,000 × .13 × 90/360 = $520 + $16,000 = $16,520 maturity***

2. Determine the amount of discount in problem 1.
$16,520 × .15 × 70/360 = $481.83 discount

3. What are the proceeds in problem 1?
$16,520 − $481.83 = $16,038.17 proceeds

4. Which firm receives the proceeds in problem 1?
Jackson Associates receives the proceeds.

5. Which firm is the maker of the note in problem 1?
Balentine and Company is the maker.

6. Which firm is the payee of the note in problem 1?
Jackson Associates is the payee.

7. What is the due date of the note in problem 1?
31 − 12 = 19 (May) + 30 (June) + 31 (July) + 10 = August 10

8. Jaber Professional Association discounted a promissory note with a maturity value of $4,560 on October 10 at 12 percent. The 120-day note was signed on September 14. What are the proceeds? (Use ordinary interest.)
$4,560 × .12 × 94/360 = $142.88
$4,560 − $142.88 = $4,417.12 proceeds

9. Smithson Corporation agreed to sign a promissory note on May 26 for their $60,000 debt to X-Ron Limited. The 60-day note carried a 15 percent interest rate. On July 21, X-Ron Limited discounted the note at 13.5 percent at Blanchard Federal Bank of Lewiston. What are the proceeds of the note? (Use ordinary interest.)
$60,000 × .15 × 60/360 = $1,500
$61,500 × .135 × 4/360 = $92.25
$61,500 − $92.25 = $61,407.75 proceeds

10. The proceeds of a promissory note with a face value of $23,400 will be used to pay current expenses for the Bay Rite Company. The 9 percent, 45-day note was dated November 10. The Bay Rite Company discounted the note on November 28 at 10.3 percent at the Commercial Bank of Union City. What are the proceeds of the note? (Use ordinary interest.)
$23,400 × .09 × 45/360 = $263.25
$23,663.25 × .103 × 27/360 = $182.80
$23,663.25 − $182.80 = $23,480.45 proceeds

11. The Bader Company received a $4,000 note from a customer who was unable to pay for merchandise purchased on account. The note was dated April 7, 2001, and was to run for 90 days without interest. On June 1, 2001, the Bader Company decided to discount the note at 12 percent for cash. What were the proceeds? (Use ordinary interest.)

$$\$4{,}000 \times .12 \times \frac{35}{360} = \$46.67$$

$4,000.00
−46.67
$3,953.33 proceeds

12. National City Bank received a check for $16,565.31 on November 1, 2002, from a promissory note customer. The customer had two notes outstanding at the time, and bank officials were not sure which one should be cancelled. Both notes had a $9\frac{1}{2}$ percent interest rate; one was dated October 17, 2002, for $16,500; and the other was dated June 17, 2002, for $16,000. Which note was the check to cover? (Use ordinary interest.)

$$\$16{,}500 \times .095 \times \frac{15}{360} = \$65.31$$

$16,500.00
+65.31
$16,565.31

$$\$16{,}000 \times .095 \times \frac{137}{360} = \$578.44$$

$16,000.00
+578.44
$16,578.44 ***The $16,500 note***

13. Sass and Company accepted a 15 percent 90-day note for $4,000 from a customer on July 2, 2002. On August 20, 2002, Mrs. Jane Sass discounted the note at 18 percent for cash at the Warren State Bank. What were the proceeds from the note? (Use ordinary interest.)

$\$4{,}000 \times .15 \times \frac{90}{360} = \$150$

$4,000
+150

$\$4{,}150 \times .18 \times \frac{41}{360} = \$85.08$

$4,150.00
−85.08
$4,064.92 proceeds

14. A promissory note held by payee Howard Gorman was discounted for cash on December 20, 2000. The face value of the note was $62,000. The note was to run for a term of 60 days at 20 percent. The date on the note was December 10, 2000. Fred Hass agreed to purchase the note by discounting it 22 percent. What were the proceeds? (Use ordinary interest.)

$\$62{,}000.00 \times .20 \times \frac{60}{360} = \$2{,}066.67$

+2,066.67

$\$64{,}066.67 \times .22 \times \frac{50}{360} = \$1{,}957.59$

−1,957.59
$62,109.08 proceeds

15. The promissory note shown here was discounted for cash on October 12, 2002. What were the proceeds if the discount rate was 15 percent? (Use ordinary interest.)

No. 546776 Date September 1, 2002, 19__ Due March 1, 2002

One hundred eighty-one days after date, I/~~We~~ promise to pay to the order of
EMPIRE NATIONAL BANK, BRONX, N.Y. 10460
the sum of Thirteen thousand-- Dollars, $13,000.00,
plus 12 % simple interest after 9/1/02.
This note secured by Office equipment
Holder, upon deeming himself insecure, may, without notice declare the unpaid balance to be immediately due and payable. Holder upon default may realize on or dispose of the collateral. Makers severally waive demand, notice and protest, and any defense due to extension of time or other indulgence by holder and to any substitution or release of collateral. This Note will bear interest on the entire unpaid balance at the rate of 12 per centum (12 %) per annum during any period of default hereunder.

Addresses: 2309 Taylor Street Signatures: Bonnie Kurtz
Pilskton Corp.

10-1 REV. 8/70

$\$13{,}000.00 \times .12 \times \frac{181}{360} = \$784.33$

+784.33

$\$13{,}784.33 \times .15 \times \frac{140}{360} = \$804.09$

−804.09
$12,980.24 proceeds

16. A firm decided to discount a note at 14 percent on July 16, 2000, that was valued at $21,880 at maturity. The note was made on June 1, 2000, to run for 120 days. What were the proceeds? (Use ordinary interest.)

$\$21{,}880 \times .14 \times \frac{75}{360} = \$638.17$

$21,880.00
−638.17
$21,241.83 proceeds

17. The Works will discount a promissory note with a face value of $20,000 and an interest rate of 10 percent on March 14, 2001. The note was made on February 28, 2001, to run for 120 days. The discount rate agreed to by the lender and The Works is 15 percent. What are the proceeds? (Use ordinary interest.)

$\$20{,}000.00 \times .10 \times \frac{120}{360} = \$666.67$

+666.67

$\$20{,}666.67 \times .15 \times \frac{106}{360} = \$912.78$

−912.78
$19,753.89 proceeds

18. The Birk Corp. received a note from Alfred Coloni for a service provided on an industrial machine. The note stated that Mr. Coloni promised to pay $1,750 within 90 days after the date of the note, June 14, 2002. The Birk Corporation discounted the note at 14 percent on July 20, 2002, for cash. What was the amount of the check Mr. Coloni used to pay off the note if he paid it off on August 18, 2002?

$1,750 (noninterest-bearing)

19. On July 14, 2003, the lawyers in the law office of Peterson, Peterson, & Snell agreed to accept a promissory note for $1,200 for services. The note interest rate was 12 percent and was to be paid off within 90 days. The office manager, Janis Tobian, decided there was a shortage of cash for payroll, and discounted the note on September 11, 2003, at a rate of 14 percent. What were the proceeds? (Use ordinary interest.)

$I = P \times R \times T$

$I = \$1{,}200 \times .12 \times \frac{90}{360}$

$I = \$36$

$1,200
+36

$\$1{,}236 \times .14 \times \frac{31}{360} = \$14.90$

$1,236.00
−14.90

$1,221.10 proceeds

20. A $4,500 face-value promissory note bearing an interest rate of 14 percent for 90 days was discounted on June 3, 2004. The note was dated May 1, 2004. The discount rate was 18 percent. What were the proceeds? (Use ordinary interest.)

$\$4{,}500.00 \times .14 \times \frac{90}{360} = \$157.50$

+157.50

$\$4{,}657.50 \times .18 \times \frac{57}{360} = \$132.74$

−132.74

$4,524.76 proceeds

21. Billie accepted a 90-day promissory note from a friend on October 22, 2003. The face value of the note was $14,000. The note was interest free. Billie discounted the note on December 1, 2003, at a rate of 14 percent. What is the amount of the discount? (Use exact interest.)

$I = P \times R \times T$

$I = \$14{,}000 \times .14 \times \frac{50}{365}$

$I = \$268.49$

22. Write a complete sentence in answer to this question. In problem 20, what would the interest have been if the note had carried an interest rate of 12 percent?

$\$4{,}500 \times .12 \times \frac{90}{365} = \$133.15$

The interest would have been $133.15 at 12 percent.

23. Write a complete sentence in answer to this question. Ted Jameson purchased a promissory note from Leonard Silk Design on May 16, 2002. Mr. Jameson purchased the $40,000 face-value note at a discount of 16 percent. The note carried an interest rate of 10 percent for the 60-day term that began on March 30, 2002. What were the proceeds? (Use exact interest.)

$\$40{,}000.00 \times .10 \times \frac{60}{365} = \$657.53$

+657.53

$\$40{,}657.53 \times .16 \times \frac{13}{365} = \$231.69$

−231.69
$40,425.84

The proceeds were $40,425.84.

★ Challenge Problem

A consumer promissory note was negotiated on January 2, 2002, for $5,100 at $9\frac{3}{4}$ percent for 120 days. On March 18, 2002, the payee discounted the note at a rate of $13\frac{1}{8}$ percent for cash at the State Bank of Burns. The proceeds were used to meet a 15 percent increase in payroll expenses. The maker of the note won a lottery prize of $10,000 on April 6, 2002, and decided to pay off the note on that date. What was the amount that the payee was obligated to pay on April 6, 2002?

$0 (payee)

MASTERY TEST Chapter 10

Name ______________________ **Date** ______________ **Score** ____________

Select the terms and correct answers using the following promissory note.

No. 463801 Date June 23, 20 02 Due One hundred twenty (120) days after date, I/We promise to pay to the order of the sum of One thousand, eight hundred fifty and no/100 Dollars, $ 1,850.00, plus 12 % simple interest after 6-23-02.

This note secured by Pilskton Corporation

Holder, upon deeming himself insecure, may, without notice declare the unpaid balance to be immediately due and payable. Holder upon default may realize on or dispose of the collateral. Makers severally waive demand, notice and protest, and any defense due to extension of time or other indulgence by holder and to any substitution or release of collateral. This Note will bear interest on the entire unpaid balance at the rate of Twelve per centum (12 %) per annum during any period of default hereunder.

Addresses: 1401 Main Street, Your City, New York

Signatures: *James White, Treasurer*

10-1 REV. 8/70

B 1. Face value of note
E 2. Maturity date
D 3. Maturity value (ordin. int.)
K 4. Date of note
C 5. Maker
A 6. Payee

a. First National Bank
b. $1,850
c. Pilskton Corp.
d. $1,924
e. October 21, 2002
f. 463801
g. 12 percent
h. September 20, 2002
i. $2,072
j. October 20, 2002
k. June 23, 2002

Dalmas Office Equipment Co. accepted a promissory note from Warren Steel for the purchase of $3,580 of office furniture on July 10, 2001. The note had an interest rate of 14 percent and was to run for ninety days. On August 7, Park Dalmas, president, discounted the note at 15 percent for cash. (Use ordinary interest.)

7. What was the term of discount? *62 days*
8. What was the discount amount? *$95.72*
9. What were the proceeds? *$3,609.58*
10. How much is the maker required to pay at maturity to the holder of the note? *$3,705.30*

11. The Peters Corp. agreed to sell $900 of industrial supplies to Franton Manufacturing if Franton would sign a promissory note for the amount of the purchase. The interest-free note was to be paid off thirty days after the November 10, 2002 purchase. On November 20, 2002 Peters discounted the note at 14 percent for cash. What were the proceeds? (Use ordinary interest.)

$I = P \times R \times T$

$I = \$900 \times .14 \times \frac{20}{360}$

$I = \$7$

$900 = Maturity and face value
− 7 = Discount
$893 = Proceeds

INSTALLMENT LOANS 11

Suggested Classroom Time for This Chapter

hours 1 2 3 4 5

Perceived Student Difficulty with Topical Material

easy moderate difficult

OBJECTIVES

After mastering the material in this chapter, you will be able to:

1. Determine the payment amount on an installment contract using a table or, when given, the amount of a loan, the time period of the loan, and the rate of interest. p. 198
2. Compute a repayment schedule for an installment contract. p. 201
3. Describe the use of an escrow account and its effect on a mortgage. p. 204
4. Describe the effect of an early payoff on an installment contract. p. 205
5. Understand and use the following terms and concepts:
   Down Payment
   Payment Amount
   Principal Balance
   Repayment Schedule
   Fixed Rate
   Variable Rate
   Fixed Payment Amount
   Variable Payment Amount
   Balloon Payment
   Points
   Closing Costs
   Escrow Account
   Rule of 78s

SPORT-A-MERICA

Kerry and Hal recognize that in order to expand the business, Sport-A-Merica needs to borrow money. "We cannot afford to discount a promissory note like we did with Shape-Up recently. The lack of working capital is going to be our next big hurdle to overcome,"commented Kerry. They need more money, not only to buy more inventory and to finance the new store, but to establish the firm with a bank so that someday they can make the big push to Eastern Mall, a regional mall two miles away.

Kerry and Hal take the records of Sport-A-Merica to Larry Mataway, a loan officer at First National Bank, to apply for a loan. Now is the time for some serious financial planning. Kerry and Hal have a business plan in which $85,000 may be needed as additional working capital over the next 12 months. Though all of the $85,000 may not be used, they want to have approval for that amount in the event it is needed. Once the merchandise they plan to buy has been sold, a portion of the loan can be repaid. The entire $85,000 will probably not be needed for the full 12 months.

Kerry has also determined that a $120,000 loan is needed for the purchase of the new space. This loan would need to be paid off over a longer period of time, probably several years. She indicates to Larry that the only other loan outstanding for Sport-A-Merica is the $26,000 short-term loan with Community National Bank for the delivery vehicle. The loan will be paid off within two months.

As they discuss the loans with Larry, most of their needs are clear. However, Larry raises some important issues in the initial discussion. Rates, points, down payment, and amortizing are some of the items that must be determined. Hal, in particular, was happy they inquired early about the loans. He will need to know the monthly payment on the big loan to help him with cash planning. There are a lot of details to work out.

Credit Purchases

There are certain circumstances in a consumer's life that result in the purchase of goods and services costing more than the consumer earns. These instances are usually in the early years and again in the later years when, for example, children enter college or are married. Goods and services such as automobiles, furniture, homes, appliances, medical services, petroleum products, and entertainment are often purchased on credit. A credit purchase allows the consumer to own and use the product when it is needed, and to pay for it while it is being used. Without such an opportunity, consumers would have to go without desired and necessary items until they have saved the money for the purchase. Debt through credit purchases is typically incurred in the consumer's early twenties (purchase of an automobile, appliances, and a home), is reduced over a period of the next fifteen to twenty years, and then increases again (weddings, college education, etc.). Most of the debt is paid off by the time the consumer retires.

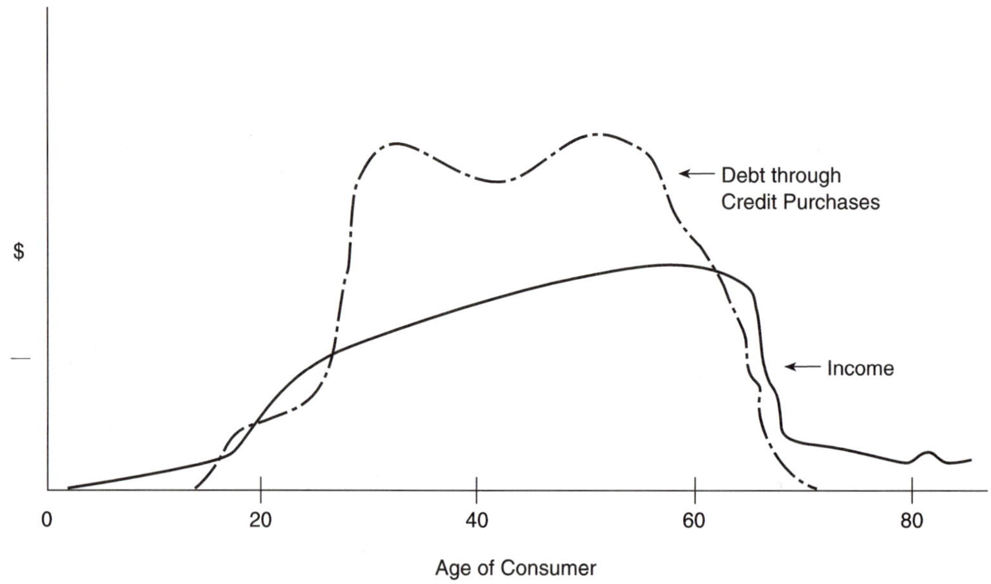

The ability to purchase on credit appeals to most Americans, as the majority use some type of credit financing. As with any freedom, there is also a degree of responsibility. The responsibility of repayment and credit management falls into the hands of the consumer. Currently, more than one million households declare bankruptcy each year, at least partly due to misuse of credit.

Consumers and business firms often make purchases that are in excess of available cash or exceed the limit of the credit card. When this happens, the bank or other lending institution provides a source of funds through an *installment loan*. Installment loans are not only used for larger amounts; they are also used when a longer time period (term) in which to pay back the loan is desired. The longer time period allows for a smaller monthly payment and lower total interest amount than would be the case if the loan were made through a promissory note. The following graph shows a comparison between a promissory note and an installment contract for $500 at 12 percent annual interest over a period of ten months.

Promissory Note Repayment

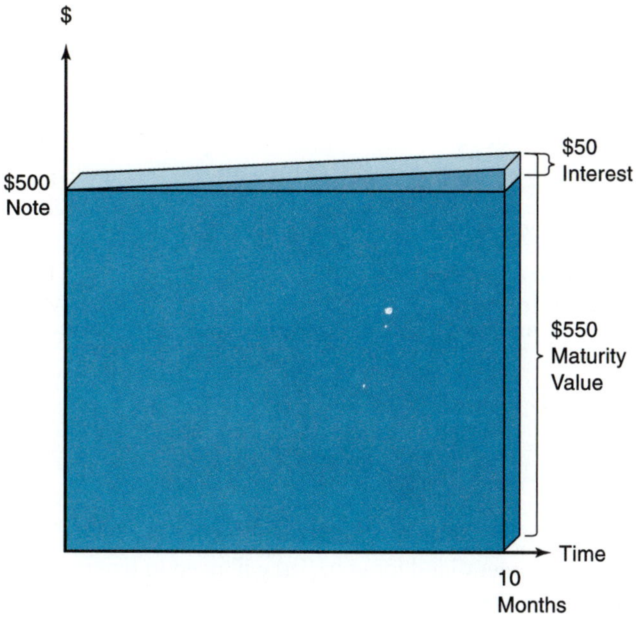

Installment Loan Repayment

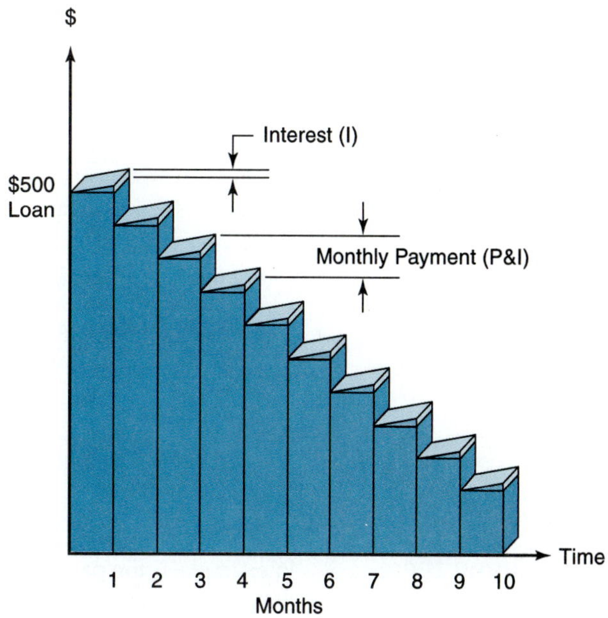

The buyer borrowed $500 and decreases the amount of the principal by making monthly payments. Because the amount of the loan is decreased over its life, the amount of interest also is less. The example shows that the buyer would repay a total of $550 on a promissory note but only a total of $527.92 if it were financed on an installment contract. (See page 201 for computations on the interest for the installment contract.) The borrower should also realize there is use of only approximately one-half of the principal over the term of the loan on an installment contract.

down payment

Unlike a promissory note, an installment loan usually requires a **down payment.** The down payment is made to secure the purchase and to prove the buyer's intent to the seller or lender. The borrower indicates the amount of down payment that can be made. The lender subtracts the down payment from the amount of the purchase. The loan is made on the balance (amount to be financed).

❍ The larger the down payment, the lower the monthly payments.

Example 1

Jenny bought a TV satellite dish for a total purchase price of $386.75. She decided to make a **down payment** of $86.75 and to finance the remainder. Jenny borrowed the $300 at 10 percent interest for four months, to be repaid at $76.60 per month. The repayment schedule is computed as follows:

Month 1	$300.00	$\times .10 \times \frac{1}{12} = \$2.50$
	+2.50	= Interest
	$302.50	
	−76.60	= Payment
Month 2	$225.90	$\times .10 \times \frac{1}{12} = \$1.88$
	+1.88	= Interest
	$227.78	
	−76.60	= Payment

(Continued on next page)

Example 1 ***continued***

Month 3	\$151.18	$\times .10 \times \frac{1}{12} = \$1.26$
	+1.26	= Interest
	\$152.44	
	−76.60	= Payment
Month 4	\$75.84	$\times .10 \times \frac{1}{12} = \$0.63$
	+.63	= Interest
	\$76.47	= Last payment

Record of Calculations (Repayment Schedule)

Payment Number	Payment Amount	Interest Amount	Payment on Principal	Principal Balance
1	\$76.60	\$2.50	\$74.10	\$225.90
2	76.60	1.88	74.72	151.18
3	76.60	1.26	75.34	75.84
4	76.47	.63	75.84	0.00

Determining the Payment Amount

Objective 1

payment amount

In example 1, the **payment amount** was determined for you. *You* will need to determine the payment amount as a businessperson. There are two circumstances that will determine which method you will need to use. If the loan is for a few months, you will use the "estimating total interest on an installment contract" method shown below. This will be especially useful if you are selling small-dollar-amount items. If you are involved in the sale of large-dollar-value items such as buildings, equipment, or vehicles, you will want to use the payment table method presented later in this chapter, which will spread the payments out over several years.

Estimating the Total Interest on an Installment Contract

The Truth-in-Lending Act requires that the total amount of finance charges and the effective annual rate be stated on a loan. In order to compute the total amount of interest on a loan and the amount of the monthly payment, it will be necessary to be able to use this formula.

$$\text{estimating total interest on an installment contract} = \frac{\text{Interest for the first month} \times (\text{the months in the contract} + 1)}{2}$$

This estimating formula is used for installment contracts of only a few months. For installment contracts that have a term of years, use the table on page 200.

Note that the formula is different than the $I = P \times R \times T$ formula used for single-payment promissory notes.

Example 2

Betty bought her husband two sport coats for his birthday gift. The merchandise totaled \$218. The men's store where the purchase was made requires 20 percent down and charges 11 percent interest. As the store manager, you are asked to determine the monthly payment required to pay the loan off within eight months. The calculations are as follows:

\$218.00	= Amount of purchase	\$218.00	
× .20		−43.60	
\$43.60	= Down payment	\$174.40	= Amount to be financed

$$\text{total interest} = \frac{\$174.40 \times .11 \times \frac{1}{12} \times (8 + 1)}{2}$$

$$= \frac{\$1.60 \times 9}{2} = \$7.20$$

\$174.40 = Amount to be financed
+7.20 = Interest on contract over 8 months
\$181.60 = Total to be repaid

\$181.60 ÷ 8 payments = \$22.70 (monthly payment amount)

Example 3 Sandy Fossum purchased kitchen furniture for her condominium from Kirkby Associates, Inc. She decided to take the credit terms that were offered by store manager Carolyn Kirkby. She will finance the balance at 11 percent for twelve months. What will be the monthly payment amount?

Kirkby Associates, Inc. — **INVOICE**

P O Box 459
Laingsburg, Mi. 48848-0459
Phone: 517-651-6302
Fax: 517-651-5134

SOLD TO: Sandy Fossum
3216 Emanuel Ave.
Muskegon, MI 49507

SHIPPED TO: Same

INVOICE NUMBER	41623
INVOICE DATE	6-16-97
OUR ORDER NUMBER	12141
YOUR ORDER NUMBER	—
TERMS	installments
SALES REP	Carolyn
SHIPPED VIA	Nomo Trucking
F.O.B.	Muskegon
PREPAID or COLLECT	Prepaid

QUANTITY	DESCRIPTION	UNIT PRICE	AMOUNT
1	Table w/two leaves	999.–	999.–
6	Chairs w/caned backs	75.–	450.–
			1,449.–
	Sales Tax		86.94
	Total		$1,535.94
	less down payment		– 192.44
	Charged		$1,343.50

THANK YOU FOR YOUR BUSINESS!

Courtesy of Kirkby Associates, Inc. Laingsburg, Michigan.

$$
\begin{aligned}
\$1{,}535.94 &= \text{Purchase amount} \\
-192.44 &= \text{Down payment} \\
\$1{,}343.50 &= \text{Amount to be financed}
\end{aligned}
$$

$$
\text{Total interest} = \frac{\$1{,}343.50 \times .11 \times \frac{1}{12} \times (12 + 1)}{2}
$$

$$
= \frac{\$12.32 \times 13}{2}
$$

$$
= \$80.08
$$

$$
\begin{aligned}
\$1{,}343.50 &= \text{Amount to be financed} \\
+80.08 &= \text{Amount of interest} \\
\$1{,}423.58 &= \text{Amount to be repaid in 12 payments} \\
&= \$118.63 \text{ monthly payment amount}
\end{aligned}
$$

Use of Payment Tables

The previous examples were for small loans for such items as office equipment, appliances, recreation equipment, furniture, and engagement rings. Installment loans are also used for larger loans—such items as machinery, buildings, land, and private homes. Because these items cost so much more, the loan amortization is usually extended over a period of years to let the buyer successfully pay off the loan. The process of computing the repayment schedule is exactly the same as in the earlier examples. The amount of the monthly payment can be determined by using the following table.

Monthly Payment Amount per $1,000 to Pay Principal and Interest on a Loan (P & I)

Term of Loan	5%	6%	7%	8%	9%	10%	11%	12%	13%	14%	15%	16%
5 years	$18.90	$19.35	$19.81	$20.28	$20.76	$21.25	$21.75	$22.25	$22.76	$23.27	$23.79	$24.32
10 years	10.61	11.11	11.62	12.14	12.67	13.22	13.78	14.35	14.94	15.53	16.14	16.76
15 years	7.89	8.43	8.99	9.56	10.15	10.75	11.37	12.01	12.66	13.32	14.00	14.69
20 years	6.60	7.17	7.76	8.37	9.00	9.66	10.33	11.02	11.72	12.44	13.17	13.92
25 years	5.82	6.43	7.07	7.72	8.40	9.09	9.81	10.54	11.28	12.04	12.81	13.59
30 years	5.38	6.01	6.66	7.34	8.05	8.78	9.53	10.29	11.07	11.85	12.65	13.45

Example 4

A boat was sold for $23,800. The buyers agreed to make a $2,800 down payment and finance the balance at 10 percent over five years. The amount of monthly payment on the boat would be found as follows:

$23,800 = Purchase price
−2,800 = Down payment
$21,000 = Amount to be financed

From the above table (Monthly Payment Amount per $1,000), the amount of $21.25 can be found (10 percent, 5 years). This amount should be multiplied by twenty-one (the number of $1,000 in the loan) to find a payment amount of $446.25.

$\$21.25 \times 21 = \$446.25$

Example 5

Sport-A-Merica will purchase its new space with a 12 percent loan over twenty years. Kerry will make a $30,000 down payment and will finance the $120,000 balance. What will the monthly payment amount be?

$11.02 = Payment per $1,000
×120 = The number of $1,000s in loan
$1,322.40 = Monthly payment amount

Exercise A

In the following problems, compute the monthly payment amount. Use the table or the estimating formula.

	Amount of Purchase	Down Payment	Rate of Interest	Term of Loan	Monthly Payment Amount
1.	$13,600	$ 3,000	7% (19.81 × 10.6 =)	5 years	$209.99
2.	40,000	6,000	13 *(12.66 × 34 =)*	15 years	*430.44*
3.	69,900	21,000	10 *(9.09 × 48.9 =)*	25 years	*444.50*
4.	3,000	500	14 *(2,500 × .14 × 1 ÷ 12 × 6 ÷ 2 =)*	5 months	*517.50*
5.	21,500	3,100	8 *(20.28 × 18.4 =)*	5 years	*373.15*
6.	500	125	12 *(375 × .12 × 1 ÷ 12 × 11 ÷ 2 =)*	10 months	*39.56*
7.	800	400	16 *(400 × .16 × 1 ÷ 12 × 16 ÷ 2 =)*	15 months	*29.51*
8.	56,500	$10,000	6 *(6.01 × 46.5 =)*	30 years	*279.47*

Computing the Principal Balance

Objective 2

Example 6

principal balance
repayment schedule

Let us use an example of a firm that wishes to purchase an office machine. The firm will finance or borrow \$500 in order to make the purchase. A lending institution, bank, or finance company has agreed to lend the \$500 at an annual rate of 12 percent for a period of ten months. A payment amount of \$52.75 is established, which will pay back the \$500 principal and the 12 percent interest over the ten months of the loan. A record is kept of payments, interest, and the amount due after each payment, which is known as the **principal balance.** This record is termed a **repayment schedule.** It is developed as follows.

Computing the Interest

$$\text{Interest} = \$500 \text{ (principal)} \times 12\% \text{ (annual rate)} \times \frac{1}{12} \text{ (time − monthly payments)}$$

$$\$5 = \$500.00 \times .12 \times \frac{1}{12}$$

+5.00 = Interest for the first month
\$505.00 = Amount owed at the end of the first month
−52.75 = Amount of monthly payment
\$452.25 = Amount of loan after the first monthly payment (principal balance)

$$\$4.52 = \$452.25 \times .12 \times \frac{1}{12}$$

+4.52 = Interest for the second month
\$456.77 = Amount owed at the end of the second month
−52.75 = Amount of monthly payment
\$404.02 = Amount of loan after the second monthly payment

$$\$4.04 = \$404.02 \times .12 \times \frac{1}{12}$$

+4.04 = Interest for the third month
\$408.06 = Amount owed at the end of the third month
−52.75 = Amount of monthly payment
\$355.31 = Amount of loan after the third monthly payment

This process is continued until the principal balance is reduced to zero. The completed repayment schedule for the \$500 loan follows.

Repayment Schedule

Payment Number	Payment Amount	Interest Amount	Payment on Principal	Principal Balance
1	\$52.75	\$5.00	\$47.75	\$452.25
2	52.75	4.52	48.23	404.02
3	52.75	4.04	48.71	355.31
4	52.75	3.55	49.20	306.11
5	52.75	3.06	49.69	256.42
6	52.75	2.56	50.19	206.23
7	52.75	2.06	50.69	155.54
8	52.75	1.56	51.19	104.35
9	52.75	1.04	51.71	52.64
10	53.17	.53	52.64	0.00

Note that the last payment may be a few cents more or less than the established payment amount.

Home Mortgages

Developed property is purchased by both business (stores, office buildings, research centers, manufacturing plants, and warehouses) and individuals (single-family homes, multi-unit complexes, and condominiums). The information and problems that follow, although applicable to both business and individuals, deal with purchasing a home—a decision relevant to most Americans.

Maximum Monthly Cost

As a guide, the maximum amount that should be spent on shelter by a family is about one-quarter of the gross income of the household. A family with a combined income of $60,000 could spend a maximum of $15,000 per year on shelter, or $1,250 per month. That amount must include the payment, taxes (if applicable), insurance, utilities, and repairs. This guide applies to those renting or buying.

Advantages

There are some advantages to both buying and renting. Examples follow.

Buying:

1. Interest and tax expenses are deductible for federal income tax purposes.
2. There may be some appreciation in the value of the property over the period that it is owned.
3. The property is being purchased. At some point in time, there will be no more mortgage payments!

Renting:

1. Little or no effort is needed to maintain the property.
2. Renter is free to move when the lease expires. There is no need to sell. This is an important factor for those who have jobs that require frequent relocation.

Many of the homes sold in the United States in recent years have been financed by the seller. This is an important factor to consider, because as a potential buyer through this method, you will need to know exactly what you are facing with respect to interest rates, terms of the loan, and how the balance is affected by each payment. The homes that are not financed by the seller are financed by savings and loan associations and commercial banks, which usually have more conservative loan terms and payment schedules.

The average single-family home costs about $150,000 with new homes somewhat higher. Most buyers will put a 10 or 20 percent down payment on the property in order to reduce the loan amount and, thereby, the monthly payment amount.

Effect of Interest Rate Difference

Interest rates are competitive, and therefore, a prospective buyer should shop around or try to negotiate a lower rate. The difference of one percentage point can result in several thousand dollars in additional payments over the term of the agreement.

Example 7

A family decides to purchase a small used home that is priced at $75,000. They put a $25,000 down payment on the home. The balance of $50,000 will be financed by a commercial bank. One bank will provide 7 percent financing on a twenty-year term. Another bank will provide 8 percent financing for the same term. What is the difference in the amount of the total payments if the loan goes to term?

$7.76	(7%, 20 years)	$8.37	(8%, 20 years)
×50	= Thousands	×50	= Thousands
$388	= Payment per month	$418.50	= Payment per month
×240	= (12 months × 20 years)	×240	= (12 months × 20 years)
$93,120	= Total repayment	$100,440	= Total repayment

$100,440	= (8%, 20 years)
−93,120	= (7%, 20 years)
$7,320	difference

The family in example 7 has a combined household income of $30,000. The mortgage payment (along with taxes and insurance) total $650 per month. This family may not be able to afford this home, because the total shelter expense exceeds the one-quarter maximum guideline (*$30,000* $\times \frac{1}{4} =$ $7,500, and $7,500 ÷ 12 = $625 per month).

Effect of Term Difference

Example 8

Another lender was found that would provide financing at 7 percent for a thirty-year term. Would the extra ten years of term allow the buyers to fit the guideline of one-quarter of gross income?

$6.66 (7%, 30 years)
×50 = Thousands
$333 = Payment per month

$388 = (7%, 20 years)
−$333 = (7%, 30 years)
$55 = Less per month

$650 = Monthly household expenses with 20-year mortgage
−55 = Reduction in monthly mortgage payment
$595 = This would allow the buyer to meet the guidelines!

Example 9

How much additional cost would be incurred over the life of the loan in example 8 for the expanded amortization period?

$333 = Payment per month
×360 = (12 months × 30 years)
$119,880 = Total repayment over 30 years
−93,120 = Total repayment over 20 years ($388 payment × 12 months × 20 years)
$26,760 = Additional cost over the term of the loan

Example 9 points out very clearly that time does cost money—in this case, $26,760 more! To put the information into proper perspective, another factor must be considered: The average family moves every six years; therefore, the family in examples 7, 8, and 9 probably won't make all the payments on the property to term. The family will be more than likely to sell the home after making only 72 payments.

Because lenders are aware that consumers are looking for a variety of financing plans to fit their individual family needs, they have formulated several plans that will fit both their needs and the needs of the borrower. Some of these variations are outlined here.

Fixed Rate

The rate on a **fixed-rate** loan remains the same over the term of the loan. This is a traditional type of loan and the type that will be used in the assigned problems unless otherwise indicated.

Variable Rate

The rate on a **variable-rate** loan (sometimes termed an adjustable rate loan) will usually change annually. The change in the rate will usually be based on some interest indicator, such as treasury bills. There may be limits on the amount of the adjustment.

Fixed Payment Amount

The traditional method of amortizing a loan involves using a **fixed payment amount,** and this is the method used in the assignments unless otherwise indicated. The payment amount remains the same over the term of the loan.

Variable Payment Amount

A loan with a **variable payment amount** begins with low payments that increase in amount as the loan begins to mature. This enables the new (and usually younger) buyer to buy a home early in life. The loan is designed to anticipate the increase in household income in future years. In this type of loan, the balance may actually increase as payments are made because they are so small. This is termed *negative amortization.*

Balloon Payment

A **balloon payment** is often used with the other loan variations. When there is a balloon payment, the total amount of the balance of the loan must be paid on the balloon date. The balloon date is usually five years after the original loan begins. The payment amount is for the loan balance at that point and is, therefore, a large amount. The loan will therefore have to be refinanced by the balloon date or paid in full.

Home Equity Loan

The Tax Reform Act of 1986 made a change in the way the federal government looked at the interest expenses of property buyers. The government disallowed the interest on what was termed *personal loans*. The law then made the *home equity loan* a way to finance family expenditures. The home buyer could use the amount of the home that was paid for, the equity, as a basis for a loan. This allowed the interest to be deductible for tax purposes. This type of loan allows some tax advantages, but also places the home in jeopardy if the buyer defaults on payments.

Points

● ***A rule of thumb on refinancing a mortgage: An interest rate of two percent lower, but with two points charged by the new lender, will bring the borrower ahead in two years. Two + two = two!***

When the buyer applies for a loan from a financial institution (commercial bank or savings and loan), there may be an application fee. The lender may also be allowed to charge a fee for a legal review of the papers, an appraisal fee, filing fees, prorated taxes, and points. A **point** is one percent of the amount of the loan. The points assessed on the loan may be paid by the buyer, by the seller, or by both. Points can be negotiable and should be reviewed by legal counsel.

Closing Costs

● ***A second guide on refinancing. How long of a period of time will it take to repay the costs of refinancing? Will the property be sold before that date?***

When the buyer makes the final financial commitment to purchase the property, the buyer and seller go through what is known as the *closing*. At the closing meeting, the appropriate papers are signed, and fees and monies are transferred to the lender, seller, and other parties. The total of these fees is termed the **closing costs,** because they are paid at closing. These costs are in excess of and not included in the amount of the loan. Often, the buyer is not aware of the exact amount of the closing costs until the day or week before the closing date.

Escrow Account

Objective 3

When a piece of property is mortgaged, the lending institution may, at the request of the buyer, collect money from the buyer to cover the yearly property taxes and fire insurance. A lump sum is collected at closing for the portion of taxes prorated to that point in the tax year. Each monthly payment thereafter will include a predetermined amount for taxes and fire insurance. This predetermined amount, which is over and above the regular monthly payment, is put into an **escrow account** for the buyer. The escrow account money, which does not earn or pay interest, is a reserve for the buyer to ensure that there will always be money to make the tax and insurance payments.

When the tax and insurance payments are due, the lender draws money from the escrow account and makes the payments for the buyer. In this way, too, the escrow account provides some protection to the lender by ensuring that the payments will be made.

Example 10

Ken and Linda purchased a home for $143,000. They made a down payment of $30,000, and financed the balance on a mortgage at 7 percent for thirty years. The lending institution stated that the home must be insured at a cost of $1,860 over three years. The property taxes were $1,470 per year. What would the monthly payment amount be in order to pay the principal (P), interest (I), taxes (T), and insurance (I) (i.e., PITI)?

$143,000 = Purchase price
−30,000 = Down payment
$113,000 = Amount of loan

$6.66 = Payment per $1,000 (from table)
113 = The number of $1,000s in loan
$752.58 = Monthly payment amount for principal and interest (P and I)

$752.58 = Monthly payment amount (P & I)
$1,470 (taxes for 1 year) ÷ 12 = 122.50 = Taxes per month (T)
$1,860 (insurance for 3 years) ÷ 36 = +51.67 = Insurance per month (I)
$926.75 = Monthly payment amount including escrow amount (PITI)

To develop a repayment schedule or to determine the balance on the loan, the payment amount without escrow is used.

Exercise B

In the following problems, compute the monthly payment amount including the amount for escrow. Use the table.

	Amount of Purchase	Down Payment	Rate of Interest	Term of Loan	Insurance Cost per Year	Taxes per Year	Monthly Payment Amount Including Escrow (i.e., PITI)
1.	$ 31,500	$ 0	9%	15 years	$147	$ 600	$381.98 (10.15 × 31.5 =)
2.	149,000	30,000	5	25	458	1,942	*892.58 (5.82 × 119 =)*
3.	18,000	3,000	7	5	92	480	*344.82 (19.81 × 15 =)*
4.	41,600	5,600	13	10	186	1,520	*680.01 (14.94 × 36 =)*
5.	194,000	40,000	6	25	—	1,800	*1,140.22 (6.43 × 154 =)*
6.	21,000	13,000	10	15	108	165	*108.75 (10.75 × 8 =)*
7.	233,500	45,000	5	20	160	478	*1,297.26 (6.60 × 188.5 =)*

Early Payoff Penalties

Objective 4

The development of a repayment schedule earlier in the chapter (examples 1 and 6) provided a complete analysis of the amount of interest, payment on the principal, and payment amount, as well as the balance after each monthly payment. This information is useful and necessary on an installment contract. Sometimes a loan is paid off before the end of the term. In this case, usually the borrower must pay some type of penalty to receive credit for complete payment. For example, on a long-term mortgage, the penalty may be a simple 1 percent of the original amount of the loan. This early payoff penalty must be spelled out in the loan agreement.

Example 11

A 25-year $42,000 loan at 8 percent has a principal balance of $7,860. The borrower has found enough money to pay off the loan before the end of the twenty-five years. After reviewing the contract, the borrower finds a penalty clause, which states that the borrower must pay a $\frac{3}{4}$ percent penalty on the current principal balance if an early payoff is made. The payoff amount would be computed as follows:

$7,860.00 = Principal balance

$\$7{,}860 \times .08 \times \frac{1}{12} =$ +52.40 = Interest for month

+58.95 = $\frac{3}{4}$% on principal balance for early payoff penalty

$7,971.35 = Payoff amount

Exercise C

Determine the payoff amount in the following exercises using the penalty information indicated.

	Amount Borrowed	Current Principal Balance	Interest Rate		Penalty	Payoff Amount
1.	$40,000	$28,960	10%	(28,960 × .005 =)	$\frac{1}{2}$% of principal balance	$29,346.13
2.	372,000	13,643	12	*(3,643 × .01 =)*	1% of principal balance over $10,000	*13,815.86*
3.	125,000	103,970	13	*(103,970 × .005 + 250 =)*	$250 if principal balance is over $80,000 plus $\frac{1}{2}$% of principal balance	*105,866.19*
4.	52,500	8,950	10	*(52,500 × .01 =)*	1% of amount borrowed	*9,549.58*
5.	185,000	13,805	7	*(13,805 × .0025 + 300 =)*	$\frac{1}{4}$% of principal balance and $300	*14,220.04*

Rule of 78s

- *Some students may have had an experience with early payoffs on their purchase of a car.*

On a shorter-term loan a bank may use the Rule of 78s. This rule goes along with the concept that more of the interest is earned in the earlier months of a loan's life. To compute the amount of payoff on a loan using the **Rule of 78s,** the number of the months in the term of the loan must be totaled. In a twelve-month loan, the total is 78; that is,

$$12 + 11 + 10 + 9 + 8 + 7 + 6 + 5 + 4 + 3 + 2 + 1 = 78$$

The bank uses $\frac{12}{78}$ of the principal to compute the first month's interest for payoff purposes; the second month uses $\frac{11}{78}$, etc. This results in a higher payoff amount than most consumers expect.

Note: The formula $\frac{n(n+1)}{2}$ can be used to find the sum of the months, where n is equal to the number of months of the loan. In a twelve-month loan, the formula gives $\frac{12(12+1)}{2} = 78$.

To find the payoff amount, the lender will

1. Determine the amount of unearned interest.
2. Determine the amount still owed (payments remaining × payment amount).
3. Subtract unearned interest from amount still owed.

Example 12

Fred borrowed \$780, which was to be repaid in eighteen monthly payments of \$47 each. The total to be repaid was \$846. After making the fifteenth payment, Fred inquired about the payoff amount. By using the Rule of 78s, the bank indicated the amount to be \$138.68. The amount was found as follows.

$$\begin{array}{rl} \$846 & = \text{Amount to be repaid } (18 \times \$47) \\ -780 & = \text{Amount of loan} \\ \hline \$66 & = \text{finance charge} \end{array}$$

Months	
18, 17, 16, 15, 14, 13, 12, 11, 10, 9, 8, 7, 6, 5, 4	first 15 months of loan (earned interest)
3, 2, +1	remaining 3 months of unearned interest total = 6
171	

$$\frac{6}{171} \times \$66 = \$2.32 \text{ Unearned interest}$$

$$\begin{array}{rl} \$47 & = \text{Payment amount} \\ \times\ 3 & = \text{Number of months remaining in loan} \\ \hline \$141.00 & = \text{Balance of loan after the fifteenth payment (still owed)} \\ -2.32 & = \text{Unearned interest} \\ \hline \$138.68 & = \text{Payoff amount} \end{array}$$

Exercise D

Determine the payoff amount in the following exercises using the Rule of 78s.

Amount Borrowed	Term (Months)	After Payment No.	Payment Amount	Payoff Amount
1. $400 (420 − 400 = 20)	10 (6 ÷ 55 × 20 =)	7	$ 42	$123.82
2. 1,600 *(1,680 − 1,600 = 80)*	7 *(10 ÷ 28 × 80 =)*	3	240	*931.43*
3. 750 *(795 − 750 = 45)*	15 *(15 ÷ 120 × 45 =)*	10	53	*259.37*
4. 1,000 *(1,080 − 1,000 = 80)*	12 *(6 ÷ 78 × 80 =)*	9	90	*263.85*
5. 625 *(666 − 625 = 41)*	6 *(3 ÷ 21 × 41 =)*	4	111	*216.14*

Computer Applications

It should be realized that even though the calculations in this chapter seem lengthy, they are very easily programmed into a computer. The repayment schedule, for example, can be developed for the full term of a 25-year loan in a matter of seconds. The computer needs only the principal, the interest rate, and the payment amount, as well as a simple set of instructions, in order to develop the complete repayment schedule.

This task is typical of many business problems that are repetitious, and therefore, it is a task that is easily automated. Computer applications also include determinations of depreciation, payroll, compound interest, and monthly billing.

Firms that have their own computers and computer service firms need trained employees to help program the computers. Individual consumers need training in the topic of installment contracts in case they purchase or sell property or merchandise on a land contract.

National bank credit cards and retail merchants that sell on credit all use a similar installment contract repayment analysis. The variations, though small, are numerous and should be understood by the consumer before authorizing a purchase on credit.

ASSIGNMENT Chapter 11

Name ______________________ **Date** ______________ **Score** ______________

Skill Problems

Using the chart within the chapter, compute the payment amount (P&1) of the following loan amounts.

1. Amount of purchase = $74,000
Rate of interest = 10%
Term of loan = 25 years
Monthly payment = ***$672.66***
(74 × 9.09 =)

2. Amount of purchase = $1,500
Rate of interest = 7%
Term of loan = 5 years
Monthly payment = ***$29.72***
(1.5 × 19.81 =)

3. Amount of purchase = $16,000
Rate of interest = 16%
Term of loan = 15 years
Monthly payment = ***$235.04***
(16 × 14.69 =)

4. Amount of purchase = $47,540
Rate of interest = 11%
Term of loan = 10 years
Monthly payment = ***$655.10***
(47.54 × 13.78 =)

In the following problems, compute the monthly payment amount (P&I). Use the table for long-term loans and the estimating formula for short-term loans.

	Amount of Purchase	Down Payment	Rate of Interest	Term of Loan	Monthly Payment Amount	
5.	$15,800	$2,000	8%	5 years	***$279.86***	***(15,800 − 2,000 = 13.8 × 20.28 =)***
6.	60,000	6,000	10	10 years	***713.88***	***(60,000 − 6,000 = 54 × 13.22 =)***
7.	750	0	18	5 months	***156.75***	***(750 × .18/12 × (5 + 1)/2 = 11.25 × 6/2 = 33.75 + 750 = 783.75 ÷ 5 =)***
8.	48,500	$10,000	13	10 months	***4,079.39***	***(38,500 × .13/12 × 10 + 1/2 = 417.08 × 11/2 = 2,293.94 + 38,500 = 40,793.94 ÷ 10 =)***
9.	64,400	5,400	7	25 years	***417.13***	***(64,400 − 5,400 = 59 × 7.07 =)***

10. Amount of purchase = $64,000
Rate of interest = 7%
Term of loan = 10 years
Monthly payment = ***743.68*** (P&I) ***(11.62 × 64 =)***

11. Amount of purchase = $50,000
Down payment = $8,000
Rate of interest = 5%
Term of loan = 25 years
Monthly payment = ***$244.44*** (P&I) ***(5.82 × 42 =)***

12. Amount of purchase = $29,500
Down payment = $4,000
Rate of interest = 13%
Term of loan = 30 years
Monthly payment = ***$282.29*** (P&I) ***(11.07 × 25.5 =)***

13. Amount of purchase = $160,000
Down payment = $12,000
Rate of interest = 12%
Term of loan = 15 years
Taxes per year = $540
Insurance for three years = $1,008
Monthly payment = ***$1,850.48*** (PITI) ***(12.01 × 148 =)***

14. Amount of purchase = $25,000
Down payment = $8,000
Rate of interest = 14%
Term of loan = 15 years
Taxes per year = $768
Insurance per year = $261
Monthly payment = ***$312.19*** (PITI) ***(13.32 × 17 =)***

15. Amount of purchase = $286
Rate of interest = $16\frac{1}{2}$%
Term of loan = 8 months
Down payment = $36
Monthly payment = ***$33.19***
(Use the Estimating Total Interest on an Installment Contract method.)
(250 × .165 = ÷ 12 =
3.44 × (8 + 1) ÷ 2 = 15.48   250 + 15.48 =
÷ 8 =)

In the following problems, compute the monthly payment amount including the amount for escrow. Use the table in the chapter.

	Amount of Purchase	Down Payment	Rate of Interest	Term of Loan	Insurance Cost per Year	Taxes per Year	Monthly Payment Amount Including Escrow (PITI)	
16.	$ 45,500	$ 0	13%	15 yr.	$225	$ 457	*$632.86*	*(45.5 × 12.66 = 576.03, (225 + 457)/12 + 576.03 =)*
17.	150,000	10,000	10	25	483	1,594	*1,445.68*	*(140 × 9.09 = 1,272.60, (483 + 1,594)/12 + 1,272.60 =)*
18.	46,750	3,250	11	10	140	0	*611.10*	*(43.5 × 13.78 = 599.43 140/12 + 599.43 =)*
19.	50,050	5,050	12	5	645	2,870	*1,294.17*	*(45 × 22.25 = 1,001.25 (645 + 2,870)/12 + 1,001.25 =)*
20.	250,000	15,000	7	30	915	3,450	*1,928.85*	*(235 × 6.66 = 1,565.10 (915 + 3,450)/12 + 1,565.10 =)*

Cost of item = $548.25
Down payment = 48.25
Loan period = 6 months
Interest rate = 10%
Complete the repayment schedule below.

Payment Amount	Interest Amount	Payment on Principal	Principal Balance	
$85.78	**21.** *$4.17*	**22.** *$81.61*	**23.** *$418.39*	*(500 × .1/12 = 85.78 − 4.17 = 500 − 81.61 =)*
85.78	**24.** *3.49*	**25.** *82.29*	**26.** *336.10*	*(418.39 × .1/12 = 85.78 − 3.49 = 418.39 − 82.29 =)*
85.78	**27.** *2.80*	**28.** *82.98*	**29.** *253.12*	*(336.10 × .1/12 = 85.78 − 2.80 = 336.10 − 82.98 =)*
85.78	**30.** *2.11*	**31.** *83.67*	**32.** *169.45*	*(253.12 × .1/12 = 85.78 − 2.11 = 253.12 − 83.67 =)*
85.78	**33.** *1.41*	**34.** *84.37*	**35.** *85.08*	*(169.45 × .1/12 = 85.78 − 1.41 = 169.45 − 84.37 =)*
85.79	**36.** *.71*	**37.** *85.08*	**38.** *0.00*	*(85.08 × .1/12 = 85.79 − .71 = 85.08 − 85.08 =)*

Cost of item = $247
Down payment = 15%
Loan period = 6 months
Interest rate = 18%

39. Down payment amount = *$37.05*
(247 × .15 =)

40. Amount to be financed = *$209.95*
(247 − 37.05 =)

41. Total interest = *$11.03*
(209.95 × .18/12 × (6 + 1) = 3.15 × 7/2 =)

42. Monthly payment = *$36.83*
(209.95 + 11.03 = 220.98 ÷ 6 =)

Cost of item = $658
Down payment = 0
Loan period = 12 months
Interest rate = $14\frac{1}{4}$%

43. Total interest for 12 months = *$50.77*
(658 × .1425/12 × (12 + 1) = 7.81 × 13/2 =)

44. Monthly payment for 12 months = *$59.06*
(658 + 50.77 = 708.77 ÷ 12 =)

45. Amount of purchase = \$8,000
Rate of interest = 12%
Monthly payment = \$180 (P&I)
Principal balance after the second monthly payment = ***\$7,799***
(8,000 × .12 ÷ 12 = 7,900 × .12 ÷ 12 =)

46. Amount of purchase = \$8,640
Rate of interest = 9%
Monthly payment = \$120 (P&I)
Interest amount for fourth month = ***\$63.55***
(64.80 64.39 63.97 8,473.16 × .09 ÷ 12 =)

47. Amount of purchase = \$26,000
Rate of interest = 8%
Monthly payment = \$305 (P&I)
Principal balance after the third monthly payment = ***\$25,602.36***
(173.33 172.46 25,735.79 × .08 ÷ 12 =)

48. Amount of purchase = \$13,100
Rate of interest = 16%
Monthly payment = \$260 (P&I)
Interest for first two months = ***\$348.20***
(174.67 13,014.67 × .16 ÷ 12 =)

49. Amount of purchase = \$26,000
Rate of interest = 9%
Monthly payment = \$390 (including \$106 for escrow account)
Principal balance after the third monthly payment = ***\$25,730.99*** ***(195 194.33 25,821.33 × .09 ÷ 12 =)***

50. Amount of purchase = \$26,000
Down payment = \$0
Rate of interest = 10%
Monthly payment = \$236.34
Term of loan = ***25*** years ***(from table) (236.34 ÷ 26 =)***

51. Amount of purchase = \$380
Monthly payments = \$48
Amount to be repaid = \$384
8 monthly payments
Amount of payoff after the sixth payment = ***\$95.67***
(Use Rule of 78s.)
(384 − 380 = 4 4 × 3 = ÷ 36 =
48 × 2 = 96 96 − .33 =)

In the following problems, compute the payoff amount.

	Amount Borrowed	Current Principal Balance	Interest Rate	Early Payoff Penalty	Payoff Amount	
52.	\$ 45,500	\$20,737	13%	½% of principal balance	***\$21,065.34***	***(20,737 × .005 = 103.69, 20,737 × .13 ÷ 12 = 224.65, 20,737 + 103.69 + 224.65 =)***
53.	150,000	7,635.60	10	1% of amount borrowed	***9,199.23***	***(150,000 × .01 = 1,500, 7,635.60 × .1 ÷ 12 = 63.63, 7,635.60 + 63.63 + 1,500 =)***
54.	46,750	7,193.16	11	\$300 and ¼% of principal balance	***7,577.08***	***(7,193.16 × .0025 = 17.98, 7,193.16 × .11 ÷ 12 = 65.94, 7,193.16 + 17.98 + 300 + 65.94 =)***
55.	50,050	48,060.00	12	½% of principal balance over \$10,000	***48,730.90***	***(38,060 × .005 = 190.30, 48,060 × .12 ÷ 12 = 480.60, 48,060 + 190.30 + 480.60 =)***
56.	\$250,000	31,302.00	7	¼% of balance	***31,562.86***	***(31,302 × .0025 = 78.26, 31,302 × .07 ÷ 12 = 182.60, 31,302 + 78.26 + 182.60 =)***

Determine the payoff amount in exercises 57–62 using the Rule of 78s.

	Amount Borrowed		Term (Months)		After Payment No.	Payment Amount	Payoff Amount	
57.	\$600	***(728 − 600 =)***	7	***(6/28 × 128 =)***	4	\$104.00	***\$284.57***	***(104 × 3 − 27.43 =)***
58.	4,600	***(5,261.20 − 4,600 =)***	14	***(10/105 × 661.20 =)***	10	375.80	***1,440.23***	***(375.8 × 4 − 62.97 =)***
59.	750	***(780 − 750 =)***	6	***(6/21 × 30 =)***	3	130.00	***381.43***	***(130 × 3 − 8.57 =)***
60.	845	***(981.20 − 845 =)***	11	***(21/66 × 136.20 =)***	5	89.20	***491.86***	***(89.2 × 6 − 43.34 =)***
61.	1,500	***(1,836 − 1,500 =)***	18	***(55/171 × 336 =)***	8	102.00	***911.93***	***(102 × 10 − 108.07 =)***
62.	678	***(795.84 − 678 =)***	24	***(210/300 × 117.84 =)***	4	33.16	***580.71***	***(33.16 × 20 − 82.49 =)***

Business Application Problems

1. A company bought a building for $82,500. The building was financed through a mortgage at 11 percent for 20 years. What is the amount of monthly payment on the mortgage loan?

82.5 × $10.33 = $852.23 payment

2. In problem 1, what is the total amount of interest charges over the life of the loan?

20 × 12 × $852.23 = $204,535.20 − $82,500 =
$122,035.20 interest

3. In problem 1, what is the balance of the mortgage after the second monthly payment?

$82,500 × .11 × 1/12 =
$756.25 + $82,500 − $852.23 =
$82,404.02 × .11 × 1/12 =
$755.37 + $82,404.02 − $852.23 =
$82,307.16 balance

4. In problem 1, how much interest has the company paid in the first two months of the mortgage?

$756.25 + $755.37 = $1,511.62 interest

5. A business has a balance of $14,540 on a $15,000, five-year installment contract. The business has made three monthly payments of $318. What is the penalty on the contract if there is an early payoff penalty clause of 1 percent of the current balance?

$14,540 × .01 = $145.40

6. In problem 5, how much interest has the company paid in the first three months?

$318 × 3 = $954 $15,000 − $14,540 = $460
$954 − $460 = $494 interest

7. What is the difference in the total interest paid on a 25-year, $30,000 mortgage at 6 percent versus at 5 percent?

$6.43 × 30 × 25 × 12 = $57,870
$5.82 × 30 × 25 × 12 = $52,380
$5,490 difference

8. What is the difference in the mortgage balance using 5 percent versus 6 percent, after the first monthly payment in problem 7?

$30,000 × .05 × 1/12 = $125 $5.82 × 30 =
$174.60 − $125 = $49.60
$30,000 × .06 × 1/12 = $150
$6.43 × 30 = $192.90 − $150 = $42.90
$49.60 − $42.90 = $6.70 difference

9. Kelly and Carl have a combined household annual income of $72,000. They have accumulated a down payment of $14,000. What is the maximum amount they can spend on shelter per month?

$72,000/4/12 = $1,500 payment

10. In problem 9, the current interest rate is 10 percent and escrow payments are estimated to be $150 per month. To the nearest thousand dollars, what is the maximum value of shelter they can purchase through a 30-year mortgage?

$1,500 − $150/$8.78 = 153.75 =
$154,000 + $14,000 = $168,000 value

11. Candy's Confectionary will add on to their existing store with a loan from the American State Bank. The loan will be for $125,000. The owner negotiated a 10 percent rate and an amortization over twenty years. What is the payment amount?

125 thousands
×$9.66 per thousand
$1,207.50 payment

12. As cash payments clerk of the Accounting Department, you are asked to develop a repayment schedule on a loan of $22,800. The rate of interest is 12 percent and the loan will run for fifteen years. Taxes on the property are $640 a year. The insurance coverage for a year will be $230. Complete the repayment schedule through the third month.

Repayment Schedule

Payment Number	(PITI) Payment Amount	Interest Amount	Payment on Principal	Principal Balance
1	*$346.33*	*$228.00*	*$45.83*	*$22,754.17*
2	*346.33*	*227.54*	*46.29*	*22,707.88*
3	*346.33*	*227.08*	*46.75*	*22,661.13*

13. The teller of the Mid-State Bank referred a customer to the loan manager to explain the payoff amount on an installment loan for $600. The amount to be repaid was $684 over eight months. The customer has paid four payments and feels that the loan balance should be $258 because four payments have been made on the $600 loan. Using the Rule of 78s, determine the correct payoff amount.

$684 total repayment amount
−600 amount of loan
$84 finance charges

$\frac{10}{36} \times \$84 = \$23.33$ *unearned interest*

$684 ÷ 8 = $85.50 monthly payment
$85.50 × 4 = $342 − $23.33
= $318.67
$318.67 payoff amount

14. Alice Baid paid off the mortgage on her home early. The amount of the mortgage was initially $93,000. The balance at the date of payoff was $56,328.40. The mortgage contract stipulated a 1 percent penalty charge on the current balance for early payoff or for payments in excess of 10 percent of the initial mortgage amount. What is the payoff amount?

$56,328.40 balance × .01 = $563.28
+563.28 1% penalty
$56,891.68 payoff amount

15. Four months ago, Predmore Associates purchased a piece of land for future development at a price of $46,000. They paid $10,000 down and financed the balance at 15 percent interest. The payments are $600 per month. No formal record or repayment schedule has been established to date. As assistant to the Budget Director, you are asked to prepare the figures for this month's interest expense on the loan. What is the interest for the fourth month?

$46,000
−10,000

$\$36{,}000 \times .15 \times \frac{1}{12} = \$450$

+450
−600

$\$35{,}850.00 \times .15 \times \frac{1}{12} = \$448.13$

+448.13
−600.00

$\$35{,}698.13 \times .15 \times \frac{1}{12} = \$446.23$

+446.23
−600.00

$\$35{,}544.36 \times .15 \times \frac{1}{12} = \$444.30$ *interest*

16. Blanche Olsen selected a condominium to purchase. She decided to pay $12,500 as a down payment on the purchase price of $79,500. She will amortize the 11 percent loan over a period of fifteen years. What will be the principal balance after the second monthly payment?

$79,500
−12,500
$67,000

67.00
×$11.37
$761.79 payment

$\$67{,}000.00 \times .11 \times \frac{1}{12} = \$614.17$ *interest*

+614.17
−761.79

$\$66{,}852.38 \times .11 \times \frac{1}{12} = \$612.81$ *interest*

+612.81
−761.79
$66,703.40 balance after the second month

17. If Blanche, in problem 16, had elected to take a 10 percent, 30-year loan with Marlin Mortgage, how much more would it have cost her over the term of the loan?

67.00
×$11.37
$761.79
×180 (12 × 15 years)
$137,122.20 payments

67.00
×$8.78
$588.26
×360 (12 × 30 years)
$211,773.60
−137,122.20
$74,651.40 difference

18. Write a complete sentence in answer to this question. Lloyd must obtain a mortgage on a home in the amount of $80,000. He has the choice of a 6-percent 30-year mortgage, or a 5-percent 25-year mortgage. What is the difference on the first month's interest?

$\$80{,}000 \times .06 \times \frac{1}{12} = \$400.00$

$\$80{,}000 \times .05 \times \frac{1}{12} = \frac{333.33}{\$66.67}$

The difference in the interest the first month is $66.67.

19. Write a complete sentence in answer to this question. What is the difference in the interest over the terms of the mortgages in problem 18?

80 × $6.01 × 360 (12 × 30 years)
= $173,088
−80,000
$93,088

80 × $5.82 × 300 (12 × 25 years)
= $139,680
−80,000
$59,680

$93,088
−59,680
$33,408
The difference over the terms of the mortgages is $33,408.

20. Write a complete sentence in answer to this question. Marin and Associates will build a new home on a piece of land. The finished home will sell for $184,900. A loan company will finance 90 percent of the value of the home. If a buyer finances the 90 percent at 7 percent interest over a twenty-year period, what will be the amount of interest for the first month?

$184,900
×.90
$\$166{,}410 \times .07 \times \frac{1}{12} = \$970.73$

The amount of interest the first month is $970.73.

★ Challenge Problem

The accountant for the Vilper Corporation must determine the principal balance of a loan negotiated six months ago on a piece of property. The property purchase price was $24,000. The rate of interest on the loan was 9 percent. The payment amount of $263 per month will pay off the loan and the $300 annual tax bill. The vacant land required a down payment of $6,000 before the bank would finance the loan. A penalty charge of 1 percent of the balance of the loan was agreed upon by Vilper should they decide to pay off the loan early. What is the principal balance?

$24,000 amount of purchase
−6,000 down payment
$18,000 amount of loan

$300 ÷ 12 = $25 per month for tax
$263 payment
−25 for tax
$238 P & I payment

$\$18{,}000 \times .09 \times \frac{1}{12} = \$135$
+135
−238
$\$17{,}897.00 \times .09 \times \frac{1}{12} = \$134.23$
+134.23
−238.00
$\$17{,}793.23 \times .09 \times \frac{1}{12} = \$133.45$
+133.45
−238.00
$\$17{,}688.68 \times .09 \times \frac{1}{12} = \$132.67$
+132.67
−238.00
$\$17{,}583.35 \times .09 \times \frac{1}{12} = \$131.88$
+131.88
−238.00
$\$17{,}477.23 \times .09 \times \frac{1}{12} = \$131.08$
+131.08
−238.00
$17,370.31 principal balance

MASTERY TEST Chapter 11

Name ______________________ Date ______________ Score ____________

Complete the following problems.

1. In computing the principal balance on an installment loan, the payment amount is first applied to the interest. Any additional amount is applied to the reduction of the principal. (True) False

2. When computing a principal balance on an installment loan, the escrow amount of the payment is added to the beginning principal before computing the interest. True (False)

3. Amount of purchase = $3,000
   Down payment = $500
   Interest rate = 7%
   Term of loan = 15 years
   Monthly payment amount = $ ***22.48*** ***(P&I)***

4. Amount of purchase = $91,000
   Down payment = $5,000
   Interest rate = 6%
   Term of loan = 25 years
   Insurance per year = $1,200
   Taxes per year = $1,890
   Monthly payment = $ ***810.48*** ***(PITI)***

5. Amount of loan = $28,000
   Payment amount = $300
   Interest rate = 9%
   Principal balance at the end of the second month = $ ***27,819.33***

6. Mr Boldt borrowed $900 for six months. The monthly payments were $158.87. At the end of the fifth month he paid off the loan. Using the Rule of 78s, what was the payoff amount?

   $158.87
   × 6
   $953.22
   − 900.00
   $ 53.22 finance charges $\times \frac{1}{21}$ ***= $2.53***

   $158.87
   − 2.53
   $156.34 payoff

7. Determine the amount of the final payment on a property loan that was: (1) originally for $64,000, (2) has a current balance of $4,907.35, and (3) had a clause in the contract that any payment in excess of 5 percent of the original loan amount be penalized at the rate of 1 percent of the current balance. The mortgage carries a 10 percent interest rate.

   $64,000
   × .05
   $ 3,200

   $4,907.35 $\times$ ***.10*** $\times \frac{1}{12}$ ***= $40.89***
   $4,907.35 × .01 = $49.07
   + 40.89 interest
   + 49.07 penalty
   $4,997.31 payoff

Consumer Credit 12

Suggested Classroom Time for This Chapter

hours 1 2 3 4 5

Perceived Student Difficulty with Topical Material

easy moderate difficult

Objectives

After mastering the material in this chapter, you will be able to:

1. Describe the obligations of buying with a credit card. p. 218
2. Determine the balance for a credit card statement. p. 219
3. Determine the annual percentage rate on a loan. p. 221
4. Describe how a business owner turns credit sales into cash. p. 224
5. Understand and use the following terms and concepts:
   Credit Card
   Store Card
   Gasoline Card
   Bank Card
   Travel or Entertainment Card
   Monthly Statement
   Minimum Payment Amount
   Amortizing
   Regulation Z
   Annual Percentage Rate

Sport-A-Merica

Sales last weekend set a new record for the store. Kerry saw firsthand the huge role that credit plays at Sport-A-Merica's checkout counter over the weekend. Nearly 70 percent of all sales made were on credit. She was convinced of the importance of the ability to accept credit cards. She does wish that the cost of credit to Sport-A-Merica were lower. The credit card company takes 4 percent off the top. That hurts! She asks herself how much Sport-A-Merica gives up on each deposit for the opportunity to sell on credit using credit cards. Kerry occasionally offers store credit to a good customer. Though this takes some figuring, it saves the store the 4 percent that the credit card company would have charged on that sale. She knows that a store credit card could be offered to the best customers. However, that would require a billing-and-collections operation. She recognizes that she is the strong sales arm in the company, and that there will be opportunity in the future to expand into offering more store credit. She thinks a lot about the cost of credit.

Credit Cards

Since World War II, the acceptance of **credit cards** has grown to the point that today more than one-half of the families in the United States use them. This widespread acceptance is based at least in part on the functions that credit cards can provide. Today, a credit card acts to identify a person, to allow for credit payment of an item or service, and to provide a means to obtain additional credit from other sources. Having one credit card with a good payment record opens up the door to other credit arrangements. There are four types of credit cards:

● *Ask the students in class to indicate if they have their own credit cards. What type of card?*

Store Card

The **store card** can be used in one store or one chain of stores.

Gasoline Card

The **gasoline card** was initially issued by gasoline companies to attempt to gain a greater market share and customer loyalty. Today, the oil and gas companies have a select group of customers, often business people, who use their card exclusively.

Bank Card

The **bank card** is the most frequently used type of credit card. It is being sought because of its versatility. It can be used to purchase merchandise; most gasoline stations will accept it; and entertainment establishments often accept it as a form of payment. This type of card is often offered through organizations other than banks.

Travel or Entertainment Card

The **travel** or **entertainment card** is issued on a more selective basis to business people or professionals who are apt to use it more often. The charges on the card per month will exceed $100 on most occasions.

Objective 1

The obligation for payment begins when a purchase is made with a credit card. The finance charge or service charge agreement is usually printed on the card or on the credit card application. The charge for the use of credit with a credit card ranges from 1 to $1\frac{3}{4}$ percent per month.

monthly statement

The credit card customer receives a **monthly statement** of purchases from the card-issuing institution. On this statement the date of the purchase and amount are itemized. Charges for credit will also be shown at the bottom of the statement. The amount owed, or balance, must be paid within a specified time period, or a **minimum payment amount** will be specified on the statement. The minimum payment amount will be a percent of the balance or a flat amount such as $10.

minimum payment amount

MINIMUM AMOUNT DUE	PAST DUE AMOUNT	PAYMENT DUE DATE	NEW BALANCE	ACCOUNT NUMBER	PLEASE WRITE AMOUNT ENCLOSED
10.00	.00	10/15/01	206.90	0000–000000–0000	

☐ IF ADDRESS AS SHOWN IS INCORRECT PLEASE INDICATE CHANGE ON BACK

Brian D. Fitzpatrick
38 Bromfield Street
Boston, Massachusetts
02144

BANKCARD CENTER
P.O. BOX 2040
BOSTON, MA 02143-9917

PLEASE DETACH AND ENCLOSE THIS PORTION WITH REMITTANCE MAKE CHECK PAYABLE TO MASTERCARD

IMPORTANT: PLEASE REPORT ANY DISCREPANCIES ON THIS BILL TO THE BANK IMMEDIATELY.

DATE OF TRANS.	DATE OF POSTING	REFERENCE NUMBER	DESCRIPTION OF TRANSACTION		AMOUNT
08/31	09/02	784853OU7JK13GBA498KLHI642	Restaurant St. Michel	Coral Gables, FL	122.10
09/08	09/09	753604BL8BD34IOS444DMKP738	Dolphins Plus, Inc.	Key Largo, FL	84.80

ACCOUNT NUMBER	TO REPORT LOST OR STOLEN CREDIT CARDS OR INQUIRIES CALL:	SEND INQUIRIES TO:	
0000–000000–0000	800–345–8571	**BANKCARD CENTER** P.O. BOX 2040 BOSTON, MA 02143-9917	PAYMENTS MUST BE RECEIVED BY 2:00 P.M. OR THEY WILL BE CREDITED AS OF THE NEXT BUSINESS DAY. PAYMENTS RECEIVED AT ANY OTHER BANK LOCATION OR ADDRESS WILL BE CREDITED PROMPTLY BUT NO LATER THAN 5 DAYS FROM RECEIPT.

PREVIOUS BALANCE	CASH ADVANCES	PURCHASES	FINANCE CHARGE	ANNUAL FEE	INSURANCE PREMIUM	CREDITS	PAYMENTS	NEW BALANCE
.00	.00	206.90	.00	.00	.00	.00	.00	206.90

CREDIT LIMIT	MINIMUM PAYMENT	BILLING DATE	PAYMENT DUE DATE			AS A PERCENT OF THE NEW BALANCE IS:	BUT NOT LESS THAN:
2000	10.00	09/20/01	10/15/01	TO AVOID ADDITIONAL **FINANCE CHARGES**, PAYMENT OF NEW BALANCE MUST BE RECEIVED BY DUE DATE.	THE MINIMUM PAYMENT	%	$ 10.00

PERIODIC RATES	ON BALANCES UP TO:	THE MONTHLY PERIODIC RATE IS	WHICH IS AN ANNUAL PERCENTAGE RATE OF:	ON BALANCES OVER:	THE MONTHLY PERIODIC RATE IS:	WHICH IS AN ANNUAL PERCENTAGE RATE OF:	AVERAGE DAILY BALANCE ON WHICH FINANCE CHARGE IS COMPUTED
	$ ALL	1.650 %	19.80%	$	%	%	

Objective 2

The finance charge is usually based on the unpaid balance at the end of the month. An example of the effects of the minimum payment amount and finance charge are shown in example 1.

Example 1

Martin Stahl has a credit card. The past four monthly statements show the following information.

Month	Previous Balance	Purchases this Month	Payments	Finance Charges	New Balance
1	$42.10	$37.20	$20	______	______
2	______	85.00	20	______	______
3	______	. . .	20	______	______
4	______	7.50	50	______	______

The minimum payment amount is $20 and the finance charge is 1 percent. What is the amount of total finance charge for the four months on Martin's credit card and what is the new balance at the end of the fourth month?

Solution

The new balance for any month is found by (1) computing the finance charge on the previous balance, (2) adding the finance charge to the previous balance, (3) adding the amount of the purchases for the month, and (4) subtracting the payments.

Month

1 $\quad \$42.10 \times .01 = \$.42$
$\$42.10 + \$.42 = \$42.52$
$\$42.52 + \$37.20 = \$79.72$
$\$79.72 - \$20 = \$59.72$—New balance

2 $\quad \$59.72 \times .01 = \$.60$
$\$59.72 + \$.60 = \$60.32$
$\$60.32 + \$85 = \$145.32$
$\$145.32 - \$20 = \$125.32$—New balance

3 $\quad \$125.32 \times .01 = \$1.25$
$\$125.32 + \$1.25 = \$126.57$
$\$126.57 + 0 = \$126.57$
$\$126.57 - \$20 = \$106.57$—New balance

4 $\quad \$106.57 \times .01 = \$1.07$
$\$106.57 + \$1.07 = \$107.64$
$\$107.64 + \$7.50 = \$115.14$
$\$115.14 - \$50 = \$65.14$—New balance

The total finance charge is $3.34 ($.42 + $.60 + $1.25 + $1.07). The new balance at the end of the fourth month is $65.14.

Some card-issuing banks and institutions are adding an annual service charge if the card is unused for a year or so. Other firms add interest charges that begin at the date of the purchase. These variations would increase the cost of a credit purchase to the consumer. Other card-issuing banks and institutions offer incentives to use their card, such as a card with no annual fee to the cardholder during the first year.

Example 2

A record of Malinda's use of her credit card is shown below. The monthly purchases are for gasoline, clothes, and jewelry. There is a 1 percent finance charge on the unpaid monthly balance, a minimum payment amount of $10, and a flat service charge of $1.00 per month. What is the amount of the new balance at the end of the third month?

Month	Previous Balance	Purchases this Month	Payments	Flat Service Charge and Finance Charges	New Balance
1	$86.40	—	$10	$1.86 ($86.40 × .01 + $1)	$78.26
2	78.26	$17.58	40	1.78 ($78.26 × .01 + $1)	57.62
3	57.62	—	10	1.58 ($57.62 × .01 + $1)	49.20

The new balance at the end of the third month is $49.20.

Exercise A

Compute the new balance in these exercises.

Month	Previous Balance	Purchases this Month	Payments	Finance Charges (1%)	New Balance
1. 1	$48.20	—	$10	$.48 ($48.20 × .01)	$38.68
2	38.68	$35.00	10	.39 ($38.68 × .01)	64.07
3	64.07	22.60	10	.64 ($64.07 × .01)	77.31
2. 1	187.00	13.00	30	*1.87*	*171.87*
2	*171.87*	—	20	*1.72*	*153.59*
3	*153.59*	70.00	50	*1.54*	*175.13*
3. 1	36.50	10.00	30	*.37*	*16.87*
2	*16.87*	46.10	10	*.17*	*53.14*
3	*53.14*	51.80	50	*.53*	*55.47*
4	*55.47*	—	10	*.55*	*46.02*
4. 1	0.00	53.00	—	*0*	*53.00*
2	*53.00*	65.50	10	*.53*	*109.03*
3	*109.03*	—	10	*1.09*	*100.12*
5. 1	56.10	—	20	*.56*	*36.66*
2	*36.66*	186.50	15	*.37*	*208.53*
3	*208.53*	21.50	20	*2.09*	*212.12*
4	*212.12*	—	30	*2.12*	*184.24*

Annual Percentage Rate (APR)

amortizing

Credit buying requires the buyer, to whom credit is extended, to first pay the amount of interest due on the loan at the end of each month, as well as some amount on the principal. The amount of payment on the principal will reduce the amount of the loan to zero over a period of time. This payment process is termed **amortizing** the loan. This differs from a promissory note that requires only one payment, which includes the face value of the note as well as the interest, at the end of the term of the loan.

The Truth-in-Lending Act states that (1) the buyer must be made aware of all of the finance charges on a loan, and that (2) the annual rate of interest on the loan must be clearly stated. A monthly rate of interest of $1\frac{1}{2}$ percent is equal to 18 percent annually, 1 percent is equal to 12 percent annually, etc. This rate of interest is determined differently on an installment contract which involves a declining amount of principal borrowed over the life of the loan.

regulation Z

Regulation Z, the implementation phase of the Truth-in-Lending Act, states that total financing charges on credit purchases must be disclosed to the borrower. The amount of financing charges is found by subtracting the amount that is borrowed from the amount that is to be paid back. The difference is the amount of financing charges.

Objective 3

● *Use a page from a local newspaper that shows auto dealer ads or bank ads to demonstrate how APR is used as a basis of comparison.*

The **annual percentage rate (APR)** on an installment contract must be stated, according to Regulation Z. The Federal Reserve Board has prepared a table that is used to determine the APR. For example, an 8-month loan for $700 has interest charges of $28, an application fee of $6, and a filing fee of $8. The borrower will pay back

$$
\begin{aligned}
\$700 &= \text{amount of loan}\\
28 &= \text{interest}\\
6 &= \text{application fee}\\
\underline{+8} &= \text{filing fee}\\
\$742 &= \text{amount to be paid back}\\
\underline{-700} &= \text{amount borrowed}\\
\$42 &= \text{finance charges}
\end{aligned}
$$

$$\frac{\text{Finance charges}}{\text{Number of hundreds in loan}} = \text{Finance charge per \$100 of loan}$$

$$\frac{\$42}{7} = \$6 \text{ per hundred}$$

By referring to the table on page 222, the annual percentage rate can be determined. Using eight payments along the left-hand edge, look across until you find the $6 amount. The annual percentage rate is 15.75 percent.

Example 3

Max charged a purchase of $500 for 10 months on store credit from Sport-A-Merica. Kerry indicated that the interest charges would be $32.50. There would also be a service charge of $3 and a filing fee of $2.50. What is the annual percentage rate on the loan?

$$
\begin{aligned}
\$500.00 &= \text{Loan}\\
32.50 &= \text{Interest}\\
3.00 &= \text{Service charge}\\
\underline{+2.50} &= \text{Filing fee}\\
\$538.00 &= \text{Amount to be paid back}\\
\underline{-500.00} &= \text{Amount borrowed}\\
\$38.00 &= \text{Total finance charges}
\end{aligned}
$$

$$\frac{\text{Total finance charges}}{\text{Number of \$100 in loan}} = \text{Finance charge per \$100 of loan}$$

$$\frac{\$38}{5} = \$7.60 \text{ finance charge per \$100 of loan} = 16.25\% \text{ from the APR table listed}$$

To find the annual percentage rate, look down the number of payments column on the table until you find the number of months for the loan. Proceed across this line until you locate the amount that was charged on the loan for each $100. When you find the amount, look at the percent column heading. This will be the annual percentage rate. In this example, the rate is 16.25%. Kerry knows that this is much less than the rates that some credit card companies charge.

The annual percentage rate table can also be used to determine the amount of finance charge, if the rate is known. The amount per hundred ($7.60 in example 3) is found in the 16.25 percent column across from the ten payments level. By multiplying $7.60 times the number of hundreds (five), the finance charge of $38.00 is found.

Annual Percentage Rate

Number of Payments	14.00%	14.25%	14.50%	14.75%	15.00%	15.25%	15.50%	15.75%	16.00%	16.25%	16.50%	16.75%
1	1.17	1.19	1.21	1.23	1.25	1.27	1.29	1.31	1.33	1.35	1.38	1.40
2	1.75	1.78	1.82	1.85	1.88	1.91	1.94	1.97	2.00	2.04	2.07	2.10
3	2.34	2.38	2.43	2.47	2.51	2.55	2.59	2.64	2.68	2.72	2.76	2.80
4	2.93	2.99	3.04	3.09	3.14	3.20	3.25	3.30	3.36	3.41	3.46	3.51
5	3.53	3.59	3.65	3.72	3.78	3.84	3.91	3.97	4.04	4.10	4.16	4.23
6	4.12	4.20	4.27	4.35	4.42	4.49	4.57	4.64	4.72	4.79	4.87	4.94
7	4.72	4.81	4.89	4.98	5.06	5.15	5.23	5.32	5.40	5.49	5.58	5.66
8	5.32	5.42	5.51	5.61	5.71	5.80	5.90	6.00	6.09	6.19	6.29	6.38
9	5.92	6.03	6.14	6.25	6.35	6.46	6.57	6.68	6.78	6.89	7.00	7.11
10	6.53	6.65	6.77	6.88	7.00	7.12	7.24	7.36	7.48	7.60	7.72	7.84
11	7.14	7.27	7.40	7.53	7.66	7.79	7.92	8.05	8.18	8.31	8.44	8.57
12	7.74	7.89	8.03	8.17	8.31	8.45	8.59	8.74	8.88	9.02	9.16	9.30
13	8.36	8.51	8.66	8.81	8.97	9.12	9.27	9.43	9.58	9.73	9.89	10.04
14	8.97	9.13	9.30	9.46	9.63	9.79	9.96	10.12	10.29	10.45	10.62	10.78
15	9.59	9.76	9.94	10.11	10.29	10.47	10.64	10.82	11.00	11.17	11.35	11.53
16	10.20	10.39	10.58	10.77	10.95	11.14	11.33	11.52	11.71	11.90	12.09	12.28
17	10.82	11.02	11.22	11.42	11.62	11.82	12.02	12.22	12.42	12.62	12.83	13.03
18	11.45	11.66	11.87	12.08	12.29	12.50	12.72	12.93	13.14	13.35	13.57	13.78
19	12.07	12.30	12.52	12.74	12.97	13.19	13.41	13.64	13.86	14.09	14.31	14.54
20	12.70	12.93	13.17	13.41	13.64	13.88	14.11	14.35	14.59	14.82	15.06	15.30
21	13.33	13.58	13.82	14.07	14.32	14.57	14.82	15.06	15.31	15.56	15.81	16.06
22	13.96	14.22	14.48	14.74	15.00	15.26	15.52	15.78	16.04	16.30	16.57	16.83
23	14.59	14.87	15.14	15.41	15.68	15.96	16.23	16.50	16.78	17.05	17.32	17.60
24	15.23	15.51	15.80	16.08	16.37	16.65	16.94	17.22	17.51	17.80	18.09	18.37
25	15.87	16.17	16.46	16.76	17.06	17.35	17.65	17.95	18.25	18.55	18.85	19.15
26	16.51	16.82	17.13	17.44	17.75	18.06	18.37	18.68	18.99	19.30	19.62	19.93
27	17.15	17.47	17.80	18.12	18.44	18.76	19.09	19.41	19.74	20.06	20.39	20.71
28	17.80	18.13	18.47	18.80	19.14	19.47	19.81	20.15	20.48	20.82	21.16	21.50
29	18.45	18.79	19.14	19.49	19.83	20.18	20.53	20.88	21.23	21.58	21.94	22.29
30	19.10	19.45	19.81	20.17	20.54	20.90	21.26	21.62	21.99	22.35	22.72	23.08
31	19.75	20.12	20.49	20.87	21.24	21.61	21.99	22.37	22.74	23.12	23.50	23.88
32	20.40	20.79	21.17	21.56	21.95	22.33	22.72	23.11	23.50	23.89	24.28	24.68
33	21.06	21.46	21.85	22.25	22.65	23.06	23.46	23.86	24.26	24.67	25.07	25.48
34	21.72	22.13	22.54	22.95	23.37	23.78	24.19	24.61	25.03	25.44	25.86	26.28
35	22.38	22.80	23.23	23.65	24.08	24.51	24.94	25.36	25.79	26.23	26.66	27.09
36	23.04	23.48	23.92	24.35	24.80	25.24	25.68	26.12	26.57	27.01	27.46	27.90
48	31.17	31.77	32.37	32.98	33.59	34.20	34.81	35.42	36.03	36.65	37.27	37.88
60	39.61	40.39	41.17	41.95	42.74	43.53	44.32	45.11	45.91	46.71	47.51	48.31

Annual Percentage Rate

Number of Payments	17.00%	17.25%	17.50%	17.75%	18.00%	18.25%	18.50%	18.75%	19.00%	19.25%	19.50%	19.75%
1	1.42	1.44	1.46	1.48	1.50	1.52	1.54	1.56	1.58	1.60	1.63	1.65
2	2.13	2.16	2.19	2.22	2.26	2.29	2.32	2.35	2.38	2.41	2.44	2.48
3	2.85	2.89	2.93	2.97	3.01	3.06	3.10	3.14	3.18	3.23	3.27	3.31
4	3.57	3.62	3.67	3.73	3.78	3.83	3.88	3.94	3.99	4.04	4.10	4.15
5	4.29	4.35	4.42	4.48	4.54	4.61	4.67	4.74	4.80	4.86	4.93	4.99
6	5.02	5.09	5.17	5.24	5.32	5.39	5.46	5.54	5.61	5.69	5.76	5.84
7	5.75	5.83	5.92	6.00	6.09	6.18	6.26	6.35	6.43	6.52	6.60	6.69
8	6.48	6.58	6.67	6.77	6.87	6.96	7.06	7.16	7.26	7.35	7.45	7.55
9	7.22	7.32	7.43	7.54	7.65	7.76	7.87	7.97	8.08	8.19	8.30	8.41
10	7.96	8.08	8.19	8.31	8.43	8.55	8.67	8.79	8.91	9.03	9.15	9.27
11	8.70	8.83	8.96	9.09	9.22	9.35	9.49	9.62	9.75	9.88	10.01	10.14
12	9.45	9.59	9.73	9.87	10.02	10.16	10.30	10.44	10.59	10.73	10.87	11.02
13	10.20	10.35	10.50	10.66	10.81	10.97	11.12	11.28	11.43	11.59	11.74	11.90
14	10.95	11.11	11.28	11.45	11.61	11.78	11.95	12.11	12.28	12.45	12.61	12.78
15	11.71	11.88	12.06	12.24	12.42	12.59	12.77	12.95	13.13	13.31	13.49	13.67
16	12.46	12.65	12.84	13.03	13.22	13.41	13.60	13.80	13.99	14.18	14.37	14.56
17	13.23	13.43	13.63	13.83	14.04	14.24	14.44	14.64	14.85	15.05	15.25	15.46
18	13.99	14.21	14.42	14.64	14.85	15.07	15.28	15.49	15.71	15.93	16.14	16.36
19	14.76	14.99	15.22	15.44	15.67	15.90	16.12	16.35	16.58	16.81	17.03	17.26
20	15.54	15.78	16.01	16.25	16.49	16.73	16.97	17.21	17.45	17.69	17.93	18.17
21	16.31	16.56	16.81	17.07	17.32	17.57	17.82	18.07	18.33	18.58	18.83	19.09
22	17.09	17.36	17.62	17.88	18.15	18.41	18.68	18.94	19.21	19.47	19.74	20.01
23	17.88	18.15	18.43	18.70	18.98	19.26	19.54	19.81	20.09	20.37	20.65	20.93
24	18.66	18.95	19.24	19.53	19.82	20.11	20.40	20.69	20.98	21.27	21.56	21.86
25	19.45	19.75	20.05	20.36	20.66	20.96	21.27	21.57	21.87	22.18	22.48	22.79
26	20.24	20.56	20.87	21.19	21.50	21.82	22.14	22.45	22.77	23.09	23.41	23.73
27	21.04	21.37	21.69	22.02	22.35	22.68	23.01	23.34	23.67	24.00	24.33	24.67
28	21.84	22.18	22.52	22.86	23.20	23.55	23.89	24.23	24.58	24.92	25.27	25.61
29	22.64	22.99	23.35	23.70	24.06	24.41	24.77	25.13	25.49	25.84	26.20	26.56
30	23.45	23.81	24.18	24.55	24.92	25.29	25.66	26.03	26.40	26.77	27.14	27.52
31	24.26	24.64	25.02	25.40	25.78	26.16	26.55	26.93	27.32	27.70	28.09	28.47
32	25.07	25.46	25.86	26.25	26.65	27.04	27.44	27.84	28.24	28.64	29.04	29.44
33	25.88	26.29	26.70	27.11	27.52	27.93	28.34	28.75	29.16	29.57	29.99	30.40
34	26.70	27.12	27.54	27.97	28.39	28.81	29.24	29.66	30.09	30.52	30.95	31.37
35	27.52	27.96	28.39	28.83	29.27	29.71	30.14	30.58	31.02	31.47	31.91	32.35
36	28.35	28.80	29.25	29.70	30.15	30.60	31.05	31.51	31.96	32.42	32.87	33.33
48	38.50	39.13	39.75	40.37	41.00	41.63	42.26	42.89	43.52	44.15	44.79	45.43
60	49.12	49.92	50.73	51.55	52.36	53.18	54.00	54.82	55.64	56.47	57.30	58.13

Annual Percentage Rate

Number of Payments	20.00%	20.25%	20.50%	20.75%	21.00%	21.25%	21.50%	21.75%	22.00%	22.25%	22.50%	22.75%
1	1.67	1.69	1.71	1.73	1.75	1.77	1.79	1.81	1.83	1.85	1.88	1.90
2	2.51	2.54	2.57	2.60	2.63	2.66	2.70	2.73	2.76	2.79	2.82	2.85
3	3.35	3.39	3.44	3.48	3.52	3.56	3.60	3.65	3.69	3.73	3.77	3.82
4	4.20	4.25	4.31	4.36	4.41	4.47	4.52	4.57	4.62	4.68	4.73	4.78
5	5.06	5.12	5.18	5.25	5.31	5.37	5.44	5.50	5.57	5.63	5.69	5.76
6	5.91	5.99	6.06	6.14	6.21	6.29	6.36	6.44	6.51	6.59	6.66	6.74
7	6.78	6.86	6.95	7.04	7.12	7.21	7.29	7.38	7.47	7.55	7.64	7.73
8	7.64	7.74	7.84	7.94	8.03	8.13	8.23	8.33	8.42	8.52	[illegible]	8.72
9	8.52	8.63	8.73	8.84	8.95	9.06	9.17	9.28	9.39	9.50	9.61	9.72
10	9.39	9.51	9.63	9.75	9.88	10.00	10.12	10.24	10.36	10.48	10.60	10.72
11	10.28	10.41	10.54	10.67	10.80	10.94	11.07	11.20	11.33	11.47	11.60	11.73
12	11.16	11.31	11.45	11.59	11.74	11.88	12.02	12.17	12.31	12.46	12.60	12.75
13	12.05	12.21	12.36	12.52	12.67	12.83	12.99	13.14	13.30	13.46	13.61	13.77
14	12.95	13.11	13.28	13.45	13.62	13.79	13.95	14.12	14.29	14.46	14.63	14.80
15	13.85	14.03	14.21	14.39	14.57	14.75	14.93	15.11	15.29	15.47	15.65	15.83
16	14.75	14.94	15.14	15.33	15.52	15.71	15.90	16.10	16.29	16.48	16.68	16.87
17	15.66	15.86	16.07	16.27	16.48	16.68	16.89	17.09	17.30	17.50	17.71	17.92
18	16.57	16.79	17.01	17.22	17.44	17.66	17.88	18.09	18.31	18.53	18.75	18.97
19	17.49	17.72	17.95	18.18	18.41	18.64	18.87	19.10	19.33	19.56	19.79	20.02
20	18.41	18.66	18.90	19.14	19.38	19.63	19.87	20.11	20.36	20.60	20.84	21.09
21	19.34	19.60	19.85	20.11	20.36	20.62	20.87	21.13	21.38	21.64	21.90	22.16
22	20.27	20.54	20.81	21.08	21.34	21.61	21.88	22.15	22.42	22.69	22.96	23.23
23	21.21	21.49	21.77	22.05	22.33	22.61	22.90	23.18	23.46	23.74	24.03	24.31
24	22.15	22.44	22.74	23.03	23.33	23.62	23.92	24.21	24.51	24.80	25.10	25.40
25	23.10	23.40	23.71	24.02	24.32	24.63	24.94	25.25	25.56	25.87	26.18	26.49
26	24.04	24.36	24.68	25.01	25.33	25.65	25.97	26.29	26.62	26.94	27.26	27.59
27	25.00	25.33	25.67	26.00	26.34	26.67	27.01	27.34	27.68	28.02	28.35	28.69
28	25.96	26.30	26.65	27.00	27.35	27.70	28.05	28.40	28.75	29.10	29.45	29.80
29	26.92	27.28	27.64	28.00	28.37	28.73	29.09	29.46	29.82	30.19	30.55	30.92
30	27.89	28.26	28.64	29.01	29.39	29.77	30.14	30.52	30.90	31.28	31.66	32.04
31	28.86	29.25	29.64	30.03	30.42	30.81	31.20	31.59	31.98	32.38	32.77	33.17
32	29.84	30.24	30.64	31.05	31.45	31.85	32.26	32.67	33.07	33.48	33.89	34.30
33	30.82	31.23	31.65	32.07	32.49	32.91	33.33	33.75	34.17	34.59	35.01	35.44
34	31.80	32.23	32.67	33.10	33.53	33.96	34.40	34.83	35.27	35.71	36.14	36.58
35	32.79	33.24	33.68	34.13	34.58	35.03	35.47	35.92	36.37	36.83	37.28	37.73
36	33.79	34.25	34.71	35.17	35.63	36.09	36.56	37.02	37.49	37.95	38.42	38.89
48	46.07	46.71	47.35	47.99	48.64	49.28	49.93	50.58	51.23	51.88	52.54	53.19
60	58.96	59.80	60.64	61.48	62.32	63.17	64.01	64.86	65.71	66.57	67.42	68.28

Annual Percentage Rate

Number of Payments	23.00%	23.25%	23.50%	23.75%	24.00%	24.25%	24.50%	24.75%	25.00%	25.25%	25.50%	25.75%
1	1.92	1.94	1.96	1.98	2.00	2.02	2.04	2.06	2.08	2.10	2.12	2.15
2	2.88	2.92	2.95	2.98	3.01	3.04	3.07	3.10	3.14	3.17	3.20	3.23
3	3.86	3.90	3.94	3.98	4.03	4.07	4.11	4.15	4.20	4.24	4.28	4.32
4	4.84	4.89	4.94	5.00	5.05	5.10	5.16	5.21	5.26	5.32	5.37	5.42
5	5.82	5.89	5.95	6.02	6.08	6.14	6.21	6.27	6.34	6.40	6.46	6.53
6	6.81	6.89	6.96	7.04	7.12	7.19	7.27	7.34	7.42	7.49	7.57	7.64
7	7.81	7.90	7.99	8.07	8.16	8.25	8.33	8.42	8.51	8.59	8.68	8.77
8	8.82	8.91	9.01	9.11	9.21	9.31	9.40	9.50	9.60	9.70	9.80	9.90
9	9.83	9.94	10.04	10.15	10.26	10.37	10.48	10.59	10.70	10.81	10.92	11.03
10	10.84	10.96	11.08	11.21	11.33	11.45	11.57	11.69	11.81	11.93	12.06	12.18
11	11.86	12.00	12.13	12.26	12.40	12.53	12.66	12.80	12.93	13.06	13.20	13.33
12	12.89	13.04	13.18	13.33	13.47	13.62	13.76	13.91	14.05	14.20	14.34	14.49
13	13.93	14.08	14.24	14.40	14.55	14.71	14.87	15.03	15.18	15.34	15.50	15.66
14	14.97	15.13	15.30	15.47	15.64	15.81	15.98	16.15	16.32	16.49	16.66	16.83
15	16.01	16.19	16.37	16.56	16.74	16.92	17.10	17.28	17.47	17.65	17.83	18.02
16	17.06	17.26	17.45	17.65	17.84	18.03	18.23	18.42	18.62	18.81	19.01	19.21
17	18.12	18.33	18.53	18.74	18.95	19.16	19.36	19.57	19.78	19.99	20.20	20.40
18	19.19	19.41	19.62	19.84	20.06	20.28	20.50	20.72	20.95	21.17	21.39	21.61
19	20.26	20.49	20.72	20.95	21.19	21.42	21.65	21.89	22.12	22.35	22.59	22.82
20	21.33	21.58	21.82	22.07	22.31	22.56	22.81	23.05	23.30	23.55	23.79	24.04
21	22.41	22.67	22.93	23.19	23.45	23.71	23.97	24.23	24.49	24.75	25.01	25.27
22	23.50	23.77	24.04	24.32	24.59	24.86	25.13	25.41	25.68	25.96	26.23	26.50
23	24.60	24.88	25.17	25.45	25.74	26.02	26.31	26.60	26.88	27.17	27.46	27.75
24	25.70	25.99	26.29	26.59	26.89	27.19	27.49	27.79	28.09	28.39	28.69	29.00
25	26.80	27.11	27.43	27.74	28.05	28.36	28.68	28.99	29.31	29.62	29.94	30.25
26	27.91	28.24	28.56	28.89	29.22	29.55	29.87	30.20	30.53	30.86	31.19	31.52
27	29.03	29.37	29.71	30.05	30.39	30.73	31.07	31.42	31.76	32.10	32.45	32.79
28	30.15	30.51	30.86	31.22	31.57	31.93	32.28	32.64	33.00	33.35	33.71	34.07
29	31.28	31.65	32.02	32.39	32.76	33.13	33.50	33.87	34.24	34.61	34.98	35.36
30	32.42	32.80	33.18	33.57	33.95	34.33	34.72	35.10	35.49	35.88	36.26	36.65
31	33.56	33.96	34.35	34.75	35.15	35.55	35.95	36.35	36.75	37.15	37.55	37.95
32	34.71	35.12	35.53	35.94	36.35	36.77	37.18	37.60	38.01	38.43	38.84	39.26
33	35,86	36.29	36.71	37.14	37.57	37.99	38.42	38.85	39.28	39.71	40.14	40.58
34	37.02	37.46	37.90	38.34	38.78	39.23	39.67	40.11	40.56	41.01	41.45	41.90
35	38.18	38.64	39.09	39.55	40.01	40.47	40.92	41.38	41.84	42.31	42.77	43.23
36	39.35	39.82	40.29	40.77	41.24	41.71	42.19	42.66	43.14	43.61	44.09	44.57
48	53.85	54.51	55.16	55.83	56.49	57.15	57.82	58.49	59.15	59.82	60.50	61.17
60	69.14	70.01	70.87	71.74	72.61	73.48	74.35	75.23	76.11	76.99	77.87	78.76

Annual Percentage Rate

Number of Payments	26.00%	26.25%	26.50%	26.75%	27.00%	27.25%	27.50%	27.75%	28.00%	28.25%	28.50%	28.75%
1	2.17	2.19	2.21	2.23	2.25	2.27	2.29	2.31	2.33	2.35	2.37	2.40
2	3.26	3.29	3.32	3.36	3.39	3.42	3.45	3.48	3.51	3.54	3.58	3.61
3	4.36	4.41	4.45	4.49	4.53	4.58	4.62	4.66	4.70	4.74	4.79	4.83
4	5.47	5.53	5.58	5.63	5.69	5.74	5.79	5.85	5.90	5.95	6.01	6.06
5	6.59	6.66	6.72	6.79	6.85	6.91	6.98	7.04	7.11	7.17	7.24	7.30
6	7.72	7.79	7.87	7.95	8.02	8.10	8.17	8.25	8.32	8.40	8.48	8.55
7	8.85	8.94	9.03	9.11	9.20	9.29	9.37	9.46	9.55	9.64	9.72	9.81
8	9.99	10.09	10.19	10.29	10.39	10.49	10.58	10.68	10.78	10.88	10.98	11.08
9	11.14	11.25	11.36	11.47	11.58	11.69	11.80	11.91	12.03	12.14	12.25	12.36
10	12.30	12.42	12.54	12.67	12.79	12.91	13.03	13.15	13.28	13.40	13.52	13.64
11	13.46	13.60	13.73	13.87	14.00	14.13	14.27	14.40	14.54	14.67	14.81	14.94
12	14.64	14.78	14.93	15.07	15.22	15.37	15.51	15.66	15.81	15.95	16.10	16.25
13	15.82	15.97	16.13	16.29	16.45	16.61	16.77	16.93	17.09	17.24	17.40	17.56
14	17.00	17.17	17.35	17.52	17.69	17.86	18.03	18.20	18.37	18.54	18.72	18.89
15	18.20	18.38	18.57	18.75	18.93	19.12	19.30	19.48	19.67	19.85	20.04	20.22
16	19.40	19.60	19.79	19.99	20.19	20.38	20.58	20.78	20.97	21.17	21.37	21.57
17	20.61	20.82	21.03	21.24	21.45	21.66	21.87	22.08	22.29	22.50	22.71	22.92
18	21.83	22.05	22.27	22.50	22.72	22.94	23.16	23.39	23.61	23.83	24.06	24.28
19	23.06	23.29	23.53	23.76	24.00	24.23	24.47	24.71	24.94	25.18	25.42	25.65
20	24.29	24.54	24.79	25.04	25.28	25.53	25.78	26.03	26.28	26.53	26.78	27.04
21	25.53	25.79	26.05	26.32	26.58	26.84	27.11	27.37	27.63	27.90	28.16	28.43
22	26.78	27.05	27.33	27.61	27.88	28.16	28.44	28.71	28.99	29.27	29.55	29.82
23	28.04	28.32	28.61	28.90	29.19	29.48	29.77	30.07	30.36	30.65	30.94	31.23
24	29.30	29.60	29.90	30.21	30.51	30.82	31.12	31.43	31.73	32.04	32.34	32.65
25	30.57	30.89	31.20	31.52	31.84	32.16	32.48	32.80	33.12	33.44	33.76	34.08
26	31.85	32.18	32.51	32.84	33.18	33.51	33.84	34.18	34.51	34.84	35.18	35.51
27	33.14	33.48	33.83	34.17	34.52	34.87	35.21	35.56	35.91	36.26	36.61	36.96
28	34.43	34.79	35.15	35.51	35.87	36.23	36.59	36.96	37.32	37.68	38.05	38.41
29	35.73	36.10	36.48	36.85	37.23	37.61	37.98	38.36	38.74	39.12	39.50	39.88
30	37.04	37.43	37.82	38.21	38.60	38.99	39.38	39.77	40.17	40.56	40.95	41.35
31	38.36	38.76	39.16	39.57	39.97	40.38	40.79	41.19	41.60	42.01	42.42	42.83
32	39.68	40.10	40.52	40.94	41.36	41.78	42.20	42.62	43.05	43.47	43.90	44.32
33	41.01	41.44	41.88	42.31	42.75	43.19	43.62	44.06	44.50	44.94	45.38	45.82
34	42.35	42.80	43.25	43.70	44.15	44.60	45.05	45.51	45.96	46.42	46.87	47.33
35	43.69	44.16	44.62	45.09	45.56	46.02	46.49	46.96	47.43	47.90	48.37	48.85
36	45.05	45.53	46.01	46.49	46.97	47.45	47.94	48.42	48.91	49.40	49.88	50.37
48	61.84	62.52	63.20	63.87	64.56	65.24	65.92	66.60	67.29	67.98	68.67	69.36
60	79.64	80.53	81.42	82.32	83.21	84.11	85.01	85.91	86.81	87.72	88.63	89.54

Exercise B

Determine the annual percentage rate or the interest from the table in the following problems.

Amount Borrowed	Interest	Other Charges		Term of Loan (in Months)	Annual Percentage Rate
1. $600	$22.42	$5.00	(22.42 + 5 = ÷ 6 =)	6	15.5%
2. 1,000	232.00	25.00	*(232 + 25 = ÷ 10 =)*	24	*23*
3. 450	*35.87*	0		9	18.75
4. 800	27.00	6.60	*(27 + 6.6 = ÷ 8 =)*	6	*14.25*
5. 2,500	859.75	31.00	*(859.75 + 31 = ÷ 25 =)*	36	*21*

Credit Card Deposits

Objective 4

When a store accepts a credit card form of payment from a customer, the sales clerk follows a standard procedure: The customer's credit card is run through an electronic reader or an imprinting machine (which has a plate that indicates the name of the store) with a blank standard charge form; the clerk completes the form by filling in the correct amounts; the clerk usually calls the credit card company to verify the card and the chargeable amount; the customer then signs the charge form, the clerk gives a copy to the customer and keeps two copies.

When the store is ready to make a deposit at the bank, they fill out a deposit slip which is presented at the bank along with one copy of each charge form. The bank computes the amount of discount charged by the credit card company, subtracts this from the deposit amount, and credits the store's account with the difference. The store automatically receives payment for the difference to their account. This may also be accomplished electronically. The credit card company has the responsibility for billing and collection.

Some credit card companies require the store to send the charge forms to them. They determine the discount and return the difference to the store. This may take longer, but the results are the same.

Though nearly all stores are charged some discount by the credit card company, the rate of discount is negotiable. Some credit card companies charge smaller stores as much as 5 percent of the credit sale.

Example 4

Kerry made out a deposit slip for Tuesday's credit sales. She took the deposit slip to the bank with copies of the charge forms and asked the bank teller to credit the account of Sport-A-Merica with the amount remaining after the discount charge was subtracted. The credit forms totaled \$2,180. Sport-A-Merica pays a discount charge of 4 percent. What is the amount of the deposit to the account after the discount charge?

\$2,180	\$2,180.00
×.04	−87.20
\$87.20 Discount charge	\$2,092.80 Deposit

Exercise C

Determine the discount charge amount and the deposit amount for the following stores.

Store	Total of Charge Forms	Discount Charge Rate	Discount Charge Amount	Deposit Amount
1. Pick-up	\$480.00	5% (480 × .05 = 480 − 24 =)	\$24.00	\$456.00
2. Sarah's T's	1,900.00	$4\frac{1}{2}$ *(1,900 × .045 =)*	*85.50*	*1,814.50*
3. Pringle & Tamold	600.65	2 *(600.65 × .02 =)*	*12.01*	*588.64*
4. The Fast Hut	1,200.00	3 *(1,200 × .03 =)*	*36.00*	*1,164.00*
5. Ann's Shop for Men	89.20	$3\frac{3}{4}$ *(89.20 × .0375 =)*	*3.35*	*85.85*

ASSIGNMENT Chapter 12

Name ______________________ Date ______________ Score ______________

Skill Problems

1. When does the obligation for payment begin when one owns a credit card? *c*
   (a) The time of the application.
   (b) The time of credit approval.
   (c) The time of a purchase.
   (d) The receipt of the monthly statement.
2. When a store accepts a credit card form of payment, what will the sales clerk do? *d*
   (a) Fill out form by hand or cash register.
   (b) Verify the card and the amount.
   (c) Run card through imprinting machine or slide magnetic strip through machine.
   (d) All of the above.
3. When the store is ready to make a deposit of credit sales at the bank, what amount does the store put in its check register? *b*
   (a) The store records the total credit sales and a sales commission from credit card companies.
   (b) The store records the total credit sales less a fee charged by the credit card company.
   (c) The store records just the total credit sales, no fees are ever charged.
   (d) None of the above.

Select the best term for the description below.

A. Credit Card B. Annual Percentage Rate C. Monthly Statement
D. Regulation Z E. Amortizing F. Minimum Payment Amount

4. ___*A*___ A purchase device issued to those with a good credit history.
5. ___*F*___ The smallest amount that can be paid on a credit card balance to maintain a good credit rating.
6. ___*D*___ This specifies that interest rates must be calculated using a set method.
7. ___*C*___ A record of the account balance from the viewpoint of the issuer.
8. ___*E*___ Paying off a loan by making monthly payments.

Compute the last balance in the following problems.

Month	Previous Balance	Purchases this Month	Payments	Finance Charges (1%)	Balance	
1	$36.50	$21.50	$10	9. *$.37*	10. *$48.37*	*(36.50 × .01 =)*
2	*48.37*	—	10	11. *.48*	12. *38.85*	*(48.37 × .01 =)*
3	*38.85*	12.00	20	13. *.39*	14. *31.24*	*(38.85 × .01 =)*
1	$24	$24.95	$40	15. *$.24*	16. *$ 9.19*	*(24 × .01 =)*
2	*9.19*	65.62	20	17. *.09*	18. *54.90*	*(9.19 × .01 =)*
3	*54.90*	—	40	19. *.55*	20. *15.45*	*(54.90 × .01 =)*
1	$1,467	$82.00	$150	21. *$14.67*	22. *$1,413.67*	*(1,467 × .01 =)*
2	*$1,413.67*	$135.00	150	23. *14.14*	24. *1,412.81*	*(1,413.67 × .01 =)*
3	*1,412.81*	50.83	300	25. *14.13*	26. *1,177.77*	*(1,412.81 × .01 =)*

Complete the following chart for the finance charge and new balance. There is a 1.8 percent finance charge on unpaid monthly balance and a minimum payment amount of $15.00.

Month	Previous Balance	Purchases this Month	Payments	Finance Charges	New Balance	
Jan.	$ 0.00	$ 79.95	$ 0.00	**27.** *$ 0.00*	**28.** *$ 79.95*	*(0 × .018 =)* *(0 + 79.95 + 0 =)*
Feb.	*79.95*	50.50	15.00	**29.** *1.44*	**30.** *116.89*	*(79.95 × .018 =)* *(79.95 + 50.50 − 15 + 1.44 =)*
Mar.	*116.89*	0.00	30.00	**31.** *2.10*	**32.** *88.99*	*(116.89 × .018 =)* *(116.89 − 30 + 2.10 =)*
Apr.	*88.99*	46.15	15.00	**33.** *1.60*	**34.** *121.74*	*(88.99 × .018 =)* *(88.99 + 46.15 − 15 + 1.60 =)*
May	*121.74*	495.21	60.00	**35.** *2.19*	**36.** *559.14*	*(121.74 × .018 =)* *(121.74 + 495.21 − 60 + 2.19 =)*
June	*559.14*	0.00	300.00	**37.** *10.06*	**38.** *269.20*	*(559.14 × .018 =)* *(559.14 − 300 + 10.06 =)*
July	*269.20*	25.31	60.00	**39.** *4.85*	**40.** *239.36*	*(269.20 × .018 =)* *(269.20 + 25.31 − 60 + 4.85 =)*
Aug.	*239.36*	0.00	200.00	**41.** *4.31*	**42.** *43.67*	*(239.36 × .018 =)* *(239.36 − 200 + 4.31 =)*
Sep.	*43.67*	15.59	15.00	**43.** *.79*	**44.** *45.05*	*(43.67 × .018 =)* *(43.67 + 15.59 − 15 + .79 =)*

Determine the annual percentage rate or the interest from the table in the following problems.

45.
$1,000 = Loan
62.00 = Interest
25.00 = Filing fee
11 = Number of payments
1,087 = Amount to be paid back
87 = Total finance charge
8.70 = Finance charge per $100 of loan
17% = APR
(1,000 + 62.00 + 25 =) ((62.00 + 25 =)/10)

46.
$500.00 = Loan
21.25 = Interest
20.00 = Filing fee
7 = Number of payments
541.25 = Amount to be paid back
41.25 = Total finance charge
8.25 = Finance charge per $100 of loan
24.25% = APR
(500 + 21.25 + 20 =) ((21.25 + 20 =)/5)

47.
$450 = Loan
13.57 = Interest
2.50 = Filing fee
4 = Number of payments
466.07 = Amount to be paid back
16.07 = Total finance charge
3.57 = Finance charge per $100 of loan
17% = APR
(450 + 13.57 + 2.5 =) ((13.57 + 2.5 =)/4.5)

48.
$10,000 = Loan
4,199 = Interest
75.00 = Filing fee
60 = Number of payments
14,274.00 = Amount to be paid back
4,274.00 = Total finance charge
42.74 = Finance charge per $100 of loan
15% = APR
(10,000 + 4,199 + 75 =) ((4,199 + 75 =)/100)

49. $500 = Loan
38.90 = Interest
10.50 = Filing fee
10 = Number of payments
549.40 = Amount to be paid back
49.40 = Total finance charge
9.88 = Finance charge per $100 of loan
21% = APR
(500 + 38.90 + 10.5 =) ((38.9 + 10.5 =)/5)

50. $100.00 = Loan
25.84 = Interest
0.00 = Filing fee
29 = Number of payments
125.84 = Amount to be paid back
25.84 = Total finance charge
25.84 = Finance charge per $100 of loan
19.25% = APR
(100 + 25.84 =) (25.84/1)

51. $105,200 = Loan
42,365.28 = Interest
125.00 = Filing fee
60 = Number of payments
147,690.28 = Amount to be paid back
42,490.28 = Total finance charge
40.39 = Finance charge per $100 of loan
14.25% = APR
(105,200 + 42,365.28 + 125 =)
((42,365.28 + 125 =)/1052)

52. $7,840 = Loan
1,877.07 = Interest
50.00 = Filing fee
28 = Number of payments
9,767.07 = Amount to be paid back
1,927.07 = Total finance charge
24.58 = Finance charge per $100 of loan
19% = APR
(7,840 + 1,877.07 + 50 =)
((1,877.07 + 50 =)/78.4)

53. Amount borrowed = $1,500
Interest charges = $99.40
Service charges = $20
Term of loan = 10 months
Annual percentage rate = *17* %
(99.40 + 20 ÷ 15 =)

54. Amount borrowed = $600
Interest charges = $44.86
Filing fee = $5
Term of loan = 12 months
Annual percentage rate = *15* %
(44.86 + 5 ÷ 6 =)

55. Amount borrowed = $4,000
Interest charges = $2,104
Service charges = $40
Filing fee = $10
Term of loan = 48 months
Annual percentage rate = *23* %
(2,104 + 40 + 10 ÷ 40 =)

56. Amount borrowed = $1,200
Interest charges = $44.92
Other fees = $26
Term of loan = 6 months
Annual percentage rate = *20* %
(44.92 + 26 ÷ 12 =)

Determine the discount charge amount and the deposit amount for the following problems.

57. $1,280 = Charge sales
3% = Discount charge rate
38.40 = Discount charge amount
1,241.60 = Deposit amount
(1,280 × .03 =) (1,280 − 38.40 =)

58. $2,040.49 = Charge sales
3 1/2% = Discount charge rate
71.42 = Discount charge amount
1,969.07 = Deposit amount
(2,040.49 × .035 =) (2,040.49 − 71.42 =)

59. $87.23 = Charge sales
5% = Discount charge rate
4.36 = Discount charge amount
82.87 = Deposit amount
(87.23 × .05 =) *(87.23 − 4.36 =)*

60. $842.11 = Charge sales
2.6% = Discount charge rate
21.89 = Discount charge amount
820.22 = Deposit amount
(842.11 × .026 =) *(842.11 − 21.89 =)*

61. $84,020 = Charge sales
3% = Discount charge rate
2,520.60 = Discount charge amount
81,499.40 = Deposit amount
(84,020 × .03 =) *(84,020 − 2,520.60 =)*

62. $674.61 = Charge sales
4% = Discount charge rate
26.98 = Discount charge amount
647.63 = Deposit amount
(674.61 × .04 =) *(674.61 − 26.98 =)*

Determine the deposit amount on credit card sales in the following problems.

Store	Total of Charge Forms	Discount Charge Rate	Deposit Amount	
63. Confectionaires	$1,853	3%	*$1,797.41*	*(1,853 × .03 =)*
64. Kelly's Pub	2,940	$2\frac{1}{2}$	*2,866.50*	*(2,940 × .025 =)*
65. Benzie's	1,560	$1\frac{3}{4}$	*1,532.70*	*(1,560 × .0175 =)*

Business Application Problems

1. Ed and Lou heard an ad on the radio indicating a $1,995 living room suite could be bought for just $59 down plus 12 monthly payments of $178.94. What is the difference in cost between buying on credit and paying cash?
$178.94 × 12 + $59 = $2,206.28 − $1,995 = $211.28 cost

2. In problem 1, what is the APR on the loan?
$211.28/19.36 ($1,995 − $59 = /$100) = $10.91 = 19 1/2% rate

3. Bilison Supply Company sold an office desk and chair for a total cash price of $570. The buyer made a down payment of $40 and three monthly payments of $183.49. What is the amount of interest and finance charges earned by Bilison Supply Company on the installment sale? ***$570 − $40 = $530***
$183.49 × 3 = $550.47 − $530 = $20.47 earned

4. What is the APR in problem 3?
$20.47/5.3 = $3.86 23% rate

5. A credit card balance of $400 was paid off in payments of $150 each month. The credit card company charged 2 percent per month. What is the amount of the finance charges on the loan?
$400 × .02 = $8 $408 − $150 = $258 × .02 = $5.16
$263.16 − $150 = $113.16 × .02 = $2.26 + $5.16 + $8 = $15.42 charges

6. What is the amount of the last monthly payment in problem 5?
$113.16 + $2.26 = $115.42 payment

7. Benny deposited the daily receipts from his restaurant at the bank. The total was $1,689.40. Cash and currency were $1,434.52 of the total. The bank charges 2 percent on credit card deposits. What is the value of the deposit to Benny?
$1,689.40 − $1,434.52 = $254.88 × .02 = $5.10
$1,689.40 − $5.10 = $1,684.30 deposit

8. Bob Marsh paid service plus finance charges of $2,703.20 on an $8,000 monthly installment loan over a three-year term. What is the annual percentage rate on the loan?
$2,703.20/80 = $33.79 = 20% APR

9. What is the amount of interest on the loan in the first month in problem 8?
$8,000 × .20 × 1/12 = $133.33 interest

10. Waterslide Resort, Inc. receipts are an average of $642 a day for the 98 days a year that the firm operates. One fourth of the receipts are credit card charges. The credit card company charges 3.5 percent for deposits at the bank. What is the cost of credit to the Waterslide Resort, Inc. over the length of the season?
$642 × 98 = $62,916 × .25 = $15,729 × .035 = $550.52 cost

11. Julie Smith purchased a video camera on a credit card. The video camera was priced at $302. She paid $50 down and financed the balance. The credit card stipulated a finance charge of 1 percent plus a flat charge of $2 per month. Her previous balance was $63. What is the balance that she will owe after a $30 payment is made on the balance that includes the video camera purchase?

$63.00 × .01 = $.63
.63 interest
+ 2.00 finance charge
$65.63
+ 252.00 purchase
$317.63
− 30.00 payment
$287.63 balance

$302
− 50
$252

12. Ricardo Perez purchased patio furniture for $860. He paid $150 down and financed the balance through the furniture store credit card at 1 percent interest per month. He paid $200 a month on the loan. What will be the finance charges on the loan?

$860 amount of purchase
− 150 down payment
$710.00 amount of loan
+ 7.10 interest
− 200.00 payment
$517.10
+ 5.17
− 200.00
$322.27
+ 3.22
− 200.00
$125.49
+ 1.25
$126.74

$7.10
5.17
3.22
+ 1.25
$16.74 finance charges

13. What will be the last monthly payment amount in problem 12?
$126.74 last payment

14. Marilyn bought a television for $580. She paid $80 down and charged the balance on her credit card. She paid $150 a month until the balance was paid off. The service charge was $1\frac{1}{2}$ percent per month. What was the amount of the last payment?

$580 amount of purchase
− 80 down payment
$500 amount of loan
$500.00 × .015 = $7.50
+ 7.50
− 150
$375.50 × .015 = $5.36
+ 5.36
− 150
$212.86 × .015 = $3.19
+ 3.19
− 150
$66.05 × .015 = $.99
+ .99
$67.04 last payment

15. What were the finance charges in problem 14?

$7.50
5.36
3.19
+ .99
$17.04 finance charge

16. Sally found a store willing to provide credit for six months on a purchase of $450. The store manager indicated that the store would charge $22.95 for interest and a $5 service fee. Sally knows that her credit card's APR is 18.5 percent. Should she borrow that money through the store credit or charge it on her credit card?

$22.95 interest
+ 5.00 service fee
$27.95 finance charges
$27.95 ÷ 4.5 = $6.21 charge per hundred
$6.21 = 21% (APR table)
Use credit card

17. Gerry had a previous balance of $653 on his credit card. He decided to not use the card and to pay off the balance as quickly as possible by making payments of $200 each month. The credit card company charges his card at the rate of 1 percent per month. What is the last payment amount?

$653. balance
+ 6.53 interest (1%)
$659.53
− 200. payment
459.53
+ 4.60 interest
$464.13

$464.13
− 200. payment
$264.13
+ 2.64 interest
266.77
− 200. payment
$66.77

$66.77
+ .67 interest
$67.44 last payment

18. Ken has a balance on his credit card of $1,426.50 from last month. The credit card company charges $1\frac{3}{4}$ percent per month on the balance from the previous month. His charges for the month were $239.95. He will make a payment of $200 this month. What will his balance be after the payment this month?

$1,426.50
+ 24.96 interest ($1\frac{3}{4}$%)
+ 239.95 charges
− 200. payment
$1,491.41 balance

19. Write a complete sentence in answer to this question. Amy took $457 in currency, $5.23 in coin, and $389 in credit card charge forms to the bank for deposit. The credit card company uses a discount charge rate of 3 percent for her firm. What is the total amount of the deposit for the firm?

$389
× .03
$11.67 discount amount

$457.
+ 5.23
+ 377.33 ($389 − $11.67)
$839.56

The total amount of the deposit is $839.56.

20. Write a complete sentence in answer to this question. Another credit card company has offered the firm in problem 19 a discount charge rate of $2\frac{1}{2}$ percent. If each day the firm makes the same amount of credit card deposits as they did in problem 19 for the 312 days that they are open each year, how much would the firm save annually?

$389
× .025
$9.73 discount amount

$11.67
− 9.73
$ 1.94 difference × 312
= $605.28

The firm would have a yearly savings of $605.28.

★ Challenge Problem

Lee had a credit card balance of $369 on her statement in September. She made monthly payments of $200. The credit card company charged $1\frac{3}{4}$ percent interest per month. On October 7, Anna made an additional purchase of $328.60. What is the balance on the statement that she received at the end of November?

$369 × .0175 = $6.46
$369 + $6.46 − $200 = $175.46 (end of September)
$175.46 × .0175 = $3.07
$175.46 + $3.07 + $328.60 − $200 = $307.13 (end of October)
$307.13 × .0175 = $5.37 $307.13 + $5.37 = $312.50 November balance

MASTERY TEST Chapter 12

Name ______________________ **Date** ______________ **Score** ____________

1. The annual percentage rate includes all charges that are used for the use of the borrowed funds. Amount paid back less amount borrowed equals finance charges.
   (True) False
2. A credit card can be a source of identification.
   (True) False
3. The APR is a banker's formula to identify interest rates.
   True (False)
4. Amortization means to increase the credit card loan limit. True (False)
5. Bill took $327 in currency, $4.10 in coins, and $824.80 in credit card slips to the bank for deposit. The charge on credit card deposits is 4%. What is the value of the deposit made by Bill?

   $824.80
   × .04
   $32.99

   $327.
   4.10
   + 791.81 ($824.80 − $32.99)
   $1,122.91 deposit

6. Kern Fredrick purchased a coat for $160. She paid $20 down and repaid the balance at $1\frac{1}{2}$ percent interest on her credit card. She made payments of $50 a month. What was the last payment amount?

   $160 cost
   − 20 down payment
   $140.00 financed × .015 = $2.10
   + 2.10
   − 50.00
   $ 92.10 × .015 = $1.38
   + 1.38
   − 50.00
   $ 43.48 × .015 = $.65

   $43.48
   + .65
   $44.13 last payment

7. Determine the APR on a loan for $750 that is paid back in eight monthly installments of $100 each.

   $100
   × 8
   $800
   − 750
   $ 50 finance charges

   $\frac{\$50}{7.5}$ *= $6.67/hundred = 17.5%*

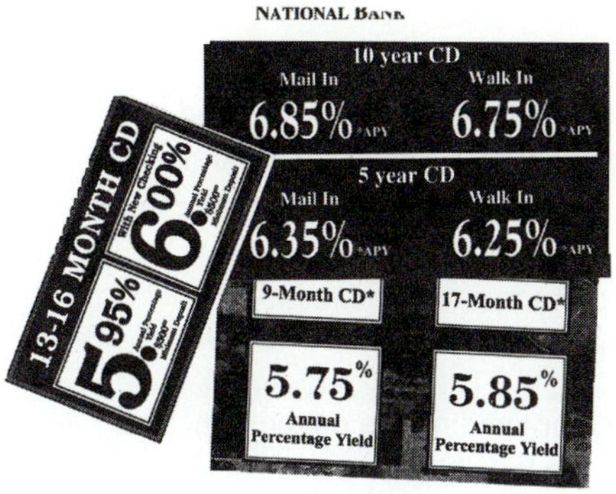

It's Time

Compound Interest and Present Value

13

Suggested Classroom Time for This Chapter

hours 1 2 3 4 5

Perceived Student Difficulty with Topical Material

easy moderate difficult

Objectives

After mastering the material in this chapter, you will be able to:

1. Compute the compounded amount in an investment problem and use compound interest tables. p. 236
2. Compute the balance on a savings account. p. 237
3. Compute the present value in an investment problem using present value tables. p. 240
4. Compute the cash value of two types of annuities in an investment problem using an annuity table. p. 244
5. Determine the investment amount in a sinking fund problem. p. 250
6. Understand and use the following terms and concepts:
   Compound Interest
   Compounded Amount
   Present Value
   Annuity
   Ordinary Annuity
   Annuity Due
   Sinking Fund

Sport-A-Merica

As Larry Mataway was talking about loan repayment at the bank a few days ago, Kerry realized that both she and Hal were very important to the success of the business. She would like Hal to realize that she knows his importance to the business. She calls Mark Barink once again. "Mark, I want to provide Hal with a benefit that will make him realize how important he is to our organization. I've budgeted $920 a year to make that point to him. Is there something that State Mutual Insurance can do to help me accomplish that goal?" Kerry continues, explaining Hal's importance, what he does in the firm, and how old he is. Mark listens carefully to Kerry's concerns. He then suggests the purchase of an annuity in Hal's name, indicating that an annuity would act as a savings plan for Hal. Deposits made over a period of time would earn compound interest, which allows the annuity to grow as time passes. Kerry tells Mark that $6,000 was placed in a savings account for a rainy day five years ago. She is sure that the value of that account has grown, but she is not sure how much it has grown. Mark explains that a savings account operates somewhat the same way as an annuity, but that it is more basic. Kerry agrees to meet Mark to discuss the matter further.

Compound Interest

Objective 1

Compound interest is another way the chapter on simple interest can be applied to solve business problems. To calculate **compound interest,** the interest from the previous period becomes part of the new principal amount; that is, the interest is added to the principal before the interest for the next period is calculated. See the Compound Interest and Present Value Model on page 241 for a visual presentation of compound interest.

Example 1

compounded amount

An example will help illustrate the dramatic effect compounding has on the principal. Under three sets of circumstances, $1,000 invested at 4 percent for a year will show a difference in the amount returned to the investor, termed the **compounded amount.**

$1,000 at 4% simple interest

Interest = $40 = $1,000 × .04 × 1

$40 = Interest
+1,000 = Principal
$1,040 = Return to investor

● *The compounded amount increases at a decreasing rate as the frequency of compounded periods increases.*

$1,000 at 4% interest compounded semiannually

Int = $20.00 = $1,000 × .04 × $\frac{1}{2}$ (first six months)
+20

Int = $20.40 = $1,020 × .04 × $\frac{1}{2}$ (second six months)
+20.40
$1,040.40 = Return to investor (compounded amount)

$1,000 at 4% interest compounded quarterly

Int = $10.00 = $1,000 × .04 × $\frac{1}{4}$ (first three months)
+10

Int = $10.10 = $1,010 × .04 × $\frac{1}{4}$ (second three months)
+10.10

Int = $10.20 = $1,020.10 × .04 × $\frac{1}{4}$ (third three months)
+10.20

Int = $10.30 = $1,030.30 × .04 × $\frac{1}{4}$ (fourth three months)
+10.30
$1,040.60 = Return to investor (compounded amount)

In this first-year period, the investor has earned an additional 40 cents by selecting an opportunity to invest at 4 percent compounded semiannually, as opposed to 4 percent at simple interest. The investor would receive 60 cents more by compounding quarterly rather than using simple interest, or 20 cents more than by compounding semiannually. Notice that the more frequently the principal is compounded in a year, the greater the return or compounded amount. Notice also that the *increase* in earnings gained from compounding interest twice a year to compounding interest four times a year is not as great as the increase in earnings gained by going from simple interest to compounding interest twice a year. The amount of increase in interest earnings declines as the principal is compounded more frequently.

Objective 2
Savings Accounts

❍ *It pays to shop for the best overall rate.*

Often one finds banks, savings and loan associations, and credit unions that compound monthly or even daily interest. Daily compounding provides a small increase in the amount of return over quarterly compounding. A consumer should not be persuaded on this fact alone. The computer has made such frequent compoundings fast and inexpensive for these firms. The resulting additional return to the investor is small compared to the additional return gained from compounding semiannually or quarterly.

Compound interest applies very well to saving accounts, money market funds, and certificates of deposit (Cds).

Use of Tables

This type of problem is repetitious and, therefore, open to a shortcut. Tables have been developed to reduce the number of these calculations.

Example 2

Kerry had placed $6,000 in a savings bank five years ago. It could be used for a rainy day for Sport-A-Merica. The bank paid 4 percent interest compounded semiannually. What is the compounded amount?

An investment of $6,000 at 4 percent compounded semiannually for a period of five years can be calculated as follows using the table.

I = P × R × T

Interest = $6,000 (principal) × 4 percent interest (annual rate) × $\frac{1}{2}$ (compounded semiannually)

❍ *What would the compounded amount have been if the interest rate was 6 percent compounded quarterly?*

Consider for a moment only the rate and time factors in this formula.

R = annual rate ÷ compoundings per year
R = 4% ÷ 2 = 2%
T = compoundings per year × number of years
T = 2 × 5 = 10

The reduced factors indicate that the investment is actually compounded at a 2 percent rate each six-month period. Therefore, the *effective rate* of interest is 2 percent *per period* for the ten 6-month periods of the five-year investment. In using the compound interest table on page 238, the effective rate of 2 percent (find across the top), and the ten periods (along the edge), the investment to be compounded is a factor of 1.21899442. This factor is the amount returned to the investor for each dollar invested. In the example, $6,000 was invested.

$6,000	= Investment	*(6,000 × 1.21899442 =)*
×1.21899442	= Factor/$1.00	
$7,313.97	= Returned to investor (compounded amount)	

Note: If you are using a calculator, enter the maximum number of digits possible.

Example 3

Let's check our answer in example 1 of an investment of $1,000 at 4 percent compounded quarterly.

R = 4% ÷ 4 = 1% effective rate
T = 4 × 1 = 4 periods

● ***Drill students on effective rates and periods before assigning exercises.***

One percent is the effective rate for the four quarter periods in one year. Using the table, 1 percent (across top) and four periods (along the edge) show a factor equal to 1.04060401 for each dollar invested.

$1,000	= Investment
×1.04060401	= Factor/$1.00
$1,040.60	= Returned to investor (compounded amount)

(1,000 × 1.04060401 =)

Compound Interest

Periods	¾%	1%	1½%	2%	2½%	3%
1	1.00750000	1.01000000	1.01500000	1.02000000	1.02500000	1.03000000
2	1.01505625	1.02010000	1.03022500	1.04040000	1.05062500	1.06090000
3	1.02266917	1.03030100	1.04567837	1.06120800	1.07689062	1.09272700
4	1.03033919	1.04060401	1.06136355	1.08243216	1.10381289	1.12550881
5	1.03806673	1.05101005	1.07728400	1.10408080	1.13140821	1.15927407
6	1.04585224	1.06152015	1.09344326	1.12616242	1.15969342	1.19405230
7	1.05369613	1.07213535	1.10984491	1.14868567	1.18868575	1.22987387
8	1.06159885	1.08285671	1.12649259	1.17165938	1.21840290	1.26677008
9	1.06956084	1.09368527	1.14338998	1.19509257	1.24886297	1.30477318
10	1.07758255	1.10462213	1.16054083	1.21899442	1.28008454	1.34391638
11	1.08566441	1.11566835	1.17794894	1.24337431	1.31208666	1.38423387
12	1.09380690	1.12682503	1.19561817	1.26824179	1.34488882	1.42576089
13	1.10201045	1.13809328	1.21355244	1.29360663	1.37851104	1.46853371
14	1.11027553	1.14947421	1.23175573	1.31947876	1.41297382	1.51258972
15	1.11860259	1.16096896	1.25023207	1.34586834	1.44829817	1.55796742
16	1.12699211	1.17257864	1.26898555	1.37278571	1.48450562	1.60470644
17	1.13544455	1.18430443	1.28802033	1.40024142	1.52161826	1.65284763
18	1.14396039	1.19614748	1.30734064	1.42824625	1.55965872	1.70243306
19	1.15254009	1.20810895	1.32695075	1.45681117	1.59865019	1.75350605
20	1.16118414	1.22019004	1.34685501	1.48594740	1.63861644	1.80611123
21	1.16989302	1.23239194	1.36705783	1.51566634	1.67958185	1.86029457
22	1.17866722	1.24471586	1.38756370	1.54597967	1.72157140	1.91610341
23	1.18750723	1.25716302	1.40837715	1.57689926	1.76461068	1.97358651
24	1.19641353	1.26973465	1.42950281	1.60843725	1.80872595	2.03279411
30	1.25127176	1.34784892	1.56308022	1.81136158	2.09756758	2.42726247

Compound Interest

Periods	8%	9%	10%	11%	12%	13%
1	1.08000000	1.09000000	1.10000000	1.11000000	1.12000000	1.13000000
2	1.16640000	1.18810000	1.21000000	1.23210000	1.25440000	1.27690000
3	1.25971200	1.29502900	1.33100000	1.36763100	1.40492800	1.44289700
4	1.36048896	1.41158161	1.46410000	1.51807041	1.57351936	1.63047361
5	1.46932808	1.53862395	1.61051000	1.68505816	1.76234168	1.84243518
6	1.58687432	1.67710011	1.77156100	1.87041455	1.97382269	2.08195175
7	1.71382427	1.82803912	1.94871710	2.07616015	2.21068141	2.35260548
8	1.85093021	1.99256264	2.14358881	2.30453777	2.47596318	2.65844419
9	1.99900463	2.17189328	2.35794769	2.55803692	2.77307876	3.00404194
10	2.15892500	2.36736367	2.59374246	2.83942099	3.10584821	3.39456739
11	2.33163900	2.58042641	2.85311671	3.15175729	3.47854999	3.83586115
12	2.51817012	2.81266478	3.13842838	3.49845060	3.89597599	4.33452310
13	2.71962373	3.06580461	3.45227121	3.88328016	4.36349311	4.89801110
14	2.93719362	3.34172703	3.79749834	4.31044098	4.88711229	5.53475255
15	3.17216911	3.64248246	4.17724817	4.78458949	5.47356576	6.25427038
16	3.42594264	3.97030588	4.59497299	5.31089433	6.13039365	7.06732553
17	3.70001805	4.32763341	5.05447028	5.89509271	6.86604089	7.98607785
18	3.99601950	4.71712042	5.55991731	6.54355291	7.68996580	9.02426797
19	4.31570106	5.14166125	6.11590904	7.26334373	8.61276169	10.19742280
20	4.66095714	5.60441077	6.72749995	8.06231154	9.64629309	11.52308776
21	5.03383372	6.10880774	7.40024994	8.94916581	10.80384826	13.02108917
22	5.43654041	6.65860043	8.14027494	9.93357404	12.10031006	14.71383077
23	5.87146365	7.25787447	8.95430243	11.02626719	13.55234726	16.62662877
24	6.34118074	7.91108317	9.84973268	12.23915658	15.17862893	18.78809051
30	10.06265689	13.26767847	17.44940227	22.89229657	29.95992212	39.11589796

Periods	3½%	4%	4½%	5%	5½%	6%	7%
1	1.03500000	1.04000000	1.04500000	1.05000000	1.05500000	1.06000000	1.07000000
2	1.07122500	1.08160000	1.09202500	1.10250000	1.11302500	1.12360000	1.14490000
3	1.10871788	1.12486400	1.14116612	1.15762500	1.17424137	1.19101600	1.22504300
4	1.14752300	1.16985856	1.19251860	1.21550625	1.23882465	1.26247696	1.31079601
5	1.18768631	1.21665290	1.24618194	1.27628156	1.30696001	1.33822558	1.40255173
6	1.22925533	1.26531902	1.30226012	1.34009564	1.37884281	1.41851911	1.50073035
7	1.27227926	1.31593178	1.36086183	1.40710042	1.45467916	1.50363026	1.60578148
8	1.31680904	1.36856905	1.42210061	1.47745544	1.53468651	1.59384807	1.71818618
9	1.36289735	1.42331181	1.48609514	1.55132822	1.61909427	1.68947896	1.83845921
10	1.41059876	1.48024428	1.55296942	1.62889463	1.70814446	1.79084770	1.96715136
11	1.45996972	1.53945406	1.62285305	1.71033936	1.80209240	1.89829856	2.10485195
12	1.51106866	1.60103222	1.69588143	1.79585633	1.90120749	2.01219647	2.25219159
13	1.56395606	1.66507351	1.77219610	1.88564914	2.00577390	2.13292826	2.40984500
14	1.61869452	1.73167645	1.85194492	1.97993160	2.11609146	2.26090396	2.57853415
15	1.67534883	1.80094351	1.93528244	2.07892818	2.23247649	2.39655819	2.75903154
16	1.73398604	1.87298125	2.02237015	2.18287459	2.35526270	2.54035168	2.95216375
17	1.79467555	1.94790050	2.11337681	2.29201832	2.48480215	2.69277279	3.15881521
18	1.85748920	2.02581652	2.20847877	2.40661923	2.62146627	2.85433915	3.37993228
19	1.92250132	2.10684918	2.30786031	2.52695020	2.76564691	3.02559950	3.61652754
20	1.98978886	2.19112314	2.41171402	2.65329771	2.91775749	3.20713547	3.86968446
21	2.05943147	2.27876807	2.52024116	2.78596259	3.07823415	3.39956360	4.14056237
22	2.13151158	2.36991879	2.63365201	2.92526072	3.24753703	3.60353742	4.43040174
23	2.20611448	2.46471554	2.75216635	3.07152376	3.42615157	3.81974966	4.74052986
24	2.28332849	2.56330416	2.87601383	3.22509994	3.61458990	4.04893464	5.07236695
30	2.80679370	3.24339751	3.74531813	4.32194238	4.98395129	5.74349117	7.61225504

Periods	14%	15%	16%	17%	18%	19%	20%
1	1.14000000	1.15000000	1.16000000	1.17000000	1.18000000	1.19000000	1.20000000
2	1.29960000	1.32250000	1.34560000	1.36890000	1.39240000	1.41610000	1.44000000
3	1.48154400	1.52087500	1.56089600	1.60161300	1.64303200	1.68515900	1.72800000
4	1.68896016	1.74900625	1.81063936	1.87388721	1.93877776	2.00533921	2.07360000
5	1.92541458	2.01135719	2.10034166	2.19244804	2.28775776	2.38635366	2.48832000
6	2.19497262	2.31306077	2.43639632	2.56516420	2.69955415	2.83976086	2.98598400
7	2.50226879	2.66001988	2.82621973	3.00124212	3.18547390	3.37931542	3.58318080
8	2.85258642	3.05902286	3.27841489	3.51145328	3.75885920	4.02138535	4.29981696
9	3.25194852	3.51787629	3.80296127	4.10840033	4.43545386	4.78544856	5.15978035
10	3.70722131	4.04555774	4.41143508	4.80682839	5.23383555	5.69468379	6.19173642
11	4.22623230	4.65239140	5.11726469	5.62398922	6.17592595	6.77667371	7.43008371
12	4.81790482	5.35025011	5.93602704	6.58006738	7.28759263	8.06424172	8.91610045
13	5.49241149	6.15278762	6.88579137	7.69867884	8.59935930	9.59644764	10.69932054
14	6.26134910	7.07570576	7.98751799	9.00745424	10.14724397	11.41977269	12.83918465
15	7.13793798	8.13706163	9.26552087	10.53872146	11.97374789	13.58952950	15.40702157
16	8.13724930	9.35762087	10.74800420	12.33030411	14.12902251	16.17154011	18.48842589
17	9.27646420	10.76126400	12.46768488	14.42645581	16.67224656	19.24413273	22.18611107
18	10.57516918	12.37545361	14.46251446	16.87895329	19.67325094	22.90051795	26.62333328
19	12.05569287	14.23177165	16.77651677	19.74837535	23.21443611	27.25161636	31.94799994
20	13.74348987	16.36653739	19.46075945	23.10559916	27.39303460	32.42942347	38.33759992
21	15.66757845	18.82151800	22.57448097	27.03355102	32.32378083	38.59101393	46.00511991
22	17.86103944	21.64474570	26.18639792	31.62925470	38.14206138	45.92330658	55.20614389
23	20.36158496	24.89145756	30.37622159	37.00622799	45.00763243	54.64873482	66.24737267
24	23.21220685	28.62517619	35.23641704	43.29728675	53.10900627	65.03199444	79.49684720
30	50.95015858	66.21177196	85.84987691	111.06465001	143.37063844	184.67531215	237.37631380

Exercise A

Complete the following problems to find the compounded amount.

	Amount Invested	Annual Interest Rate	Term of Investment (Years)	Compound Frequency	Compounded Amount
1.	$ 1,400	5%	5	Semi. (1,400 × 1.28008454 =)	$1,792.12
2.	4,000	5	12	Semi. *(4,000 × 1.80872595 =)*	*7,234.90*
3.	5,500	4	8	Semi. *(5,500 × 1.37278571 =)*	*7,550.32*
4.	8,950	4	4	Qtr. *(8,950 × 1.17257864 =)*	*10,494.58*
5.	20,000	4	20	Annual *(20,000 × 2.19112314 =)*	*43,822.46*
6.	7,500	3	5	Semi. *(7,500 × 1.16054083 =)*	*8,704.06*
7.	19,500	6	$5\frac{1}{2}$	Qtr. *(19,500 × 1.38756370 =)*	*27,057.50*
8.	5,000	4	6	Qtr. *(5,000 × 1.26973465 =)*	*6,348.67*
9.	5,000	5	18	Annual *(5,000 × 2.40661923 =)*	*12,033.10*
10.	9,000	4	9	Semi. *(9,000 × 1.42824625 =)*	*12,854.22*

Present Value

Objective 3

Money has a "time value." That is, a given amount of money can be invested today with the expectation that the compounded amount or future value will be worth more in the future. That was just proven in exercise A. There is also a converse situation. We could ask, "What is the current value of an amount of money to be received at some date in the future?" The current, now or today, value of the future amount is termed its **present value.** Present value is the exact opposite of future value, or compound interest. We are asking, "What amount of money today is equal to the amount that will be received in the future, if today's amount (whatever it is) could be invested now and allowed to grow to that future (known) amount?" See the Compound Interest and Present Value Model on page 241 for a visual presentation of this concept.

present value

❍ A back to the future viewpoint of the value of your investment!

Example 4

Acrodin, Incorporated, will receive the sum of $15,600 three years from this date. The treasurer realizes that money can safely be invested at an 8 percent annual interest rate compounded semiannually. What is the equivalent amount (present value) of $15,600 to be received three years from today?

The present value table will aid in the solution to the example problem in very much the same way as did the compound interest table.

R = 8% ÷ 2 = 4% effective rate
T = 3 × 2 = 6 periods

The effective rate for the six 6-month periods is 4 percent. Using the present value table, 4 percent (across top) and six periods (along edge) gives a factor of .79031453.

$15,600 = Future value
×.79031453 = Factor/$1.00
$12,328.91 = Present value

(15,600 × .79031453 =)

This means that \$12,328.91 received today is equal to \$15,600 received three years from now if the \$12,328.91 can be invested at 8 percent compounded semiannually for three years. This can be checked by using the compound interest table.

\$12,328.91 × 1.26531902 (4 percent, six periods) = \$15,600

Example 5

What is the present value of \$2,680 to be received five years from today if an investment can be made at 6 percent compounded quarterly?

R = 6 % ÷ 4 = $1\frac{1}{2}$% effective rate
T = 5 × 4 = 20 periods

The effective rate for twenty periods is $1\frac{1}{2}$ percent. Using the table, the factor equals .74247042.

\$2,680	=	Expected sum
×.74247042	=	Factor/\$1.00
\$1,989.82	=	Present value

(2,680 × .74247042 =)

Compound Interest and Present Value Model

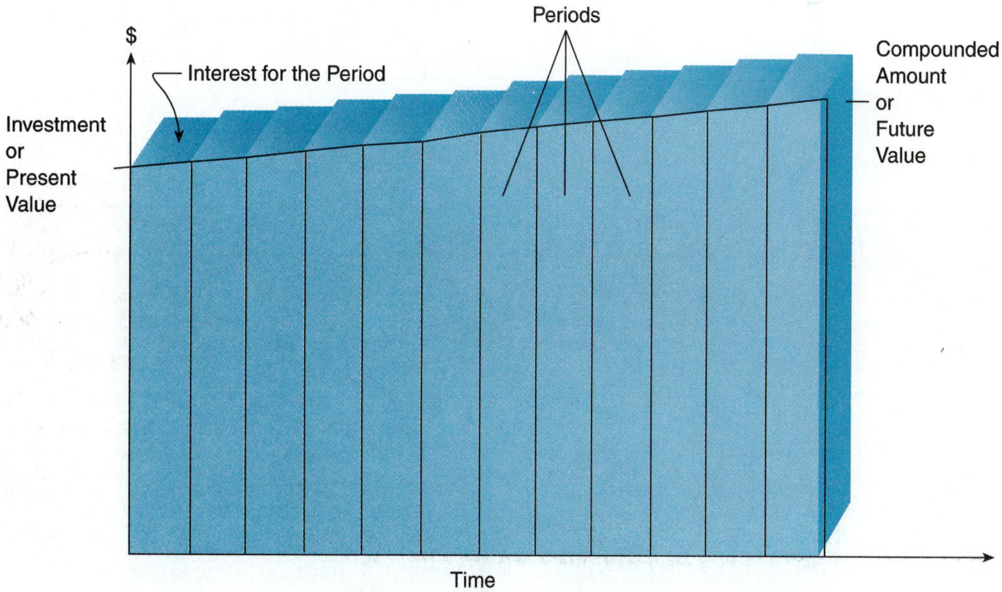

Present Value of 1

Periods	¾%	1%	1½%	2%	2½%	3%
1	0.99255583	0.99009901	0.98522167	0.98039216	0.97560976	0.97087379
2	0.98516708	0.98029605	0.97066175	0.96116878	0.95181440	0.94259591
3	0.97783333	0.97059015	0.95631699	0.94232233	0.92859941	0.91514166
4	0.97055417	0.96098034	0.94218423	0.92384543	0.90595064	0.88848705
5	0.96332920	0.95146569	0.92826033	0.90573081	0.88385429	0.86260878
6	0.95615802	0.94204524	0.91454219	0.88797138	0.86229687	0.83748426
7	0.94904022	0.93271805	0.90102679	0.87056018	0.84126524	0.81309151
8	0.94197540	0.92348322	0.88771112	0.85349037	0.82074657	0.78940923
9	0.93496318	0.91433982	0.87459224	0.83675527	0.80072836	0.76641673
10	0.92800315	0.90528695	0.86166723	0.82034830	0.78119840	0.74409391
11	0.92109494	0.89632372	0.84893323	0.80426304	0.76214478	0.72242128
12	0.91423815	0.88744923	0.83638742	0.78849318	0.74355589	0.70137988
13	0.90743241	0.87866260	0.82402702	0.77303253	0.72542038	0.68095134
14	0.90067733	0.86996297	0.81184928	0.75787502	0.70772720	0.66111781
15	0.89397254	0.86134947	0.79985150	0.74301473	0.69046556	0.64186195
16	0.88731766	0.85282126	0.78803104	0.72844581	0.67362493	0.62316694
17	0.88071231	0.84437749	0.77638526	0.71416256	0.65719506	0.60501645
18	0.87415614	0.83601731	0.76491159	0.70015937	0.64116591	0.58739461
19	0.86764878	0.82773992	0.75360747	0.68643076	0.62552772	0.57028603
20	0.86118985	0.81954447	0.74247042	0.67297133	0.61027094	0.55367575
21	0.85477901	0.81143017	0.73149795	0.65977582	0.59538629	0.53754928
22	0.84841589	0.80339621	0.72068763	0.64683904	0.58086467	0.52189250
23	0.84210014	0.79544179	0.71003708	0.63415592	0.56669724	0.50669175
24	0.83583140	0.78756613	0.69954392	0.62172149	0.55287535	0.49193374
30	0.79918690	0.74192292	0.63976243	0.55207089	0.47674269	0.41198676

Present Value of 1

Periods	8%	9%	10%	11%	12%	13%
1	0.92592593	0.91743119	0.90909091	0.90090090	0.89285714	0.88495575
2	0.85733882	0.84167999	0.82644628	0.81162243	0.79719388	0.78314668
3	0.79383224	0.77218348	0.75131480	0.73119138	0.71178025	0.69305016
4	0.73502985	0.70842521	0.68301346	0.65873097	0.63551808	0.61331873
5	0.68058320	0.64993139	0.62092132	0.59345133	0.56742686	0.54275994
6	0.63016963	0.59626733	0.56447393	0.53464084	0.50663112	0.48031853
7	0.58349040	0.54703424	0.51315812	0.48165841	0.45234922	0.42506064
8	0.54026888	0.50186628	0.46650738	0.43392650	0.40388323	0.37615986
9	0.50024897	0.46042778	0.42409762	0.39092477	0.36061002	0.33288483
10	0.46319349	0.42241081	0.38554329	0.35218448	0.32197324	0.29458835
11	0.42888286	0.38753285	0.35049390	0.31728331	0.28747610	0.26069765
12	0.39711376	0.35553473	0.31863082	0.28584082	0.25667509	0.23070589
13	0.36769792	0.32617865	0.28966438	0.25751426	0.22917419	0.20416450
14	0.34046104	0.29924647	0.26333125	0.23199482	0.20461981	0.18067655
15	0.31524170	0.27453804	0.23939205	0.20900435	0.18269626	0.15989075
16	0.29189047	0.25186976	0.21762914	0.18829220	0.16312166	0.14149624
17	0.27026895	0.23107318	0.19784467	0.16963262	0.14564434	0.12521791
18	0.25024903	0.21199374	0.17985879	0.15282218	0.13003959	0.11081231
19	0.23171206	0.19448967	0.16350799	0.13767764	0.11610678	0.09806399
20	0.21454821	0.17843089	0.14864363	0.12403391	0.10366677	0.08678229
21	0.19865575	0.16369806	0.13513057	0.11174226	0.09255961	0.07679849
22	0.18394051	0.15018171	0.12284597	0.10066870	0.08264251	0.06796327
23	0.17031528	0.13778139	0.11167816	0.09069252	0.07378796	0.06014448
24	0.15769934	0.12640494	0.10152560	0.08170498	0.06588210	0.05322521
30	0.09937733	0.07537114	0.05730855	0.04368282	0.03337792	0.02556505

Periods	*3½%*	*4%*	*4½%*	*5%*	*5½%*	*6%*	*7%*
1	0.96618357	0.96153846	0.95693780	0.95238095	0.94786730	0.94339623	0.93457944
2	0.93351070	0.92455621	0.91572995	0.90702948	0.89845242	0.88999644	0.87343873
3	0.90194271	0.88899636	0.87629660	0.86383760	0.85161366	0.83961928	0.81629788
4	0.87144223	0.85480419	0.83856134	0.82270247	0.80721674	0.79209366	0.76289521
5	0.84197317	0.82192711	0.80245105	0.78352617	0.76513435	0.74725817	0.71298618
6	0.81350064	0.79031453	0.76789574	0.74621540	0.72524583	0.70496054	0.66634222
7	0.78599096	0.75991781	0.73482846	0.71068133	0.68743681	0.66505711	0.62274974
8	0.75941156	0.73069021	0.70318513	0.67683936	0.65159887	0.62741237	0.58200910
9	0.73373097	0.70258674	0.67290443	0.64460892	0.61762926	0.59189846	0.54393374
10	0.70891881	0.67556417	0.64392768	0.61391325	0.58543058	0.55839478	0.50834929
11	0.68494571	0.64958093	0.61619874	0.58467929	0.55491050	0.52678753	0.47509280
12	0.66178330	0.62459705	0.58966386	0.55683742	0.52598152	0.49696936	0.44401196
13	0.63940415	0.60057409	0.56427164	0.53032135	0.49856068	0.46883902	0.41496445
14	0.61778179	0.57747508	0.53997286	0.50506795	0.47256937	0.44230096	0.38781724
15	0.59689062	0.55526450	0.51672044	0.48101710	0.44793305	0.41726506	0.36244602
16	0.57670591	0.53390818	0.49446932	0.45811152	0.42458109	0.39364628	0.33873460
17	0.55720378	0.51337325	0.47317639	0.43629669	0.40244653	0.37136442	0.31657439
18	0.53836114	0.49362812	0.45280037	0.41552065	0.38146590	0.35034379	0.29586392
19	0.52015569	0.47464242	0.43330179	0.39573396	0.36157906	0.33051301	0.27650833
20	0.50256588	0.45638695	0.41464286	0.37688948	0.34272896	0.31180473	0.25841900
21	0.48557090	0.43883360	0.39678743	0.35894236	0.32486158	0.29415540	0.24151309
22	0.46915063	0.42195539	0.37970089	0.34184987	0.30792567	0.27750510	0.22571317
23	0.45328563	0.40572633	0.36335013	0.32557131	0.29187267	0.26179726	0.21094688
24	0.43795713	0.39012147	0.34770347	0.31006791	0.27665656	0.24697855	0.19714662
30	0.35627841	0.30831867	0.26700002	0.23137745	0.20064402	0.17411013	0.13136712

Periods	*14%*	*15%*	*16%*	*17%*	*18%*	*19%*	*20%*
1	0.87719298	0.86956522	0.86206897	0.85470085	0.84745763	0.84033613	0.83333333
2	0.76946753	0.75614367	0.74316290	0.73051355	0.71818443	0.70616482	0.69444444
3	0.67497152	0.65751623	0.64065767	0.62437056	0.60863087	0.59341581	0.57870370
4	0.59208028	0.57175325	0.55229110	0.53365005	0.51578888	0.49866875	0.48225309
5	0.51936866	0.49717674	0.47611302	0.45611115	0.43710922	0.41904937	0.40187757
6	0.45558655	0.43232760	0.41044225	0.38983859	0.37043154	0.35214233	0.33489798
7	0.39963732	0.37593704	0.35382953	0.33319538	0.31392503	0.29591792	0.27908165
8	0.35055905	0.32690177	0.30502546	0.28478237	0.26603816	0.24867052	0.23256804
9	0.30750794	0.28426241	0.26295298	0.24340374	0.22545607	0.20896683	0.19380670
10	0.26974381	0.24718471	0.22668360	0.20803738	0.19106447	0.17560238	0.16150558
11	0.23661738	0.21494322	0.19541690	0.17780973	0.16191904	0.14756502	0.13458799
12	0.20755910	0.18690715	0.16846284	0.15197413	0.13721953	0.12400422	0.11215665
13	0.18206939	0.16252796	0.14522659	0.12989242	0.11628773	0.10420523	0.09346388
14	0.15970999	0.14132866	0.12519534	0.11101916	0.09854893	0.08756742	0.07788657
15	0.14009648	0.12289449	0.10792701	0.09488817	0.08351604	0.07358606	0.06490547
16	0.12289165	0.10686477	0.09304053	0.08110100	0.07077630	0.06183703	0.05408789
17	0.10779969	0.09292589	0.08020735	0.06931709	0.05997992	0.05196389	0.04507324
18	0.09456113	0.08080512	0.06914427	0.05924538	0.05083044	0.04366713	0.03756104
19	0.08294836	0.07026532	0.05960713	0.05063708	0.04307664	0.03669507	0.03130086
20	0.07276172	0.06110028	0.05138546	0.04327955	0.03650563	0.03083619	0.02608405
21	0.06382607	0.05313068	0.04429781	0.03699107	0.03093698	0.02591277	0.02173671
22	0.05598778	0.04620059	0.03818776	0.03161630	0.02621778	0.02177544	0.01811393
23	0.04911209	0.04017443	0.03292049	0.02702248	0.02221845	0.01829869	0.01509494
24	0.04308078	0.03493428	0.02837973	0.02309614	0.01882920	0.01537705	0.01257912
30	0.01962702	0.01510305	0.01164824	0.00900376	0.00697493	0.00541491	0.00421272

Exercise B

Complete the following problems to find the present value.

Investment Opportunity

	Future Value or Expected Sum	Annual Interest Rate	Compound Frequency	Term before Receipt	Present Value
1.	$1,600	4%	Semi.	4 years (1,600 × .85349037 =)	$1,365.58
2.	2,000	6	Qtr.	3 *(2,000 × .83638742 =)*	*1,672.77*
3.	5,000	5	Semi.	5 *(5,000 × .78119840 =)*	*3,905.99*
4.	2,800	9	Annually	7 *(2,800 × .54703424 =)*	*1,531.70*
5.	1,000	8	Semi.	6 *(1,000 × .62459705 =)*	*624.60*
6.	16,000	6	Semi.	8 *(16,000 × .62316694 =)*	*9,970.67*
7.	6,000	5	Semi.	2 *(6,000 × .90595064 =)*	*5,435.70*
8.	8,500	10	Qtr.	$3\frac{1}{2}$ *(8,500 × .70772720 =)*	*6,015.68*
9.	10,000	5	Semi.	8 *(10,000 × .67362493 =)*	*6,736.25*
10.	12,500	4	Semi.	$5\frac{1}{2}$ *(12,500 × .80426304 =)*	*10,053.29*

Annuities

Objective 4

The concepts that concern an annuity are very similar to the concept of compound interest. Money invested over a period of time accumulates interest at a compound rate on the investment. The important difference is that the **annuity** involves a *series* of investments, usually of an equal amount and at regular intervals. The value of an annuity will grow at the stated compound rate and the amount of regular investment.

❍ *Compound interest involves one investment. An annuity involves several equal investments.*

Annuities are used by business, religious, and financial firms to save money to pay off a bond issue, a large note, or a piece of property. Individuals use annuities to save money for their children's education, for a planned trip, or for retirement. An annuity can be purchased from an insurance agent or a securities broker.

There are two types of annuities. An *annuity certain* is the type in which the number of payments is stated in the annuity agreement. A *contingent annuity* involves an open-ended agreement, and the investor can make as many payments as he or she chooses to make.

ordinary annuity
annuity due

The point at which the payment is made in the time period further classifies an annuity. In an **ordinary annuity,** the payment is made at the end of the period. In an **annuity due,** the payment is made at the beginning of the period. This can make a difference of one period of interest when using the annuity table.

Ordinary Annuity

An annuity can be calculated over a 4-year period at 4 percent, as is shown in example 6.

Example 6

An ordinary annuity for $200 at 4 percent for four years will result in what amount or cash value?

Investment	End of First Year Value	End of Second Year Value	End of Third Year Value	End of Fourth Year Value
Year 1 $200	(no interest) $200			
Year 2 $200		$200 × 1.04 = $208 (no interest) +200 $408		
Year 3 $200			$408 × 1.04 = $424.32 (no interest) +200.00 $624.32	
Year 4 $200				$624.32 × 1.04 = $649.29 (no interest) +200.00 $849.29 (cash value)

Growth of an Annuity

The compound interest table can be used to compute the same cash value.

Investment
Year 1 $200 invested for 3 years
at 4% compounded annually = 1.12486400 (factor) × $200 = $224.97
Year 2 $200 invested for 2 years
at 4% compounded annually = 1.08160000 (factor) × $200 = 216.32
Year 3 $200 invested for 1 year
at 4% compounded annually = 1.04000000 (factor) × $200 = 208.00
Year 4 $200 invested for 0 years
at 4% compounded annually = 1.00000000 (factor) × $200 = 200.00
4.24646400 = factor or return/$1.00 $849.29

The ordinary annuity table simplifies the problem, as does the table in the compound interest and present value problems.

The effective interest rate (across top) of 4 percent and four periods (along the edge) give you a factor of 4.24646400.

$200 = Payment
×4.24646400 = Factor or return/$1.00
$849.29 = Cash value

Example 7

Kerry is willing to set $460 aside every six months in an annuity for Hal. The annuity will earn 4 percent over the next six years, according to Mark Barink of State Mutual. This should encourage Hal to stay with Sport-A-Merica for at least six years.

What will be the cash value of an ordinary annuity for $460 at 4 percent compounded for six years if payments are made semiannually?

The effective interest rate (across top) is 2 percent, and using 12 periods (along the edge) will show a factor of 13.41208973.

$460 = Payment
×13.41208973 = Factor
$6,169.56 = Cash value

(460 × 13.41208973 =)

Annuity Due

The cash value of an annuity due can also be found by using an ordinary annuity table. Remember that an annuity due requires that the payment be made at the beginning of the period. This will make a difference in the cash value compared to that of an ordinary annuity.

Example 8

Using example 6, an ordinary annuity for $200 at 4 percent interest with four annual payments, it is possible to show the comparison to an annuity due.

Investment
Year 1 $200 invested for 4 years
at 4% compounded annually = 1.16985856 (factor)
Year 2 $200 invested for 3 years
at 4% compounded annually = 1.12486400 (factor)
Year 3 $200 invested for 2 years
at 4% compounded annually = 1.08160000 (factor)
Year 4 $200 invested for 1 year
at 4% compounded annually = 1.04000000 (factor)
4.41632256 = factor or return/$1.00

$200 = Payment
×4.41632256 = Factor or return/$1.00
$883.26 = Cash value

(200 × 4.41632256 =)

❍ Add one period, subtract one for an annuity due.

By referring to the ordinary annuity table, you will find the effective interest rate of 4 percent (across top) and five periods (along the edge). Five periods is used because the payment is being made at the beginning of the period. The effect is the same as starting an ordinary annuity one period earlier. The factor is 5.41632256. Subtract 1.00000000 from this factor, which will compensate for *not* making the fifth payment with no interest, and the factor 4.41632256 will be found. This compares to the sum of the factors found by using the compound interest table (4%, 1, 2, 3, & 4 periods, i.e., 1.04000000 + 1.08160000 + 1.12486400 + 1.16985856 = 4.41632256).

Example 9

Find the cash value of an annuity due for $380 at 3 percent with annual payments for eleven years. The effective interest rate (across top) is 3 percent, and there are 12 periods (along the edge), so the factor is 14.19202956.

*❍ An annuity **due** means you **do** subtract 1.00000000!*

14.19202956
−1.00000000
13.19202956 = Factor or return/$1.00
×$380 = Payment
$5,012.97 = Cash value

(13.19202956 × 380 =)

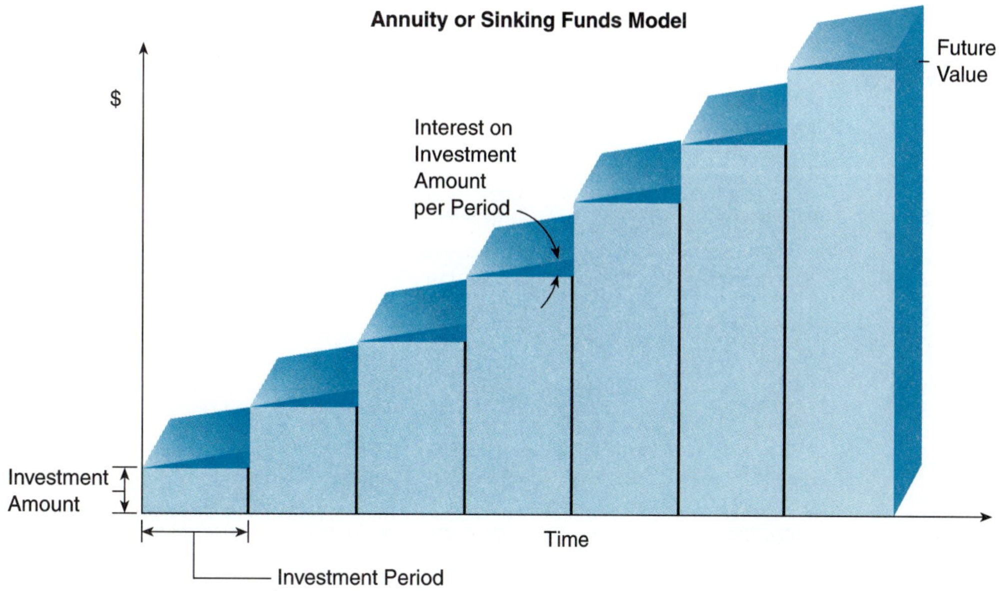

Example 10

○ *In an annuity due, you add one period and subtract 1.00000000!*

To make the distinction between an ordinary annuity and annuity due even clearer, this example will solve the answer to both using the same information.

Find the cash value of an annuity for $200 at 7 percent with annual payments for five years.

Annuity Due

7.15329074	Factor for 7%,
−1.00000000	6 periods
6.15329074	
×$200	
$1,230.66 cash value	

Ordinary Annuity

5.75073901	Factor for 7%,
×$200	5 periods
$1,150.15 cash value	

(6.15329074 × 200 =) *(5.75073901 × 200 =)*

Example 11

Find the cash value of an annuity for $350 at 16 percent with semiannual payments for six years.

Annuity Due

21.49529658	Factor for 8%,
−1.00000000	13 periods
20.49529658	
×$350	
$7,173.35 cash value	

Ordinary Annuity

18.97712646	Factor for 8%,
×$350	12 periods
$6,641.99 cash value	

(20.49529658 × 350 =) *(18.97712646 × 350 =)*

Ordinary Annuity Table Rate for Each Interest Period

Periods	¾%	1%	1½%	2%	2½%	3%
1	1.00000000	1.00000000	1.00000000	1.00000000	1.00000000	1.00000000
2	2.00750000	2.01000000	2.01500000	2.02000000	2.02500000	2.03000000
3	3.02255625	3.03010000	3.04522500	3.06040000	3.07562500	3.09090000
4	4.04522542	4.06040100	4.09090338	4.12160800	4.15251563	4.18362700
5	5.07556461	5.10100501	5.15226693	5.20404016	5.25632852	5.30913581
6	6.11363135	6.15201506	6.22955093	6.30812096	6.38773673	6.46840988
7	7.15948358	7.21353521	7.32299419	7.43428338	7.54743015	7.66246218
8	8.21317971	8.28567056	8.43283911	8.58296905	8.73611590	8.89233605
9	9.27477856	9.36852727	9.55933169	9.75462843	9.95451880	10.15910613
10	10.34433940	10.46221254	10.70272167	10.94972100	11.20338177	11.46387931
11	11.42192194	11.56683467	11.86326249	12.16871542	12.48346631	12.80779569
12	12.50758636	12.68250301	13.04121143	13.41208973	13.79555297	14.19202956
13	13.60139325	13.80932804	14.23682960	14.68033152	15.14044179	15.61779045
14	14.70340370	14.94742132	15.45038205	15.97393815	16.51895284	17.08632416
15	15.81367923	16.09689554	16.68213778	17.29341692	17.93192666	18.59891389
16	16.93228183	17.25786449	17.93236984	18.63928525	19.38022483	20.15688130
17	18.05927394	18.43044314	19.20135539	20.01207096	20.86473045	21.76158774
18	19.19471849	19.61474757	20.48937572	21.41231238	22.38634871	23.41443537
19	20.33867888	20.81089504	21.79671636	22.84055863	23.94600743	25.11686844
20	21.49121897	22.01900399	23.12366710	24.29736980	25.54465761	26.87037449
21	22.65240312	23.23919403	24.47052211	25.78331719	27.18327405	28.67648572
22	23.82229614	24.47158598	25.83757994	27.29898354	28.86285590	30.53678030
23	25.00096336	25.71630183	27.22514364	28.84496321	30.58442730	32.45288370
24	26.18847059	26.97346485	28.63352080	30.42186247	32.34903798	34.42647022
25	27.38488412	28.24319950	30.06302361	32.03029972	34.15776393	36.45926432
26	28.59027075	29.52563150	31.51396896	33.67090572	36.01170803	38.55304225
27	29.80469778	30.82088781	32.98667850	35.34432383	37.91200073	40.70963352
28	31.02823301	32.12909669	34.48147867	37.05121031	39.85980075	42.93092252
29	32.26094476	33.45038766	35.99870085	38.79223451	41.85629577	45.21885020
30	33.50290184	34.78489153	37.53868137	40.56807921	43.90270316	47.57541571

Ordinary Annuity Table Rate for Each Interest Period

Periods	8%	9%	10%	11%	12%	13%
1	1.00000000	1.00000000	1.00000000	1.00000000	1.00000000	1.00000000
2	2.08000000	2.09000000	2.10000000	2.11000000	2.12000000	2.13000000
3	3.24640000	3.27810000	3.31000000	3.34210000	3.37440000	3.40690000
4	4.50611200	4.57312900	4.64100000	4.70973100	4.77932800	4.84979700
5	5.86660096	5.98471061	6.10510000	6.22780141	6.35284736	6.48027061
6	7.33592904	7.52333456	7.71561000	7.91285957	8.11518904	8.32270579
7	8.92280336	9.20043468	9.48717100	9.78327412	10.08901173	10.40465754
8	10.63662763	11.02847380	11.43588810	11.85943427	12.29969314	12.75726302
9	12.48755784	13.02103644	13.57947691	14.16397204	14.77565631	15.41570722
10	14.48656247	15.19292972	15.93742460	16.72200896	17.54873507	18.41974915
11	16.64548746	17.56029339	18.53116706	19.56142995	20.65458328	21.81431654
12	18.97712646	20.14071980	21.38428377	22.71318724	24.13313327	25.65017769
13	21.49529658	22.95338458	24.52271214	26.21163784	28.02910926	29.98470079
14	24.21492030	26.01918919	27.97498336	30.09491800	32.39260238	34.88271190
15	27.15211393	29.36091622	31.77248169	34.40535898	37.27971466	40.41746444
16	30.32428304	33.00339868	35.94972986	39.18994847	42.75328042	46.67173482
17	33.75022569	36.97370456	40.54470285	44.50084281	48.88367407	53.73906035
18	37.45024374	41.30133797	45.59917313	50.39593551	55.74971496	61.72513819
19	41.44626324	46.01845839	51.15909045	56.93948842	63.43968075	70.74940616
20	45.76196430	51.16011964	57.27499949	64.20283215	72.05244244	80.94682896
21	50.42292144	56.76453041	64.00249944	72.26514368	81.69873554	92.46991672
22	55.45675516	62.87333815	71.40274939	81.21430949	92.50258380	105.49100590
23	60.89329557	69.53193858	79.54302433	91.14788353	104.60289386	120.20483667
24	66.76475922	76.78981305	88.49732676	102.17415072	118.15524112	136.83146543
25	73.10593995	84.70089623	98.34705943	114.41330730	133.33387006	155.61955594
26	79.95441515	93.32397689	109.18176538	127.99877110	150.33393446	176.85009821
27	87.35076836	102.72313481	121.09994191	143.07863592	169.37400660	200.84061098
28	95.33882983	112.96821694	134.20993611	159.81728587	190.69888739	227.94989040
29	103.96593622	124.13535646	148.63092972	178.39718732	214.58275388	258.58337616
30	113.28321111	136.30753855	164.49402269	199.02087793	241.33268434	293.19921506

Periods	3½%	4%	4½%	5%	5½%	6%	7%
1	1.00000000	1.00000000	1.00000000	1.00000000	1.00000000	1.00000000	1.00000000
2	2.03500000	2.04000000	2.04500000	2.05000000	2.05500000	2.06000000	2.07000000
3	3.10622500	3.12160000	3.13702500	3.15250000	3.16802500	3.18360000	3.21490000
4	4.21494287	4.24646400	4.27819112	4.31012500	4.34226638	4.37461600	4.43994300
5	5.36246588	5.41632256	5.47070973	5.52563125	5.58109103	5.63709296	5.75073901
6	6.55015218	6.63297546	6.71689166	6.80191281	6.88805103	6.97531854	7.15329074
7	7.77940751	7.89829448	8.01915179	8.14200845	8.26689384	8.39383765	8.65402109
8	9.05168677	9.21422626	9.38001362	9.54910888	9.72157300	9.89746791	10.25980257
9	10.36849581	10.58279531	10.80211423	11.02656432	11.25625951	11.49131598	11.97798875
10	11.73139316	12.00610712	12.28820937	12.57789254	12.87535379	13.18079494	13.81644796
11	13.14199192	13.48635141	13.84117879	14.20678716	14.58349825	14.97164264	15.78359932
12	14.60196164	15.02580546	15.46403184	15.91712652	16.38559065	16.86994120	17.88845127
13	16.11303030	16.62683768	17.15991327	17.71298285	18.28679814	18.88213767	20.14064286
14	17.67698636	18.29191119	18.93210937	19.59863199	20.29257203	21.01506593	22.55048786
15	19.29568088	20.02358764	20.78405429	21.57856359	22.40866350	23.27596988	25.12902201
16	20.97102971	21.82453114	22.71933673	23.65749177	24.64113999	25.67252808	27.88805355
17	22.70501575	23.69751239	24.74170689	25.84036636	26.99640269	28.21287976	30.84021730
18	24.49969130	25.64541288	26.85508370	28.13238467	29.48120483	30.90565255	33.99903251
19	26.35718050	27.67122940	29.06356246	30.53900391	32.10267110	33.75999170	37.37896479
20	28.27968181	29.77807858	31.37142277	33.06595410	34.86831801	36.78559120	40.99549232
21	30.26947068	31.96920172	33.78313680	35.71925181	37.78607550	39.99272668	44.86517678
22	32.32890215	34.24796979	36.30337795	38.50521440	40.86430965	43.39229028	49.00573916
23	34.46041373	36.61788858	38.93702996	41.43047512	44.11184669	46.99582769	53.43614090
24	36.66652821	39.08260412	41.68919631	44.50199887	47.53799825	50.81557735	58.17667076
25	38.94985669	41.64590829	44.56521015	47.72709882	51.15258816	54.86451200	63.24903772
26	41.31310168	44.31174462	47.57064460	51.11345376	54.96598051	59.15638272	68.67647036
27	43.75906024	47.08421440	50.71132361	54.66912645	58.98910943	63.70576568	74.48382328
28	46.29062734	49.96758298	53.99333317	58.40258277	63.23351045	68.52811162	80.69769091
29	48.91079930	52.96628630	57.42303316	62.32271191	67.71135353	73.63979832	87.34652927
30	51.62267728	56.08493775	61.00706966	66.43884750	72.43547797	79.05818622	94.46078632

Periods	14%	15%	16%	17%	18%	19%	20%
1	1.00000000	1.00000000	1.00000000	1.00000000	1.00000000	1.00000000	1.00000000
2	2.14000000	2.15000000	2.16000000	2.17000000	2.18000000	2.19000000	2.20000000
3	3.43960000	3.47250000	3.50560000	3.53890000	3.57240000	3.60610000	3.64000000
4	4.92114400	4.99337500	5.06649600	5.14051300	5.21543200	5.29125900	5.36800000
5	6.61010416	6.74238125	6.87713536	7.01440021	7.15420976	7.29659821	7.44160000
6	8.53551874	8.75373844	8.97747702	9.20684825	9.44196752	9.68295187	9.92992000
7	10.73049137	11.06679920	11.41387334	11.77201245	12.14152167	12.52271273	12.91590400
8	13.23276016	13.72681908	14.24009307	14.77325456	15.32699557	15.90202814	16.49908480
9	16.08534658	16.78584195	17.51850797	18.28470784	19.08585477	19.92341349	20.79890176
10	19.33729510	20.30371824	21.32146924	22.39310817	23.52130863	24.70886205	25.95868211
11	23.04451641	24.34927597	25.73290432	27.19993656	28.75514419	30.40354584	32.15041853
12	27.27074871	29.00166737	30.85016901	32.82392578	34.93107014	37.18021955	39.58050224
13	32.08865353	34.35191748	36.78619605	39.40399316	42.21866276	45.24446127	48.49660269
14	37.58106503	40.50470510	43.67198742	47.10267200	50.81802206	54.84090891	59.19592323
15	43.84241413	47.58041086	51.65950541	56.11012623	60.96526603	66.26068160	72.03510787
16	50.98035211	55.71747249	60.92502627	66.64884769	72.93901392	79.85021111	87.44212945
17	59.11760141	65.07509336	71.67303048	78.97915180	87.06803642	96.02175122	105.93055534
18	68.39406560	75.83635737	84.14071536	93.40560761	103.74028298	115.26588395	128.11666640
19	78.96923479	88.21181097	98.60322981	110.28456090	123.41353392	138.16640190	154.73999969
20	91.02492766	102.44358262	115.37974658	130.03293626	146.62797002	165.41801826	186.68799962
21	104.76841753	118.81012001	134.84050604	153.13853542	174.02100463	197.84744173	225.02559955
22	120.43599598	137.63163801	157.41498700	180.17208644	206.34478546	236.43845566	271.03071946
23	138.29703542	159.27638372	183.60138492	211.80134114	244.48684684	282.36176223	326.23686335
24	158.65862038	184.16784127	213.97760651	248.80756913	289.49447928	337.01049706	392.48423602
25	181.87082723	212.79301747	249.21402355	292.10485588	342.60348554	402.04249150	471.98108322
26	208.33274304	245.71197009	290.08826732	342.76268138	405.27211294	479.43056488	567.37729986
27	238.49932707	283.56876560	337.50239009	402.03233722	479.22109327	571.52237221	681.85275984
28	272.88923286	327.10408044	392.50277250	471.37783454	566.48089006	681.11162293	819.22331180
29	312.09372546	377.16969250	456.30321610	552.51206642	669.44745027	811.52283129	984.06797417
30	356.78684702	434.74514638	530.31173068	647.43911771	790.94799132	966.71216923	1181.88156900

Exercise C

Using the ordinary annuity table, determine the cash value in the following problems. Be careful to distinguish the type of annuity that is specified!

	Payment Amount	Payment Frequency	Annuity Term	Type of Annuity	Interest Rate	Cash Value
1.	$3,000	Semi.	6 yr	Ordin.	6%	$42,576.09 (3,000 × 14.19202956 =)
2.	500	Annual	10	Due	5	*6,603.39 (500 × 13.20678716 =)*
3.	1,800	Semi.	4	Ordin.	4	*15,449.34 (1,800 × 8.58296905 =)*
4.	2,000	Semi.	5	Due	4	*22,337.43 (2,000 × 11.16871542 =)*
5.	750	Annual	8	Ordin.	6	*7,423.10 (750 × 9.89746791 =)*
6.	1,500	Semi.	3	Ordin.	6	*9,702.61 (1,500 × 6.46840988 =)*
7.	3,600	Annual	3	Due	4	*11,687.27 (3,600 × 3.24646400 =)*
8.	5,000	Annual	12	Due	3	*73,088.95 (5,000 × 14.61779045 =)*
9.	875	Semi.	5	Ordin.	4	*9,581.01 (875 × 10.94972100 =)*
10.	3,750	Semi.	3	Due	6	*24,984.23 (3,750 × 6.66246218 =)*

Sinking Funds

Objective 5

When a company has plans to replace worn equipment, expand their facilities, or retire a bond issue, they often establish a sinking fund. An investment in the form of a series of payments is made at the end of each period into a fund that pays interest. These payments, along with interest, accumulate over several periods in order to provide the desired amount. The money set aside in the form of payments and its accumulation is termed a **sinking fund.**

sinking fund

A sinking fund is very much like an annuity, except:

1. Sinking fund problems require a solution to the payment question, not the maturity or future value question.
2. Sinking fund problems usually require payments at the end of each period and are therefore similar to ordinary annuities.

To compute the payment amount in a sinking fund problem the following information is needed:

1. the future amount desired
2. the interest rate
3. the frequency that the investment is to be compounded
4. the length of time before the desired amount is required

Example 12

What payment amount will be required if $58,000 is needed ten years from now, and the interest rate is 6 percent compounded annually?

$58,000	= Amount desired
×.07586796	= (6%, 10 periods)—sinking funds table
$4,400.34	= Payment at the end of each year for ten years

(58,000 × .07586796 =)

Example 13

What payment amount will be required every six months if a $150,000 office addition will be needed in three years, and the interest rate is 8 percent compounded semiannually?

$150,000	= Amount desired
×.15076190	= (4%, 6 periods)
$22,614.29	= Payment at the end of each six months for three years

(150,000 × .15076190 =)

Exercise D

Using the sinking funds table, determine the payment amount in the following problems.

	Required Future Value	Payment Frequency	Sinking Fund Term	Interest Rate	Payment Amount
1.	$43,000	Annual	14 yr	6%	$2,046.15 (43,000 × .04758491 =)
2.	16,500	Semi.	5	6	*1,439.30 (16,500 × .08723051 =)*
3.	122,000	Annual	12	5	*7,664.70 (122,000 × .06282541 =)*
4.	42,500	Semi.	6	4	*3,168.78 (42,500 × .07455960 =)*
5.	65,000	Semi.	10	6	*2,419.02 (65,000 × .03721571 =)*
6.	25,000	Annual	3	3	*8,088.26 (25,000 × .32353036 =)*
7.	140,000	Annual	12	5	*8,795.56 (140,000 × .06282541 =)*
8.	75,000	Semi.	7	4	*4,695.14 (75,000 × .06260197 =)*
9.	62,000	Annual	4	5	*14,384.73 (62,000 × .23201183 =)*
10.	57,500	Semi.	12	6	*1,670.23 (57,500 × .02904742 =)*

Sinking Funds Table

Periods	¾%	1%	1½%	2%	2½%	3%
1	1.00000000	1.00000000	1.00000000	1.00000000	1.00000000	1.00000000
2	0.49813200	0.49751244	0.49627792	0.49504950	0.49382716	0.49261084
3	0.33084579	0.33002211	0.32838296	0.32675467	0.32513717	0.32353036
4	0.24720501	0.24628109	0.24444479	0.24262375	0.24081788	0.23902705
5	0.19702242	0.19603980	0.19408932	0.19215839	0.19024686	0.18835457
6	0.16356891	0.16254837	0.16052521	0.15852581	0.15654997	0.15459750
7	0.13967488	0.13862828	0.13655616	0.13451196	0.13249543	0.13050635
8	0.12175552	0.12069029	0.11858402	0.11650980	0.11446735	0.11245639
9	0.10781929	0.10674036	0.10460982	0.10251544	0.10045689	0.09843386
10	0.09667123	0.09558208	0.09343418	0.09132653	0.08925876	0.08723051
11	0.08755094	0.08645408	0.08429384	0.08217794	0.08010596	0.07807745
12	0.07995148	0.07884879	0.07667999	0.07455960	0.07248713	0.07046209
13	0.07352188	0.07241482	0.07024036	0.06811835	0.06604827	0.06402954
14	0.06801146	0.06690117	0.06472332	0.06260197	0.06053652	0.05852634
15	0.06323639	0.06212378	0.05994436	0.05782547	0.05576646	0.05376658
16	0.05905879	0.05794460	0.05576508	0.05365013	0.05159899	0.04961085
17	0.05537321	0.05425806	0.05207966	0.04996984	0.04792777	0.04595253
18	0.05209766	0.05098205	0.04880578	0.04670210	0.04467008	0.04270870
19	0.04916740	0.04805175	0.04587847	0.04378177	0.04176062	0.03981388
20	0.04653063	0.04541531	0.04324574	0.04115672	0.03914713	0.03721571
21	0.04414543	0.04303075	0.04086550	0.03878477	0.03678733	0.03487178
22	0.04197748	0.04086372	0.03870332	0.03663140	0.03464661	0.03274739
23	0.03999846	0.03888584	0.03673075	0.03466810	0.03269638	0.03081390
24	0.03818474	0.03707347	0.03492410	0.03287110	0.03091282	0.02904742
30	0.02984816	0.02874811	0.02663919	0.02464992	0.02277764	0.02101926

Sinking Funds Table

Periods	8%	9%	10%	11%	12%	13%
1	1.00000000	1.00000000	1.00000000	1.00000000	1.00000000	1.00000000
2	0.48076923	0.47846890	0.47619048	0.47393365	0.47169811	0.46948357
3	0.30803351	0.30505476	0.30211480	0.29921307	0.29634898	0.29352197
4	0.22192080	0.21866866	0.21547080	0.21232635	0.20923444	0.20619420
5	0.17045645	0.16709246	0.16379748	0.16057031	0.15740973	0.15431454
6	0.13631539	0.13291978	0.12960738	0.12637656	0.12322572	0.12015323
7	0.11207240	0.10869052	0.10540550	0.10221527	0.09911774	0.09611080
8	0.09401476	0.09067438	0.08744402	0.08432105	0.08130284	0.07838672
9	0.08007971	0.07679880	0.07364054	0.07060166	0.06767889	0.06486890
10	0.06902949	0.06582009	0.06274539	0.05980143	0.05698416	0.05428956
11	0.06007634	0.05694666	0.05396314	0.05112101	0.04841540	0.04584145
12	0.05269502	0.04965066	0.04676332	0.04402729	0.04143681	0.03898608
13	0.04652181	0.04356656	0.04077852	0.03815099	0.03567720	0.03335034
14	0.04129685	0.03843317	0.03574622	0.03322820	0.03087125	0.02866750
15	0.03682954	0.03405888	0.03147378	0.02906524	0.02682424	0.02474178
16	0.03297687	0.03029991	0.02781662	0.02551675	0.02339002	0.02142624
17	0.02962943	0.02704625	0.02466413	0.02247148	0.02045673	0.01860844
18	0.02670210	0.02421229	0.02193022	0.01984287	0.01793731	0.01620085
19	0.02412763	0.02173041	0.01954687	0.01756250	0.01576300	0.01413439
20	0.02185221	0.01954648	0.01745962	0.01557564	0.01387878	0.01235379
21	0.01983225	0.01761663	0.01562439	0.01383793	0.01224009	0.01081433
22	0.01803207	0.01590499	0.01400506	0.01231310	0.01081051	0.00947948
23	0.01642217	0.01438188	0.01257181	0.01097118	0.00955996	0.00831913
24	0.01497796	0.01302256	0.01129978	0.00978721	0.00846344	0.00730826
30	0.00882743	0.00733635	0.00607925	0.00502460	0.00414366	0.00341065

Periods	3½%	4%	4½%	5%	5½%	6%	7%
1	1.00000000	1.00000000	1.00000000	1.00000000	1.00000000	1.00000000	1.00000000
2	0.49140049	0.49019608	0.48899756	0.48780488	0.48661800	0.48543689	0.48309179
3	0.32193418	0.32034854	0.31877336	0.31720856	0.31565407	0.31410981	0.31105167
4	0.23725114	0.23549005	0.23374365	0.23201183	0.23029449	0.22859149	0.22522812
5	0.18648137	0.18462711	0.18279164	0.18097480	0.17917644	0.17739640	0.17389069
6	0.15266821	0.15076190	0.14887839	0.14701747	0.14517895	0.14336263	0.13979580
7	0.12854449	0.12660961	0.12470147	0.12281982	0.12096442	0.11913502	0.11555322
8	0.11047665	0.10852783	0.10660965	0.10472181	0.10286401	0.10103594	0.09746776
9	0.09644601	0.09449299	0.09257447	0.09069008	0.08883946	0.08702224	0.08348647
10	0.08524137	0.08329094	0.08137882	0.07950457	0.07766777	0.07586796	0.07237750
11	0.07609197	0.07414904	0.07224818	0.07038889	0.06857065	0.06679294	0.06335690
12	0.06848395	0.06655217	0.06466619	0.06282541	0.06102923	0.05927703	0.05590199
13	0.06206157	0.06014373	0.05827535	0.05645577	0.05468426	0.05296011	0.04965085
14	0.05657073	0.05466897	0.05282032	0.05102397	0.04927912	0.04758491	0.04434494
15	0.05182507	0.04994110	0.04811381	0.04634229	0.04462560	0.04296276	0.03979462
16	0.04768483	0.04582000	0.04401537	0.04226991	0.04058254	0.03895214	0.03585765
17	0.04404313	0.04219852	0.04041758	0.03869914	0.03704197	0.03544480	0.03242519
18	0.04081684	0.03899333	0.03723690	0.03554622	0.03391992	0.03235654	0.02941260
19	0.03794033	0.03613862	0.03440734	0.03274501	0.03115006	0.02962086	0.02675301
20	0.03536108	0.03358175	0.03187614	0.03024259	0.02867933	0.02718456	0.02439293
21	0.03303659	0.03128011	0.02960057	0.02799611	0.02646478	0.02500455	0.02228900
22	0.03093207	0.02919881	0.02754565	0.02597051	0.02447123	0.02304557	0.02040577
23	0.02901880	0.02730906	0.02568249	0.02413682	0.02266965	0.02127848	0.01871393
24	0.02727283	0.02558683	0.02398703	0.02247090	0.02103580	0.01967900	0.01718902
30	0.01937133	0.01783010	0.01639154	0.01505144	0.01380539	0.01264891	0.01058640

Periods	14%	15%	16%	17%	18%	19%	20%
1	1.00000000	1.00000000	1.00000000	1.00000000	1.00000000	1.00000000	1.00000000
2	0.46728972	0.46511628	0.46296296	0.46082949	0.45871560	0.45662100	0.45454545
3	0.29073148	0.28797696	0.28525787	0.28257368	0.27992386	0.27730789	0.27472527
4	0.20320478	0.20026535	0.19737507	0.19453311	0.19173867	0.18899094	0.18628912
5	0.15128355	0.14831555	0.14540938	0.14256386	0.13977784	0.13705017	0.13437970
6	0.11715750	0.11423691	0.11138987	0.10861480	0.10591013	0.10327429	0.10070575
7	0.09319238	0.09036036	0.08761268	0.08494724	0.08236200	0.07985490	0.07742393
8	0.07557002	0.07285009	0.07022426	0.06768989	0.06524436	0.06288506	0.06060942
9	0.06216838	0.05957402	0.05708249	0.05469051	0.05239482	0.05019220	0.04807946
10	0.05171354	0.04925206	0.04690108	0.04465660	0.04251464	0.04047131	0.03852276
11	0.04339427	0.04106898	0.03886075	0.03676479	0.03477639	0.03289090	0.03110379
12	0.03666933	0.03448078	0.03241473	0.03046558	0.02862781	0.02689602	0.02526496
13	0.03116366	0.02911046	0.02718411	0.02537814	0.02368621	0.02210215	0.02062000
14	0.02660914	0.02468849	0.02289797	0.02123022	0.01967806	0.01823456	0.01689306
15	0.02280896	0.02101705	0.01935752	0.01782209	0.01640278	0.01509191	0.01388212
16	0.01961540	0.01794769	0.01641362	0.01500401	0.01371008	0.01252345	0.01143614
17	0.01691544	0.01536686	0.01395225	0.01266157	0.01148527	0.01041431	0.00944015
18	0.01462115	0.01318629	0.01188485	0.01070600	0.00963946	0.00867559	0.00780539
19	0.01266316	0.01133635	0.01014166	0.00906745	0.00810284	0.00723765	0.00646245
20	0.01098600	0.00976147	0.00866703	0.00769036	0.00681998	0.00604529	0.00535653
21	0.00954486	0.00841679	0.00741617	0.00653004	0.00574643	0.00505440	0.00444394
22	0.00830317	0.00726577	0.00635264	0.00555025	0.00484626	0.00422943	0.00368962
23	0.00723081	0.00627839	0.00544658	0.00472141	0.00409020	0.00354156	0.00306526
24	0.00630284	0.00542983	0.00467339	0.00401917	0.00345430	0.00296727	0.00254787
30	0.00280279	0.00230020	0.00188568	0.00154455	0.00126431	0.00103443	0.00084611

ASSIGNMENT Chapter 13

Name ______________________ **Date** ______________ **Score** ____________

Skill Problems

True or False. Circle the best answer.

T F 1. A present value problem involves a series of investments, usually of an equal amount and at regular intervals. *F*

T F 2. The money set aside in the form of payments and its accumulation is termed an annuity due. *F*

T F 3. To calculate compound interest, the interest from the previous period is subtracted from the principal amount. *F*

T F 4. The total amount returned to the investor is termed the investment amount. *F*

T F 5. An annuity due requires that the payment be made at the end of the period. *T*

T F 6. Money has no "time value." *F*

T F 7. The current value of the future amount is termed the future value. *F*

T F 8. In an ordinary annuity, the payment is made at the end of the period. *T*

T F 9. The computer has made compounding interest inexpensive for the banking industry. *T*

T F 10. When a company has plans to replace worn equipment, they often establish a trust. *F*

Compute the investment balance when interest is compounded using the charts in the chapter.

Amount Invested	Annual Interest Rate	Term of Investment (Years)	Compound Frequency	Interest Rate Per Period	Number of Periods Paid	Compounded Amount	
$ 4,500	6%	5	Semi.	11. *3%*	12. *10*	13. *$6,047.62*	*(4,500 × 1.34391638 =)*
14,500	4	$7\frac{1}{2}$	Qtr.	14. *1*	15. *30*	16. *19,543.81*	*(14,500 × 1.34784892 =)*
58,700	5	6	Annually	17. *5*	18. *6*	19. *78,663.61*	*(58,700 × 1.34009564 =)*
75,840	10	3	Qtr.	20. *2.5*	21. *12*	22. *101,996.37*	*(75,840 × 1.34488882 =)*
6,000	9	2	Monthly	23. *3/4*	24. *24*	25. *7,178.48*	*(6,000 × 1.19641353 =)*
42,000	8	7	Semi.	26. *4*	27. *14*	28. *72,730.41*	*(42,000 × 1.73167645 =)*

Compute the compounded quarterly balance in the following eighteen-month savings account investment.

Quarter Ending Date	Beginning Balance	5% Annual Interest Pmt. for Quarter	Balance at End of Quarter	
June. 30, 01	$4,000.00	29. *$50.00*	30. *$4,050.00*	*(4,000.00 × .05 ×,25 = 50 + 4,000.00 =)*
Sep. 30, 01	31. *4,050.00*	32. *50.63*	33. *4,100.63*	*(4,050. × .05 × .25 = 50.63 + 4,050 =)*
Dec. 31, 01	34. *4,100.63*	35. *51.26*	36. *4,151.88*	*(4,100.63 × .05 × .25 = 51.26 + 4,100.63 =)*
Mar. 31, 02	37. *4,151.88*	38. *51.90*	39. *4,203.78*	*(4,151.88 × .05 × .25 = 51.90 + 4,151.88 =)*
June 30, 02	40. *4,203.78*	41. *52.55*	42. *4,256.33*	*(4,203.78 × .05 × .25 = 52.55 + 4,203.78 =)*
Sep. 30, 02	43. *4,256.33*	44. *53.20*	45. *4,309.53*	*(4,256.33 × .05 × .25 = 53.20 + 4,256.33 =)*

Complete the following problems to find the present value.

	Future Value	Annual Interest Rate	Compound Frequency	Term before Receipt	Effective Rate Per Period	Number of Periods Paid	Present Value	
46.	$41,000	10%	Qtr.	3 years	*2.5%*	*12*	*$30,485.79*	*(41,000 × .74355589 =)*
47.	54,500	6	Semi.	5	*3*	*10*	*40,553.12*	*(54,500 × .74409391 =)*
48.	16,750	5	Annually	7	*5*	*7*	*11,903.91*	*(16,750 × .71068133 =)*
49.	20,000	8	Semi.	15	*4*	*30*	*6,166.37*	*(20,000 × .30831867 =)*
50.	1,000	4	Qtr.	6	*1*	*24*	*787.57*	*(1,000 × .78756613 =)*
51.	16,000	18	Monthly	2	*1.5*	*24*	*11,192.70*	*(16,000 × .69954392 =)*

Fill in the missing information in problems 52 through 61. Use the compound interest and present value tables.

Present Value	Annual Rate of Interest	Term of Investment	Compound Frequency	Future Value
52. $23,500.00	6%	6 years	Semi.	*$33,505.38 (23,500 × 1.42576089 =)*
53. 41,000.00	4	6	Qtr.	*52,059.12 (41,000 × 1.26973465 =)*
54. *8,444.55*	8	5	Semi.	12,500.00 *(12,500 × .67556417 =)*
55. *16,159.49*	4	9	Annual	23,000.00 *(23,000 × .70258674 =)*
56. 5,860.00	10	2	Qtr.	*7,139.84 (5,860 × 1.21840290 =)*
57. *6,565.97*	5	4	Semi.	8,000.00 *(8,000 × .82074657 =)*
58. 26,000.00	9	6	Annual	*43,604.60 (26,000 × 1.67710011 =)*
59. *3,683.48*	5	10	Annual	6,000.00 *(6,000 × .61391325 =)*
60. 4,000.00	10	3	Qtr.	*5,379.56 (4,000 × 1.34488882 =)*
61. 61,000.00	4	4	Qtr.	*71,527.29 (61,000 × 1.17257864 =)*

Complete the following problems to find the cash value of the annuity.

	Investment Amount	Payment Frequency	Annuity Term	Type of Annuity	Interest Rate	Effective Rate Per Period	Number of Periods Paid	Future Value	
62.	$700	Semi.	10 years	due	4%	*2%*	*21*	*$17,348.32*	*(25.78331719 − 1 × 700 =)*
63.	640	Qtr.	$7\frac{1}{2}$	ordinary	6	*1.5*	*30*	*24,024.76*	*(37.53868137 × 640 =)*
64.	50	Monthly	2	ordinary	9	*$\frac{3}{4}$*	*24*	*1,309.42*	*(26.18847059 × 50 =)*
65.	654	Semi.	8	due	4	*2*	*17*	*12,433.89*	*(20.01207096 − 1 × 654 =)*
66.	3,600	Annual	28	ordinary	6	*6*	*28*	*246,701.20*	*(68.52811162 × 3,600 =)*
67.	9,500	Annual	12	due	5	*5*	*13*	*158,773.32*	*(17.71298285 − 1 × 9,500 =)*

Fill in the missing information in problems 68 through 77. Use the ordinary annuity table or sinking funds table.

	Present Value—Investment Amount	Payment Frequency	Annuity Term	Sinking Fund or Type of Annuity	Interest Rate	Cash Value—Future Value
68.	$1,000.00	Annual	10 years	Ordin. *(1,000 × 12.57789254 =)*	5%	*$12,577.89*
69.	5,600.00	Semi.	7	Due *(5,600 × 17.59891389 =)*	6	*98,553.91*
70.	*1,017.24*	Semi.	5	Sink. fund *(12,500 × .08137882 =)*	9	12,500.00
71.	*879.56*	Annual	12	Sink. fund *(14,000 × .06282541 =)*	5	14,000.00
72.	6,250.00	Annual	3	Due *(6,250 × 3.18362700 =)*	3	*19,897.67*
73.	*7,806.16*	Semi.	4	Sink. fund *(67,000 × .11650980 =)*	4	67,000.00
74.	1,250.00	Semi.	6	Ordin. *(1,250 × 14.19202956 =)*	6	*17,740.04*
75.	1,250.00	Annual	9	Ordin. *(1,250 × 11.02656432 =)*	5	*13,783.21*
76.	*10,083.60*	Annual	13	Sink. fund *(190,400 × .05296011 =)*	6	190,400.00
77.	*4,138.10*	Annual	11	Sink. fund *(53,000 × .07807745 =)*	3	53,000.00

Using the sinking funds table, determine the payment amount in the following problems.

	Future Value	Payment Frequency	Sinking Fund Term	Interest Rate	Effective Rate Per Period	Number of Periods Paid	Investment Amount	
78.	$15,400	Annual	13 years	8%	*8%*	*13*	*$716.44*	*(15,400 × .04652181 =)*
79.	40,000	Semi.	5	7	*3½*	*10*	*3,409.65*	*(40,000 × .08524137 =)*
80.	78,000	Qtr.	4	12	*3*	*16*	*3,869.65*	*(78,000 × .04961085 =)*
81.	$170,000	Annual	8	5½	*5½*	*8*	*17,486.88*	*(170,000 × .10286401 =)*
82.	11,500	Semi.	2	9	*4½*	*4*	*2,688.05*	*(11,500 × .23374365 =)*
83.	8,000	Annual	10	11	*11*	*10*	*478.41*	*(8,000 × .05980143 =)*

Business Application Problems

1. Hampe Associates purchased a building for $95,000, giving the seller a note earning 8 percent compounded quarterly and full repayment after three years. What is the amount of the payoff at the end of three years?
$95,000 × 1.26824179 = $120,482.96 payoff

2. The Wellar Corporation must pay a debt of $121,000 in five years. If the corporation knows that other investment opportunities pay 6 percent compounded semiannually, what is the current value of the debt?
$121,000 × .74409391 = $90,035.36 value

3. What amount should Wellar Corporation, in problem 2, deposit in a 6 percent sinking fund at the end of each quarter to be able to pay off their debt?
$121,000 × .04324574 = $5,232.73 deposit

4. Bill just made the sixth annual payment of $500 to an ordinary annuity that pays 7 percent compounded annually. What is the value of the annuity?
$500 × 7.15329074 = $3,576.65 value

5. Bill, in problem 4, makes no more payments to the annuity for three years. He will let the current balance grow at 7 percent compounded annually. What will be the new balance in the account?
$3,576.65 × 1.22504300 = $4,381.55 balance

6. Bill, in problem 5, withdrew the balance of the annuity at the end of three years of no deposits. What is the amount of interest earned over the nine-year life of the annuity?
$4,381.55 − $3,000 = $1,381.55 interest

7. Suli Ahmed has two certificates of deposit that pay 6 percent semiannually. He opened the first CD for $4,000 three years ago and the second CD for $1,000 one year ago. What is the value of his CDs in total?
$4,000 × 1.19405230 = $4,776.21
$1,000 × 1.06090000 = $1,060.90
$4,776.21 + $1,060.90 = $5,837.11 value

8. An ordinary annuity was purchased five years ago. The annuity pays 6 percent compounded quarterly. The quarterly payments have been $500. What is the amount of interest earned on the annuity to date?
$500 × 23.12366710 = $11,561.83 − ($500 × 20) = $1,561.83 interest

9. Alicia Hemez placed $162,000 she inherited in a savings account earning 3 percent compounded quarterly. She needs $200,000 to enroll in law school. She intends to enroll in two years. What will be the balance in the savings account in two years?
$162,000 × 1.06159885 = $171,979 balance

10. In problem 9, what is the additional amount that Alicia will need to deposit at the end of each quarter in a 6 percent quarterly compounded investment to reach her $200,000 goal for law school?
$200,000 − $171,979 = $28,021 × .11858402 = $3,322.84 deposit

11. Kimpton and Company placed $5,000 in a CD for a business deal five and a half years ago. This CD paid 6 percent, compounded quarterly. What would be the value of the CD today?
$5,000
×1.38756370 ($1\frac{1}{2}$%—22 periods)
$6,937.82 value

12. The Dalton Company wishes to pay off a promissory note due in eight years. The firm decides to establish a sinking fund and invest money at 11 percent compounded annually. The payments will be made once a year. The maturity value of the promissory note is $140,000. What will be the payment amount?
$140,000
×.08432105 (11%—8 periods)
$11,804.94 payment amount

13. A promissory note for $16,000 including interest is due in two years to the Bulster Company. What is the present value of the note if money can be invested at 6 percent compounded semiannually?
$16,000
×.88848705 (3%—4 periods)
$14,215.79 present value

14. A 6 percent contingent ordinary annuity was purchased by James D. Armour seven years ago. The semiannual payment was $600. What is the cash value of the annuity today?
$600
×17.08632416 (3%—14 periods)
$10,251.79 cash value

15. The parents of Jamez Hotz wished to save some money for her college education. They felt that if they could save $20,000 by the time she became 18 years old the balance of the education cost could be earned by part-time jobs, possible scholarships, etc. Jamez will soon have her twelfth birthday. What will be the payment amount on an ordinary annuity at a 6 percent rate if payments are made annually?
$20,000
×.14336263 (6%—6 periods)
$2,867.25 payment amount

16. Phil Orel deposited $1,700 in his savings account on July 1, 1986. The bank paid 4 percent compounded semiannually. On June 30, 1992, he deposited an additional $3,100 in his account. The bank began compounding deposits quarterly on January 1, 1996. On June 30, 1997, Phil withdrew the total amount in his savings account to take a trip to Europe. What was the value of the withdrawal?

$1,700
×1.26824179 (2%—12 periods) July 1, 1986, to June 30, 1992
$2,156.01
+3,100.00 June 30, 1992
$5,256.01
×1.14868567 (2%—7 periods) July 1, 1992, to December 31, 1995
$6,037.50
×1.06152015 (1%—6 periods) January 1, 1996, to June 30, 1997
$6,408.93 value of withdrawal

17. Genny Portan was recently widowed. She and her husband had made twelve annual payments of $2,500 on an annuity due at 6 percent. What is the value of the annuity before the next payment is due?

$2,500
×17.88213767 (18.88213767 − 1.00000000)
$44,705.34 value

18. The P.H. Rickton Corporation borrowed $67,500 from Villance Trust Company with a promissory note at 6 percent for five years. The Rickton Corporation negotiated an ordinary annuity at 4 percent on the same date. Payments will be made at six-month intervals. What will the payment amount be in order to repay the maturity value of the note to Villance Trust Company in five years? (Use the sinking funds table.)

$67,500 × .06 × $\frac{5}{1}$ = $20,250 interest
+20,250
$87,750 due at maturity
×.09132653 (2%—10 periods)
$8,013.90 payment

19. Benzol Products will need to retire (pay off) a $165,000 debt in eight years. In order to prepare for the payoff, the firm has decided to set aside some amount at the end of each year. The insurance firm they have been speaking to said that the amount set aside would earn 7 percent. What is the amount that will need to be set aside each year?

$165,000
×.09746776 (7%—8 periods)
$16,082.17 set aside

20. Jebson and Associates invested $60,000 five years ago at 8 percent compounded semiannually. What is the compounded amount?

$60,000
×1.48024428 (4%—10 periods)
$88,814.65 compounded amount

21. The Wycoff Corporation will need to retire a bond debt that is due to mature in six years. The bond issue has a total face value of $6,300,000. The treasurer has recommended that a sinking fund be set up to retire the debt. The funds can be invested at 7 percent compounded semiannually. What will be the amount of the semiannual payment to the sinking fund?

$6,300,000
×.06848395 ($3\frac{1}{2}$%—12 periods)
$431,448.57 semiannual payment

22. Write a complete sentence in answer to this question. Lester placed $5,000 in a cash account with his stockbroker. The account pays an annual rate of 9 percent compounded monthly. After five months, he decided to purchase 300 shares of a stock at a price of $15.50 per share. The broker charged a fee of $\frac{1}{2}$ percent of the total purchase price. Lester used the money from his cash account to pay for the stock. How much was left in the account after the shares were bought and the broker had been paid?

$5,000
×1.03806673 ($\frac{3}{4}$%—5 periods)
$5,190.33
−4,650.00 (300 × $15.50) stock price
−23.25 ($4,640 × .005) broker fee
$517.08
There was $517.08 left in the account.

23. Write a complete sentence in answer to this question. A savings and loan association is paying depositors 4 percent interest on savings deposits, and compounds the interest every six months. The funds are then lent to home buyers at 9 percent for home mortgages (monthly repayments). This savings and loan association has deposits and loans totaling $51,000,000. The overhead for office operations is maintained at 1 percent of deposits. What is the amount of profit for a one-year period?

$51,000,000 × .09 = $4,590,000 income
$51,000,000 × .04040000 (2%—2 periods) = $2,060,400 interest to depositors
$51,000,000 × .01 = $510,000 overhead
$4,590,000 = income
−2,060,400 = interest expense
−510,000 = overhead
$2,019,600
The profit was $2,019,600 for one year.

★ Challenge Problem

Silvia Ryan is a widow. Her husband left $20,000 for her living expenses. She placed the $20,000 in a savings account that paid 4 percent compounded semiannually. At the beginning of each year, she transferred $3,000 from her savings account to her checking account. Each month, she wrote herself a check for $250. How many monthly $250 checks will she be able to use before the $20,000 is used up? She made the first transfer at the end of the first year.

Year 2 = $20,000 × 1.04040000 =
$20,808 − $3,000 = $17,808
3 = $17,808 × 1.04040000 =
$18,527.44 − $3,000 = $15,527.44
4 = $15,527.44 × 1.04040000 =
$16,154.75 − $3,000 = $13,154.75
5 = $13,154.75 × 1.04040000 =
$13,686.20 − $3,000 = $10,686.20
6 = $10,686.20 × 1.04040000 =
$11,117.92 − $3,000 = $8,117.92
7 = $8,117.92 × 1.04040000 =
$8,445.88 − $3,000 = $5,445.88
8 = $5,445.88 × 1.04040000 =
$5,665.89 − $3,000 = $2,665.89
9 = $2,665.89 × 1.04040000 = $2,773.59

7 years × 12 = 84 + 11 in the 9th year
= <u>95 monthly checks</u>

MASTERY TEST Chapter 13

Name ____________________ Date ____________ Score ____________

Complete the following problems. Use the appropriate tables:

Present Value	Annual Rate of Interest	Term of Investment	Compound Frequency	Future Value
1. $ 2,000.00	8%	4 years	Qtrly.	$ *2,745.57*
2. *2,981.34*	9	6	Annual	5,000.00
3. 19,000.00	10	7	Semi.	*37,618.70*
4. *7,490.68*	8	3	Qtrly.	9,500.00

Investment Amount	Payment Frequency	Annuity Term	Type of Annuity or Sinking Fund	Interest Rate	Cash Value—Future Value
5. $ 400.00	Annual	7 years	Due	9%	$ *4,011.39*
6. *2,498.73*	Semi.	5	Sink. Fund	8	30,000.00
7. 375.00	Semi.	6	Ordinary	10	*5,968.92*
8. *24,271.84*	Annual	2	Sink. Fund	6	50,000.00

9. Dolores Heine placed $300 in a new savings account that paid 5 percent interest compounded semiannually. Her money has been on deposit for $4\frac{1}{2}$ years. What is the value of Ms. Heine's account at the present time if no other deposits or withdrawals have been made?
$300 × 1.24886297 = $374.66 present value

10. The Blackburn Corp. purchased a contingent annuity due eight years ago with payments of $750 annually. The rate of interest on the annuity was 7 percent. The treasurer of the firm decided to terminate the annuity and purchase bonds with the cash value. What was the cash value of the annuity?
$750 × 10.97798875
(7%—8 periods = 7%—9 periods—1.0000000) =
$8,233.49 cash value

DISCOUNTS 14

Suggested Classroom Time for This Chapter

hours 1 2 3 4 5

Perceived Student Difficulty with Topical Material

easy moderate difficult

OBJECTIVES

After mastering the material in this chapter, you will be able to:

1. Compute the selling price of merchandise subject to a trade discount. p. 264
2. Compute a single equivalent trade discount. p. 266
3. Compute the net price of merchandise subject to a cash discount using several methods. p. 266
4. Compute the net price in problems that include cash discounts and transportation charges using F.O.B. or C.O.D. p. 271
5. Understand and use the following terms and concepts:
   Trade Discount
   List Price
   Selling Price
   Direct Method
   Equivalent Trade Discount
   Cash Discount
   Terms of Trade
   Net Price
   E.O.M.
   R.O.G.
   F.O.B.
   F.O.B. Destination
   F.O.B. Shipping Point
   C.O.D.

SPORT-A-MERICA

In the new store, the additional space will allow for expansion of the merchandise now carried by Sport-A-Merica. Kerry knows that in order to continue to grow, the store should include new items. When the sales agent for Tourist Bicycles calls on her, she sees the possibility of adding racing models to the current line. The addition of a fine line of racing bicycles and related equipment would mean that Sport-A-Merica could sponsor road races. The store could use the two delivery vehicles as escorts for the cyclists, gain some media exposure for the sponsorship of the event, and promote public relations through the cyclists and their friends. Most importantly, the addition of the racing line would add to the profits of the firm.

The price offered by Tourist Bicycles, Inc. is dependent on the volume of sales and the shipping arrangements. Specifically, the agent for Tourist Bicycles offered a volume discount of 20 percent off the list price and another 5 percent for the handling of the freight. The agent also specified terms of trade of 2/10, n/30. Kerry considers placing an order for 20 bicycles with a list price of $1,295 each.

The discount would make a substantial difference in the net price to Sport-A-Merica and, therefore, in the amount of profit. Kerry decides to discuss this with Hal. It would be important to make this addition to the product line before the beginning of the next selling season for bicycles.

Trade Discounts

Objective 1

- ***Trade discounts are offered by one firm to another, not to final consumers. Final consumers may receive a "sale price" through a markdown.***

Manufacturers, wholesalers, and other middlemen often offer their customers discounts based on the buyer's providing some marketing functions. These discounts are termed **trade discounts.** The seller offers this price reduction if the buyer purchases in large quantities, transports or stores the merchandise for the seller, promotes the product to the final consumer, gathers sales information for the seller, or simply negotiates successfully. The number of marketing functions performed will determine the size of the trade discounts offered. The size of the discounts varies from seller to seller.

Finding the Selling Price

list price

selling price

The trade discount is computed from a **list price** established by the seller, and given in a catalog or on a price sheet. Since not all buyers will perform the same marketing functions, the trade discounts are not listed. The trade discount is subtracted from the list price in order to find the **selling price.** This is the amount that the seller is willing to accept for the merchandise when a buyer provides the agreed-upon marketing functions.

Example 1

A piece of office equipment is listed at $1,460. The seller agrees to offer the buyer a 15 percent trade discount if the buyer transports and installs the equipment. What is the selling price of the equipment?

$1,460 = List price (base)
×.15 = Trade discount percent (rate)
$219 = Trade discount amount (percentage)

$1,460 = List Price
−219 = Trade discount amount
$1,241 = Selling price

(1,460 × .15 =)

The same problem can be solved by using another shorter approach: Subtract the trade discount percent from 100 percent and then multiply.

Example 2

$1,460 = List price
×.85 = (100% − 15% trade discount)
$1,241 = Selling price

(1,460 × .85 =)

Example 3

If the selling firm in the previous example were to offer an additional 10 percent trade discount for allowing on-site inspections of the equipment by other prospective customers, the new selling price would be calculated as follows.

$1,460.00 = List price
−219.00 = Trade discount amount (15%)
$1,241.00 = (Base)
×.10 = Trade discount percent (rate)
$124.10 = Trade discount amount (10%) (percentage)

$1,241.00
−124.10 = Trade discount amount (10%)
$1,116.90 = Selling price

(1,241 × .10 =)

When two or more trade discounts are offered, it is termed a *chain discount,* or *series of trade discounts*. Notice that the series of trade discounts, 15 percent and 10 percent, is *not* the same as a 25 percent discount.

$1,460	$1,460
×.25	−365
$365	$1,095

Direct Method

The **direct method** is a much quicker and effective way to find the selling price. Each discount is subtracted from 100 percent, and each successive difference is then used to multiply the successively discounted prices, beginning with the list price, to find the selling price.

Example 4

The 20 percent and 5 percent trade discounts offered to Sport-A-Merica by Tourist Bicycles' agent on a group of 20 bicycles would have a significant effect on the selling price to Sport-A-Merica. The list price to the retailer of each bicycle is \$1,295. What is the selling price of the group of 20 bicycles?

\$1,295 Each
×20
\$25,900 Total

100%	100%
−20%	−5%
80%	95%

\$25,900 (list price) × .80 = \$20,720
\$20,720 × .95 = \$19,684 (selling price)

Note: This method will be used to determine the answers to the problems in this chapter.

(1,295 × 20 = 25,900 × .80 = 20,720 × .95 =)

Example 5

Monson Distributing purchased 50 computers that would be connected with the existing server. The list price per computer was \$600. J. Bonner, the distributor, agreed to allow the buyer a series of 10, 5, and 20 percent discounts for quantity, promotional considerations, and user feedback reasons. What will be the selling price of the lot of 50 computers?

\$600
×50
\$30,000 Total list price

\$30,000 × .90 = \$27,000
\$27,000 × .95 = \$25,650
\$25,650 × .80 = \$20,520 selling price

(600 × .50 = 30,000 × .90 = 27,000 × .95 = 25,650 × .80 =)

Exercise A

Find the total selling price in each of the following problems.

	List Price per Unit	Units Purchased	Trade Discount Offered	Total Selling Price
1.	\$ 64.00	100	20% (6,400 × .80 =)	\$5,120.00
2.	1,200.00	40	15 ***(48,000 × .85 =)***	***40,800.00***
3.	59.00	17	19 ***(1,003 × .81 =)***	***812.43***
4.	96.00	60	25 ***(5,760 × .75 =)***	***4,320.00***
5.	189.00	12	10 ***(2,268 × .90 =)***	***2,041.20***
6.	48.00	12	25 ***(576 × .75 =)***	***432.00***
7.	120.00	144	10 ***(17,280 × .90 =)***	***15,552.00***
8.	18.00	50	5 and 20 ***(900 × .95 = × .8 =)***	***684.00***
9.	185.00	3	10 and 10 ***(555 × .90 = × .90 =)***	***449.55***
10.	69.00	100	20 and 10 ***(6,900 × .80 = × .90 =)***	***4,968.00***

Objective 2
equivalent trade discount

There is a method that can be used to find one single discount percent equal to a series of trade discounts. This is termed an **equivalent trade discount.** It is found by subtracting the product of two trade discounts from their sum. This method is useful when the same series of discounts are used over and over.

Example 6

An 8 percent and a 20 percent discount are offered to the A. J. Rislone Corporation for purchasing 1,500 vertical hydrators listed at $46 each. A & K Distributing, the seller, offered the discounts because of the size of the purchase and because Rislone will be penetrating a new market area. The single equivalent trade discount and the selling price are computed as follows:

20%		20% =	.20
+8%		×8% =	×.08
28%	= Sum of trade discounts		.0160 = 1.6% product of trade discounts
−1.6%	= Product of trade discounts (.20 × .08)		
26.4%	= Single equivalent discount (20% and 8%)		

The single equivalent discount can be used to find the total trade discount amount.

$69,000	(1,500 × $46) = Total list price
× .264	= Equivalent trade discount percent
$18,216	= Total trade discount amount

By subtracting the single equivalent trade discount from 100%, the difference can be multiplied by the list price to find the selling price.

100%	$69,000 = Total list price
−26.4%	×.736
73.6%	$50,784 = Total selling price

(.20 + .08 = .20 × .08 =)

When there are three or more trade discounts offered, a single equivalent trade discount can still be found. The equivalent discount of two of the trade discounts with the third discount will provide an equivalent of all three discounts.

Exercise B

Find the total selling price of the following problems.

	List Price per Unit	Units Purchased	Trade Discount Offered	Total Selling Price	
1.	$45.00	200	5 and 8%	$7,866.00	(9,000 × .874 =)
2.	60.00	100	10 and 15	**4,590.00**	***(6,000 × .765 =)***
3.	25.00	25	5 and 12	**522.50**	***(625 × .836 =)***
4.	525.00	144	10 and 20	**54,432.00**	***(75,600 × .72 =)***
5.	1,200.00	65	20 and 5	**59,280.00**	***(78,000 × .76 =)***
6.	80.00	100	7, 5, and 10	**6,361.20**	***(8,000 × .79515 =)***
7.	42.00	500	10, 10, and 20	**13,608.00**	***(21,000 × .648 =)***
8.	130.00	25	5, 8, and 30	**1,988.35**	***(3,250 × .6118 =)***
9.	125.00	20	5, 10, and 10	**1,923.75**	***(2,500 × .7695 =)***
10.	7.50	50	10, 5, and 17	**266.12**	***(375 × .70965 =)***

Cash Discounts

Objective 3

Manufacturers, wholesalers, and other middlemen also offer their customers **cash discounts.** These discounts are not based on the customer's providing marketing functions, but are designed to motivate the customer to pay for the merchandise promptly. The motivating tool is a reduction in the selling price if the payment is made at an early date. The date (or dates) is specified in the **terms of trade** by the seller. Examples of terms of trade are 3/10,n/30 and 2/30,n/60; and these are shown on the invoice or bill. In each of

terms of trade

Cash discounts are offered to firms by other firms, not to final consumers.

the previous examples, the first number represents a percent discount. The second number following the slash mark represents the last day after the invoice or billing date that the discount is available.

Example 7

Terms of trade 1/10,n/60 mean that a 1 percent cash discount off the selling price of the merchandise is offered if the invoice is paid within the 10 days following the invoice date. The n/60 means that the customer has 60 days following the invoice date to pay the bill: In the first 10 days, a 1 percent discount is offered; in the next 50 days, no discount is offered. Sixty days after the invoice date, the unpaid bill may be placed in the hands of a collection agent. Firms that have their accounts placed in the hands of a collection agent will incur a financial penalty and the loss of a favorable status with the seller. The seller probably will no longer extend credit terms to the buyer, therefore requiring that future purchases be made on a cash basis (C.O.D.). Most new firms begin on a cash purchase basis until the seller is convinced that the firm is creditworthy.

A cash discount of $31.39 to WBC Systems, Inc., for prompt payment, based on the terms of 2/10,n30.

WORK RIGHT®

MASTER DESIGNED SHOWER DOORS AND BATH ENCLOSURES
4615 WORK RIGHT CIRCLE, LAKEPORT CA 95453-9302
Call (707) 263-0290
FAX (707) 263-4048

REMIT TO:
WORK RIGHT PRODUCTS, INC.
P.O.BOX 638
NOVATO, CA 94948-0638

INVOICE

INVOICE NUMBER	INVOICE DATE	PAGE
1971	June 15	1

BILL TO: WBC Systems, Inc.
1000 Nagle
Chicago, IL 60606

SHIP TO: Same

CUSTOMER PURCHASE ORDER NO.	TERMS	SHIP VIA	F.O.B. POINT
800-1982	2/10, n/30	UDIC	Lakeport

ORDERED BY	SALES REPRESENTATIVE	ORDER DATE	OUR ORDER NO.	CUSTOMER I.D.
Wally	Mike	6-2	15603-42	A

CL	DL	QUANTITY ORDERED	QUANTITY SHIPPED	PART NUMBER	DESCRIPTION / COMMENTS	UNIT	UNIT PRICE	EXTENDED PRICE
		5	5	DISTALL	STALL DOOR WRAP	EA	195.69	978.45
		5	5	DI-R-L-24	BOTTOM RAIL & DRIP	EA	49.84	249.20
		1	1	TE1400-60	TUB ENCLOSURE	EA	341.95	341.95
					TOTAL			$1569.60
					LESS 2% CASH DISCOUNT			−31.39
					NET			$1,538.21

Courtesy of Work Right Products, Inc., Lakeport, CA.

Example 8

An invoice listing goods purchased on May 26, with terms of trade 3/5,2/30,n/60, would give the buyer until May 31 to receive a 3 percent discount from the selling price, until June 25 to receive a 2 percent discount from the selling price, and until July 25 to receive no discount with no penalty on the selling price.

Finding the Net Price

net price

When the cash discount is subtracted from the selling price, the balance is termed the **net price.** If there is no cash discount available in the terms of the sale, the selling price may also be called the net price. The net price paid to the seller will complete the financial obligation.

Example 9

On October 17, Sport-A-Merica purchased merchandise from Pratt Body Builders, Inc., at a selling price of \$1,650 with terms of trade of 2/10,1/30,n/90. What is the net price if payment is made on November 6?

\$1,650	\$1,650.00 = Selling price
×.01	−16.50 = Cash discount
\$16.50	\$1,633.50 = Net price

or

\$1,650 = Selling price
×.99 = Percent of net price (100% − 1%)
\$1,633.50 = Net price

Note that Sport-A-Merica did not pay for the merchandise by October 27 and therefore did not qualify for the 2 percent cash discount.

(1,650 × .99 =)

Cash discounts are an important financial consideration, even though the discount rates sound small. Terms of trade of 2/10,n/30 equal a potential 36 percent annual discount rate. The proof is shown by the following: 2 percent discount can be taken if payment is made within ten days. By not paying on the tenth day, the money is in effect *borrowed* for the next 20-day period of the year. There are at least 18 such 20-day periods in a full year. Eighteen periods times 2 percent equals 36 percent annually (if the process were to repeat itself for a full year).

What would the net price be to Haeck & Co. if payment was made by November 9th?

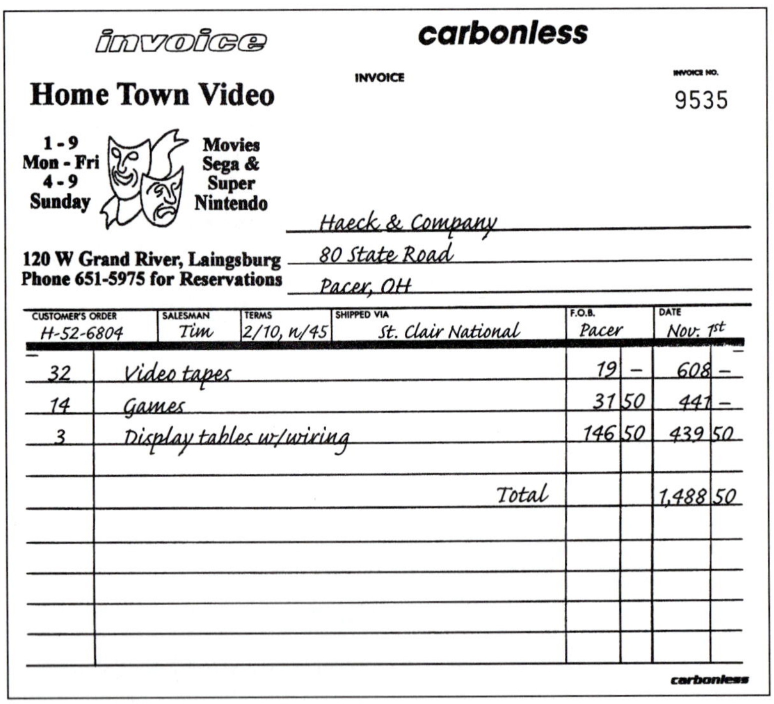
invoice carbonless

INVOICE — INVOICE NO. 9535

Home Town Video

1-9 Mon - Fri, 4-9 Sunday — Movies, Sega & Super Nintendo

120 W Grand River, Laingsburg
Phone 651-5975 for Reservations

Haeck & Company
80 State Road
Pacer, OH

CUSTOMER'S ORDER	SALESMAN	TERMS	SHIPPED VIA	F.O.B.	DATE
H-52-6804	Tim	2/10, n/45	St. Clair National	Pacer	Nov. 1st

Qty	Description	Price	Amount
32	Video tapes	19 –	608 –
14	Games	31 50	441 –
3	Display tables w/wiring	146 50	439 50
	Total		1,488 50

carbonless

Courtesy of Home Town Video, Laingsburg, MI.

Exercise C

Find the net price in the following problems. Note the dates.

	Selling Price	Terms of Trade	Invoice Date	Payment Date	Net Price (Payment Amount)
1.	\$675.00	3/15,n/30	July 26	Aug. 8	\$654.75 (675 × .97 =)
2.	450.00	2/10,n/30	Nov. 10	Nov. 17	*441.00 (450 × .98 =)*

Exercise C Continued

	Selling Price	Terms of Trade	Invoice Date	Payment Date	Net Price (Payment Amount)
3.	$ 200.00	1/10,n/90	Sept. 16	Oct. 30	*200.00 (NA)*
4.	65.00	2/10,1/30,n/60	Aug. 20	Sept. 12	*64.35 (65 × .99 =)*
5.	175.00	1/30,n/60	May 13	July 1	*175.00 (NA)*
6.	1,400.00	n/30	Sept. 15	Oct. 10	*1,400.00 (NA)*
7.	70.00	2/15,1/60,n/90	Jan. 3	Feb. 16	*69.30 (70 × .99 =)*
8.	500.00	2/10,n/120	June 30	July 2	*490.00 (500 × .98 =)*
9.	120.00	1/20,n/45	Aug. 1	Aug. 13	*118.80 (120 × .99 =)*
10.	89.00	1/30,n/60	Dec. 2	Dec. 30	*88.11 (89 × .99 =)*

End of Month (E.O.M.)

A variation in terms of trade is the end-of-month **(E.O.M.)** stipulation. Terms of trade of 2/10,n/60,E.O.M. indicate that the buyer computes the days of discount beginning at the end of the month in which the purchase was made. If the purchase is made after the twenty-fifth of the month, then an additional month is allowed in the discount time period.

Example 10

A $595 (selling price) purchase of merchandise was made on June 12, with terms of trade of 2/10,n/60,E.O.M. What is the last date that a cash discount is available?

In this example, E.O.M. means the end of the month of June (month in which the purchase is made). Ten days after the end of the month of June is July 10. The 2 percent discount is offered until July 10.

Example 11

An invoice for $69.80 is dated March 28 with terms of 3/20,n/30,E.O.M. What is the last date that a cash discount is available?

Here, E.O.M. would normally mean the end of the month of March. However, because the purchase was made after the twenty-fifth of the month, an additional month is added (April). Twenty days after the end of the month of April is May 20. The 3 percent discount is offered until May 20.

If payment is made before May 20, the net price on this purchase would be $67.71.

(69.80 × .97 =)

Exercise D

Find the net price in the following problems.

	Selling Price	Terms of Trade	Invoice Date	Payment Date	Net Price (Payment Amount)
1.	$ 190.00	2/10,n/30,E.O.M.	Oct. 15	Nov. 20	$190.00 (NA)
2.	150.00	1/20,n/60,E.O.M.	July 20	Sept. 26	*150.00 (NA)*
3.	60.00	1/10,n/30,E.O.M.	April 30	June 10	*59.40 (60 × .99 =)*
4.	750.00	2/20,E.O.M.	Feb. 28	April 16	*735.00 (750 × .98 =)*
5.	600.00	1/15,n/30,E.O.M.	May 10	June 30	*600.00 (NA)*
6.	260.00	3/10,n/60,E.O.M.	Aug. 31	Oct. 7	*252.20 (260 × .97 =)*
7.	550.00	2/10,n/20,E.O.M.	Jan. 16	Mar. 28	*550.00 (NA)*
8.	1,530.00	1/30,E.O.M.	Dec. 23	Jan. 25	*1,514.70 (1,530 × .99 =)*
9.	32.50	4/10,2/30,E.O.M.	June 30	Aug. 3	*31.20 (32.50 × .96 =)*
10.	700.00	2/20,1/30,E.O.M.	July 3	Aug. 25	*693.00 (700 × .99 =)*

Receipt of Goods (R.O.G.)

Another variation in terms of trade is the receipt-of-goods **(R.O.G.)** stipulation. This, like E.O.M., also specifies an extension of the period during which the buyer can still receive the cash discount. Terms of trade of 3/10,n/30,R.O.G. indicate that the buyer computes the days of discount beginning with the date that the merchandise is actually received by the buyer. The R.O.G. stipulation is often used when goods are not available for immediate shipment by the seller. By specifying R.O.G., the buyer also has an opportunity to inspect the goods received before making payment.

Example 12

A $1,385 purchase of equipment on August 20 is received on October 6. The terms of trade are 2/5,n/60,R.O.G. What is the latest date that a discount is available and what is the net price?

R.O.G. means that the discount period will be computed starting on the date that the goods are received: October 6 plus five days equals October 11. The net price equals:

$1,385 = Selling price (base)
×.02 = Cash discount rate (rate)
$27.70 = Cash discount amount (percentage)

$1,385.00 = Selling price
−27.70 = Cash discount amount
$1,357.30 = Net price, due on October 11

(1,385 × .02 =)

Exercise E

Find the net price in the following problems.

	Selling Price	Terms of Trade	Invoice Date	Date Received	Payment Date	Net Price (Payment Amount)
1.	$620.00	3/10,n/30,R.O.G.	Aug.6	Sept. 10	Sept. 17	$601.40 (620 × .97 =)
2.	85.00	2/5,n/10,R.O.G.	June 25	July 20	July 22	*83.30 (85 × .98 =)*
3.	120.00	1/10,n/60,R.O.G.	Jan. 16	Feb. 1	Feb. 13	*120.00 (NA)*
4.	450.00	2/5,n/30,R.O.G.	April 3	May 30	June 16	*450.00 (NA)*
5.	327.00	1/10,n/20,R.O.G.	Sept. 1	Sept. 3	Sept. 11	*323.73 (327 × .99 =)*
6.	700.00	1/30,n/60,R.O.G.	July 5	July 27	Aug. 13	*693.00 (700 × .99 =)*
7.	175.00	2/10,n/45,R.O.G.	Oct. 26	Nov. 30	Dec. 7	*171.50 (175 × .98 =)*
8.	78.00	2/20,n/30,R.O.G.	May 1	June 16	July 3	*76.44 (78 × .98 =)*
9.	325.00	1/20,R.O.G.	Feb. 28	March 26	March 31	*321.75 (325 × .99 =)*
10.	802.00	1/5,n/30,R.O.G.	Nov. 2	Nov. 16	Nov. 30	*802.00 (NA)*

Combination of Cash and Trade Discounts

When both cash and trade discounts are offered, the trade discount is subtracted before the cash discount is computed. The model below shows the relationship.

list price
− trade discounts
selling price
− cash discount
net price

Example 13

Sport-A-Merica purchased 125 baseball mitts from the O and W Manufacturing Company at a list price of $32 each. Sport-A-Merica was offered a series of trade discounts of 10 percent and 20 percent, on terms of trade of 3/10,n/60,E.O.M. The purchase was made on July 12, delivery was made on July 18, and payment was made on July 21. What was the net price on the purchase?

$32 = Price per unit at list price
×125 = Units purchased
$4,000 = Total list price
×.28 = Single equivalent discount (10% and 20%)
$1,120 = Trade discount amount

$4,000 = Total list price
−1,120 = Trade discount amount
$2,880 = Total selling price
×.03 = Cash discount rate
$86.40 = Cash discount amount

$2,880.00 = Total selling price
−86.40 = Cash discount
$2,793.60 = Net price

(4,000 × .28 = 2,880 × .03 =)

Exercise F

Find the net price in the following problems.

	Total List Price	Trade Discounts	Terms of Trade	Invoice Date	Date Received	Payment Date	Net Price
1.	$ 582	10 and 5%	1/10,n/30,E.O.M.	July 5	July 7	July 12	$492.63 (582 × .855 = × .99 =)
2.	1,600	10 and 7	2/10,n/60	May 3	May 6	June 5	*1,339.20* *(1,600 × .837 =)*
3.	270	5 and 7	2/10,n/60,R.O.G.	March 1	March 19	March 25	*233.77* *(270 × .8835 = × .98 =)*
4.	950	15 and 10	3/5,n/60,E.O.M.	July 29	Aug. 17	Sept. 3	*704.95* *(950 × .765 = × .97 =)*
5.	2,800	10 and 10	1/10,n/90	Dec. 2	Dec. 9	Dec. 11	*2,245.32* *(2,800 × .81 = × .99 =)*

Transportation Charges

Objective 4

The price of merchandise varies due to discounts made available by the seller. Much of the discount variation depends on the negotiating ability of the two parties. Another factor included in the negotiations that can influence the final price is the transportation charges.

Whatever the method of delivery, there is a transportation cost that must be paid by one of the parties. Either the seller will include the transportation cost in the quoted price, or the buyer will have to pay the cost in addition to the negotiated price. The transportation cost is variable; it may be as high as the cost of the merchandise itself, depending on the distance traveled, the means of delivery, and the weight. The businessperson should consider this cost as an important price factor.

F.O.B.

F.O.B. is an abbreviation for "free on board." It means that the seller will include in the quoted price the service of placing the merchandise on the transportation device—truck, boat, plane, train, etc.—for "free." From a cost standpoint, F.O.B. does not mean very much. The transportation device will be located near the seller's place of business. The meaningful portion of F.O.B. follows the abbreviation by indicating the location to which the merchandise is free of transportation costs.

F.O.B. Destination

When the buyer and seller agree that the merchandise will be shipped to a location specified by the buyer, and all of the transportation costs will be paid by the seller, the term **F.O.B. destination** is used. This means that the seller will ship the merchandise free to the buyer's place of business. The price per unit quotation will always include transportation. The buyer will usually not be aware of the transportation costs.

Example 14

Downson & Patrick sold 12 pairs of women's slacks to the Tops & Bottoms Shop in Palo Alto, California. The price of each pair of slacks was $26. The merchandise will be shipped via truck by Baldwin Transportation Co. for a charge of $43.72. Downson & Patrick agreed to ship F.O.B. Palo Alto (the destination) from New York. How much will the Tops & Bottoms Shop pay the seller (terms of 2/10,n/60) if payment is within the discount period?

$26	Price per pair of slacks
×12	Pairs of slacks
$312	Total selling price
×.02	(2% cash discount)
$6.24	

$312.00 = Total selling price
−6.24 = Cash discount
$305.76 = Net price

The net price of $305.76 will be paid by Tops & Bottoms to Downson & Patrick for the merchandise, including transportation. Downson & Patrick owes Baldwin Transportation Co. $43.72 for the transportation.

(26 × 12 = 312 × .02 =)

Sometimes the buyer specifies several locations if the merchandise is to be shipped to all of the buyer's plants, retail locations, or warehouses. The result is the same if *F.O.B. destination* is specified. The seller pays the freight costs to the multiple locations that the buyer specifies.

F.O.B. Shipping Point

❍ *F.O.B. charges can amount to a large sum of money. Be sure you know who is paying when you negotiate price!*

When the buyer and seller agree that the price of the merchandise does not include the transportation charges, **F.O.B. shipping point** is specified. Under these terms, the buyer is responsible for all of the costs associated with moving the merchandise from the seller's location. In this case, F.O.B. indicates that the seller will place the merchandise on the transportation device free. From that point on, the costs are covered by the buyer. This allows the buyer to negotiate the transportation method and cost separately. This may aid in reducing costs and speeding up delivery. Each cost can be identified using F.O.B. shipping point. Transportation charges will not be part of the seller's invoice amount.

Example 15

Sport-A-Merica purchased 42 exercise sets for their stock. Each set was priced at $268, F.O.B. shipping point, 3/10, n/30. Built Well, Inc., the seller, indicated that the best method of shipment would be by truck at $38 a set. What would be the total cost of the 42 sets delivered if payment is made within the discount period?

$268 = Price per set
+38 = Transportation cost per set
$306 = Total cost per set delivered
×42 = Sets
$12,852 = Total cost/42 sets delivered

The amount paid to Built Well, Inc. would be:

$268 = Selling price per set
×42 = Sets
$11,256 = Selling price of 42 sets
×.03
$337.68 = Cash discount

$11,256.00 = Selling price
−337.68 = Cash discount
$10,918.32 = Net price/42 sets

The amount paid to the common carrier would be:

$38 = Price of delivery per set
×42 = Sets
$1,596 = Total cost of delivery/42 sets

$10,918.32 = Cost of 42 sets
+1,596.00 = Cost of delivery
$12,514.32 = Total cost of 42 sets delivered

Note that the 3/10, n/30 terms apply to the merchandise only. The common carrier did not offer a cash discount on the cost of the transportation.

(268 + 38 = 306 × 42 =) (268 × 42= 11,256 × .03 = 38 × 42 =)

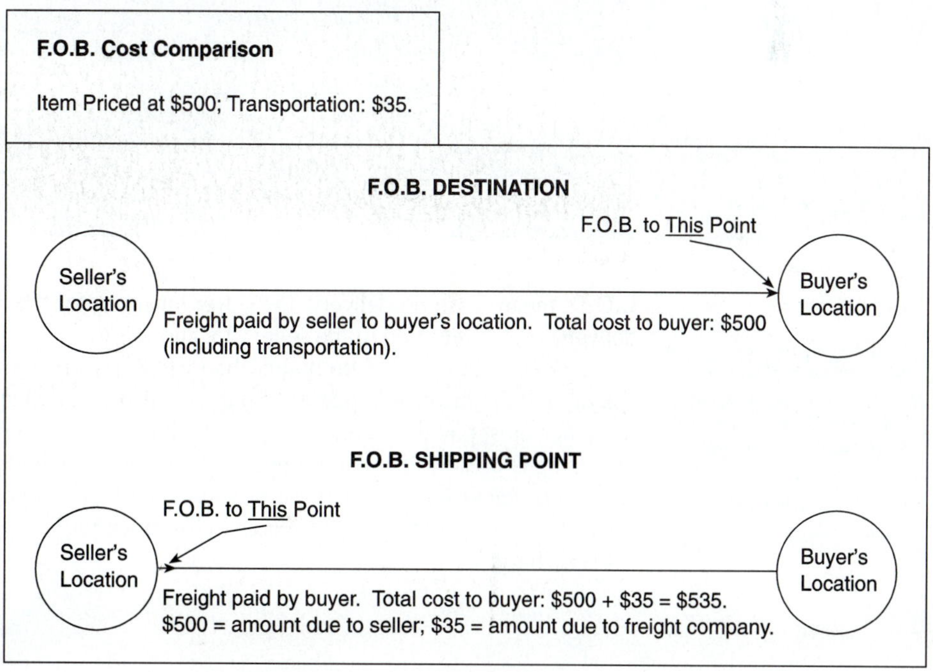

Exercise G

Fill in the missing amount in the problems below. Assume that the payment was made within the discount period and that the terms of trade apply to the sales price of the merchandise.

	Sales Price per Unit	Units Purchased	Terms of Trade	Freight Cost F.O.B. Destination	Cost to Buyer (total)
1.	$14.65	144	2/10,n/30	$ 16.46	$2,067.41
					(14.65 × 144 = × .98 =)
2.	*8.50*	90	n/60	135.68	765.00
					(765 ÷ 90 =)
3.	18.00	150	3/10,n/30	72.00	*2,619.00*
					(18 × 150 = × .97 =)
					Amount Due Seller
4.	6.00	30	—	27.00	*$180.00*
					(6 × 30 =)
5.	60.00	*30*	1/10,n/60	32.00	1,782.00
					(1,782 ÷ 60 = ÷ .99 =)

Exercise H

Fill in the missing amount in the following problems. Assume that payment was made within the discount period and that the terms of trade apply to the sales price of the merchandise.

	Sales Price per Unit	Units Purchased	Freight Cost F.O.B. Shipping Point	Terms of Trade	Cost to Buyer (total)
1.	$ 1.44	1,500	$ 162.50	3/10,n/20	$2,257.70
					(1.44 × 1,500 = × .97 = +162.50 =)
2.	15.00	80	2,359.00	n/60	*3,559.00*
					(15 × 80 = + 2,359 =)
3.	1,950.00	12	493.51	1/10,n/30	*23,659.51*
					(1,950 × 12 = × .99 = + 439.51 =)
					Amount Due Seller
4.	84.90	*150*	125.00	n/60	$12,735.00 *(12,735 ÷ 84.90 =)*
5.	*76.00*	71	792.00	3/10,n/45	5,234.12
					(5,234.12 ÷ .97 = 5,396 ÷ 71 =)

C.O.D.

● *New firms usually must pay for their purchases in cash until credit is established. This makes it difficult to be financially successful in the early years of a business.*

C.O.D. means cash on delivery. The buyer must pay for the merchandise in cash upon delivery, or it is returned to the seller's location.

C.O.D. prices tend to be higher than F.O.B. destination prices, because the seller is risking that the buyer will pay for the merchandise at delivery. If the buyer refuses to pay, the seller must pay the transportation cost of both directions and lose the sale. This type of pricing is used when selling impulse consumer merchandise, such as that sold through Internet or television merchandise programs, and when dealing with middlemen with poor or new credit ratings. The merchandise price will be quoted "C.O.D." or "plus C.O.D. charges."

Example 16

The Palderson Company sold Mrs. Alma Dickinson four bolts of material for her new upholstery business. The bolts were quoted at $268 per bolt C.O.D. What is the amount that the delivery firm will collect when the merchandise is accepted by Mrs. Dickinson?

$268 = Price per bolt/C.O.D.
×4 = Bolts
$1,072 = Total price for 4 bolts

(268 × 4 =)

Example 17

The Watson family was watching television and noticed an announcement that stated an item could be purchased for $29.95 plus C.O.D. charges. The family decided to order one. When the delivery man came to the home, he indicated that the cost of delivery was $4.50. How much is due the delivery man?

$29.95 = Price of merchandise
+4.50 = C.O.D. charges
$34.45 = Total amount due

(29.95 + 4.50 =)

ASSIGNMENT Chapter 14

Name ______________________ **Date** ______________ **Score** ____________

Skill Problems

Find the total selling price in each of the following problems

	List Price per Unit	Units Purchased	Trade Discount Offered	Total Selling Price	
1.	$64.50	200	15%	*$10,965.*	*(64.5 × 200 × .85 =)*
2.	995.00	31	26	*22,825.30*	*(995 × 31 × .74 =)*
3.	5.45	1,412	10	*6,925.86*	*(5.45 × 1,412 × .90 =)*
4.	39.95	120	10	*4,314.60*	*(39.95 × 120 × .90 =)*
5.	629.00	50	15	*26,732.50*	*(629 × 50 × .85 =)*
6.	95.00	30	10	*2,565.*	*(95 × 30 = × .90 =)*
7.	65.00	144	10 and 20	*6,739.20*	*(65 × 144 = × .72 =)*
8.	125.00	25	5	*2,968.75*	*(125 × 25 = × .95 =)*
9.	65.00	30	5 and 15	*1,574.63*	*(65 × 30 = × .8075 =)*
10.	41.50	144	7, 20, and 5	*4,223.84*	*(41.50 × 144 = × .93 = × .80 = × .95 =)*

Find the total selling price in each of the following problems using the direct method.

	List Price per Unit	Units Purchased	Trade Discount Offered	Total Selling Price	
11.	$145,100.00	2	5 and 8%	*$253,634.80*	*(145,100 × 2 × .95 = × .92 =)*
12.	367.00	47	10 and 15	*13,195.49*	*(367 × 47 × .90 = × .85 =)*
13.	19.95	651	10, 5, and 3	*10,771.14*	*(19.95 × 651 × .90 = × .95 = × .97 =)*
14.	10,483.00	8	20 and 10	*60,382.08*	*(10,483 × 8 × .80 = × .90 =)*
15.	150.00	245	5 and 12	*30,723.00*	*(150 × 245 × .95 = × .88 =)*

Find the total selling price in each of the following problems using the equivalent trade discount.

	List Price per Unit	Units Purchased	Trade Discount Offered	Equivalent Trade Discount		Total Selling Price	
16.	$ 842.00	411	10 and 10%	*19%*	*(.1 + .1 − .01 =)*	*$280,310.22*	*(842 × 411 × .81 =)*
17.	8,164.00	203	17 and 5	*21.15*	*(.17 + .05 − .0085 =)*	*1,306,774.74*	*(8,164 × 203 × .7885 =)*
18.	29.60	764	15, 9, and 5	*26.5175*	*(.15 + .09 − .0135 = .2265 + .05 − .011325 =)*	*16,617.63*	*(29.6 × 764 × .734825 =)*
19.	8.45	120	27 and 20	*41.6*	*(.27 + .20 − .054 =)*	*592.18*	*(8.45 × 120 × .584 =)*
20.	60.00	200	14 and 5	*18.3*	*(.14 + .05 − .007 =)*	*9,804.00*	*(60 × 200 × .817 =)*

Find the last date that the cash discount is available in the following problems.

	Terms of Trade	Invoice Date	Last Date Discount Available
21.	3/10,n/30 E.O.M.	June 1	*July 10*
22.	1/20,n/45 E.O.M.	Nov. 10	*December 20*
23.	2/10,n/120 E.O.M.	Dec. 27	*February 10*
24.	3/15,n/90 E.O.M.	July 18	*August 15*

	Terms of Trade	Invoice Date	Date Goods Received	Last Date Discount Available
25.	2/10,n/30,R.O.G.	June 20	Nov. 6	*November 16*
26.	3/15,n/60,R.O.G.	Sept. 1	Oct. 29	*November 13*
27.	1/5,n/45,R.O.G.	Jan. 1	Apr. 9	*April 14*
28.	1/30,n/90,R.O.G.	Oct. 25	Nov. 26	*December 26*

Find the net price in the following problems. Note the dates.

	Selling Price	Terms of Trade	Invoice Date	Payment Date	Net Price	
29.	$939.00	1/30,n/60	June 30	July 16	*$929.61*	*(939 × .99 =)*
30.	1,237.00	2/10,n/120	Nov. 14	Nov. 30	*1,237*	*n/a*
31.	429.95	1/20,n/45	Dec. 9	Dec. 24	*425.65*	*(429.95 × .99 =)*
32.	40.00	2/15,1/60,n/90	May 4	July 1	*39.60*	*(40 × .99 =)*
33.	1,930.00	n/30	March 1	March 2	*1,930*	*n/a*

Find the net price in the following problems.

	Total List Price	Trade Discounts	Terms of Trade	Invoice Date	Date Received	Payment Date	Net Price	
34.	$900	10 and 7%	1/10,n/90	Mar. 3	Apr. 2	Mar. 11	*$745.77*	*(900 × .9 × .93 × .99 =)*
35.	295	5 and 10	3/5,n/60,E.O.M.	Feb. 7	Mar. 9	Mar. 10	*252.23*	*(295 × .95 × .9 =)*
36.	940	10 and 10	2/10,n/30,R.O.G.	Oct. 4	Nov. 10	Nov. 15	*746.17*	*(940 × .9 × .9 × .98 =)*
37.	300	5 and 8	1/10,n/90	Sep. 8	Nov. 22	Sep. 16	*259.58*	*(300 × .95 × .92 × .99 =)*

Fill in the missing amount in the following problems. Assume that payment was made within the discount period and that the terms apply to the sales price for the merchandise.

	Selling Price	Terms of Trade	Invoice Date	Date Received	Payment Date	Net Price
38.	$259.00	3/10,n/20	July 12	July 15	July 20	*$251.23 (259 × .97 =)*
39.	890.00	1/20,n/90,E.O.M.	Aug. 29	Sept. 2	Oct. 15	*881.10 (890 × .99 =)*
40.	59.50	2/15,n/60,R.O.G.	Oct. 21	Nov. 3	Nov. 16	*58.31 (59.50 × .98 =)*
41.	120.00	2/10,n/60,E.O.M.	Feb. 12	March 1	March 7	*117.60 (120 × .98 =)*
42.	856.00	n/30	April 22	April 26	May 10	*856.00 (NA)*

	Total List Price	Trade Discounts	Terms of Trade	Invoice Date	Date Received	Payment Date	Net Price
43.	$3,260.00	5 and 10%	2/10,n/30	July 6	July 9	July 20	*$2,787.30 (3,260 × .855 =)*
44.	695.00	12	2/5,n/30,R.O.G.	Jan. 5	Feb. 16	Feb. 20	*599.37 (695 × .88 = × .98 =)*

Sales Price Per Unit	Units Bought	Trade Discounts	Terms of Trade	Transportation Costs	Total Cost to Buyer	Amount Due Seller	
$19.95	122	5 and 10%	3/10,n/20	$95.00 F.O.B. destination	**45.** *$2,018.55*	**46.** *$2,018.55*	*(19.95 × 122 ×.95 × .9 × .97 =)*
9.70	148	12	2/15,n/60, E.O.G.	850.00 F.O.B. shipping point	**47.** *2,088.06*	**48.** *1,238.06*	*((9.7 × 148 × .88 × .98) + 850 =) (2,088.06 − 850 =)*
5.05	35	10 and 10	n/30	65.00 F.O.B. shipping point	**49.** *208.17*	**50.** *143.17*	*((5.05 × 35 × .9 × .9) + 65 =) (208.17 − 65 =)*
87.45	48	5, 4, and 10	2/5,n/90, R.O.G.	44.50 F.O.B. destination	**51.** *3,376.48*	**52.** *3,376.48*	*(87.45 × 48 × .95 × .96 × .9 × .98 =)*
6.82	912	—	3/10,n/90	78.75 F.O.B. shipping point	**53.** *6,111.99*	**54.** *6,033.24*	*((6.82 × 912 × .97) + 78.75 =)*
7.49	74	15 and 10	—	29.99 C.O.D.	**55.** *453.99*	**56.** *453.99*	*((7.49 × 74 × .85 × .9) + 29.99 =)*
55.00	35	20 and 8	5/30,n/90	64.95 F.O.B. destination	**57.** *1,345.96*	**58.** *1,345.96*	*(55 × 35 × .8 × .92 × .95 =)*
184.25	341	4 and 7	1/10,n/30 R.O.G.	984.44 F.O.B. shipping point	**59.** *56,517.45*	**60.** *55,533.01*	*((184.25 × 341 × .96 × 93 × .99) + 984.44 =) (56,517.45 − 984.44 =)*

Determine the missing amount in the following problems. Assume that the payment was made within the discount period and that the terms of trade apply to the invoice amount.

	Amount of Invoice	Terms of Trade	Transportation Agreement	Transportation Costs		Amount Due Seller
61.	$ 453.80	2/10,n/60	F.O.B. destination	$32.80	*(453.80 × .98 =)*	*$444.72*
62.	980.00	—	C.O.D. included	62.68	*(NA)*	*980.00*
63.	1,698.00	1/10,n/45	F.O.B. shipping point	—	*(1,698 × .99 =)*	*1,681.02*

Business Application Problems

1. Anter Company sold 140 units to Kass Furniture at a list price of $345 each. The invoice indicated a quantity discount of 15 percent. What is the selling price for the total sale?
$345 × 140 × .85 = $41,055 sale

2. Junior Devins purchased 50 TVs for the retirement complex his company is building. Each TV is priced at $439. The invoice indicates terms of 2/15, n/30. If payment is made on the tenth day following the invoice date, what is the total net amount due?
$439 × 50 × .98 = $21,511 due

3. A multistation exercise unit was purchased by Realife Body at a list price of $1,250. A trade discount of 17 percent was negotiated. The terms on the invoice indicated 2/10,n/40. The purchase was made on July 14th. Payment was made on July 20th. What is the amount of the trade discount?
$1,250 × .17 = $212.50 trade discount

4. In problem 3, what is the last day that the cash discount is available to Realife Body?
July 14th + 10 = July 24th

5. In problem 3, what is the amount that Realife Body will need to remit to pay off the invoice in full on July 20th?
$1,250 × .83 × .98 = $1,016.75 payoff

6. Thurston's, a retail store, ordered 300 gas grills listed at $239 each from Mentrail Manufacturing on April 3. The seller agreed to a 10 percent discount due to the size of the order. They also agreed to a 15 percent discount because Thurston's was arranging and paying for shipping. The seller also agreed to terms of 1/15,n/30, R.O.G. The goods were shipped on April 12 and received on April 16. The invoice was paid on April 28. What was the total list price of the order?
$239 × 300 = $71,700 list

7. What is the total selling price of the order in problem 6?
$239 × 300 × .9 × .85 = $54,850.50 selling price

8. What is the total net price of the order in problem 6?
$239 × 300 × .9 × .85 × .99 = $54,302 net

9. If the shipping cost in problem 6 were $450, what is the total cost of each unit to Thurston's?
$239 × .9 × .85 × .99 = $181.01 + $450/300 = $182.51 cost

10. Kelsey, Inc. sold three sets of office furniture to Jambor Rhe Associates for $1,649 per set, list, F.O.B. destination. The agreement stipulated a trade discount of 12 percent and terms of 2/10,E.O.M. The purchase was made on March 12. What is the amount that Kelsey should expect in full payment on April 20?
$1,649 × 3 × .88 = $4,353.36 payment

11. The Sales Furniture Store ordered 20 dining sets at a list price of $230 each. The manufacturer offered a 12 percent discount for the size of the order. What is the sales price of the total order?

$230
×20
$4,600 × .88 = $4,048 sales price

12. Billings and Alpert Company received a price quotation from two firms for 900 units of a processing tool used in the canning business. A representative of the Bernwalther Corporation quoted a price of $3.91 per unit F.O.B. destination, with terms of trade of 2/10,n/60. A spokesperson for the Altron Digital Corporation quoted a price of $3.78 per unit F.O.B. shipping point, with terms of trade 3/15,n/45. Total shipping costs were expected to be $93.60 for the 900 units. Which price quotation should be accepted, considering the price of the merchandise, transportation costs, and cash discounts?

Bernwalther	***Altron***
$3.91	***$3.78***
×900	***×900***
$3,519	***$3,402 × .97 = $3,299.94***
×.98	***+ 93.60***
$3,448.62	***$3,393.54***

Accept Altron

13. Kelly bought 50 units of raw materials for her manufacturing firm at a list price of $17 per unit. Terms of trade were 3/10,n/20,E.O.M. The purchase was made on March 12, the goods were delivered on April 6, and payment was made on April 16. What was the amount of the check sent by Kelly's company?

$17
×50
$850 check

14. Karen Liscomb was instructed to pay for a purchase made on September 1 for 30 pieces of office furniture that were listed at $174 each. A trade discount of 20 percent was offered, as well as terms of trade of 1/10,n/30. What is the net price if payment is made on September 9?

174
×30
$5,220 × .80 × .99 = $4,134.24 payment

15. The Aikman Office Furniture Company ordered three desks with matching chairs for their inventory. The items were priced at $1,900 for each set (desk and chair). The manufacturer offered a 10 and a 15 percent discount if the Aikman Office Furniture Company would handle and ship the merchandise themselves. What is the selling price of a set to the buyer?
$1,900 × .9 × .85 = $1,453.50 per set

16. The Fredrick Insurance Agency purchased six new computers listed at $480 each. They received a 10 percent discount because they purchased more than five, and they received an additional 5 percent discount because they took immediate delivery. Terms of trade were 2/5,n/30. What was the net price if payment was made within the discount period?

$480
×6
$2,880 × .9 × .95 × .98
= $2,413.15 net price

17. A check for the net amount of $84.28 is received from a customer who was extended terms of 2/10,n/30. The check is past the discount period. You must contact the customer and indicate the correct amount due. What is the amount due?

$84.28 ÷ .98 = $86 due

18. Sharon Sadlowski, payments clerk and secretary for the Tog and Slumber Shop, made out a check to the P. K. Worth Company for a purchase made on July 14. The purchase was for 30 travel wear sets priced at $31 per set. The P. K. Worth Company extended terms of 1/10,n/45, E.O.M. The price quoted was F.O.B. shipping point. Freight on the purchase was $21.80. The check will have the date of August 5. What should be the amount of the check to P. K. Worth Company?

$31
×30
$930 × .99 = $920.70 check

19. Peter Frank ordered 20 men's jackets at the Apparel Center in Chicago for his men's shop. The jackets were priced at $90 each, with terms of trade of 2/20,1/30,n/45,R.O.G. The purchase was made on March 12 for fall delivery. The merchandise was received on August 5 and payment was made on August 22. What was the net price of the lot of 20 jackets?

$90
×20
$1,800 × .98 = $1,764 net price

20. Using the invoice below, show the purchase of seven refrigerators from Wilton's by Petti and Barret Construction Company on July 17, 2001. Each refrigerator will cost $722 after a trade discount has been considered. There will be a $184 C.O.D. charge added to the invoice. Find the total amount due at delivery.

$722
×7
$5,054
+ 184
$5,238 total due

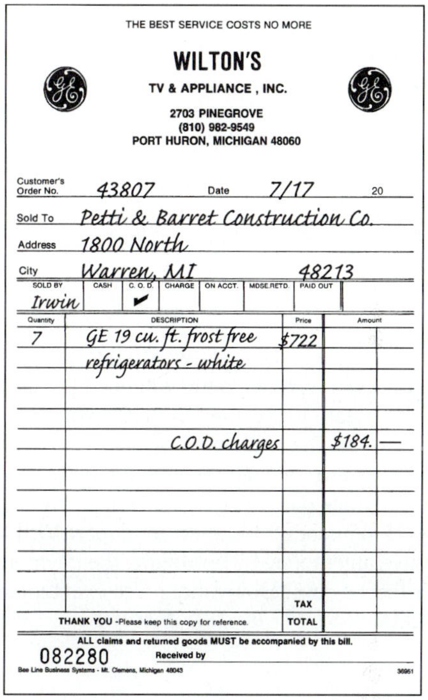
THE BEST SERVICE COSTS NO MORE

WILTON'S
TV & APPLIANCE , INC.
2703 PINEGROVE
(810) 982-9549
PORT HURON, MICHIGAN 48060

Customer's Order No. 43807 Date 7/17 20
Sold To Petti & Barret Construction Co.
Address 1800 North
City Warren, MI 48213

SOLD BY	CASH	C. O. D.	CHARGE	ON ACCT.	MDSE.RETD.	PAID OUT
Irwin		✔				

Quantity	DESCRIPTION	Price	Amount
7	GE 19 cu. ft. frost free	$722	
	refrigerators - white		
	C.O.D. charges		$184. —
		TAX	
THANK YOU -Please keep this copy for reference.		TOTAL	

ALL claims and returned goods MUST be accompanied by this bill.
082280 Received by
Bee Line Business Systems - Mt. Clemens, Michigan 48043 36951

Courtesy of Wilton's TV & Appliance, Inc., Port Huron, MI.

21. Pulver and Company sold three video systems to Tec-Rec, Inc. The price was $1,680 per system F.O.B. the buyer's locations in St. Paul, Boston, and Atlanta. Terms of 3/15,n/90 were offered. The transportation costs totaled $567.80. How much did Pulver and Company receive for the systems after the cash discount and transportation costs were considered?

$1,680
×3
$5,040 × .97 = $4,888.80 received

22. Patio Encounters paid $382.50 for a table-and-four-chair set after the supplier provided a 15 percent trade discount. What was the list price of the set?

$382.50 ÷ .85 = $450 original price

23. Write a complete sentence in answer to this question. Randy Marsh paid an invoice with a check for $509.25. The vendor had agreed to a 25 percent trade discount and terms of 3/10,n/60. Randy paid the invoice within the first ten days. What was the list price of the goods before the trade and cash discounts?
$509.25 ÷ .75 ÷ .97 = $700
The list price was $700.

24. Write a complete sentence in answer to this question. A check for $1,162.80 was sent on August 30 for merchandise purchased from Eron Corporation. Eron Corporation allowed a series of 15 percent, 10 percent, and 5 percent trade discounts, as well as terms of 3/10,E.O.M. What was the list price of the merchandise? The invoice was dated July 12.
$1,162.80 ÷ .85 ÷ .9 ÷ .95
= $1,600
The list price was $1,600.

25. Write a complete sentence in answer to this question. The Jackson and Fulmer Company will need to borrow the money necessary to pay the $56,850 invoice from Acme Corporation. The terms of trade are 2/10,n/30. The company will receive enough revenue on the twenty-fifth day after the invoice date to repay the lender. The lender will charge 17 percent for the short-term loan. Will the company save any money by borrowing the payment amount for the 15 days? How much? Use exact interest.
$56,850 × .02 = $1,137.00 cash discount
$56,850
−1,137
$55,713 borrowed $\times .17 \times \frac{15}{365}$ *= $389.23 interest*

$1,137.00 discount
−389.23 interest expenses
$747.77
The company will save $747.77.

★ Challenge Problem

Deanna purchased 1,200 pounds of a food additive for the Ideal Food Company at a list price of $9 per pound. The purchase was made on February 12. Delivery was made on March 31. Deanna negotiated a series of trade discounts of 10 percent, 10 percent, and 7 percent. Terms of trade were 3/10,1/30,n/45,R.O.G. Because of an overpurchase, $1,300 of the additive was returned to the supplier on April 16 for credit at list price. A partial payment of $500 was made on May 7, and the balance was paid on May 28. What was the amount of the payment on May 28? The supplier uses E.O.M.
1,200
×$9
$10,800 total list price
−1,300 returned
$9,500 total purchase × .9 × .9 × .93
= $7,156.35 selling price
$7,156.35
−515.46 effect of $500 payment ($500 = .97 of $515.46)
$6,640.89
×.99
$6,574.48 amount of payment

MASTERY TEST Chapter 14

Name ______________________ **Date** ______________ **Score** ____________

Select the best phrase or term.

Answer	Item		Phrases/Terms
L	**1.** Equivalent trade discount	A.	Discount period begins at end of the month
		B.	Three times per year
J	**2.** Cash discount	C.	C.O.D.
		D.	Abbreviation of cash discount agreement
E	**3.** Trade discount	E.	Offered for buyer aiding the seller
H	**4.** Seller pays transportation costs	F.	List price less trade discounts
		G.	Selling, storing, gathering information
O	**5.** Chain discount	H.	F.O.B. destination
		I.	F.O.B. shipping point
C	**6.** Buyer pays the transportation costs to the seller, who pays the transportation firm	J.	A reduction in the selling price
		K.	Usually found in a catalogue or price sheet
		L.	Equal to two or more trade discounts
D	**7.** Terms of trade	M.	Discount period begins when the goods are received
M	**8.** R.O.G.	N.	Amount due after cash discount is subtracted
N	**9.** Net price	O.	More than one trade discount
G	**10.** Marketing functions		
I	**11.** Buyer pays the transportation costs to the transportation firm		
A	**12.** E.O.M.		
K	**13.** List price		
F	**14.** Selling price		

Compute the selling price in the following problems.

List Price	Trade Discounts Offered	Selling Price
15. $600	18%	*$492.*
16. 890	10 and 20%	*640.80*

Compute the net price in the following problems. Assume discounts are taken.

Amount of Invoice	Transportation Agreement	Terms of Trade	Transportation Costs	Amount Due Seller
17. $900	F.O.B. destination	1/10,n/60	$12.00	*$891.*
18. 748	F.O.B. shipping point	2/10,n/30,R.O.G.	—	*733.04*

Compute the net price in the following problems. Assume discounts are taken.

	Total List Price	Trade Discounts	Terms of Trade	Invoice Date	Payment Date	Date Rec'd	Net Price
19.	$ 800	10 and 20%	2/10,n/30,R.O.G.	Dec. 3	Jan. 5	Dec. 28	*$564.48*
20.	1,000	5, 5, and 10%	1/20,n/60	May 10	May 31	May 12	*812.25*
21.	650	5, 10, and 8%	2/15,n/60,E.O.M.	July 16	Aug. 12	July 20	*501.06*

22. The accounts payable clerk of the A-K Marketing Company made out a check for the amount due on the purchase of ten file cabinets. The list price of the cabinets was $92 each. The firm had received an 18 percent trade discount on terms of 2/30,n/60. What is the amount of the check if payment is made within the discount period?

$92
× 10
$920/.82 × .98 = $739.31 check

23. What is the amount due on a $600 purchase of merchandise bought on April 22? There is a 10 and 12 percent trade discount offered on terms of 2/5,1/20,n/60,R.O.G. The goods were received on May 2 and the payment is made on May 18.
$600 × .9 × .88 × .99 = $470.45 due

Markup

15

Suggested Classroom Time for This Chapter

hours 1 2 3 4 5

Perceived Student Difficulty with Topical Material

easy moderate difficult

Objectives

After mastering the material in this chapter, you will be able to:

1. Calculate sales based on cost or sales. p. 286
2. Calculate cost based on cost or sales. p. 289
3. Calculate markup based on cost or sales. p. 289
4. Calculate the markdown or new selling price. p. 290
5. Convert a markup on cost to a markup on sales. p. 291
6. Convert a markup on sales to a markup on cost. p. 291
7. Determine wholesale catalog prices when a trade discount and a markup are to be considered. p. 292
8. Understand and use the following terms:
   Markup
   Cost Price
   Selling Price
   List Price

Sport-A-Merica

Hal has ordered a new line of exercise equipment for Sport-A-Merica. The items, including rowing machines, stationary walking machines, and electronic stationary bikes, will arrive at the end of the month. Using the feedback he receives from customers and information from the trade association, he knows the area is a good market for the equipment. Sales of this equipment will offset the sluggish sales during the slower months when the weather turns bad. The major buyers of this equipment are expected to be either senior citizens who want to stay inside where it is safe and close to a phone, or exercise enthusiasts who want to set up an exercise area of their own to avoid joining a gym. Though these items will not sell in the volume Sport-A-Merica is used to selling, say, tennis balls, the fact that they are big-ticket items will allow for substantial profits.

Hal knows that the price marked on the item must cover the cost of the item *and* a portion of the overhead, just to break even. The exercise machine that he plans to purchase for $420 will be marked up by 20 percent, for example. That markup will cover his overhead and provide a profit, especially when he sells all of the units that he plans to order.

At the same time, he has some items in the tennis area that are not moving very fast, including a few of the remaining tennis shirts that he purchased in the spring. He is considering a 40 percent markdown just to see if he can move the inventory. In order to make room for the newest products in the equipment area, he feels Sport-A-Merica should advertise a markdown in order to sell them quickly. All of this price concern is an ongoing question for the firm. It is a question that Hal continues to struggle with, recalculate, and adjust.

In order for a merchandising firm to stay in business, it must sell its merchandise for more than it pays for it. The merchant must increase the price enough to cover overhead, such as the cost of utilities, rent, taxes, wages, and insurance. There must be an incentive to risk the capital investment that is required to go into business. This incentive, or profit, is a return on the capital investment, a reward for the risk taken, and, in some cases, a compensation for hours worked by the owners. Overhead and profit are grouped together and termed **markup** by merchants. Markup is the difference between the amount that is paid for the merchandise and the amount that it is sold for by the merchant. Markup is used by all members of the channel of distribution, such as retailers, wholesalers, distributors, manufacturers, and independent salespeople. Any firm or person dealing with the flow of goods or materials through the channel of distribution comes into contact with markup. Members of the channel of distribution are often termed *middlemen*. This term is used because their activities are between the manufacturer of the product and its use or consumption by the final consumer.

markup

cost price

The price the firm pays for an item is termed the **cost price.** Cash and trade discounts must be subtracted to find the cost price, which is the net price paid for merchandise.

selling price

The **selling price** is the amount for which merchandise is sold to the customers of the firm. The selling price is found by adding a markup to a cost price (i.e., cost price + markup = selling price).

Example 1

A merchant purchased a desk for \$165 and marked it up by \$70. What is the selling price on the desk?

\$165 = Cost price
+70 = Markup
\$235 = Selling price

(165 + 70 =)

Often the markup is not a specific dollar amount but a percent. The percent is determined by considering the amount of overhead and profit that the firm must earn to stay in business. The percent of markup must be based on one of the two factors in the formula: cost price + markup = selling price.

Markup on Cost

Middlemen often mark up goods based on cost. The cost price of the merchandise is increased by the markup rate. In this case, the markup rate is based on the cost price of the merchandise. The cost price is, therefore, 100 percent (the base). The selling price percent is the total of the cost price (100 percent) and the markup rate.

Objective 1

Example 2

A gift wholesaler purchased a carload of decorative radios for a price of \$15 per item. The goods were priced to sell using the company's standard markup rate of 40 percent. What is the selling price?

100% +	40% =	140%
cost price +	markup =	selling price
\$15		

The selling price in this example is the amount. The 140 percent is the rate and the \$15 cost price is base. The amount is found by:

A = B × R
A = \$15 × 1.40 (100% + 40%)
A = \$21 (selling price)

(15 × 1.40 =)

❍ *Consider using the percent model in markup problems.*

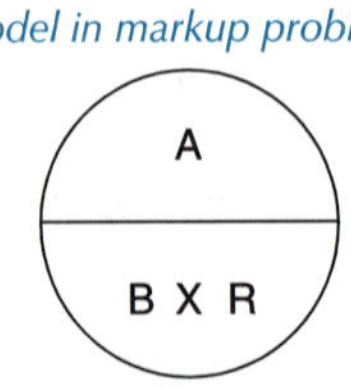

❍ *Solving Algebraically*
S = Selling Price
S = 1.4 × C
S = 1.4 × \$15
S = \$21

Example 3

Sport-A-Merica marks up pool merchandise 55 percent on cost. A pool cleaner is purchased for $8. What is the selling price?

100% + 55% = 155%
cost price + markup = selling price
$8

A = B × R
A = $8 × 1.55
A = $12.40 (selling price)

(8 × 1.55 =)

Exercise A

Determine the selling price and markup amount for the following cost prices. Round off your answer to the nearest cent.

	Cost Price	Markup	Markup Percent on Cost	Selling Price
1.	$ 40.00	$20.00	50%	$60.00 (40 × 1.50 =)
2.	60.00	***12.00***	20	***72.00 (60 × 1.20 =)***
3.	100.00	***35.00***	35	***135.00 (100 × 1.35 =)***
4.	125.00	***50.00***	40	***175.00 (125 × 1.40 =)***
5.	68.00	***15.64***	23	***83.64 (68 × 1.23 =)***

Markup on Sales

Goods can also be marked up based on selling price. If the markup is based on the selling price or sales, then the selling price is 100 percent (the base).

Example 4

● ***Have the students find the selling price of an item with a cost price of $20 and a markup of 50%. Solve for a markup on cost, then a markup on sales.***
(on cost = $20 × $1.50 = $30)
(on sales = $20 ÷ .5 = $40)

Using the information in example 2, what is the selling price of an item that cost $15 if it will be marked up 40 percent on sales?

60% + 40% = 100%
cost price + markup = selling price
$15

The selling price in this example is the base (100 percent). The 60 percent is the rate, and $15 is the amount. The base is found by:

B = A ÷ R
B = $15 ÷ .60
B = $25 (selling price)

(15 ÷ .60 =)

○ *Solving Algebraically*
S = Selling Price
S = C + .4S
.6S = C
S = C ÷ .6
S = $15 ÷ 6
S = $25

Notice that a problem that indicates a markup on sales is handled differently *and* results in a different answer than one that indicates markup on cost.

Example 5

Hal marked up a $420 piece of exercise equipment 20 percent on sales. What was the selling price that Sport-A-Merica will mark on the price tag?

80% + 20% = 100%
cost price + markup = selling price
$420

Base = Amount ÷ Rate
Base = $420 ÷ .80
Base = $525 (selling price)

(420 ÷ .80 =)

○ *A given percent markup on sales is different than a markup on cost. Which provides more dollars in markup?*

Exercise B

Determine the selling price and markup amount. Round off your answer to the nearest cent.

	Cost Price	Markup	Markup Percent on Sales	Selling Price
1.	$150	$37.50	20%	$187.50 (150 ÷ .80 =)
2.	250	*107.14*	30	*357.14 (250 ÷ .70 =)*
3.	80	*26.67*	25	*106.67 (80 ÷ .75 =)*
4.	48	*112.00*	70	*160.00 (48 ÷ .30 =)*
5.	13	*10.64*	45	*23.64 (13 ÷ .55 =)*

Exercise C

Determine the selling price. Note the basis of the markup rate. Round off your answers to the nearest cent.

	Cost Price	Markup Rate	Markup on	Selling Price
1.	$90.00	10%	Cost Price	$99.00 (90 × 1.10 =)
2.	30.00	40	Cost Price	*42.00 (30 × 1.40 =)*
3.	160.00	25	Selling Price	*213.33 (160 ÷ .75 =)*
4.	146.00	60	Selling Price	*365.00 (146 ÷ .40 =)*
5.	12.00	50	Selling Price	*24.00 (12 ÷ .50 =)*
6.	14.50	40	Cost Price	*20.30 (14.50 × 1.40 =)*
7.	6.50	60	Cost Price	*10.40 (6.50 × 1.60 =)*
8.	780.00	45	Cost Price	*1,131.00 (780 × 1.45 =)*
9.	19.00	25	Selling Price	*25.33 (19 ÷ .75 =)*
10.	46.00	70	Selling Price	*153.33 (46 ÷ .30 =)*

Finding Cost, Sales, Markup, or Markdown

There are occasions when a middleman needs to know an answer other than the selling price in a markup problem. The information needed may be the cost price, the markup amount, the markup percent, the percent of cost to sales, etc. There are many possible questions. These questions require the student to understand the markup formula and the way the percent works. It would be difficult to memorize every procedure for every problem type. The following examples represent some types of markup problems. You should determine if the problem is a markup on cost or one on sales, then solve for either base, rate, or amount.

Notice that in all of the examples, a basic procedure is followed: (1) the markup formula is used, (2) the markup base is identified with 100 percent, (3) the dollar values available are included in the formula, (4) the percent values available are included in the formula, and (5) the unknown factor is determined. This procedure is an excellent method to solve markup problems. It is recommended that the student adopt this procedure until another method proves to be more workable.

Objective 2

Example 6 To find the cost price when the markup is on cost—What is the cost price of an article priced to sell at $40, if the merchant has marked it up 20 percent on cost?

100% + 20% = 120%
cost price + markup = selling price
$40

Base = Amount ÷ Rate
Base = $40 ÷ 1.20
Base = $33.33 (answer)

(40 ÷ 1.20 =)

Example 7 To find the cost price when the markup is on sales—What is the cost price of an item priced to sell at $65, if the markup is 30 percent on sales?

70% + 30% = 100%
cost price + markup = selling price
$65

Amount = Base × Rate
Amount = $65 × .70
Amount = $45.50 (answer)

(65 × .70 =)

Objective 3

Example 8 To find the markup when the markup percent is on sales—What is the markup on an item priced to sell at $50, if the markup percent is 20 percent on sales?

80% + 20% = 100%
cost price + markup = selling price
$50

Amount = Base × Rate
Amount = $50 × .20
Amount = $10 (answer)

(50 × .20 =)

Example 9 To find the markup when the markup percent is on cost—What is the markup on an item priced to sell at $80, if the markup percent is 30 percent on cost?

100% + 30% = 130%
cost price + markup = selling price
$80

Base = Amount ÷ Rate
Base = $80 ÷ 1.3
Base = $61.54 (cost price)

Amount = Base × Rate
Amount = $61.54 × .30
Amount = $18.46 (answer)

(80 ÷ 1.30 = 61.54 × .3 =)

Example 10

To find the percent of cost to sales—What is the percent of cost to sales of an item that has a 25 percent markup on cost? The item is priced to sell at $80.

100% + 25% = 125%
cost price + markup = selling price
$80

Base = Amount ÷ Rate
Base = $80 ÷ 1.25
Base = $64 (cost price)
Rate = Amount ÷ Base
Rate = $64 ÷ $80
Rate = 80% (answer)

(80 ÷ 1.25 = 64 ÷ 80 =)

Exercise D

Fill in the blank lines below. Note the markup base—cost or sales. Round off your answers to the nearest percent or cent.

	Cost Price	Selling Price	Markup Percent	Markup on	Markup
1.	$128.00	$160.00	20%	Sales	$32.00 (160 × .20 =)
2.	*35.00*	50.00	30	Sales	*15.00 (50 × .30 =)*
3.	40.00	*68.00*	70	Cost	*28.00 (40 × .70 =)*
4.	93.00	*111.60*	20	Cost	*18.60 (93 × .20 =)*
5.	*29.90*	46.00	35	Sales	*16.10 (46 × .35 =)*
6.	*33.25*	47.50	30	Sales	*14.25 (47.50 × .30 =)*
7.	22.00	39.95	*82*	Cost	*17.95 (17.95 ÷ 22 =)*
8.	45.00	65.50	*31*	Sales	*20.50 (20.50 ÷ 65.50 =)*
9.	*171.00*	*228.00*	25	Sales	57.00 *(57 ÷ .25 =)*
10.	*228.57*	*308.57*	35	Cost	80.00 *(80 ÷ .35 =)*
11.	20.00	48.00	*140*	Cost	*28.00 (28 ÷ 20 =)*
12.	36.00	48.00	*25*	Sales	*12.00 (12 ÷ 48 =)*
13.	126.00	200.00	*37*	Sales	*74.00 (74 ÷ 200 =)*
14.	*44.80*	56.00	20	Sales	*11.20 (56 × .20 =)*
15.	42.00	*189.00*	350	Cost	*147.00 (42 × 3.50 =)*

Objective 4
Markdown or Loss

There are times that a retailer may decide to mark down the current inventory in order to move the merchandise. The retailer usually decides on a rate of markdown and then marks the price tags accordingly. The markdown is always based on the previous selling price.

Example 11

Hal has decided to mark down some of the tennis shirts that had been priced at $59.95. The markdown will be 40 percent. What will be the sale price?

60% + 40% = 100%
sale price + markdown = selling price (previous)
$59.95

○ *How sale merchandise is marked!*

Amount = Base × Rate
Amount = $59.95 × .60
Amount = $35.97 (sale price)

(30 × .60 =)

● *Your local newspaper is a great source of example problems in markdown!*

When merchandise is sold at an actual loss (less than the cost price), the percent of loss must be calculated. The loss is always based on the cost price.

Example 12

Find the percent of loss of an item that cost \$50 and was sold for \$40.

100%
cost price + loss = selling price
\$50 − \$10 = \$40

Rate = Amount ÷ Base
Rate = \$10 ÷ \$50
Rate = 20% (loss)

(10 ÷ 50 =)

Exercise E

Determine the answer in the exercises below. Round off your answer to the nearest percent or cent.

	Selling Price	Markdown	Sale Price
1.	\$ 85	30%	\$59.50 (85 × .70 =)
2.	17	40	*10.20 (17 × .60 =)*
3.	40	35	*26.00 (40 × .65 =)*
4.	17.95	20	*14.36 (17.95 × .80 =)*
	Cost Price	**Selling Price**	**Loss (%)**
5.	\$80	\$60	*25% (20 ÷ 80 =)*
6.	35	20	*43 (15 ÷ 35 =)*
7.	80	68	*15 (12 ÷ 80 =)*
8.	129	89	*31 (40 ÷ 129 =)*

Conversion of Markup Rate
Objectives 5 & 6

A markup of 20 percent on sales is not the same amount as a markup of 20 percent on cost. When a merchant wishes to compare various markups, the markups must be stated using the same base. This will allow for a direct comparison of markup rates. The markup rate on one base can be changed to another base by using the same markup formula.

To change the markup rate on one base (cost or sales) to another base, divide the current markup percent by the new base percent.

Example 13

What is the markup percent on sales if the markup percent on cost is 30 percent?

100% + 30% = 130%
cost price + markup = selling price

$$\text{Markup percent on sales} = \frac{30\% \text{ (current markup percent)}}{130\% \text{ (new base percent } - \text{ sales)}}$$

Markup percent on sales = 23% (answer)

The new markup percent on sales in the above example can be used in the markup formula as follows.

77% + 23% = 100%
cost price + markup = selling price

(Continued on next page)

Example 13 continued

The two markup values (30 percent on cost and 23 percent on sales) are equal. To prove this, use a cost price of $3 in each case to determine the selling price.

on Cost

100% + 30% = 130%
cost price + markup = selling price
$3
Amount = Base × Rate
Amount = $3 × 1.30
Amount = $3.90 (the selling price)

on Sales

77% + 23% = 100%
cost price + markup = selling price
$3
Base = Amount ÷ Rate
Base = $3 ÷ .77
Base = $3.90 (the selling price)

(.30 ÷ 1.30 =)

Example 14

What is the markup percent on cost if the markup percent on sales is 20 percent?

80% + 20% = 100%
cost price + markup = selling price

$$\text{Markup percent on cost} = \frac{20\% \text{ (current markup percent)}}{80\% \text{ (new base percent} - \text{cost)}}$$

Markup percent on cost = 25% (answer)

(.20 ÷ .80 =)

In the previous example, a markup of 20 percent on sales is equal to a markup of 25 percent on cost. Proof: What is the selling price of a product that cost $85.60 (using 20 percent on sales and 25 percent on cost)?

on Cost

100% + 25% = 125%
cost price + markup = selling price
$85.60
Amount = Base × Rate
Amount = $85.60 × 1.25
Amount = $107 (answer)

on Sales

80% + 20% = 100%
cost price + markup = selling price
$85.60
Base = Amount ÷ Rate
Base = $85.60 ÷ .80
Base = $107 (answer)

Exercise F

Determine the markup percent on the new base in the exercises below. Round off your answers to the nearest percent.

Markup Percent on Cost	Markup Percent on Sales	Markup Percent on Cost	Markup Percent on Sales
1. 40%	29% (.40 ÷ 1.40 =)	2. 50	33 (.50 ÷ 1.50 =)
3. 67	40 (.40 ÷ .60 =)	4. 100	50 (.50 ÷ .50 =)
5. 70	41 (.70 ÷ 1.70 =)	6. 30	23 (.30 ÷ 1.30 =)
7. 67	40 (.40 ÷ .60 =)	8. 20	17 (.20 ÷ 1.20 =)
9. 45	31 (.45 ÷ 1.45 =)	10. 25	20 (.20 ÷ .80 =)

Catalog Pricing

Objective 7

Merchandise sold by wholesalers or manufacturers is often priced in a catalog and may be subject to a discount (see chapter 14). Trade discounts are available for larger purchases and for the buyer performing marketing functions for the seller. The seller must determine the price to be shown in a catalog in order to allow for trade discounts. This price is known as the **list price.** Trade discounts are subtracted from the list price to find the selling price. The relationship of cost price, selling price, and list price is shown here.

list price

(Cost Price + Markup) = Selling Price
Selling Price + Trade Discount = List Price

Example 15 What is the list price on an item that cost a wholesaler $17 if the wholesaler will mark it up 20 percent on sales and will allow for a 10 percent trade discount?

80% + 20% = 100%
cost price + markup = selling price
$17

Base = Amount ÷ Rate
Base = $17 ÷ .80
Base = $21.25 (the selling price)

90% + 10% = 100%
selling price + trade discount = list price
$21.25

Base = Amount ÷ Rate
Base = $21.25 ÷ .90
Base = $23.61 (list price)

(17 ÷ .80 = 21.25 ÷ .90 =)

Exercise G

Determine the list prices in the exercises below. Round off your answers to the nearest cent.

	Cost Price	Markup Percent	Markup on	Trade Discount	List Price
1.	$ 40	150%	Cost (40 × 2.50 =)	15% (100 ÷ .85 =)	$117.65
2.	25	35	Sales *(25 ÷ .65 =)*	10 *(38.46 ÷ .90 =)*	*42.74*
3.	163	40	Sales *(163 ÷ .60 =)*	15 and 10 *(271.67 ÷ .765 =)*	*355.12*
4.	47	60	Cost *(47 × 1.60 =)*	21 *(75.20 ÷ .79 =)*	*95.19*
5.	80	55	Cost *(80 × 1.55 =)*	17 *(124 ÷ .83 =)*	*149.40*
6.	65	7	Sales *(65 ÷ .93 =)*	20 and 15 *(69.89 ÷ .68 =)*	*102.78*
7.	72	42	Sales *(72 ÷ .58 =)*	15 *(124.14 ÷ .85 =)*	*146.04*
8.	1,530	80	Sales *(1,530 ÷ .20 =)*	10 and 15 *(7,650 ÷ .765 =)*	*10,000.00*
9.	96	12	Cost *(96 × 1.12 =)*	10 *(107.52 ÷ .90 =)*	*119.47*
10.	120	17	Sales *(120 ÷ .83 =)*	20 *(144.58 ÷ .80 =)*	*180.72*

Additional Trade Discounts

When a middleman finds another source for a product at a lower price, the middleman may decide not to issue a new catalog immediately. The lower-cost price would provide a higher profit for the seller. However, in order to stay competitive, the middleman may decide to offer an additional trade discount, possibly through an E-mail communication, passing along to the customer the cost price reduction. The percent of additional trade discount that can be passed on is found by (1) determining the difference between the old and new cost prices, and (2) dividing the difference by the old cost price.

Example 16 The cost price of a toy robot is $190. A new supplier can reduce the cost price to $172. What will be the additional trade discount the seller can offer if the markup rate and trade discounts remain the same?

$190 Old cost price
−172 New cost price
$18 Difference

$$\frac{\$18 \text{ difference}}{\$190 \text{ old cost price}} = 9\% \text{ additional trade discount}$$

(18 ÷ 190 =)

Exercise H

Determine the list price and additional trade discount. Round off your answer to the nearest percent or cent.

	Cost Price	Markup Percent	Markup on	Trade Discount	List Price	New Cost Price	Additional Trade Discount Percent
1.	$127.00	15%	Sales	10% (127 ÷ .85 =)	$166.01 (149.41 ÷ .90 =)	$120.00	6% (7 ÷ 127 =)
2.	427.00	35	Cost	20 *(427 × 1.35 =)*	*720.56* *(576.45 ÷ .80 =)*	360.70	*16* *(66.30 ÷ 427 =)*
3.	320.00	20	Cost	10 and 15 *(320 × 1.20 =)*	*501.96* *(384 ÷ .765 =)*	268.00	*16* *(52 ÷ 320 =)*
4.	95.50	38	Sales	$7\frac{1}{2}$ *(95.50 ÷ .62 =)*	*166.52* *(154.03 ÷ .925 =)*	76.00	*20* *(19.50 ÷ 95.50 =)*
5.	147.00	80	Sales	22 *(147 ÷ .20 =)*	*942.31* *(735 ÷ .78 =)*	138.60	*6* *(8.40 ÷ 147 =)*

ASSIGNMENT Chapter 15

Name ______________________ Date ______________ Score ____________

Skill Problems

Determine the selling price and markup amount. Round off your answer to the nearest cent.

Cost Price	Markup		Markup Percent on Cost	Selling Price	
$ 95.00	1. $52.25	(95 × .55 =)	55%	2. $147.25	(95 × 1.55 =)
700.00	3. 245.00	(700 × .35 =)	35	4. 945.00	(700 × 1.35 =)
55.40	5. 27.70	(55.4 × .5 =)	50	6. 83.10	(55.4 × 1.5 =)
1,746.50	7. 401.70	(1,746.5 × .23 =)	23	8. 2,148.20	(1,746.5 × 1.23 =)
49.00	9. 34.30	(49 × .7 =)	70	10. 83.30	(49 × 1.7 =)

Cost Price	Markup		Markup Percent on Sales	Selling Price	
$ 80.00	11. $97.78	(177.78 × .55 =)	55%	12. $177.78	(80 ÷ .45 =)
120.00	13. 64.62	(184.62 × .35 =)	35	14. 184.62	(120 ÷ .65 =)
55.40	15. 55.40	(110.8 × .5 =)	50	16. 110.80	(55.4 ÷ .5 =)
1,746.50	17. 521.68	(2,268.18 × .23 =)	23	18. 2,268.18	(1,746.5 ÷ .77 =)
48.50	19. 113.17	(161.67 × .7 =)	70	20. 161.67	(48.5 ÷ .30 =)

Determine the selling price. Note the basis of the markup rate. Round off your answers to the nearest cent.

Cost Price	Markup Rate	Markup on	Selling Price	
$600.00	10%	Cost Price	21. 660	(600 × 1.1 =)
369.00	25	Selling Price	22. 492	(369 ÷ .75 =)
84.60	50	Selling Price	23. 169.20	(84.6 ÷ .5 =)
10.95	70	Cost Price	24. 18.62	(10.95 × 1.7 =)
381.50	34	Selling Price	25. 578.03	(381.50 ÷ .66 =)

Fill in the blank lines below. Note the markup base—cost or sales. Round off your answers to the nearest percent or cent.

Cost Price	Selling Price	Markup Percent	Markup on	Markup	
26. $129.15	$184.50	30%	Sales	27. $55.35	(184.5 × .30 =)
50.00	28. 67.50	35	Cost	29. 17.50	(50 × .35 =)
30. 832.51	1,040.64	20	Sales	31. 208.13	(1,040.64 × .20 =)
648.12	32. 2,160.40	70	Sales	33. 1,512.28	(648.12 ÷ .30 =)
25.00	40.00	34. 60	Cost	35. 15.00	(15.00 ÷ 25 =)
$140.00	36. $189.00	35	Cost	37. $49.00	(140 × 1.35 =)
47.50	62.00	38. 31	Cost	39. 14.50	(14.50 ÷ 47.50 =)
60.00	85.00	40. 29	Sales	41. 25.00	(25 ÷ 85 =)
42. 156.67	188.00	20	Cost	43. 31.33	(188 ÷ 1.20 =)
44. 150.00	45. 210.00	40	Cost	60.00	(60 ÷ .40 =)
46. 168.00	47. 240.00	30	Sales	72.00	(72 ÷ .30 =)
72.00	48. 180.00	60	Sales	49. 108.00	(72 ÷ .40 =)
50. 16.50	27.50	40	Sales	51. 11.00	(27.50 × .40 =)
248.00	52. 322.40	30	Cost	53. 74.40	(248 × 1.30 =)
54. 5.82	12.80	120	Cost	55. 6.98	(12.80 ÷ 2.20 =)

Determine the answer in the exercises below. Round off your answer to two decimal places.

Cost Price	Selling Price	Markdown	Sale Price		Loss (%)	
$74.00	$95.50	40%	56. *$57.30*	*(95.50 × .60 =)*	57. *22.57%*	*(74 − 57.30 = 16.70 ÷ 74 =)*
80.20	90.00	30	58. *63.00*	*(90. × .7 =)*	59. *21.45*	*(80.20 − 63. = 17.20 ÷ 80.20 =)*
46.81	53.83	30	60. *37.68*	*(53.83 × .7 =)*	61. *19.50*	*(46.81 − 37.68 = 9.13 ÷ 46.81 =)*
650.00	910.00	65	62. *318.50*	*(910 × .35 =)*	63. *51.00*	*(650 − 318.50 = 331.50 ÷ 650 =)*
891.65	936.23	25	64. *702.17*	*(936.23 × .75 =)*	65. *21.25*	*(891.65 − 702.17 = 189.48 ÷ 891.65 =)*

Determine the markup percent on the new base in the exercises below. Round off your answers to the nearest percent.

	Markup Percent on Cost	Markup Percent on Sales	
66.	46%	*32%*	*(.46 ÷ 1.46 =)*
67.	*30*	23	*(.23 ÷ .77 =)*
68.	80	*44*	*(.80 ÷ 1.8 =)*
69.	*150*	60	*(.6 ÷ .4 =)*
70.	125	*56*	*(1.25 ÷ 2.25 =)*

Complete the following problems. Round off your answers to the nearest cent.

Cost Price	Markup Percent		Markup on	Selling Price	Trade Discount		Catalog List Price
$30.00	40%	*(30 ÷ .60 =)*	Sales	71. *$50.00*	15%	*(50 ÷ .85 =)*	72. *$58.82*
73. *71.65*	75	*(125.38 ÷ 1.75 =)*	Cost	74. *125.38*	15	*(147.50 × .85 =)*	147.50
46.00	30	*(46 × 1.30 =)*	Cost	75. *59.80*	6	*(59.80 ÷ .94 =)*	76. *63.62*
25.50	30	*(25.50 ÷ .70 =)*	Sales	77. *36.43*	17	*(36.43 ÷ .83 =)*	78. *43.89*
79. *38.40*	20	*(48 × .80 =)*	Sales	80. *48.00*	20	*(60 × .80 =)*	60.00

Determine the list prices and the additional trade discount percent in the exercises below. Round off your answers to the nearest percent or cent.

Cost Price	Markup Percent	Markup on	Trade Discount	Catalog List Price	New Cost Price	Additional Trade Discount Percent
$50.00	125%	Cost *(50 × 2.25 =)*	20% *(112.50 ÷ .8 =)*	81. *$140.63*	$ 45.00	82. *10%* *(5 ÷ 50 =)*
612.50	15	Sales *(612.5 ÷ .85 =)*	5 *(720.59 ÷ .95 =)*	83. *758.51*	600.00	84. *2* *(12.5 ÷ 612.5 =)*
84.40	20	Cost *(84.4 × 1.2 =)*	25 *(101.28 ÷ .75 =)*	85. *135.04*	75.50	86. *11* *(8.9 ÷ 84.4 =)*
78.85	38	Cost *(78.85 × 1.38 =)*	15 *(108.81 ÷ .85 =)*	87. *128.02*	75.35	88. *4* *(3.5 ÷ 78.85 =)*
340.60	62	Sales *(340.6 ÷ .38 =)*	20 *(896.32 ÷ .80 =)*	89. *1,120.40*	295.00	90. *13* *(45.6 ÷ 340.6 =)*
42.00	20	Cost *(42 × 1.20 =)*	12 *(50.40 ÷ .88 =)*	91. *57.27*	37.50	92. *11* *(4.50 ÷ 42 =)*
23.60	70	Cost *(23.60 × 1.70 =)*	10 *(40.12 ÷ .90 =)*	93. *44.58*	21.90	94. *7* *(1.70 ÷ 23.60 =)*
126.00	35	Sales *(126 ÷ .65 =)*	15 and 10 *(193.85 ÷ .765 =)*	95. *253.39*	93.70	96. *26* *(32.30 ÷ 126 =)*

Business Application Problems

Round off your answer to nearest percent or cent.

1. Price Wise Stores purchased a lawn mower for $259. The store will need to make a markup of 40 percent on cost. At what price will the store sell the lawn mower?

$259
C + M = S *$259 × 1.4 = $362.60 price*
100% + 40% = 140%

2. In problem 1, what is the amount of the markup?

$259
C + M = S *$259 × .4 = $103.60 markup*
100% + 40%

3. If Price Wise Stores, in problem 1, would have used a markup of 40 percent on the selling price, what would be the price of the lawn mower?

$259
C + M = S *$259/.6 = $431.67 price*
60% + 40% = 100%

4. Professional Kitchens, Inc. bought a piece of used stainless steel preparation equipment for $275. They were able to resell the equipment for $400 a week later. What markup did they earn on cost?

$400 − $275 = $125/$275 = 45% markup

5. What was the rate of markup on the selling price in problem 4?

$400 − $275 = $125/$400 = 31% markup

6. All the merchandise at Steve's store is priced to sell at $49.95 or lower. Steve uses a markup of 35 percent on cost for all of his merchandise. If he is to buy items to sell in his store, what is the most he can pay for them and maintain the current maximum price level?

$49.95
C + M = S *$49.95/1.35 = $37 maximum*
100% + 35% = 135%

7. Kevin's Kernels sells popcorn for $4.75 a bag. The supplier indicated that Kevin could produce 45 popped bags from the box of kernels he purchased for $7. What is the potential amount of markup on the purchase?

$4.75 × 45 = $213.75 − $7 = $206.75 markup

8. What is the rate of markup on cost that Kevin could earn in problem 7?

$4.75 × 45 = $213.75 − $7 = $206.75/$7 = 2,954% markup

9. A leaf blower sold for $189 at the Bayview Hardware Store. The unit was purchased for $83 through a supplier in Mexico. What is the rate of markup on the selling price?

$83 + $106 = $189
C + M = S *$106/$189 = 56% markup*
100%

10. Dawn Donuts produces the best pecan rolls in Medford. The total cost of ingredients for a batch of 100 rolls is $31.25. Each of the pecan rolls is sold for $1.25 by 10:00 a.m. each day. What is the rate of markup on the ingredients?

$31.25/100 = $.3125
$1.25 − $.3125 = $.9375/$.3125 = 300% markup

11. The Carnes Gift Shop purchased an antique for $40 and resold it for $90. To the nearest percent, what is the markup percent on sales?

$90
−40
$50 markup

$$\frac{\$50}{\$90} = 56\% \text{ markup}$$

12. Arrigo Dodge prices all repair parts to sell at a 60 percent markup based on cost. The counter-parts salesperson must price out the lamp assembly to be sold to Carl's Storage shown in the following invoice. The cost price listed on the video display unit is $17.90. Place the selling price in the list column on the invoice below.

100% + 60% = 160%
C + M = S
$17.90

$17.90
×1.6
$28.64 selling price

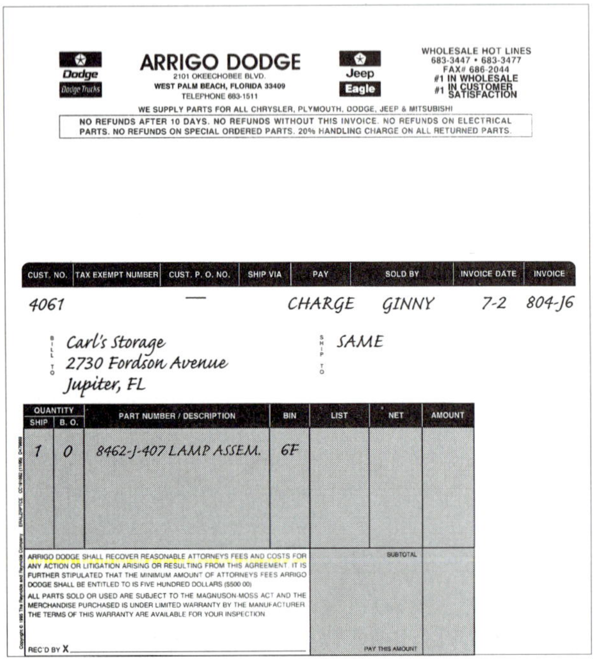

ARRIGO DODGE
2101 OKEECHOBEE BLVD.
WEST PALM BEACH, FLORIDA 33409
TELEPHONE 683-1511

WHOLESALE HOT LINES
683-3447 • 683-3477
FAX# 686-2044
#1 IN WHOLESALE
#1 IN CUSTOMER SATISFACTION

WE SUPPLY PARTS FOR ALL CHRYSLER, PLYMOUTH, DODGE, JEEP & MITSUBISHI

NO REFUNDS AFTER 10 DAYS. NO REFUNDS WITHOUT THIS INVOICE. NO REFUNDS ON ELECTRICAL PARTS. NO REFUNDS ON SPECIAL ORDERED PARTS. 20% HANDLING CHARGE ON ALL RETURNED PARTS.

CUST. NO.	TAX EXEMPT NUMBER	CUST. P. O. NO.	SHIP VIA	PAY	SOLD BY	INVOICE DATE	INVOICE
4061		—		CHARGE	GINNY	7-2	804-J6

BILL TO: Carl's Storage, 2730 Fordson Avenue, Jupiter, FL
SHIP TO: SAME

QUANTITY SHIP	QUANTITY B. O.	PART NUMBER / DESCRIPTION	BIN	LIST	NET	AMOUNT
1	0	8462-J-407 LAMP ASSEM.	6F			

ARRIGO DODGE SHALL RECOVER REASONABLE ATTORNEYS FEES AND COSTS FOR ANY ACTION OR LITIGATION ARISING OR RESULTING FROM THIS AGREEMENT. IT IS FURTHER STIPULATED THAT THE MINIMUM AMOUNT OF ATTORNEYS FEES ARRIGO DODGE SHALL BE ENTITLED TO IS FIVE HUNDRED DOLLARS ($500.00)

ALL PARTS SOLD OR USED ARE SUBJECT TO THE MAGNUSON-MOSS ACT AND THE MERCHANDISE PURCHASED IS UNDER LIMITED WARRANTY BY THE MANUFACTURER. THE TERMS OF THIS WARRANTY ARE AVAILABLE FOR YOUR INSPECTION

SUBTOTAL

REC'D BY X

PAY THIS AMOUNT

Courtesy of Arrigo Dodge, West Palm Beach, FL.

13. Sarah marked some out-of-season boots to sell for $31 a pair. The boots cost the store $56 a pair and had been marked up 23 percent on sales. Rita asked Sarah to determine the percent of markdown on the original selling price. Round off your answer to the nearest percent.

$56
C + M = S
77% + 23% = 100%
B = A ÷ R
B = $56 ÷ .77 = $72.73
$ 72.73 original selling price
−31.00 sale price
$41.73 markdown

$$\frac{\$41.73}{\$72.73} = \underline{57\%\ \text{markdown}}$$

14. Keeline Wilcox sold 20 Kentia plants to the Colwell Nursing Home, a nonprofit organization. The plants cost the firm $17.90 each. They were marked to sell for $22.90 each. Keeline Wilcox offered to give the buyer a discount as shown on the following invoice. How much markup was made on the sale?

$17.90
×20
$358 cost

$412.20
−358.00
$54.20 markup

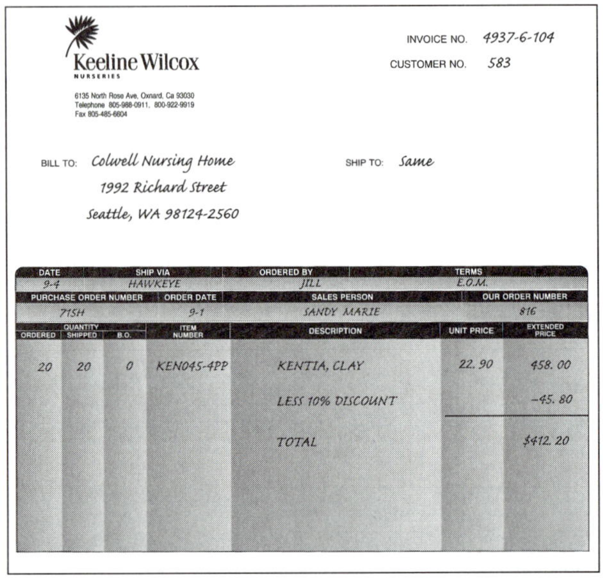

Keeline Wilcox
NURSERIES
6135 North Rose Ave, Oxnard, Ca 93030
Telephone 805-988-0911, 800-922-9919
Fax 805-485-6604

INVOICE NO. 4937-6-104
CUSTOMER NO. 583

BILL TO: Colwell Nursing Home, 1992 Richard Street, Seattle, WA 98124-2560
SHIP TO: Same

DATE	SHIP VIA	ORDERED BY	TERMS
9-4	HAWKEYE	JILL	E.O.M.

PURCHASE ORDER NUMBER	ORDER DATE	SALES PERSON	OUR ORDER NUMBER
71SH	9-1	SANDY MARIE	816

QUANTITY ORDERED	QUANTITY SHIPPED	B.O.	ITEM NUMBER	DESCRIPTION	UNIT PRICE	EXTENDED PRICE
20	20	0	KEN045-4PP	KENTIA, CLAY	22.90	458.00
				LESS 10% DISCOUNT		-45.80
				TOTAL		$412.20

Courtesy of Keeline Wilcox, Oxnard, CA.

15. The Stork Shop realized a 42 percent markup on cost on all products sold last year. The merchandise bought for resale cost \$37,860. There was \$9,650 worth of overhead during the year, including utilities, rent, and insurance. What was the amount of the owner's profit for the year? (overhead + profit = markup)

\$37,860
C + M = S
100% + 42% = 142%
P = B × R
P = \$37,860 × 1.42 = \$53,761.20

\$53,761.20 selling price
−37,860.00 cost price
−9,650.00 overhead
\$6,251.20 profit

16. Whitican & Associates, art dealers, planned to mark down a painting 50 percent. The painting was purchased from an artist for \$40. They wish to maintain a 35 percent markup on sales after the markdown. What was the pre-marked-down price?

\$40
C + M = S
65% + 35% = 100%
B = P ÷ R
B = \$40 ÷ .65 = \$61.54

\$61.54
S + D = L
50% + 50% = 100%
B = P ÷ R
B = \$61.54 ÷ .5
= \$123.08 pre-marked-down price

17. Linda's Natural Foods buys dried fruit for resale. A single package of dried apples costs \$1.80. The package is marked up 60 percent on sales. How many packages must be sold for the store to realize a total markup of \$240?

\$1.80
C + M = S
40% + 60% = 100%
B = P ÷ R
B = \$1.80 ÷ .4 = \$4.50

\$4.50 selling price
−1.80 cost price
\$2.70 markup

$\frac{\$240}{\$2.70} = 89 \text{ packages}$

18. Janis has a markup of 100 percent on cost in her dress shop. A business associate, Cyndi, indicates that she does better than Janis because she earns a 60 percent markup on sales. Who earns more per sale, Janis or Cyndi?

C + M = S
100% + 100% = 200%
$\frac{100\%}{200\%} = 50\%$ *on sales for Janis*
Cyndi earns a higher markup.

19. For what price must a merchant sell a desk that cost him \$300 in order to earn a markup of 30 percent on sales after a 10 percent trade discount has been offered to his customers?

\$300
C + M = S
70% + 30% = 100%
B = P ÷ R
B = \$300 ÷ .70 = \$428.57 selling price

20. The Cabinet Corporation bought a buffet for \$600 from the Stillo Manufacturing Company on July 17. The Cabinet Corporation operates under a policy of a 40 percent markup on sales. The corporation allows a 10 percent trade discount to its customers. What will be the list price on the buffet that costs \$600?

\$600
C + M = S
60% + 40% = 100%
B = A ÷ R
B = \$600 ÷ .6 = \$1,000

\$1,000
S + D = L
90% + 10% = 100%
B = A ÷ R
B = \$1,000 ÷ .9 = \$1,111.11 list price

21. A dress marked "25 percent off" is on sale for \$169. The dress was originally marked up 150 percent on cost. What was the cost of the dress?

\$169 ÷ .75 = \$225.33
\$225.33
C + M = S
100% + 150% = 250%
B = A ÷ R
B = \$225.33 ÷ 2.50 = \$90.13 cost

22. A diamond ring is marked \$890. The diamond cost the jeweler \$200 and is marked up 250 percent on cost. The setting cost the jeweler \$114. What is the percent of markup on cost on the setting? Round off your answer to the nearest percent.

\$200
C + M = S
100% + 250% = 350%
A = B × R
A = \$200 × 3.50 = \$700

\$700
+ 114
\$814

\$890
−814
\$76

$\frac{\$76}{\$114} = 67\% \text{ markup}$

23. Write a complete sentence in answer to this question. If the markup on food items at a convenience store is 12 percent on cost, how much merchandise must be sold to have total markups of \$200,000 per month?

\$200,000
C + M = S
100% + 12% = 112%
B = A ÷ R
B = \$200,000 ÷ .12 = \$1,666,666.67

\$1,666,666.67 costs
+ 200,000 markups
\$1,866,666.67

The store must sell \$1,866,666.67 per month.

24. Write a complete sentence in answer to this question. Angelo sells popcorn at all of the athletic games at Bucken Stadium. His popcorn sells for \$5 a bag. Angelo marks up the ingredients (popcorn, oil, salt, etc.) 1,000 percent on cost. What is the cost of the ingredients?

\$5
C + M = S
100% + 1,000% = 1,100%
B = A ÷ R
B = \$5 ÷ 11 = 45¢
His ingredients cost 45¢.

25. Write a complete sentence in answer to this question. Desmond and Associates have taken an inventory of the materials in their storage facility. The value of the items at cost is \$853,980. The accountant said that these items will be sold at an average markup of 22 percent of the selling price. If Desmond and Associates sell the complete inventory, what will they receive in revenue?

\$853,980
C + M = S
78% + 22% = 100%
B = A ÷ R
B = \$853,980 ÷ .78 = \$1,094,846.10
They will receive \$1,094,846.10 in revenue.

★ Challenge Problem

The Cronton & Trion Company purchased 70 special timers at a list price of \$37.80 each. The Burrs Corporation allowed a 15 percent trade discount, as well as terms of 2/15,n/30, E.O.M. The purchase was made on July 18. The timers were backordered and actually arrived on September 29. Four were found to be defective and were returned. The invoice was paid on October 17. The timers were sold through the Cronton & Trion catalog. The markup rate of 40 percent on cost was made.

A second purchase of 70 timers was made on November 2, C.O.D., at \$31 each with no trade discount. They were marked up in the same way as the first purchase. What is the total dollar amount of markup realized by Cronton & Trion on the two purchases? Round your answer off to the nearest cent.

70
−4
66 units
×\$37.80
\$2,494.80 list price
−374.22 trade discount
\$2,120.58 selling price

\$2,120.58
C + M = S
100% + 40% = 140%
P = A × R
P = \$2,120.58 × 1.4 = \$2,968.81 catalog price

70
×\$31
\$2,170 second purchase

\$2,170
C + M = S
100% + 40% = 140%
P = A × R
P = \$2,170 × 1.4 = \$3,038 catalog price

\$2,120.58
+2,170.00
\$4,290.58 total cost

\$2,968.81
+3,038.00
\$6,006.81 total list price

\$6,006.81
−4,290.58
\$1,716.23 markup

MASTERY TEST Chapter 15

Name ______________________ **Date** ______________ **Score** ____________

Complete the problems below. Round off your answer to the nearest cent or percent.

1. Markup = 30% on cost
Cost price = $50
Selling = $ ***65***

2. Markup = 60% on sales
Cost price = $38
Selling price = $ ***95***

3. Markup = 40% on cost
Selling price = $200
Cost price = $ ***142.86***

4. Markup = 50% on sales
Selling price = $47
Cost price = $ ***23.50***

5. Selling price = $60
Cost price = $25
Markup = ***140*** % on cost

6. Selling price = $35
Cost price = $20
Markup = ***43*** % on sales

7. Markup = 40% on sales
Markup = ***67*** % on cost

8. Markup = 50% on cost
Markup = ***33*** % on sales

9. Markup = $80
Markup = 12% on cost
Selling price = $ ***746.67***

10. Markup = $30
Markup = 8% on sales
Cost price = $ ***345***

11. Cost price = $120
Markup = 38% on sales
Trade discount = 25%
List price = $ ***258.06***

12. List price = $380
Trade discount = 15 and 22%
Markup = 25% on sales
Cost price = $ ***188.96***

13. Out-of-Doors, Inc. purchased a tent for $280. The tent was then marked up 40 percent on cost. What is the selling price of the tent?

$280
C + M = S
100% + 40% = 140%

A = B × R
A = $280 × 1.4 = $392 selling price

14. The Shipe Ice Cream Parlour marks up half-gallon containers 20 percent on cost. What is the highest cost the firm can pay for the item and sell it for $2.29? Round off your answer to the nearest cent.

100% + 20% = 120%
C + M = S
$2.29

B = A ÷ R
B = $2.29 ÷ 1.2 = $1.91 highest cost

15. Richard Yake's men's store purchased suits for $105 each. They were marked to sell for $249. The last of the suits were put on sale for $89. What is the percent of loss on the last suits sold?

$105
− 89
$ 16 loss

$$\frac{\$16}{\$105} = 15\% \text{ loss}$$

Inventory and Turnover 16

Suggested Classroom Time for This Chapter

hours 1 2 3 4 5

Perceived Student Difficulty with Topical Material

easy moderate difficult

Objectives

After mastering the material in this chapter, you will be able to:

1. Determine the value of ending inventory using the L.I.F.O., F.I.F.O., average cost, specific identification, or retail inventory estimating method. p. 304
2. Determine the cost of goods sold using the L.I.F.O., F.I.F.O., average cost, specific identification, or retail inventory estimating method. p. 304
3. Determine the rate of merchandise turnover for a firm. p. 313
4. Understand and use the following terms or concepts:
   Beginning Inventory
   Purchases
   Goods Available for Sale
   Ending Inventory
   Cost of Goods Sold
   L.I.F.O.
   F.I.F.O.
   Average Cost
   Retail Inventory Estimating Method
   Specific Identification Method
   Turnover

Sport-A-Merica

It's Saturday, and Kerry and Hal are sitting down to discuss how Sport-A-Merica has been doing in the recent past. It is important to focus on the overall financial picture at least once a month. They know that markup is an important part of how profits are generated; they also know that the more items they are able to sell over a period of time, the better the total financial picture. Some sports items in the store are sold every day. Things like tennis balls and running shoes are constantly in demand. Other items are "slow movers," but bring a high profit when they are sold. Things like sleeping bags and clay pigeon throwers are sold only occasionally. These items contribute a great deal to the year-end profit and to the image of a complete sports store; though they sit for some time and bring a limited profit, they do contribute to the bottom line for the store.

Both Kerry and Hal know that a comparison to a previous period of time would help them to determine how well the store is doing. The time of year the comparison is to be made will also make a difference. They know, for example, that the spring and the holiday seasons are busy for them. If a comparison is to be made, it should be made at the same time each year.

Hal has kept records of the stock amounts on hand since the store first opened. He thinks about the face mask item often. It sells well and must be re-ordered several times a year. Keeping track of how many are in stock, how many have been sold, and how many were received in the last order is very time-consuming. In addition, the price of this item keeps changing, and the records must reflect these changes. Each time he has ordered the face mask, the price has gone up one dollar. He must determine the current value of the stock that he has in the store. This will give them a head start in making comparisons with the most recent months.

Kerry and Hal proceed with the analysis. They will have a much better grasp on how Sport-A-Merica is doing by the end of the day.

Inventory Methods

For firms that buy merchandise for resale, the value of the merchandise on hand at tax time or inventory time is very important. Several methods of determining that value will be presented in this chapter. Each method has its own advantage and special purpose. Knowledge of these methods is useful for students studying for merchandising, distribution, or accounting positions.

beginning inventory

purchases

goods available for sale

At the beginning of an accounting period, a firm has an inventory of merchandise on hand to sell. This amount is termed **beginning inventory,** the value of inventory that the firm begins with in a period. During that period, the firm will buy more merchandise to add to the inventory. The total value bought is termed **purchases.** Beginning inventory plus purchases equal **goods available for sale,** all of the merchandise that can be sold during this period; not necessarily all available at one time in that period.

Objectives 1 & 2

ending inventory

cost of goods sold

The value of the goods on hand at the end of the period is termed **ending inventory.** By subtracting the ending inventory from the goods available for sale, the value of the goods that were sold during the period can be found. The cost price of the goods that were sold during the period is termed **cost of goods sold.** The word merchandise can be substituted for goods to create "merchandise available for sale" and "cost of merchandise sold." Note that (1) the period covered may be a year, six months, a month, or even one day; (2) all of the values in the example are at cost price, not selling price; (3) the ending inventory from one period becomes the beginning inventory for the next period; and (4) this inventory model repeats itself every period.

❍ *Purchase orders such as this one are used to replenish the stock in the firm. Work Right is ordering 144 pieces from Pace International, Inc., to add to their stock.*

❍ *Bar coding [officially known as the Uniform Product Code (UPC)] on most packaged goods plus scanning devices provide a very efficient method to take a physical inventory.*

WORK RIGHT
MASTER DESIGNED SHOWER ENVIRONMENTS
4615 WORK RIGHT CIRCLE, LAKEPORT CA 95453-9302
Call (707) 263-0290
FAX (707) 263-0261

PURCHASE ORDER

PURCHASE ORDER 317-4112 REV

BILL TO: SAME

ORDER DATE *February 10th*

PAGE NUMBER *1 of 1*

TO: *Pace International, Inc.*
102 152nd Street
New York, NY
11102

SHIP TO: SAME

VENDOR	CONFIRMING	TERMS	SHIP VIA	COLLEC	PPD	FOB	TAXABLE	INSP
51		*1/10, n/30*	*Brady Co.*			*NYC*	*No*	

BUYER	VENDOR CONTACT	REQUISITIONER	SHIP INSTRUCTIONS	REMARKS
Brad	*Brooks*	—	*A.S.A.P.*	—

L/N	DL	QUANTITY	UNIT	PART NUMBER	REV	DESCRIPTION	UNIT PRICE	EXTENSION	DUE DATE
		144	*EA*	*58-42-6114*		*SILVER ALLUMINIUM TRACKING*	*17.50*	*$2,520.~*	

THE TERMS AND CONDITIONS HEREIN SUPERSEDE ANY TERMS AND CONDITIONS OF SELLER UNLESS AGREED TO, IN WRITING, BY WORK RIGHT PRODUCTS, INC. SELLER AGREES TO DELIVER SUPPLIES OR SERVICES HEREIN SUBJECT TO THE TERMS AND CONDITIONS ON THE FACE AND REVERSE HEREOF.

M. G. Sanson
AUTHORIZED SIGNATURE

Courtesy of Work Right Products, Inc., Lakeport, CA.

Example 1

- *Students should know this model. They will refer to it many times in this chapter.*

Inventory model

	Beginning inventory
+	Purchases
	Goods available for sale
−	Ending inventory
	Cost of goods sold

Example 2

- *Ask your students if any of them have been employed to take a physical inventory.*

The Dosen Paper Company had 147,000 units of inventory on hand valued at $123 each on June 1. During the month, another 46,000 units were purchased at the same price. On June 30, the Accounting Department found that there were $15,922,350 of goods on hand. What is the value of the goods that were sold?

$18,081,000	Beginning inventory (147,000 units × $123/unit)—June 1
+5,658,000	Purchases (46,000 units × $123/unit)
$23,739,000	Goods available for sale
−15,922,350	Ending inventory—June 30
$7,816,650	Cost of goods sold—in the month of June

(147,000 × 123 = + 5,658,000 = − 15,922,350 =)

In this example, the cost price per unit of beginning inventory and of purchases was the same amount.

For income tax purposes, the cost value of ending inventory or cost of goods sold is very important. A firm can legally increase the cost of the goods sold, thereby reducing gross income. Income taxes are reduced because the taxable base, net income before taxes, is reduced.

	Revenue from sales
−	Cost of goods sold
	Gross income
−	Operating and selling expenses
	Net income before taxes
×	Tax rate
	Taxes

Inventory Valuation

- *Emphasize the flow-of-cost concept to students. They will try to think in terms of physical flow of merchandise.*

Valuation of inventory costs considers units of the same product being purchased at different unit-cost prices during the period. In order to determine the value of inventory, you must first determine the unit prices to be assigned to the items on hand at the end of the period (ending inventory). It is necessary to assume an arbitrary flow of costs of merchandise through the company. The three most frequently used methods in valuation of inventory are

1. L.I.F.O.
2. F.I.F.O.
3. Average Cost

L.I.F.O.

L.I.F.O. is an abbreviation for last in–first out. The **L.I.F.O.** concept considers that the most recently purchased goods (costs incurred) are sold first and their costs should be charged against revenue. The units remaining in the ending inventory come from the earliest purchases at their respective costs.

L.I.F.O
Flow of Costs

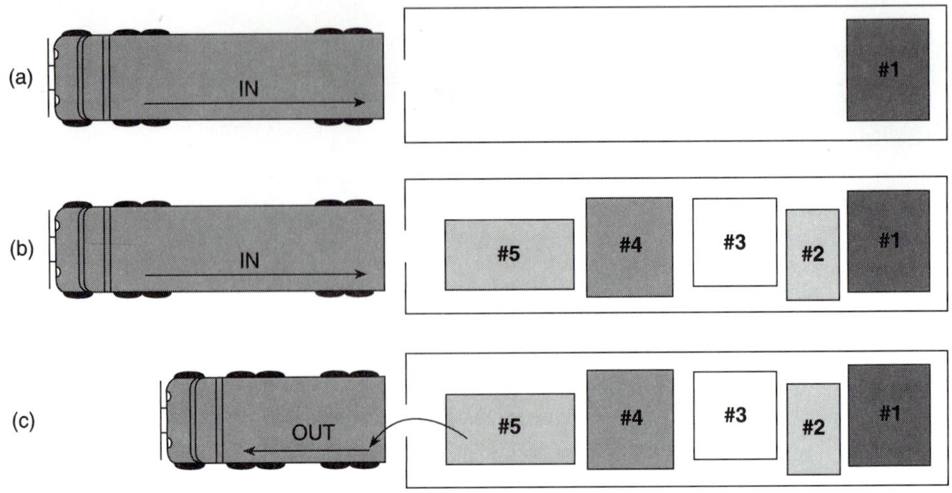

Last In–First Out (L.I.F.O.) = #5

The L.I.F.O. method will reduce gross income in times of rising prices (because the unit cost of the most recently acquired merchandise will be higher than the unit cost of the earlier purchases). This higher unit cost is used in calculating the cost of goods sold.

Example 3

Hal had made the following purchases of a face mask for Sport-A-Merica over the past year.

January	10	10 units at \$5 each
March	3	20 units at \$6 each
May	30	15 units at \$7 each
September	3	5 units at \$8 each

Hal has determined that there were 12 units of the item on hand on December 31.

A. What were the quantity and value of units available for sale?

	Units	Price per Unit	Total Price
	10	× \$5	= \$50
	20	× 6	= 120
	15	× 7	= 105
	+5	× 8	= +40
Total quantity and dollar value of units available for sale =	50 units (quantity)		\$315 (value)

Consider introdicing Last in-first out, First in-still there (LIFO-FIST) to your students to help them understand the LIFO concept.

Remember, the assignment of costs is independent of the actual physical movement of merchandise.

B. What was the value of ending inventory of the 12 units, using L.I.F.O.?

January	10	10 units at \$5 each =	\$50
March	3	+2 units at \$6 each =	+12
Total		12 (quantity)	\$62 (value)

C. What is the value of the cost of goods sold?

Goods available for sale	50 units	\$315
− Ending inventory	−12 units	− 62
Cost of goods sold	38 units	\$253 (value)

or

September	3	5 units	at \$8 each =	\$40
May	30	15 units	at 7 each =	105
March	3	18 units	at 6 each =	108
Cost of goods sold		38 units		\$253 (value)

Example 4

A new firm, Websters, Incorporated, made the following purchases during the first quarter year of operations.

January	16	450 units at $21.70 each
	21	600 units at 22.00 each
	28	250 units at 22.00 each
February	5	300 units at 21.90 each
	17	600 units at 22.10 each
	23	225 units at 22.10 each
March	6	730 units at 22.15 each
	11	400 units at 21.95 each
	16	350 units at 22.20 each
	28	600 units at 22.20 each
	31	400 units at 22.21 each

The firm sold 2,600 units during the quarter.

A. How many units were on hand at the end of the quarter?

Total purchases	4,905 units	
− Units sold	−2,600 units	(4,905 − 2,600 =)
Ending inventory	2,305 units	

Notice that the inventory model was used with alterations.

B. What was the value of ending inventory of 2,305 units, using L.I.F.O.?

Purchase of January 16 of 450 units at $21.70 each = (first purchase)	$9,765.00
January 21 600 units at $22.00 each =	13,200.00
January 28 250 units at $22.00 each =	5,500.00
February 5 300 units at $21.90 each =	6,570.00
February 17 600 units at $22.10 each =	13,260.00
February 23 105 units at $22.10 each = (of the 225 units)	+ 2,320.50
Value of ending inventory of 2,305 units, using L.I.F.O. =	$50,615.50

C. What is the value of cost of goods sold at 2,600 units, using L.I.F.O.?

Purchase on March 31 of 400 units at $22.21 each = (most recent purchase)	$8,884.00
March 28 of 600 units at $22.20 each =	13,320.00
March 16 of 350 units at $22.20 each =	7,770.00
March 11 of 400 units at $21.95 each =	8,780.00
March 6 of 730 units at $22.15 each =	16,169.50
February 23 of 120 units at $22.10 each = (of the 225 units)	2,652.00
Value of cost of goods sold of 2,600 units, using L.I.F.O. =	$57,575.50

Example 5

The Ulmac and Sterver Co. made the following purchases during the month of June.

June	4	120 units at $9.02 each
	16	200 units at $9.05 each
	21	310 units at $9.06 each
	30	150 units at $9.10 each

The beginning inventory on June 1 of 175 units was made up of 80 units purchased in May at $8.90 each and 95 units purchased in April at $8.55 each; 750 units were sold during June.

(Continued on next page)

Example 5 *continued*

A. How many units were on hand at the end of the month?

175 units—Beginning inventory (June 1)
+780 units—Purchases during June
955 units—Goods available for sale
−750 units—Sold during June
205 units—Ending inventory (June 30)

B. What is the value of ending inventory of 205 units, using L.I.F.O.?

Value of Ending Inventory, Using L.I.F.O.

95 units of beginning inventory purchased in April at $8.55 each	= $812.25
80 units of beginning inventory purchased in May at $8.90 each	= 712.00
30 units purchased June 4 at $9.02 each	= 270.60
Value of ending inventory of 205 units, using L.I.F.O.	= $1,794.85

C. What is the value of the cost of goods sold, 750 units, using L.I.F.O.?

Value of Cost of Goods Sold, Using L.I.F.O.

Purchase on June 30 of 150 units at $9.10 each (most recent purchase)		= $1,365.00
June 21	310 units at $9.06 each =	2,808.60
16	200 units at $9.05 each =	1,810.00
4	90 units at $9.02 each =	811.80
Value of cost of goods sold of 750 units, using L.I.F.O.		= $6,795.40

F.I.F.O.

F.I.F.O is an abbreviation for first in–first out. The **F.I.F.O.** concept considers that the goods are sold in order of their purchase. Therefore, the units remaining in the ending inventory come from the most recent purchases at their respective costs.

○ Consider introducing First in–first out, Last in–still there (FIFO-LIST) to your students to help them understand the FIFO concept.

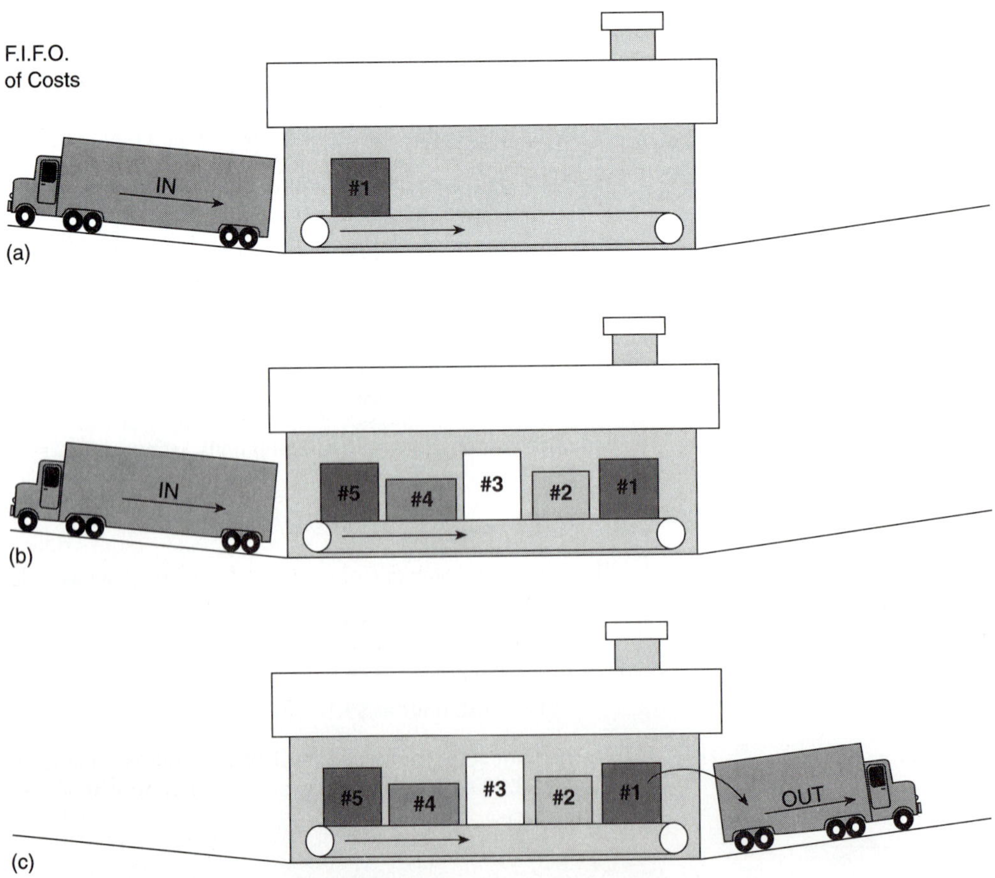

First In–First Out (F.I.F.O.) = #1

Example 6

Using the data from Sport-A-Merica in example 3:

A. What were the quantity and value of units available for sale? The answer to this question is the same as it would be in example 3.

Total quantity 50 units Dollar value $315

B. What was the value of ending inventory of the twelve units, using F.I.F.O.?

September 3	5 units at $8 each =	$40	
May 30	+ 7 units at $7 each =	+$49	
Total	12 quantity	$89 value	

C. What is the value of the cost of goods sold?

Goods available for sale	50 units	$315
− Ending inventory	−12 units	− 89
Cost of goods sold	38 units	$226 value

or

January 10	10 units at $5 each =	$50
March 3	20 units at $6 each =	120
May 30	8 units at $7 each =	56
Cost of goods sold	38 units	$226 value

Example 7

Using the purchase data from Websters, Incorporated in example 4:

A. What was the value of ending inventory of 2,305 units using F.I.F.O.?

Value of Ending Inventory, Using F.I.F.O.

Purchase on March 31 of 400 units at $22.21 each =	$ 8,884.00
(most recent purchase)	
March 28 of 600 units at $22.20 each =	13,320.00
16 of 350 units at $22.20 each =	7,770.00
11 of 400 units at $21.95 each =	8,780.00
6 of 555 units at $22.15 each =	+12,293.25
(of the 730 units)	
Value of 2,305 units, using F.I.F.O. =	$51,047.25

B. What is the value of the cost of goods sold, 2,600 units, using F.I.F.O.?

Value of Cost of Goods Sold, Using F.I.F.O.

Purchase on January 16 of 450 units at $21.70 each =	$9,765.00
(first purchase)	
January 21 of 600 units at $22.00 each =	13,200.00
28 of 250 units at $22.00 each =	5,500.00
February 5 of 300 units at $21.90 each =	6,570.00
17 of 600 units at $22.10 each =	13,260.00
February 23 of 225 units at $22.10 each =	4,972.50
March 6 of 175 units at $22.15 each =	+3,876.25
(of the 730 units)	

Value of cost of goods sold of 2,600 units, using F.I.F.O. = $57,143.75

Exercise A

Determine the answers in the exercises below.

	1.	2.	3.	4.
Beginning Inventory (units)	38	300	140	650
Cost per unit	$1.27	$4.75	$40.00	$12.50
Purchases				
January 16	600 at $1.28	75 at $4.90	20 at $35.00	130 at $10.00
February 3	350 at $1.28	200 at $4.88	35 at $34.60	150 at $10.00
15	90 at $1.30	150 at $4.95	60 at $34.70	275 at $9.59
22	400 at $1.30	200 at $4.95	40 at $35.60	300 at $10.05
March 2	250 at $1.35	225 at $5.00	75 at $36.10	500 at $10.25
Goods available for sale (units)	1,728	*1,150*	*370*	*2,005*
Goods sold (units)	1,428	*800*	*190*	*1,273*
Ending inventory (units)	300	350	180	732
Inventory method	L.I.F.O.	F.I.F.O.	L.I.F.O.	F.I.F.O.
Ending inventory	$383.62	*$1,743.75*	*$6,992.00*	*$7,456.60*
Cost of goods sold	$1,855.14	*$3,882.25*	*$6,732.50*	*$14,245.65*
	(38 × 1.27 =) *(262 × 1.28 =)*	*(225 × 5.00 =)* *(125 × 4.95 =)*	*(140 × 40 =)* *(20 × 35 =)* *(20 × 34.60 =)*	*(500 × 10.25 =)* *(232 × 10.05 =)*

Average Cost

Average cost is another method of determining inventory values. This method does not identify any specific units that were sold, but instead uses a weighted average of all units available for sale. The **average cost** is calculated by dividing the total dollar value of units available for sale by the total number of units available.

Example 8

Using the data from Sport-A-Merica in example 3:

A. What was the quantity and value of the units available for sale? The answer to this question is the same as it would be in example 3.

Total quantity: 50 units Dollar value: $315

B. What was the value of ending inventory of the 12 units, using average cost?

$$\frac{\text{Total dollar value}}{\text{Units available}} = \frac{\$315}{50} \times 12 \text{ units} = \$75.60$$

Round off at final answer only.

C. What is the value of cost of goods sold?

Goods available for sale	50 units	$315.00
− Ending inventory	−12 units	− 75.60
Cost of goods sold	38 units	$239.40

or

$$\frac{\text{Goods available for sale}}{\text{Units available}} \quad \frac{\$315.00}{50} \times 38 \text{ (units sold)} = \$239.40$$

(315 ÷ 50 = × 12 =) *(315 − 75.60 =)*

Example 9

Durst & Associates had 400 units of a file separator in stock at the end of the year. At the beginning of the year, there were 350 units in stock, at an average cost of 87 cents each. Two purchases were made during the year: one on July 16, for 800 units at 93 cents each; and one on November 12, for 250 units at 94 cents each.

A. How many units were sold during the year?

350	Beginning inventory (units)
+1,050	Purchases (800 + 250 units)
1,400	Goods available for sale (units)
−400	Ending inventory (units)
1,000	Cost of goods sold (units)

1,000 units were sold during the year.

B. What is the value of ending inventory, using average cost?

Value of ending inventory, using average cost

350 Units of beginning inventory	at 87¢ each =	$304.50
800 Units purchased	at 93¢ each =	744.00
+ 250 Units purchased	at 94¢ each =	+ 235.00
1,400 Total units available	Total cost =	$1,283.50

$$\frac{\$1{,}283.50}{1{,}400} = \frac{\text{Total cost}}{\text{Total units available}} \times \text{400 units ending inventory} = \$366.71 \text{ value of ending inventory}$$

C. What is the value of the cost of goods sold, using average cost?

Value of cost of goods sold, using average cost

$$\frac{\$1{,}283.50}{1{,}400} = \frac{\text{Total cost}}{\text{Total units available}} \times \text{1,000 units of goods sold} = \$916.79 \text{ cost of goods sold}$$

Exercise B

Notice that the average cost method considers the number of units each purchase, not the number of purchases. Determine the answers in the exercises below using the average cost method.

	1.	2.
Beginning inventory—units	720	300
	(18,136 ÷ 1,970 =) (× 730 =)	*(22,461 ÷ 17,000 =)* *(× 200 =)*
Beginning inventory (Cost per unit)	$9.80	$1.27
Purchases		
July 24	400 at $7.90/unit	7,200 at $1.30/unit
August 10	250 at $8.40/unit	4,000 at $1.31/unit
November 26	600 at $9.70/unit	5,500 at $1.36/unit
Ending inventory—units	730	200
Ending inventory value	$6,720.45	*$264.25*
Cost of goods sold value	$11,415.55	*$22,196.75*

(Continued on next page)

Exercise B Continued

	3.	4.
Beginning inventory—units	60	550
	(10,381.25 ÷ 325 =) *(× 50 =)*	*(44,230 ÷ 2,350 =)* *(× 400 =)*
Beginning inventory (Cost per unit)	$32.00	$18.50
Purchases		
July 24	90 at $32.00/unit	650 at $19.00/unit
August 10	50 at $32.25/unit	400 at $19.20/unit
November 26	125 at $31.75/unit	750 at $18.70/unit
Ending inventory—units	50	400
Ending inventory value	*$1,597.12*	*$7,528.51*
Cost of goods sold value	*$8,784.13*	*$36,701.49*

Retail Inventory Estimating Method

Occasionally, the manager may need only an estimate of the value of inventory. The **retail inventory estimating method** provides that estimate. This approach replaces the need for an actual physical inventory of the merchandise to estimate its value. In order for this method to be close to the actual value of inventory, the markup rate for all of the firm's products must be similar and consistent. The percent of cost to selling price is used with the sales for the period to estimate the cost price of goods sold and ending inventory.

The comparison of cost to sales is made from goods available for sale. All sales are made from goods available for sale. With a consistent markup rate, an estimate of inventory value at cost can be made. The information in the inventory model in example 1 must be available at selling price. Again, the inventory model has been altered to find ending inventory.

Example 10

	Cost Price	*Selling Price*
Beginning inventory	$62,580	$97,427
+ Purchases	+158,690	+229,352
Goods available for sale	$221,270	$326,779
− Sales		−261,408
Ending inventory		$65,371

Find the value of ending inventory at cost price from the above information.

$$\text{Percent of cost to sales of goods available for sale} = \frac{\text{Cost price}}{\text{Selling price}} = \frac{\$221,270}{\$326,779} = 67.7\% \text{ (rounded)}$$

67.7 percent of the ending inventory of $65,371 at selling price must be the cost price.

$65,371—Ending inventory at selling price
×.677—Percent of cost to sales
$44,256.17—Ending inventory at cost price (estimate)

Exercise C

Determine the answers to the exercises using the retail inventory estimating method. Round off the percent of cost to sales to the nearest tenth of a percent.

		Cost Price		Selling Price
1.	Beginning inventory	$4,762		$9,827
	+ Purchases	+3,846		+7,842
	Goods available for sale	$8,608		$17,669
	− Sales			−$13,931
	Ending inventory	$1,820.41	(8,608 ÷ 17,669 = × 3,738 =)	$3,738

		Cost Price		Selling Price
2.	Beginning Inventory	$8,662		$10,421
	+ Purchases	+37,356		+43,930
	Goods available for sale	*$46,018*		*$54,351*
	− Sales			−$35,567
	Ending inventory	*$15,910.05*	*(46,018 ÷ 54,351 = × 18,784 =)*	*$18,784*

		Cost Price		Selling Price
3.	Beginning Inventory	42,618		$60,247
	+ Purchases	+38,065		+57,348
	Goods available for sale	*$80,683*		*$117,595*
	− Sales			−$103,900
	Ending inventory	*$9,394.77*	*(80,683 ÷ 117,595 = × 13,695 =)*	*$13,695*

		Cost Price		Selling Price
4.	Beginning Inventory	$3,421		$5,608
	+ Purchases	+132,641		+215,648
	Goods available for sale	*$136,062*		*$221,256*
	− Sales			−$21,432
	Ending inventory	*$122,891.76*	*(136.062 ÷ 221,256 = × 199.824 =)*	*$199,824*

Specific Identification Method

When the number of sales are few, the individual merchandise items may be marked with the date and cost of the purchase. This method allows the middleman to identify which items were sold or are part of ending inventory. The **specific identification method** is used when the price of the merchandise is high or when only a few items are sold.

Example 11

Hal located the following units of a ball thrower in the Sport-A-Merica storage room at the end of last month.

3 units purchased on July 16 at $82 each
1 unit purchased on April 23 at $78.64 each
5 units purchased on September 7 at $77.90 each
2 units purchased on June 30 at $81.45 each

What is the value of ending inventory?

3 units × $82.00 each =	$246.00
1 unit × $78.64 each =	78.64
5 units × $77.90 each =	389.50
+ 2 units × $81.45 each =	+ 162.90
11 units purchased at	$877.04 = ending inventory value

Turnover

Objective 3

Turnover is the number of times the average inventory is sold during the year.

$$\text{Turnover} = \frac{\text{Sales at cost}}{\text{Average inventory}}$$

The average inventory is found by totaling the value of all inventories taken during the year and dividing that sum by the number of inventories taken. Both average inventory value and sales value must be at cost price. The answer is a number, not a dollar amount or a percent, although the number can be changed to a percent.

Example 12

A firm had sales of $66,000 during the year, with a markup of 20 percent on cost. Inventory taken at cost on January 1 was valued at $16,400; on June 30, it was $18,600; and on December 31, it was $13,000. What was the turnover?

The answer to a turnover problem is a number, not a dollar amount or a percent.

Average inventory

$16,400
18,600
+13,000
$48,000 ÷ 3 inventories = $16,000

$66,000
Cost price + markup = selling price
100% + 20% = 120%

Cost price = 1.20)$66,000 = $55,000 (sales at cost or cost or goods sold)

$$\text{Turnover} = \frac{\text{sales (at cost)}}{\text{average inventory (at cost)}} = \frac{\$55{,}000}{\$16{,}000} = 3.4 \text{ times}$$

Exercise D

Determine the turnover rate at cost price in the following problems. Round off answer to nearest tenth.

	1.	2.	3.	4.	5.
Inventories	$47,682.00 53,802.00 49,681.00 57,304.00	$23,050.00 65,400.00 61,325.00	$40,600.00 42,400.00 49,000.00 50,600.00	$2,800.00 4,380.00 3,670.00	$3,490.00 6,660.00 4,658.00 5,360.00
Sales	$426,380.00	$117,430.00	$172,480.00	$24,000.00	$31,680.00
Markup %	23	30	40	35	50
Markup on	Cost	Cost	Sales	Cost	Sales
Cost of sales	$346,650.41	*$90,330.77*	*$103,488.00*	*$17,777.78*	*$15,840.00*
Average inventory	$52,117.25	*$49,925.00*	*$45,650.00*	*$3,616.67*	*$5,042.00*
Turnover	6.7	*1.8*	*2.3*	*4.9*	*3.1*
	(346,650.41 ÷ 52,117.25 =)	*(90,330.77 ÷ 49,925 =)*	*(103,488 ÷ 45,650 =)*	*(17,777.78 ÷ 3,616.67 =)*	*(15,840 ÷ 5,042 =)*

ASSIGNMENT Chapter 16

Name ______________________ Date ______________ Score ____________

Skill Problems

Select the best term for the definition below.

A. Average Cost B. Beginning Inventory C. Cost of Goods Sold D. Ending Inventory
E. F.I.F.O. F. Goods Available for Sale G. L.I.F.O. H. Purchases
I. Retail Inventory Estimating Method J. Specific Identification Method
K. Turnover

1. ___*A*___ An accounting approach used to determine the value of inventory calculated by dividing the total dollar value of units available for sale by the total number of units available.
2. ___*E*___ The units in this ending inventory come from the most recent purchases at their respective costs.
3. ___*K*___ The number of times the average inventory is sold during the year.
4. ___*I*___ This approach replaces the need for an actual physical inventory of the merchandise.

Complete the following inventory model with the terms above.

Inventory Model:

Beginning inventory
+ ____________ 5. ___*H*___
= ____________ 6. ___*F*___
− ____________ 7. ___*D*___
= ____________ 8. ___*C*___

Determine the value of ending inventory in the problems below.

Beginning Inventory at Cost	Purchases at Cost	Net Sales	Markup on Cost	Cost of Goods Sold	Ending Inventory at Cost
$13,560	$47,390	$57,300	40% *(57,300 ÷ 1.40 =)*	9. *$40,928.57*	10. *$20,021.43*
2,658	12,256	13,560	35 *(13,560 ÷ 1.35 =)*	11. *10,044.44*	12. *4,869.56*
91,420	51,200	63,420	45 *(63,420 ÷ 1.45 =)*	13. *43,737.93*	14. *98,882.07*

Determine the answers in the exercises below.

Beginning inventory (units)	55		897	
Cost per unit	$3.45		$29.95	
Purchases:				
February 28	500 at 3.56		120 at 30.00	
March 5	200 at 3.57		200 at 31.15	
March 17	600 at 3.60		205 at 32.35	
March 30	600 at 3.75		100 at 33.10	
April 4	120 at 3.80		205 at 31.50	
Goods available for sale (units)	15. *2,075*	*(55 + 500 + 200 + 600 + 600 + 120 =)*	19. *1,727*	*(897 + 120 + 200 + 205 + 100 + 205 =)*
Goods sold (units)	16. *1,870*	*(2,075 − 205 =)*	20. *1,477*	*(1,727 − 250 =)*
Ending inventory (units)	205		250	
Inventory method	L.I.F.O.		F.I.F.O.	
Ending inventory	17. *$723.75*	*(55 × 3.45 =)* *(150 × 3.56 =)*	21. *$7,947.00*	*(205 × 31.50 =)* *(45 × 33.10 =)*
Cost of goods sold	18. *$6,826.00*	*(7,549.75 − 723.75 =)*	22. *$45,147.40*	*(53,094.40 − 7,947 =)*

Beginning inventory		85 units at 2.25	191.25
Purchases January 8		140 units at 2.35	329.00
January 16		200 units at 2.10	420.00
January 28		110 units at 2.45	269.50 (191.25 + 329 + 420 + 269.50 = 1,209.75)
Ending inventory		152 units	
Ending inventory F.I.F.O.	23.	*$357.70*	*((110 × 2.45) + (42 × 2.10) =)*
Cost of goods sold F.I.F.O.	24.	*$852.05*	*(1,209.75 − 357.70 =)*
Ending inventory L.I.F.O.	25.	*$348.70*	*((85 × 2.25) + (67 × 2.35) =)*
Cost of goods sold L.I.F.O.	26.	*$861.05*	*(1,209.75 − 348.70 =)*
Ending inventory Average costs (Do not round unit cost)	27.	*$343.70*	*(152 × (1,209.75 ÷ 535) =)*
Cost of goods sold Average costs	28.	*$866.05*	*(1,209.75 − 343.70 =)*

Use for problems 29–34.
Beginning inventory = 420 units at $3.60 each

Purchases =
November 2 1,500 units at $3.70 each
10 2,360 units at $3.75 each
23 1,500 units at $3.72 each
28 1,700 units at $3.76 each

Ending inventory = 360 units

Cost of goods sold
29. L.I.F.O. = *$26,588.00*

Ending inventory
30. L.I.F.O. = *$1,296.00* *(360 × 3.60 =)*

Cost of goods sold
31. F.I.F.O. = *$26,530.40*

Ending inventory
32. F.I.F.O. = *$1,353.60* *(360 × 3.76 =)*

Cost of goods sold
33. Average cost = *$26,541.99*

Ending inventory
34. Average cost = *$1,342.01* *(360 × 3.7278075 =)*

Use for problems 35–40.
Beginning inventory = 47 units at $1.20 each

Purchases = April 3 150 units at $1.23 each
26 250 units at $1.26 each
May 7 300 units at $1.28 each
30 200 units at $1.28 each
June 13 350 units at $1.29 each
19 200 units at $1.29 each

Ending inventory = 320 units

Cost of goods sold
35. F.I.F.O. = *$1,492.60*

Ending inventory
36. F.I.F.O. = *$412.80* *(200 × 1.29 = 120 × 1.29 =)*

Cost of goods sold
37. L.I.F.O. = *$1,509.52*

Ending inventory
38. L.I.F.O. = *$395.88*
(47 × 1.20 = 150 × 1.23 = 123 × 1.26 =)

Cost of goods sold
39. Average cost = *$1,498.10*

Ending inventory
40. Average cost = *$407.30* *(320 × 1.2728122 =)*

Use for problems 41–46.
Beginning inventory = 1,366 units at $14 each

Purchases =
January 5 600 units at $13.96 each
18 4,500 units at $13.95 each
22 3,100 units at $14.26 each
26 8,450 units at $14.35 each

Ending inventory = 1,580 units

Cost of goods sold
41. Average cost = *$233,310.28*

Ending inventory
42. Average cost = *$22,428.22*
(1,580 × 14.19507659 =)

Cost of goods sold
43. L.I.F.O. = *$233,627.06*

Ending inventory
44. L.I.F.O. = *$22,111.44*
(214 × 13.96 =) (1,366 × 14 =)

Cost of goods sold
45. F.I.F.O. = *$233,065.50*

Ending inventory
46. F.I.F.O. = *$22,673.00* *(1,580 × 14.35 =)*

Determine the answers in the exercises below using the average cost method. Round off unit cost to nearest cent.

Beginning inventory (units)	840	400
Beginning inventory (cost per unit)	$10.75	$1.40
Purchases June 20	300 at 9.85	650 at 1.35
July 15	200 at 10.50	500 at 1.25
August 17	300 at 11.10	900 at 1.50
Ending inventory (units)	760	300
Ending inventory value	47. *$8,071.20* *(17,415 ÷ 1,640 = × 760 =)*	49. *$417.00* *(3,412.5 ÷ 2,450 = × 300 =)*
Cost of goods sold value	48. *$9,345.60* *(880 × 10.62 =)*	50. *$2,988.50* *(2,150 × 1.39 =)*

Determine the value of ending inventory at cost in problems 51 through 70 using the retail inventory estimating method. Use a cost to sales percent rounded off to the nearest tenth of a percent.

	Cost Price		Selling Price
Beginning inventory	$368.40		$605.00
Purchases	943.30		1,736.80
Goods available for sale	51. *$1,311.70*		53. *$2,341.80*
Sales			$2,141.60
Ending inventory	52. *$112.11*	*(1,311.70 ÷ 2,341.80 = × 200.20 =)*	54. *$200.20*
	Cost Price		**Selling Price**
Beginning inventory	$46,356.00		$51,840.00
Purchases	117,540.00		136,983.00
Goods available for sale	55. *$163,896.00*		57. *$188,823.00*
Sales			$102,421.00
Ending inventory	56. *$74,996.94*	*(163,896 ÷ 188,823 = × 86,402 =)*	58. *$86,402.00*
	Cost Price		**Selling Price**
Beginning inventory	$40,100.00		$108,560.00
Purchases	54,980.00		126,950.00
Goods available for sale	59. *$95,080.00*		61. *$235,510.00*
Sales			$180,670.00
Ending inventory	60. *$22,155.36*	*(95,080 ÷ 235,510 = × 54,840 =)*	62. *$54,840.00*
	Cost Price		**Selling Price**
Beginning inventory	$5,870.50		$8,218.70
+ Purchases	+5,604.40		+7,846.16
Goods available for sale	63. *$11,474.90*		65. *$16,064.86*
− Sales			−9,405.48
Ending inventory	64. *$4,754.08*	*(11,474.90 ÷ 16,064.86 = .714 × 6,658.38 =)*	66. *$6,658.38*
	Cost Price		**Selling Price**
Beginning inventory	$987.40		$1,234.25
+ Purchases	+1,105.85		+1,382.31
Goods available for sale	67. *$2,093.25*		69. *$2,616.56*
− Sales			−1,744.37
Ending inventory	68. *$697.75*	*(2,093.25 ÷ 2,616.56 = .800 × 872.19 =)*	70. *$872.19*

Determine the turnover rate at cost price in the following problems. Round off answers to nearest tenth.

Inventories	$4,085.50 3,045.60 5,784.55 4,563.40	$40,875.00 65,087.00 105,040.00
Sales	$15,487.55	$85,706.00
Markup %	20	50
Markup on	Cost	Sales
Cost of sales	71. ***$12,906.29*** ***(15,487.55 ÷ 1.2 =)***	74. ***$42,853.00*** ***(85,706 × .5 =)***
Average inventory	72. ***$4,369.76*** ***(17,479.05 ÷ 4 =)***	75. ***$70,334.00*** ***(211,002 ÷ 3 =)***
Turnover	73. ***3.0*** ***(12,906.29 ÷ 4,369.76 =)***	76. ***0.6*** ***(42,853 ÷ 70,334 =)***

77. Inventories taken: July 1 = $13,560
October 1 = 14,388
December 31 = 13,840

Sales at cost = $38,645

$38,645 / $13,929.33 = 2.8 (38,645 ÷ 13,929.33 =)

78. Inventories taken: January 1 = $68,542
March 31 = 72,840
June 15 = 65,473
August 20 = 69,568
October 31 = 73,869

Sales = $630,470
Markup = 35% on cost

$467,014.81 / $70,058.40 = 6.7 (467,014 ÷ 70,058.40 =)

79. Inventories taken: April 2 = $32,480
9 = 36,400
16 = 35,931
28 = 33,560

Sales = $149,680
Markup = 40% on sales

$89,808.00 / $34,592.75 = 2.6 (89,808 ÷ 34,592.75 =)

Business Application Problems

1. The value of the cost of merchandise available for sale was $428,500 for the year. The average inventory was $68,400. To the nearest tenth, what was the turnover for the year?
$428,500/$68,400 = 6.264 = 6.3 turnover

2. The average inventory value for Pedmix Company was $43,976 over the past 12-month period. The markup rate averages 27 percent of sales. If the turnover rate is 4.1, what is the value of sales for the 12-month period?
$43,976 × 4.1 = $180,301.60
C + M = S
73% + 27% = 100% $180,301.60/.73 =
$246,988.49 sales

3. The beginning inventory of Fredmore Associates was $36,720. Purchases for the period were $75,908. The value of the cost of merchandise sold was $40,860 for the period. What is the value of ending inventory?
$36,720 + $75,908 − $40,860 = $71,768 value

4. In problem 3, what is the value of sales if the markup rate is 27 percent on sales?
$40,860
C + M = S
73% + 27% = 100% $40,860/.73 =
$55,972.60 sales

5. In problem 3, to the nearest tenth, what is the turnover rate?
$36,720 + $75,908 − $40,860 = $71,768 + $36,720 =
$108,488/2 = $54,244 $40,860/$54,244 = .753 =
.8 turnover

6. The Balcom Corporation enjoyed a low inventory level of $52,930 at the end of the year. The cost value of sales was $340,817 for the year. The turnover rate was 4.2. What was the average inventory level for the year?
$340,817/4.2 = $81,146.90 average

7. In problem 6, what was value of purchases for the period if the beginning inventory was $91,560?
$340,817 + $52,930 − $91,560 = $302,187 purchases

8. In problem 6, the markup rate was 38 percent on cost. What is the value of sales for the period?
$340,817
C + M = S
100% + 38% = 138% $340,817 × 1.38 =
$470,327.46 sales

9. Hopson Corporation reduced their investment in inventory this year, while maintaining the same 40 percent level of markup on cost. Last year the average inventory level was $45,890. This year it was $41,984. Turnover was 6.5 last year and 7.2 this year. By what amount did Hopson Corporation increase or decrease sales this year?
$45,890 × 6.5 = $298,285
$41,984 × 7.2 = $302,284.80 − $298,285 =
$3,999.80 × 1.4 = $5,599.72 increase

10. Gulf Tire Company had a beginning inventory of $106,390. They ended the period with an inventory value of $89,600. During the period they made two purchases, one for $32,640 and one for $98,600. What is the value of the goods that were available for sale in the period?
$106,390 + $32,640 + $98,600 = $237,630 available

11. Booth and Burtis, Inc., has a turnover rate of 5.4 and a markup of 40 percent on sales. Sales are $321,840 for the year. What is the average inventory that the firm carries in order to maintain this level of sales?
$321,840
C + M = S
60% + 40% = 100%
A = B × R
A = $321,840 × .6 = $193,104
$193,104 ÷ 5.4 = $35,760

12. Arsenau and Co. found that the value of inventory on January 1 was $147,368; and on March 31, it was $153,820. Sales for the period were $413,875. Goods were marked up 40 percent on cost. What was the turnover to the nearest tenth?
$147,368
+ 153,820
$301,188 ÷ 2 =
$150,594 average inventory

$413,875
C + M = S
100% + 40% = 140%
B = A ÷ R
B = $413,875 ÷ 1.40 = $295,625 C.O.G.S.

$$\frac{\$295{,}625}{\$150{,}594} = 2.0 \text{ turnover}$$

13. The Jones Corporation had a beginning inventory for the year of 275 units valued at $14,307.50. At the end of the year, there were 265 units remaining. A purchase was made on March 16 for 360 units at $65 each, and on September 1 for 800 units at $71.40 each. What is the amount of sales for the period if the firm used a F.I.F.O. system and a 30 percent markup on cost?

B.I.	*= 275*
+ Purch.	*360*
	+ 800
Units avail.	*= 1,435*
− E.I.	*= −265*
Units sold	*= 1,170*

275 units = B.I.	*= $14,307.50*
360 units × $65.00	*= 23,400.00*
535 units × $71.40	*= 38,199.00*
	$75,906.50 C.O.G.S.

$75,906.50
C + M = S
100% + 30% = 130%
A = B × R
A = $75,906.50 × 1.3 = $98,678.45 sales

14. Dawn's Discount enjoyed a fourth quarter sales record of $69,800 at a markup of 40 percent on sales. The purchases for the period were $32,590. The ending inventory was $21,730. What was the beginning inventory?
$69,800
C + M = S
60% + 40% = 100%
A = B × R
A = $69,800 × .6 = $41,880 C.O.G.S.

E.I.	*= $21,730*
+ C.O.G.S.	*= + 41,880*
Goods avail.	*= $63,610*
− Purch.	*= −32,590*
B.I.	*= $31,020*

15. Barton Wholesalers had a fire that damaged a portion of their warehouse on June 23. However, inventory had been taken on June 1, and it totaled $85,642. Only one purchase for $38,650 was made in June. Sales for the first 23 days of the month were $95,391 at a 42 percent markup on sales. A physical inventory of the undamaged merchandise totaled $41,365. What is the approximate value of the loss?

B.I. = $ 85,642.00
+ Purchase = + 38,650.00
Goods avail. = $124,292.00
− C.O.G.S. = − 55,326.78
$ 68,965.22
$68,965.22 − $41,365.00 = $27,600.22 value of loss

$95,391
C + M = S
58% + 42% = 100%
A = B × R
A = $95,391 × .58 = $55,326.78 C.O.G.S.

16. Benson and Co. started the firm with a purchase of ten units at $14 each. During the month, they purchased an additional 20 units at $14.50 each. In the second month, they purchased 50 units at a total cost of $727.50. At the end of the second month, they had 17 units left in stock. If the firm elects to use a F.I.F.O. system, what is the value of ending inventory?

$727.50 ÷ 50 = $14.55
×17
$247.35 ending inventory F.I.F.O.

17. If Benson and Co. in problem 16 elects a L.I.F.O. system, what is the value of the cost of goods sold?

13 at $14.50 = $188.50
+ 727.50
$916.00 C.O.G.S. L.I.F.O.

18. In problem 16, Benson and Co. sells the merchandise at a markup of 30 percent on the selling price. What was the turnover for the two-month period using F.I.F.O.?

10 at $14 = $140.00
20 at $14.50 = 290.00
33 at $14.55 = 480.15
$910.15 = C.O.G.S.

$$\frac{\$910.15}{\$193.68} = 4.7 \text{ turnover}$$

$140 (10 × $14)
+247.35 (17 × $14.55)
$ 387.35 ÷ 2 = $193.68

19. Write a complete sentence in answer to this question. The Sterling Corporation had sales of $589,630 for the month. At the beginning of the month, the inventory was $184,980. At the end of the month, the inventory was $203,568. The firm marks up on sales, and they use a 40 percent rate. To the nearest tenth, what is the turnover for the month?

$184,980
+203,568
$388,548 ÷ 2 = $194,274 Av. Inv.
$589,630
C + M = S
60% + 40% = 100%
A = B × R
A = $589,630 × .6 = $353,778 C.O.G.S.

$$\frac{\$353,778}{\$194,274} = 1.8$$

The turnover is 1.8 for the month.

20. Write a complete sentence in answer to this question. Dalton Distributors uses a L.I.F.O. system for inventory. The firm had a beginning inventory of $190,000. Their purchases for the month totaled 8,500 units at $16.20 each. The purchases brought the total units available for sale to 20,120. The end-of-the-month inventory, as displayed on the computer printout, said there were 5,890 units available. A physical inventory of the warehouse showed there were only 5,530 units available. The accountant decided to write off the missing units as "stock shrink." What is the amount of the write-off?

20,120
−8,500
11,620 units B.I.

$$\frac{\$190,000}{11,620} = \$16.35/\text{unit B.I.}$$

5,890
−5,530
360 units of stock shrink × $16.35 = $5,886
The amount of the write-off is $5,886.

★ Challenge Problem

The Saranac Producers Association had a beginning inventory of 682 units of a feeder on June 13. The inventory had cost Saranac $947.30. During the next six months, the following purchases were made:

September	1	200 units at $1.47 each; terms 3/10,n/30
October	21	450 units at $1.51 each; terms 3/10,n/30
November	13	600 units at $1.52 each; terms 2/10,n/60

All discounts were taken.

1,640 units were sold during the six months at $2.39 each.

What is the value of ending inventory using L.I.F.O.?

B.I. = 682 units
+ Purch. = 200 units
= 450 units
= 600 units
Units avail. = 1,932
− Units sold = 1,640
Units left = 292

$$\frac{\$947.30}{682} \times 292 = \$405.59 \text{ ending inventory}$$

MASTERY TEST Chapter 16

Name ______________________ **Date** ______________ **Score** ____________

True or False

T (F) **1.** Goods available for sale less ending inventory equals purchases.

T (F) **2.** Ending inventory equals the previous period's beginning inventory.

(T) F **3.** Turnover measures the number of times that the average inventory is sold.

(T) F **4.** Specific identification is a method of determinimg the actual value of inventory.

Complete the following matrix for the missing values.

	Beginning Inventory	Purchases	Ending Inventory	C.O.G.S.
5.	$ 42,000	$175,000	$32,000	***$185,000***
6.	***158,455***	243,980	56,835	345,600
7.	39,800	156,000	***17,800***	178,000

Determine the value of ending inventory in the problems below using the method indicated.

Beginning inventory = 890 units at $41.00 each
Purchases = January 17 400 units at $41.20 each
May 30 650 units at $41.20 each
September 16 300 units at $41.30 each
November 3 650 units at $41.50 each
Ending inventory = 970

8. Ending Inventory using L.I.F.O. = ***$39,786***

9. Ending Inventory using F.I.F.O. = ***$40,189***

10. Ending Inventory using average cost = ***$39,979.77***

11. Sales are made at a 25 percent markup on sales. What is the turnover in the problem just given if L.I.F.O. is used? Round off the answer to the nearest tenth.

BI = $36,490
EI = + 39,786
$76,276 ÷ 2 = $38,138 ave. inv.

$79,329 = C.O.G.S.

$$\frac{\$79{,}329}{\$38{,}138} = 2.1 \text{ turnover}$$

12. Estimate the value of ending inventory at cost using the Retail Inventory Estimating Method.

	Cost Price	Selling Price
Beginning Inventory	$ 38,500	$ 84,600
Purchase	153,750	301,230
Goods Available for Sale	***$ 192,250***	***$385,830***
Sales		$260,900
Ending Inventory	***$62,215.14***	***$124,930***

13. Humbert and Associates found the value of inventory on July 1 to be $96,800. An additional inventory taken on August 1 totaled $42,690. The sales for the period were $120,970 at a markup of 40 percent on sales. What was the turnover to the nearest tenth?

$ 96,800
+ 42,690
$139,490 ÷ 2 =
$69,745 ave. inv.

$$\frac{\$72{,}582}{\$69{,}745} = \underline{1.0 \text{ turnover}}$$

$120,970
C + M = S

60% + 40% = 100%
A = B × R
A = $120,970 × .6
= $72,582 C.O.G.S.

14. The Woodard Company took inventory on January 1 and found that they had 345,587 meters of their wire product. The value was set at $251,873. The warehouse was broken into on January 12. All of the inventory was stolen. The value of the sales for the twelve days was $34,980 at a 40 percent markup on sales. What is the value of merchandise that was stolen? There were no purchases during the period.

$34,980
C + M = S
60% + 40% = 100%
A = B × R
A = $34,980 × .6 = $20,988 C.O.G.S.

$251,873
− 20,988
$230,885 stolen

DEPRECIATION 17

Suggested Classroom Time for This Chapter

hours 1 2 3 4 5

Perceived Student Difficulty with Topical Material

easy moderate difficult

OBJECTIVES

After mastering the material in this chapter, you will be able to:

1. Explain the concept of depreciation and depletion. p. 324
2. Calculate book value, accumulated depreciation, and depreciation using four methods. p. 324
3. Explain the advantages of each of the depreciation methods. p. 324
4. Calculate the depreciation amount for a partial year using three methods. p. 335
5. Understand and use the following terms and concepts:
   Depreciation
   Expected Life
   Total Cost
   Salvage Value
   Total Depreciation
   Accumulated Depreciation
   Book Value
   Market Value
   Straight-Line Method
   Depreciation Schedule
   Units-of-Production Method
   Sum-of-Years'-Digits (S.O.Y.D.) Method
   Declining-Balance Method
   Modified Accelerated Cost Recovery System (M.A.C.R.S.) Method
   Depletion
   Time Line

SPORT-A-MERICA

Hal is looking at the calendar after the Saturday session on the financial picture of Sport-A-Merica with Kerry. He realizes it is time to prepare the tax information for the accountant. This always means a lot of digging into the files. The records of the assets bought the past year will have to be reviewed for any changes that need to be brought to the attention of the accountant.

Another area that always needs reviewing is the schedule of depreciation on the fixed assets. Sport-A-Merica has purchased a new delivery vehicle this year. The cost of the vehicle, along with the cost of the necessary optional equipment, will need to be gathered. Hal knows the vehicle's useful life will be shorter than the dealer had estimated because Sport-A-Merica will be keeping the vehicle very busy with the increase in sales. Hal will talk to Kerry about which method she wishes to use to depreciate the vehicle. She had earlier suggested a method that would recognize the actual use of the vehicle instead of a method that involves merely the passage of time. Hal thinks that she is right on target. After all, the total cost of the vehicle was $28,400—no small amount! He will talk to her today so that he can be ready for the meeting with the accountant.

Objective 1 *depreciation*

Depreciation is an estimate of the amount of decrease in market value of an item over a period of time. A *depreciation expense* is allowed by taxing authorities, such as the federal government, for the decline in the item's value. The depreciation expense, in effect, reduces a firm's income tax. The Economic Recovery Tax Act of 1981 was seen in part as an incentive for business to invest in new machinery, plants, and equipment. These investments would then result in a favorable tax break for the business—a faster depreciation of the asset. This allows for a greater depreciation expense per year in the early years of the asset's life. Fixed assets, such as buildings, office equipment, trucks, and furniture, decline in value due to wear, an outdating of the style, or obsolescence. An office machine may have a useful life of only two or three years, whereas a building's life may exceed fifty years. The useful life of an asset is termed the **expected life.** The expected life will be stated by the selling firm or provided through industry or government guidelines. Expected life is usually expressed in years or units of usefulness.

expected life

The amount of costs to be depreciated includes all of those costs necessary to make the item operational. Item costs, transportation costs, and installation costs make up the **total cost** of an item. Consider an item such as a computer that was purchased at a price that included transportation and installation. If the buyer had agreed instead to purchase the computer without transportation or installation, those costs would still have been incurred by the buyer before the computer became operative. In the final analysis, the total costs would have been approximately the same.

total cost

All of the total costs are depreciable over the expected life, unless the firm feels that the item can be sold at the end of its expected life. The sales price at the end of the expected life of the item is termed the **salvage value.** Terms such as *scrap value, trade-in value, residual value,* and *end value* are also used. The selling firm can provide an estimate of the salvage value.

salvage value

The **total depreciation,** or the amount that the item is depreciated over its expected life, is found by subtracting the salvage value from the total cost. A portion of the total depreciation is subtracted from the total cost each year. This amount is termed *depreciation*. By totaling the depreciation to date, the amount of **accumulated depreciation** can be determined. By subtracting the accumulated depreciation from the total cost, the amount of book value can be found. **Book value** is an estimate of an item's market value. The book value is the amount at which the item is valued on the firm's accounting records.

total depreciation

accumulated depreciation

book value

❍ *Book value is not limited to the value of a vehicle!*

Objective 2 *market value*

The **market value** is the amount that the item can be sold for at a given point in its expected life.

The terminology of depreciation is rather extensive; and at the same time, it is important to the understanding of depreciation.

Straight-Line Method

Objective 3

There are four basic methods that are used to determine depreciation. The **straight-line method** is the easiest to understand. This method establishes an equal amount of depreciation per year for an item. It is found by dividing the total depreciation by the years of expected life. Total depreciation is found by subtracting the salvage value from the total cost.

● *Students often confuse depreciation, accumulated depreciation, and total depreciation.*

The following figures may help you to visualize these terms and the concepts they represent.

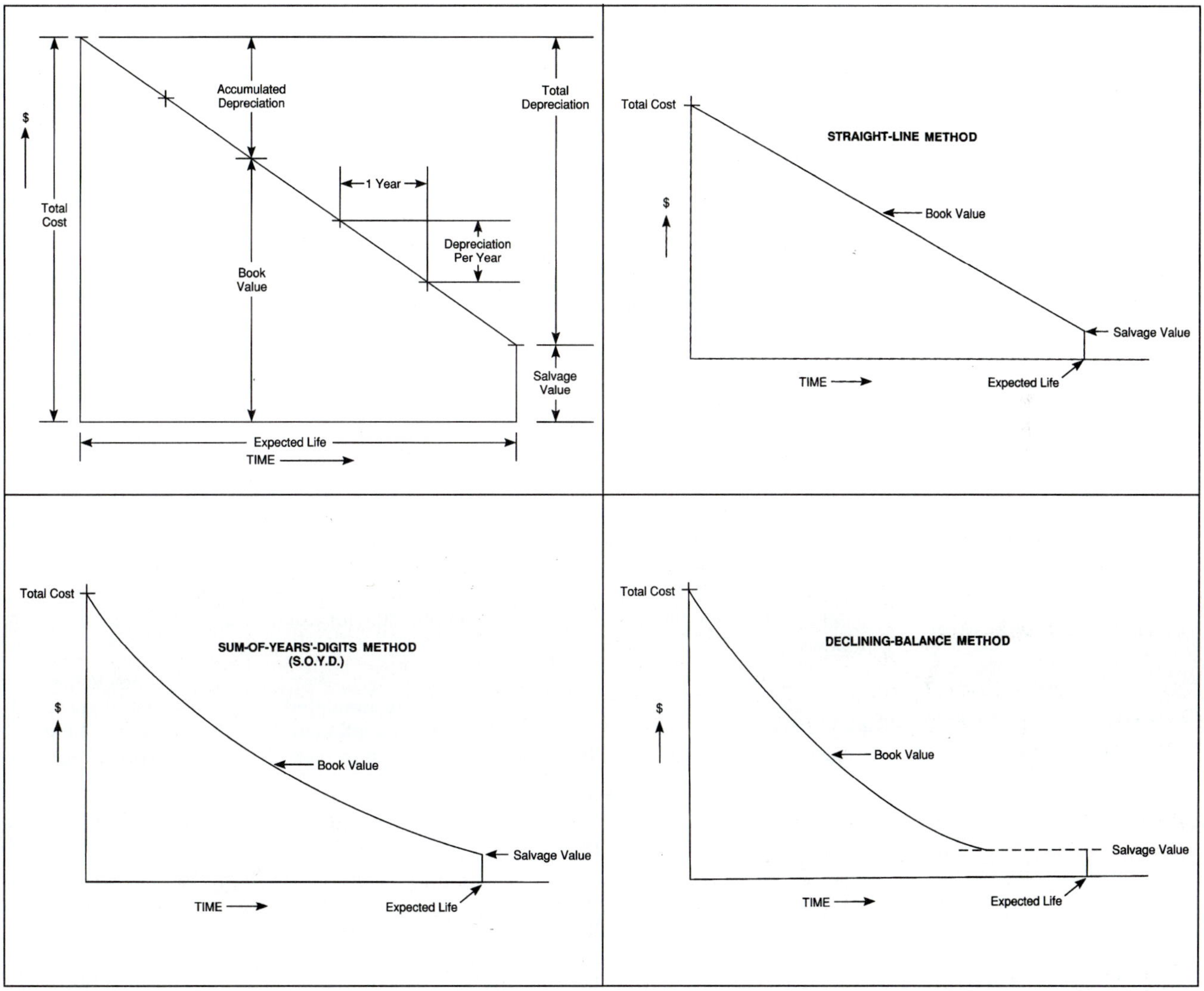

Example 1

A communications system purchased for $600 has an expected life of eight years and a salvage value of $100. What is the depreciation per year using the straight-line method?

$$\text{Depreciation} = \frac{\text{total depreciation}}{\text{expected life}} \text{ or } \frac{\text{total depreciation}}{n}$$

(n = the number of years of expected life)

\$600 − \$100 = \$500 total depreciation

$$\text{Depreciation} = \frac{\$500}{8} = \$62.50 \text{ per year}$$

(500 ÷ 8 =)

The book value at the end of the second year in example 1 is found by subtracting the accumulated depreciation ($62.50 per year × 2 years = $125) from the total cost ($600). Book value = $600 − $125 = $475.

Example 2

Sport-A-Merica purchased an asset that originally cost $300. In addition, the firm paid $14 for transportation and $26 for installation. The asset is expected to have a useful life of ten years before it will be traded in for an estimated value of $60. What will the book value be at the end of six years?

$300 Item cost
+14 Transportation cost
+26 Installation cost
$340 Total Cost
−60 Salvage value
$280 Total depreciation

$$\text{Depreciation} = \frac{\$280 \text{ (total depreciation)}}{10 \text{ years expected life}} = \$28 \text{ per year}$$

$28 = Per year depreciation
×6 = Years
$168 = Accumulated depreciation at the end of the sixth year

$340 = Total cost
−168 = Accumulated depreciation
$172 = Book value, sixth year

(280 ÷ 10 = × 6 =)

Depreciation Schedule

The **depreciation schedule** shows the effects of depreciation on the book value of the asset over the expected life. The depreciation schedule for example 2 would be as follows.

End of Year Number	Depreciation/Year	Accumulated Depreciation at the End of Year	Book Value (Cost − Accum. Depr.) at the End of Year
1	$28	$ 28	$312
2	28	56	284
3	28	84	256
4	28	112	228
5	28	140	200
6	28	168	172
7	28	196	144
8	28	224	116
9	28	252	88
10	28	280	60

Exercise A

Determine the missing values in the exercises below using the straight-line method.

	Item Cost	Transportation Cost	Installation Cost	Salvage Value	Expected Life (Years)	Year No.	Depreciation per Year	Accumulated Depreciation	Book Value
1.	$14,350	$95 (27,820 ÷ 12 =)	$16,585	$3,210	12	4	$2,318.33	$9,273.32	$21,756.68
2.	19,000	0 *(19,500 ÷ 30 =)*	1,757	1,257	30	7	*650.00*	*4,550.00*	*16,207.00*
3.	625	26 *(700 ÷ 7 =)*	249	200	7	3	*100.00*	*300.00*	*600.00*
4.	42,600	0 *(51,000 ÷ 17 =)*	10,380	1,980	17	2	*3,000.00*	*6,000.00*	*46,980.00*
5.	965	85 *(800 ÷ 8 =)*	75	325	8	3	*100.00*	*300.00*	*825.00*

Exercise B

Determine the missing values in the exercises below using the straight-line method and the formula

$$\text{Depreciation/year} = \frac{\text{Total depreciation}}{\text{Expected life}}.$$

	Total Cost	Expected Life (Years)	Salvage Value	Year No.	
1.	$70,000	10	$20,000	3	$55,000 book value (50,000 ÷ 10 =)
2.	10,000	8	2,000	*7*	$3,000 book value *(7,000 ÷ 1,000 =)*
3.	4,000	*10*	1,000	6	$1,800 accumulated depreciation *(3,000 ÷ 300 =)*
4.	20,000	10	*5,000*	2	$17,000 book value *(3,000 ÷ 2 = × 10 =)*
5.	*20,000*	7	6,000	3	$2,000 depreciation per year *(2,000 × 7 =)*

Units-of-Production Method

In the straight-line method, the depreciation expense is the same amount every year. Some assets are not used an equal amount every year. Their use varies, and therefore, the depreciation amount should also vary. The **units-of-production method** acknowledges an unequal usage. The amount of depreciation is equal for each unit produced, but it is not necessarily the same total amount each year. This method is also termed the *hours-of-production method*.

Example 3

The delivery vehicle Sport-A-Merica purchased is expected to provide approximately 100,000 trouble-free miles over its life. The vehicle, including all of the add-ons the firm needed, cost $28,400. This year, Sport-A-Merica drove the vehicle 21,000 miles. The salvage value of the vehicle after 100,000 miles have been driven is estimated to be $2,400. What is the depreciation for this year?

$$\text{Depreciation} = \frac{\text{Total depreciation}}{\text{Expected total output capacity over life of asset}} \times \text{Output this period}$$

$$\text{Depreciation} = \frac{\$28{,}400 - \$2{,}400}{100{,}000 \text{ miles}} \times 21{,}000 \text{ miles} = \$5{,}460 \text{ for } 21{,}000 \text{ miles}$$

(26,000 ÷ 100,000 = × 21,000 =)

The book value at the end of 49,000 miles would be as follows:

$$\frac{\$28{,}400 - \$2{,}400}{100{,}000 \text{ miles}} \times 49{,}000 \text{ miles} = \$12{,}740 \text{ for } 49{,}000 \text{ miles}$$

$ 28,400 = Total cost
−12,740 = Accumulated depreciation
$ 15,660 = Book value after 49,000 miles

A *depreciation schedule* for an asset depreciated by the units-of-production method could be created only if the usage rate could be known for the periods in the future.

Exercise C

Determine the missing values in the exercises below using the units-of-production method.

	Total Cost	Salvage Value	Expected Life (in Units)	Units This Period	Units to Date	Depreciation This Period	Book Value
1.	$43,000 (41,500 ÷ 140,000 =)	$1,500	140,000	1,980	35,923	$586.93	$32,351.40
2.	975 *(910 ÷ 1,400 =)*	65	1,400	350	895	*227.50*	*393.25*
3.	23,680 *(22,500 ÷ 30,000 =)*	1,180	30,000	3,500	19,600	*2,625.00*	*8,980.00*
4.	3,600 *(3,350 ÷ 16,000 =)*	250	16,000	950	2,860	*198.91*	*3,001.19*
5.	19,780 *(19,350 ÷ 21,000 =)*	430	21,000	610	13,548	*562.07*	*7,296.49*

Sum-of-Years'-Digits (S.O.Y.D.) Method

The market values of some assets do not decline by an equal amount each year. In order to develop a book value that approximates the market value of an item, the **sum-of-years'-digits (S.O.Y.D.) method** is sometimes used. This method accelerates the depreciation by allowing for the greater decline in the market value in the first years of an asset's life and a lesser decline in the later years of its life. It is similar to the straight-line method: The total depreciation is multiplied by a fraction to determine the depreciation for the year. However, each year, the fraction becomes smaller so that the depreciation amount also becomes smaller. The fraction's denominator is the sum of the years of expected life, and its numerator is the number of useful years remaining.

Example 4

An asset has an expected life of eight years. What is the fraction used to determine the depreciation for the second year using the S.O.Y.D. method?

Years of Expected Life		*Year Number*	
8		1	
7	←	2 ←	Beginning year 2 the asset had seven years of useful life left. The fraction would be $\frac{7}{36}$.
6		3	
5		4	
4		5	
3		6	
2		7	
+1		8	
36 = Sum of the years of expected life			

Example 5

An asset has an expected life of twelve years. What is the fraction used to determine the depreciation for the fifth year using the S.O.Y.D. method?

Years of Expected Life	*Year Number*	
12	1	
11	2	
10	3	
9	4	
8 ←	5 ←	Beginning year 5 the asset had eight years of useful life left. The fraction would be $\frac{8}{78}$.
7	6	
6	7	
5	8	
4	9	
3	10	
2	11	
+1	12	
78 = Sum of the years of expected life		

The sum of the years can also be found as follows:

$$\frac{n \times (n+1)}{2} \quad n = \text{the number of years of expected life}$$

Example 6

The expected life of an asset is 30 years. What is the sum of the years' digits?

$$\frac{n \times (n+1)}{2} = \frac{30 \times (30+1)}{2} = 465$$

(30 × 31 = ÷ 2 =)

This formula is very useful for assets with long life expectancies.

Exercise D

Determine the sum-of-years' fraction to be used in the following exercises.

Years of Expected Life	Sum of Years	Year Number	Fraction to be Used
1. 14 (14 × 15 = ÷ 2 =)	105	3	$\frac{12}{105}$
2. 20 *(20 × 21 = ÷ 2 =)*	***210***	2	$\frac{19}{210}$
3. 13 *(13 × 14 = ÷ 2 =)*	***91***	4	$\frac{10}{91}$
4. 15 *(15 × 16 = ÷ 2 =)*	***120***	6	$\frac{10}{120}$
5. 10 *(10 × 11 = ÷ 2 =)*	***55***	10	$\frac{1}{55}$

To determine the depreciation amount for a year, the fraction is multiplied by the total depreciation. Each year the fraction will change, and the new fraction will provide a different depreciation amount.

Example 7

A building purchased for \$42,000 five years ago has an expected life of 20 years. The salvage value is estimated to be \$3,000. What is the depreciation for each of the five years and what is the book value using the sum-of-years' method?

\$42,000 − \$3,000 = \$39,000 (total depreciation)

	Years of Expected Life	*Year Number*
(sum of values for first five years) 90	20	1
	19	2
	18	3
	17	4
	16	5
	.	
	.	
	.	
	+1	
	210 = Sum of the years of expected life	

Depreciation

$$\text{Year 1} = \$39{,}000 \times \frac{20}{210} = \$3{,}714.29$$

$$\text{Year 2} = \$39{,}000 \times \frac{19}{210} = 3{,}528.57$$

$$\text{Year 3} = \$39{,}000 \times \frac{18}{210} = 3{,}342.86 \quad \textit{or } \frac{90}{210} \times \$39{,}000 = \$16{,}714.29$$

$$\text{Year 4} = \$39{,}000 \times \frac{17}{210} = 3{,}157.14$$

$$\text{Year 5} = \$39{,}000 \times \frac{16}{210} = 2{,}971.43$$

Accumulated depreciation = \$16,714.29 years 1–5

Book value = \$42,000 − \$16,714.29 = \$25,285.71

(20 × 21 = ÷ 2 =)

Have the students compare depreciation schedules for S.O.Y.D. and straight line.

The *depreciation schedule* for the first five years of the building in example 7 using the S.O.Y.D. method would be as follows.

End of Year Number	Depreciation/Year	Accumulated Depreciation at the End of Year	Book Value at the End of Year
1	\$3,714.29	\$ 3,714.29	\$38,285.71
2	3,528.57	7,242.86	34,757.14
3	3,342.86	10,585.72	31,414.28
4	3,157.14	13,742.86	28,257.14
5	2,971.43	16,714.29	25,285.71

Exercise E

Determine the missing values in the exercises below using the S.O.Y.D. method.

	Total Cost	Salvage Value	Expected Life (Years)	Year Number	Depreciation	Accumulated Depreciation	Book Value
1.	\$ 48,500	\$ 4,000	20	10	\$ 2,330.95	\$32,845.24	\$15,654.76
							(44,500 × 11 = ÷ 210 =)
2.	21,000	600	14	4	*2,137.14*	*9,714.29*	*11,285.71*
							(20,400 × 11 = ÷ 105 =)

Exercise E Continued

Determine the missing values in the exercises below using the S.O.Y.D. method.

	Total Cost	Salvage Value	Expected Life (Years)	Year Number	Depreciation	Accumulated Depreciation	Book Value
3.	135,500	10,000	15	6	*10,458.33*	*78,437.50*	*57,062.50* (125,500 × 10 = ÷ 120 =)
4.	8,900	1,000	10	9	*282.27*	*7,756.36*	*1,143.64* (7,900 × 2 = ÷ 55 =)
5.	7,600	200	20	15	*211.43*	*6,871.43*	*728.57* (7,400 × 6 = ÷ 210 =)

Declining-Balance Method

The **declining-balance method** is similar to the sum-of-years'-digits method because both provide an accelerated amount of depreciation in the early years of an asset's life. This method is used more frequently and is even more accelerated than the sum-of-years'-digits method. The term *declining-balance* is very descriptive of the procedure used to determine the depreciation amount.

This method also has some similarity to the straight-line method. In the declining-balance method, the straight-line method rate of depreciation is doubled. This rate is then multiplied by the current book value. The depreciation amount is subtracted from the previous period book value to find the current year book value, or declined balance. The book value must not drop below the salvage value.

Note that the total cost the first year and the book value or declined value in subsequent years, not the total depreciation, are used to determine the amount of depreciation for the current year.

Example 8

A shop machine cost $840 and is expected to be used for an expected life of eight years. The salvage value is $150. What is the depreciation for each of the eight years using the declining-balance method?

$$\frac{1}{8 \text{ years of expected life} \times 2 \text{ (double)}} = \frac{2}{8} = 25\%$$

$840 = Total cost × .25 = $210 depreciation for first year
−210 = Depreciation for first year
$630 = Book value year 1 or declined value
×.25
$157.50 = Depreciation for second year

$630.00 = Book value year 1
−157.50 = Depreciation year 2
$472.50 = Book value year 2 × .25 = $118.13

$472.50 = Book value year 2
−118.13 = Depreciation for third year
$354.37 = Book value year 3

$354.37 = Book value year 3 × .25 = $88.59
−88.59 = Depreciation for year 4
$265.78 = Book value year 4 × .25 = $66.45
−66.45 = Depreciation for year 5
$199.33 = Book value year 5 × .25 = $49.83

If $49.83 was subtracted from $199.33, the book value would fall below $150. The depreciation for year six is, therefore, $49.33, the difference between book value year 5 and the salvage value ($199.33 − $150.00 = 49.33). The depreciation amount for years 7 and 8 is zero.

The *depreciation schedule* for the eight years of the machine in example 8 using the declining-balance method would be as follows.

End of Year Number	Depreciation/Year	Accumulated Depreciation at the End of Year	Book Value at the End of Year
1	$210.00	$210.00	$630.00
2	157.50	367.50	472.50
3	118.13	485.63	354.37
4	88.59	574.22	265.78
5	66.45	640.67	199.33
6	49.33	690.00	150.00
7	0	690.00	150.00
8	0	690.00	150.00

Exercise F

Determine the missing values in the exercises below using the declining-balance method. Use a factor rounded off to three decimal places.

Total Cost	Salvage Value	Expected Life (Years)	Year Number	Depreciation	Accumulated Depreciation	Book Value
1. $3,800	$ 500	4	2	$950.00	$2,850.00	$950.00
		(3,800 × .5 =)	(1,900 × .5 =)			
2. 45,000	0	5	2	*10,800.00*	*28,800.00*	*16,200.00*
		(45,000 × .4 =)	*(27,000 × .4 =)*			
3. 16,900	600	8	3	*2,376.56*	*9,770.31*	*7,129.69*
		(16,900 × .25 =)	*(12,675 × .25 =)*			*(9,506.25 × .25 =)*
4. 10,000	1,000	6	4	*988.15*	*8,020.74*	*1,979.26*
		(10,000 × .333 =)	*(6,670 × .333 =)*	*(4,448.89 × .333 =)*		*(2,967.41 × .333 =)*
5. 85,400	3,400	10	5	*6,995.97*	*57,416.13*	*27,983.87*
	(85,400 × .2 =)	*(68,320 × .2 =)*	*(54,656 × .2 =)*	*(43,724.80 × .2 =)*		*(34,979.84 × .2 =)*

Modified Accelerated Cost Recovery System (M.A.C.R.S.) Method

To improve the economy, the Economic Recovery Tax Act of 1981 included an incentive for business to invest in new equipment and other capital assets. This incentive was modified in 1986, and is now titled the **Modified Accelerated Cost Recovery System (M.A.C.R.S.)** (also known as the General Depreciation System, GDS). The M.A.C.R.S. method must be used for tax purposes and shortens the expected life of items placed in service after 1986. The salvage value is not considered in this method of computing depreciation. The shorter life and the elimination of scrap value combine to provide a larger depreciation amount, thereby reducing taxes for business. Items in service before 1986 must be depreciated for tax purposes by the A.C.R.S. methods (for years 1981 through 1986) or by the methods explained earlier in the chapter. Furthermore, a firm may not switch to the M.A.C.R.S. method on older items. Depreciable items may be classified by government regulation into eight categories.

Property Categories

Three-year items include over-the-road tractor units and race horses over two years old when placed in service.

Five-year items include autos, taxis, buses, trucks, computers and peripheral equipment, office machines, property used in research and experimentation, breeding cattle, and dairy cattle.

Seven-year items include office furniture and fixtures, and property that has not been designated in another class.

Ten-year items include vessels, barges, tugs, similar water transportation equipment, single-purpose agricultural or horticultural structure, and any tree or vine bearing fruit or nuts.

Fifteen-year property includes improvements made directly to land or added to it; items such as shrubbery, fences, roads, and bridges.

Twenty-year property includes farm buildings other than agricultural or horticultural structures.

Twenty-seven and one-half-year property includes residential real property.

Thirty-one and one-half-year items include nonresidential real property.

The M.A.C.R.S. method stipulates a 200 percent declining-balance method for items in the three-, five-, seven-, and ten-year categories, and a 150 percent declining-balance method for the fifteen- and twenty-year categories. The fifteen- and twenty-year categories use a straight-line method for real property.

The M.A.C.R.S. method also specifies that an item put into service in any of the categories from three to twenty years will use one-half of the normal depreciation for the first year.

Items in the last two categories (real property) are allowed a one-half month deduction for the month the item is put into service.

3-Year, 5-Year, 7-Year, 10-Year, 15-Year, and 20-Year Property (Half-year Convention)

	Depreciation Rate for Recovery Period					
Year	3-Year	5-Year	7-Year	10-Year	15-Year	20-Year
1	33.33%	20.00%	14.29%	10.00%	5.00%	3.750%
2	44.45	32.00	24.49	18.00	9.50	7.219
3	14.81	19.20	17.49	14.40	8.55	6.677
4	7.41	11.52	12.49	11.52	7.70	6.177
5		11.52	8.93	9.22	6.93	5.713
6		5.76	8.92	7.37	6.23	5.285
7			8.93	6.55	5.90	4.888
8			4.46	6.55	5.90	4.522
9				6.56	5.91	4.462
10				6.55	5.90	4.461
11				3.28	5.91	4.462
12					5.90	4.461
13					5.91	4.462
14					5.90	4.461
15					5.91	4.462
16					2.95	4.461
17						4.462
18						4.461
19						4.462
20						4.461
21						2.231

Residential Rental Property (27.5-Year) (Mid-month Convention)

	Use the Column for the Month of Taxable Year Placed in Service											
Year	1	2	3	4	5	6	7	8	9	10	11	12
1	3.485	3.182	2.879	2.576	2.273	1.970	1.667	1.364	1.061	0.758	0.455	0.152%
2–9	3.636	3.636	3.636	3.636	3.636	3.636	3.636	3.636	3.636	3.636	3.636	3.636%

Nonresidential Real Property (31.5-Year) (Mid-month Convention)

Use the Column for the Month of Taxable Year Placed in Service

Year	1	2	3	4	5	6	7	8	9	10	11	12
1	3.042	2.778	2.513	2.249	1.984	1.720	1.455	1.190	0.926	0.661	0.397	0.132%
2–7	3.175	3.175	3.175	3.175	3.175	3.175	3.175	3.175	3.175	3.175	3.175	3.175%

The depreciation amount is found by multiplying the cost by the percent in the table.
To find the book value, the accumulated depreciation is subtracted from the cost.

Example 9 Allen Bayes noted that the Muskegon Corporation purchased a research machine two years ago for $42,600. The machine will have a scrap value of about $750 at the end of its expected life. The material furnished by the I.R.S. indicates that this asset can be depreciated over a five-year period. What is the depreciation for this year and what is the book value?

$42,600 × .32 (for second year) = $13,632 depreciation for the second year

$42,600 × .20 (for first year) = $8,520

$42,600 Cost
−8,520 Depreciation for first year
−13,632 Depreciation for second year
$20,448 Book value
(Scrap value is not used.)

The *depreciation schedule* for the machine in example 9 for the first six years using the M.A.C.R.S. method would be as follows.

End of Year Number	Depreciation/Year	Accumulated Depreciation at the End of Year	Book Value at the End of Year
1	$ 8,520.00	$ 8,520.00	$34,080.00
2	13,632.00	22,152.00	20,448.00
3	8,179.20	30,331.20	12,268.80
4	4,907.52	35,238.72	7,361.28
5	4,907.52	40,146.24	2,453.76
6	2,453.76	42,600.00	0.00

Exercise G

Determine the missing values in the following exercises using the M.A.C.R.S. method.

	Cost	Category (Years)	Year	Depreciation	Accumulated Depreciation	Book Value
1.	$ 58,000	10	3	$8,352.00	$24,592.00	$33,408.00
2.	8,740	5	1	*1,748.00*	*1,748.00*	*6,992.00*
3.	32,844	31.5	1, 3rd mo.	*825.37*	*825.37*	*32,018.63*
4.	198,500	15	2	*18,857.50*	*28,782.50*	*169,717.50*
5.	3,900	3	2	*1,733.55*	*3,033.42*	*866.58*

Depletion

In the previous examples, the items were assets in a usable state tangible assets, such as machinery, buildings, and equipment. Other assets, such as natural resources, are not usually in a usable state. The conversion of these natural resources, termed *wasting* by the owner, is accounted for by a **depletion** allowance similar to depreciation.

The conversion of assets such as coal, timber, oil, natural gas, and minerals (when pumped, mined, or cut) reduces the balance of the value of the asset. The amount of depletion is determined by estimating the percent of the asset that has been used or "wasted" at cost.

Example 10 An oil field was purchased for $650,000. The field has an estimated 320,000 barrels of oil production possible. If 60,000 barrels were pumped in a year, what would be the depletion for the oil field?

$$\frac{60{,}000 = \text{Barrels pumped}}{320{,}000 = \text{Barrels available}} = 18.75\%$$

$650,000 = Cost of oil field
×.1875 = Percent of depletion
$121,875 = Depletion amount

(60,000 ÷ 320,000 = × 650,000 =)

Exercise H

Determine the amount of depletion in the following exercises. Round off the factor to the nearest hundredth of a percent.

Cost	Total Available	Wasted This Period	Depletion This Period
1. $450,800 (15,000 ÷ 70,000 =) (× 450,800 =)	70,000 tons	15,000 tons	$96,606.44
2. $1,600,000 *(90,000 ÷ 80,000,000 =) (× 1,600,000 =)*	80,000,000 cubic feet	90,000 cubic feet	*1,760.00*
3. 680,000 *(800,000 ÷ 6,000,000 =) (× 680,000 =)*	6,000,000 board feet	800,000 board feet	*90,644.00*
4. 230,000 *(8,000 ÷ 50,000 =) (× 230,000 =)*	50,000 barrels	8,000 barrels	*36,800.00*
5. 750,000 *(450,000 ÷ 1,840,000 =) (× 750,000 =)*	1,840,000 tons	450,000 tons	*183,450.00*

Partial Year
Objective 4

A firm must use the M.A.C.R.S. method of depreciation for tax purposes for assets put into service after 1981–1986. A firm may elect to use one of the other methods of depreciation for their *internal* records, as it may be felt that another method more closely fits real values. When the M.A.C.R.S. method is used, the owner may use a full year's depreciation no matter when the asset was purchased during the year. The other systems require that an adjustment be made to the amount of depreciation for the period. Not all assets are purchased at the beginning or end of a fiscal year; this concerns all assets whose asset life overlaps the fiscal year. There is little problem when the straight-line method is used, because the depreciation amount is the same each year. The S.O.Y.D. and declining-balance methods have a changing amount of depreciation per year and must be adjusted accordingly.

When an asset is put into service on or before the fifteenth of the month, the complete month is used to compute the amount of depreciation for the period. When it is put into service after the fifteenth, no depreciation is claimed for that month.

time line A **time line** is useful for determining the correct month and year in partial-year depreciation problems. It will help you visualize the relationship of the calendar to the fiscal year of the firm, and to the placed-in-service date of the asset. This basic tool can be very beneficial in selecting the correct number of months of the correct year to use in solving the problem. It only takes a few moments to construct the time line and place the facts on it.

Example 11 Time line

Asset life

Fiscal year

❍ *The time line can save you time (and will keep the facts straight!).*

Example 12 A desk was purchased on August 1, 2000, for $870. The expected life is five years and the salvage value is $150. The firm has a fiscal year of June 1 to May 31.

What is the depreciation amount for the first two fiscal years ending May 31, 2002, using the straight-line method?

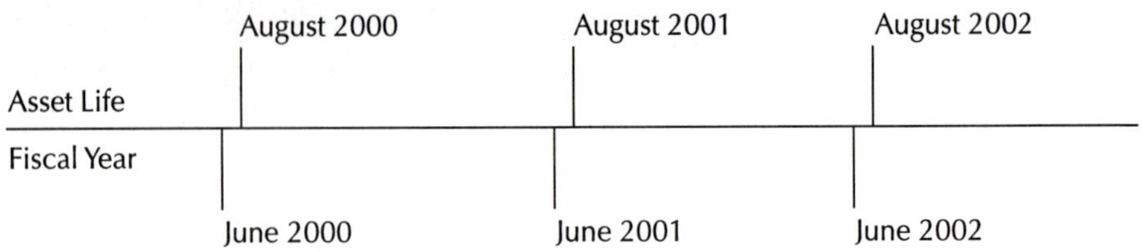

There are ten months in the first fiscal year (from August 2000 to June 2001), and a full year in the second fiscal year (June 2001 to June 2002).

The full-year depreciation for any year is $\frac{\$720}{5} = \$144$.

The first fiscal year depreciation would be $\frac{10}{12} \times \$144 = \$120$.

The depreciation for the first two fiscal years would total $264 ($144 + $120) using the straight-line method.

What is the depreciation amount for the first two fiscal years ending May 31, 2002, using the S.O.Y.D. method?

Years of Expected Life		*Year Number*
5	←	1
4	←	2
3	←	3
2		
1		

The S.O.Y.D. method requires that the asset life depreciation amounts be proportioned to the fiscal year of the firm.

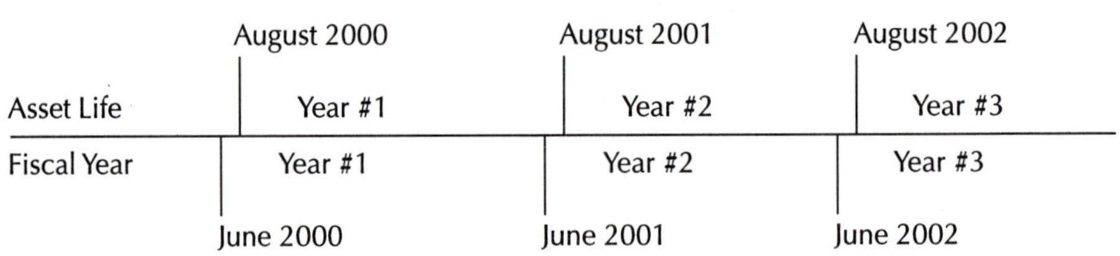

The first fiscal year (ending June 2001) would be computed as shown below.

$$\$720 \times \frac{5}{15} \times \frac{10}{12} = \$200$$

The second fiscal year (ending June 2002) would be computed as shown below.

$$\$720 \times \frac{5}{15} \times \frac{2}{12} = \$40 \text{ plus } \$720 \times \frac{4}{15} \times \frac{10}{12} = \$160$$

The depreciation for the first two full fiscal years would be

$200 + $40 + $160 = $400 using the S.O.Y.D. method

Notice that the partial year of the first year is carried over into the computations of the second year. This process would be continued if the question were to ask for the depreciation for the third year, etc. In that case, the portion of the second year would be carried over into the computations of the third year.

Example 12 *continued*

What is the depreciation amount for the first two fiscal years ending May 31, 2002, using the declining-balance method?

The declining-balance method requires that the ten months of the first fiscal year be used to adjust the balance for the next year. After the first partial year, the succeeding years are computed as full years (without any adjustment for ten months/two months, etc.).

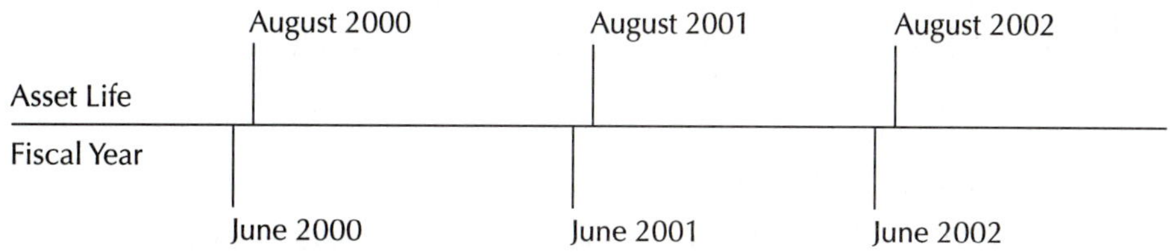

\$870 total cost $\times \frac{2}{5}$ = \$348 (full-year depreciation)
\$348 $\times \frac{10}{12}$ = \$290 for the fiscal year ending June 2001

The depreciation for the second year would be computed using the declined-balance and a full-year factor of $\frac{2}{5}$.

\$870 − \$290 = \$580 (book value June 2001)
\$580 $\times \frac{2}{5}$ = \$232 for the year ending June 2002

The depreciation for the first two full fiscal years would be \$290 + \$232 = \$522 using the declining-balance method.

Another example of the S.O.Y.D. and declining-balance methods should clear up any questions that might remain.

Example 13

A sign was purchased on July 10, 1999, for \$7,100. The expected life is ten years, with a \$1,200 salvage value. What is the depreciation for the fiscal year ending March 31, 2002, if the firm uses the S.O.Y.D. method?

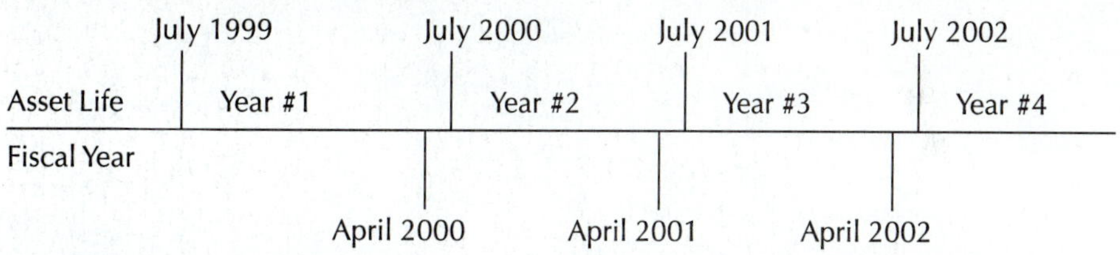

Years of Expected Life		*Year Number*
10		1
9	←———	2
8	←———	3
7		4
6		5
5		6
4		7
3		8
2		9
+1		10
55		

\$7,100 Total cost
−1,200 Salvage value
\$5,900 Total depreciation

$$\$5{,}900 \times \frac{9}{55} \times \frac{3}{12} = \$241.36 \text{ (from April 2001 to July 2001)}$$

$$\$5{,}900 \times \frac{8}{55} \times \frac{9}{12} = \underline{\$643.64} \text{ (from July 2001 to April 2002)}$$

\$885.00 depreciation for the fiscal year ending March 31, 2002

Example 14 If the firm in example 13 had used the declining-balance method, what would be the amount of depreciation for the fiscal year ending March 31, 2002?

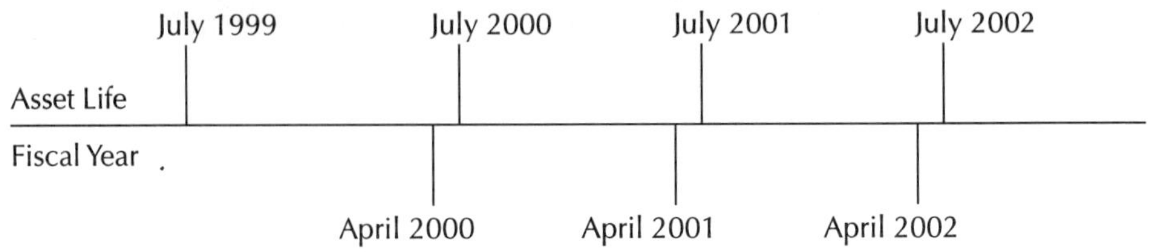

$\$7{,}100$ total cost $\times \frac{2}{10} \times \frac{9}{12} = \$1{,}065$ for the year ending March 31, 2000

$\$7{,}100 - \$1{,}065 = \$6{,}035$ book value April 2000

$\$6{,}035 \times \frac{2}{10} = \$1{,}207$ $\quad \$6{,}035 - \$1{,}207 = \$4{,}828$ book value April 2001

$\$4{,}828 \times \frac{2}{10} = \$965.60$ depreciation for the year ending March 31, 2002

Exercise I

Determine the depreciation amount in the exercises below. Use the depreciation method indicated.

Total Cost	Expected Life (Years)	Salvage Value	Purchase Date	Fiscal Year Ended	Depreciation Method	Depreciation Amount
1. $650	4	$50	August 15, 1999	January 15, 2001	S.O.Y.D. $\left(600 \times \frac{4}{10} \times \frac{7}{12} =\right)$	$215.00 $\left(600 \times \frac{3}{10} \times \frac{5}{12} =\right)$
2. 1,300	8	100	June 1, 1999	January 1, 2002	Declining-Balance $\left(1{,}300 \times .25 \times \frac{7}{12} \times .75 \times .25 =\right)$	*208.20*
3. 900	4	300	September 9, 1999	January 1, 2000	S.O.Y.D. $\left(600 \times \frac{4}{10} \times \frac{4}{12} =\right)$	*80.00*
4. 700	5	100	May 1, 1997	August 1, 2001	S.O.Y.D. $\left(600 \times \frac{2}{15} \times \frac{9}{12} =\right)$	*70.00* $\left(600 \times \frac{1}{15} \times \frac{3}{12} =\right)$
5. 630	4	0	June 1, 2000	March 1, 2002	Declining-Balance	*196.88* $\left(630 \times .5 \times \frac{9}{12} =\right)$
6. 2,800	5	300	September 23, 2000	January 1, 2001	Straight-Line *(2,500 ÷ 5 =)*	*125.00* $\left(500 \times \frac{3}{12} =\right)$
7. 750	10	50	July 3, 2000	March 1, 2002	S.O.Y.D. $\left(700 \times \frac{10}{55} \times \frac{4}{12} =\right)$	*118.78* $\left(700 \times \frac{9}{55} \times \frac{8}{12} =\right)$
8. 3,260	5	250	January 12, 2002	July 1, 2002	Declining-Balance	*652.00* $\left(3{,}260 \times .4 \times \frac{6}{12} =\right)$
9. 4,500	4	300	August 3, 2000	June 1, 2002	Declining-Balance	*1,312.50* $\left(4{,}500 \times .5 \times \frac{10}{12} =\right)$
10. 700	8	50	July 8, 1998	January 1, 2002	Straight-Line	*81.25* *(650 ÷ 8 =)*

ASSIGNMENT Chapter 17

Name ______________________ Date ______________ Score ____________

Skill Problems

Determine the missing values in the exercises below using the straight-line method and the following formula:

Depreciation/year = Total depreciation/Expected life

	Total Cost	Expected Life (yr)	Salvage Value	Year No.			
1.	$80,000	20	$10,000	4	*$66,000*	book value	*(70,000 ÷ 20 × 4 = 80,000 − 14,000 =)*
2.	4,000	*10*	800	6	1,920	accumulated depreciation	*(1,920 ÷ 6 = 3,200 ÷ 320)*
3.	*14,990*	7	500	4	2,070	depreciation per year	*(2,070 × 7 = 14,490 + 500 =)*

Determine the missing values in the exercises below using the straight-line method.

Item Cost	Trans. Cost	Install. Cost	Salvage Value	Expected Life (yr)	Year No.	Deprec. per Yr	Accum. Deprec.	Book Value
$15,600	$95	$7,050	$3,400	10	8	**4.** *$1,934.50*	**5.** *$15,476.00*	**6.** *$7,269.00*
						(15,600 + 95 + 7,050 − 3,400 ÷ 10 =) (1,934.50 × 8 =) (15,600 + 95 + 7,050 − 15,476 =)		
20,050	0	1,640	0	15	12	**7.** *1,446.00*	**8.** *17,352.00*	**9.** *4,338.00*
						(20,050 + 1,640 ÷ 15 =) (1,446 × 12 =) (20,050 + 1,640 − 17,352 =)		
750	48	150	150	17	8	**10.** *46.94*	**11.** *375.52*	**12.** *572.48*
						(750 + 48 + 150 − 150 ÷ 17 =) (46.94 × 8 =) (750 + 48 + 150 − 375.52 =)		
15,700	0	0	2,000	8	6	**13.** *1,712.50*	**14.** *10,275.00*	**15.** *5,425.00*
						(15,700 − 2,000 ÷ 8 =) (1,712.5 × 6 =) (15,700 − 10,275 =)		
894	105	355	300	4	3	**16.** *263.50*	**17.** *790.50*	**18.** *563.50*
						(894 + 105 + 355 − 300 ÷ 4 =) (263.50 × 3 =) (894 + 105 + 355 − 790.50 =)		

Determine the missing values in the exercises below using the units-of-production method. Use a four-decimal-place factor.

	Total Cost	Salvage Value	Expected Life (in units)	Units This Period	Depreciation This Period	
19.	$ 1,950	$ 50	1,400	60	*$81.43*	*(1,900 ÷ 1,400 = × 60 =)*
20.	1,650	350	90,000	4,000	*57.78*	*(1,300 ÷ 90,000 = × 4,000 =)*
21.	5,300	300	19,500	900	*230.77*	*(5,000 ÷ 19,500 = × 900 =)*
22.	3,950	200	80,000	650	*30.47*	*(3,750 ÷ 80,000 = × 650 =)*
23.	41,600	0	156,000	3,500	*933.33*	*(41,600 ÷ 156,000 = × 3,500 =)*

Total Cost	Salvage Value	Expected Life (in units)	Units This Period	Units to Date	Deprec. This Period	Book Value	
$45,000	2,500	150,000	2,870	45,972	24. *$813.07*	25. *$31,976.13*	*((45,000 − 2,500) ÷ 150,000 = .2833 × 2,870 = 45,000 − (45,972 × .2833) =)*
1,000	0	25,500	840	22,451	26. *32.93*	27. *119.92*	*(1,000 ÷ 25,500 = .0392 × 840 = 1,000 − (22,451 × .0392) =)*
15,540	2,500	200,000	10,500	190,000	28. *684.60*	29. *3,152*	*((15,540 − 2,500) ÷ 200,000 = .0652 × 10,500 = 15,540 − (190,000 × .0652) =)*

Determine the sum-of-years fraction to be used in the following exercises.

Years of Expected Life	Sum of Years		Year Number	Fraction to be Used
16	30. *136*	*(16 × 17 ÷ 2 =)*	4	31. *13/136*
15	32. *120*	*(15 × 16 ÷ 2 =)*	8	33. *8/120*
42	34. *903*	*(42 × 43 ÷ 2 =)*	21	35. *22/903*
12	36. *78*	*(12 × 13 ÷ 2 =)*	7	37. *6/78*

Determine the missing values in the exercises below using the S.O.Y.D. method.

Total Cost	Salvage Value	Expected Life (yr)	Year No.	Deprec.	Accum. Deprec.	Book Value	
$57,400	$ 2,000	12	8	38. *$3,551.28*	39. *$48,297.44*	40. *$9,102.56*	*(57,400 − 2,000 × 5 ÷ 78 = 55,400 × 8 ÷ 78 = 57,400 − 48,297.44 =)*
20,843	1,500	25	6	41. *1,190.34*	42. *8,034.78*	43. *12,808.22*	*(20,843 − 1,500 × 20 ÷ 325 = 19,343 × 135 ÷ 325 = 20,843 − 8,034.78 =)*
164,800	20,000	20	4	44. *11,721.90*	45. *51,024.76*	46. *113,775.24*	*(164,800 − 20,000 × 17 ÷ 210 = 144,800 × 74 ÷ 210 = 164,800 − 51,024.76 =)*
8,400	0	7	5	47. *900.00*	48. *7,500.00*	49. *900.00*	*(8,400 × 3 ÷ 28 = 8,400 × 25 ÷ 28 = 8,400 − 7,500 =)*

Determine the missing values in the exercises below using the declining-balance method. Use factor rounded off to three decimal places.

Total Cost	Salvage Value	Expected Life (yr)	Year No.	Deprec.	Accum. Deprec.	Book Value	
$ 4,800	$200	5	2	50. *$1,152.00*	51. *$3,072.00*	52. *$1,728.00*	*(4,800 × .4 = 2,880 × .4 =) (1,920 + 1,152 =) (4,800 − 3,072 =)*
9,870	500	8	3	53. *1,387.97*	54. *5,706.10*	55. *4,163.90*	*(9,870 × .25 = 7,402.5 × .25 = 5,551.88 × .25 = (2,467.5 + 1,850.63 + 1,387.97 =) (9,870 − 5,706.1 =)*
8,425	0	7	2	56. *1,720.42*	57. *4,129.97*	58. *4,295.03*	*(8,425 × .286 = 6,015.45 × .286 =) (2,409.55 + 1,720.42 =) (8,425 − 4,129.97 =)*
95,200	4,400	10	4	59. *9,748.48*	60. *56,206.08*	61. *38,993.92*	*(95,200 × .2 = 76,160 × .2 = 60,928 × .2 = 48,742.4 × .2 =) (19,040 + 15,232 + 12,185.6 + 9,748.48 =) (95,200 − 56,206.08 =)*

Complete the problems below. Use the method indicated. Round off declining-balance factor to the nearest tenth of a percent.

Total Cost	Salvage Value	Expected Life (Years)	Year Number	Depreciation	Accumulated Depreciation	Book Value	Method
$7,000	$500	20 (6,500 ÷ 20 =)	8	62. $325.00	63. $2,600.00	64. $4,400.00	Straight-Line (325 × 8 =)
8,600	600	8 $\left(8{,}000 \times \frac{5}{36} =\right)$	4	65. 1,111.11	66. 5,777.78	67. 2,822.22	S.O.Y.D. $\left(8{,}000 \times \frac{26}{36} =\right)$
35,700	1,000	15 (35,700 × 1.33 =)	3	68. 3,569.09	69. 12,433.80	70. 23,266.20	Declining-Balance (30,951.90 × .133 =)
1,250	50	20 (1,250 × .10 =)	3	71. 101.25	72. 338.75	73. 911.25	Declining-Balance (1,125 × .10 =) (1,012.5 × .10 =)
9,800	500	15 $\left(9{,}300 \times \frac{14}{120} =\right)$	2	74. 1,085.00	75. 2,247.50	76. 7,552.50	S.O.Y.D. $\left(9{,}300 \times \frac{29}{120} =\right)$
950	150	10 (800 ÷ 10 =)	3	77. 80.00	78. 240.00	79. 710.00	Straight-Line (80 × 3 =)
5,600	0	10 (5,600 × .2 =)	4	80. 573.44	81. 3,306.24	82. 2,293.76	Declining-Balance (4,480 × .2 =) (3,584 × .2 =)
695	95	12 (600 ÷ 12 =)	7	83. 50.00	84. 350.00	85. 345.00	Straight-Line (50 × 7 =)
1,350	350	5 $\left(1{,}000 \times \frac{4}{15} =\right)$	2	86. 266.67	87. 600.00	88. 750.00	S.O.Y.D. $\left(1{,}000 \times \frac{9}{15} =\right)$
856	56	15 $\left(800 \times \frac{13}{120} =\right)$	3	89. 86.67	90. 280.00	91. 576.00	S.O.Y.D. $\left(800 \times \frac{42}{120} =\right)$

Complete the following depreciation schedule for a 4-year-old race horse purchased for $25,000. Use the M.A.C.R.S. method and charts provided in your chapter.

End of Year No.	Deprec. Rate (% from chart)	Deprec. per Year	Accum. Deprec. at the End of Year	Book Value at the End of Year	
1	92. 33.33%	93. $8,332.50	94. $8,332.50	95. $16,667.50	(25,000 × .3333 =)
2	96. 44.45	97. 11,112.50	98. 19,445.00	99. 5,555.00	(16,667.5 × .4445 =)
3	100. 14.81	101. 3,702.50	102. 23,147.50	103. 1,852.50	(5,555 × .1481 =)

Determine the missing values in the following exercises using the M.A.C.R.S. method.

Cost	Category (years)	Year	Deprec.	Accum. Deprec.	Book Value
$ 80,600	20	2	104. $5,818.51	105. $8,841.01	106. $71,758.99
9,462	7	3	107. 1,654.90	108. 5,324.27	109. 4,137.73
174,600	31.5	1, 4 mo	110. 3,926.75	111. 3,926.75	112. 170,673.25
1,348	15	1	113. 67.40	114. 67.40	115. 1,280.60
9,000	3	2	116. 4,000.50	117. 7,000.20	118. 1,999.80
15,900	5	3	119. 3,052.80	120. 11,320.80	121. 4,579.20
85,000	10	1	122. 8,500.00	123. 8,500.00	124. 76,500.00

Determine the amount of depletion in the following exercises.

	Cost	Total Available	Wasted This Period	Depletion This Period	
125.	$640,600	100,000 tons	14,000 tons	*$89,684.00*	*(14,000/100,000 × 640,600 =)*
126.	2,460,000	8,700 cubic ft	1,940 cu. ft	*548,551.72*	*(1,940/8,700 × 2,460,000 =)*
127.	245,400	40,000 barrels	10,500 barrels	*64,417.50*	*(10,500/40,000 × 245,400 =)*

Determine the partial year depreciation amount as indicated in the exercises below. Use the depreciation method indicated. Use a three-decimal-place factor.

	Total Cost	Expected Life (yr)	Salvage Value	Purchase Date	Fiscal Year Ended	Deprec. Method	Deprec. Amount	
128.	$2,000	5 years	$250	5/1/01	7/1/03	S.O.Y.D.	*$447.22*	$\left(1{,}750 \times \frac{4}{15} \times \frac{10}{12} =\right)$
129.	1,680	4	300	9/23/00	3/1/02	Declining-Balance	*665.00*	$\left(1{,}680 \times .5 \times \frac{5}{12} =\right)$
130.	840	8	100	8/4/99	1/31/02	S.O.Y.D.	*133.61*	*((740 × 7/36 × 6/12 = 71.94) + (740 × 6/36 × 6/12 = 61.67) =)*
131.	2,500	4	0	9/9/99	6/30/01	Declining-Balance	*1,041.67*	*(2,500 × .5 × 10/12 =)*
132.	900	5	50	5/23/00	12/31/02	Straight-Line	*170.00*	*(850 ÷ 5 = 170)*
133.	6,405	10	600	7/8/02	9/30/03	Declining-Balance	*1,216.95*	*((6,405 × .2 × 3/12 = 320.25) + (6,084.75 × .2) =)*
134.	3,460	3	200	5/1/01	12/31/01	S.O.Y.D.	*1,086.67*	*(3,260 × 3/6 × 8/12 =)*
135.	680	15	50	4/5/02	7/1/02	Straight-Line	*10.50*	*(630 ÷ 15 × 3/12 =)*
136.	2,500	6	100	1/31/03	6/30/04	Declining-Balance	*716.99*	*(2,500 × .333 × 5/12 = 346.88) + (2,153.12 × .333) =*

Business Application Problems

1. A machine purchased five years ago is fully depreciated and will be sold for the salvage value of $500. The original cost was $6,500. What was the depreciation per year using the straight-line method?
$6,500 − $500 = $6,000/5 = $1,200 depreciation

2. In problem 1, what would have been the depreciation for the second year using the S.O.Y.D. method?
$6,500 − $500 = $6,000 × 4/15 = $1,600 depreciation

3. The electronic component of an installation has an expected life of 5,000 hours. The component, installed, cost $8,000 and has a salvage value of $200. The component was used 800 hours in the most recent period. What is the amount of depreciation for the period?
$8,000 − $200 = $7,800 × 800/5,000 = $1,248 depreciation

4. What is the depreciation cost per hour in problem 3?
$8,000 − $200 = $7,800/5,000 = $1.56 per hour

5. Kendron Corporation bought a wall cleaning unit for $6,800. The unit will be used for eight years and then salvaged for $400. The straight-line method will be used to determine depreciation. What is the depreciation per year?
$6,800 − $400 = $6,400/8 = $800 depreciation

6. In problem 5, what would be the book value of the cleaning unit at the end of three years?
$800 × 3 = $2,400 $6,800 − $2,400 = $4,400 book value

7. If the Kendron Corporation in problem 5 would have used the declining-balance method to determine depreciation, what would have been the amount of depreciation for the second year?
$6,800 × (2 × 1/8) = $1,700 $6,800 − $5,100 = $5,100 × (2 × 1/8) = $1,275 depreciation

8. A delivery trailer was purchased for $6,200. The trailer will be used for a period of ten years. At the end of the ten-year period it will be sold for $1,000. The firm will use the straight-line method to determine depreciation. What will be the accumulated depreciation amount at the end of the first three years?

$6,200 − $1,000 = $5,200/10 = $520 × 3 = $1,560 accumulated

9. Verton Academy bought a 15-passenger van for $26,000. They will depreciate the van using the M.A.C.R.S. system as a five-year item. The salvage value will be $850. What is the depreciation amount for the third year?

$26,000 × .192 = $4,992 depreciation

10. In problem 9, what is the accumulated depreciation at the end of the third year?

$26,000 × .712 = $18,512 accumulated

Blecher and Associates purchased a building for $139,000 for office and storage space. The building is expected to have a useful life of 20 years. The salvage value at the end of that period is estimated to be $20,000. The owner of the firm has asked that a comparison be made at the end of the tenth year using three methods of depreciation. What is the depreciation for the tenth year?

11. Straight-line method: ***$5,950.00***

(139,000 − 20,000 = 119,000 ÷ 20 =)

12. S.O.Y.D. method: ***$6,233.33***

(119,000 × 11 = ÷ 210 =)

13. Declining-balance method: ***$5,385.15***

(139,000 × .9 = × .9 = × .9 = × .9 = × .9 = × .9 = × .9 = × .9 = × .9 = × .1 =)

14. Demmer Manufacturing uses a shear to cut thick metal pieces. The shear cost $129,000 when it was new 15 years ago. The shear is expected to have a useful life of 35,000 hours. It was used 1,500 hours this year. If the scrap value is expected to be $5,000, what is the depreciation amount for this year?

$$(\$129{,}000 - \$5{,}000) \times \frac{1{,}500}{35{,}000}$$

= $5,314.29 depreciation

15. A heating system was purchased by the Fossum Import Company of New York. The installed system cost $89,000 when it was purchased 12 years ago. What is the difference in the amount of depreciation for this year if the system will have a scrap value of only $3,000 at the end of 15 years? Compare the straight-line and S.O.Y.D. methods.

$$\frac{\$89{,}000 - \$3{,}000}{15} = \$5{,}733.33$$

$$(\$89{,}000 - \$3{,}000) \times \frac{4}{120} = \$2{,}866.67$$

$2,866.67 difference

16. If an item is depreciated at the rate of 5 percent per year using the straight-line method, what was the original cost if the depreciation amount per year is $265 and the salvage value is $800?

5% = 20 years

$265
×20

$5,300 + $800 = $6,100 original cost

17. The Felder Co. purchased an office building on March 1, 1992, for $60,000. The building is expected to last for 20 years. On March 1, 2003, a real estate appraiser stated the building's value was probably $46,000. Mr. Felder, company president, has asked that the appraisal be compared to the book value. Which method of depreciation would come closest to the appraisal?

$$\frac{\$60{,}000}{20} \times 11 = \$33{,}000$$

$60,000 − $33,000 = $27,000 straight-line

$$\$60{,}000 \times \frac{165}{210} = \$47{,}142.86$$

$60,000 − $47,142.86 = $12,857.14 sum-of-years'-digits

18. An asset was purchased two years ago for $4,000. The expected life is five years with an expected salvage value of $500. The firm will depreciate the asset using the sum-of-the-years'-digits method. What is the book value of the asset?

$$(\$4{,}000 - \$500) \times \frac{9}{15} = \$2{,}100$$

$4,000
−2,100
$1,900 book value

19. The Morrison Company has a dispute with the I.R.S. over the classification of a piece of property. Morrison feels that the item should be categorized (M.A.C.R.S.) as a five-year item, and the I.R.S. indicates that it should be a ten-year item. The dispute has been going on for two full years. The item cost $8,900 and will scrap out at $1,000 at the end of the expected life. What is the difference in the book values using the five- and ten-year categories?

Five Year

$8,900 × .20 = $1,780
$8,900 × .32 = $2,848
$4,628

Ten Year

$8,900 × .10 = $890
$8,900 × .18 = $1,602
$2,492

$4,628
−2,492
$2,136 difference

20. The Forsythe Interdynamics Corporation purchased a specialized computer for $5,800 two years ago. The computer is in the five-year M.A.C.R.S. category. The firm is now interested in trading in the computer for a new unit that costs $3,790. The firm will receive a $2,000 allowance for the trade-in. What is the difference in the trade-in value and the book value of the computer?

$5,800 × .20 = $1,160
$5,800 × .32 = $1,856
$3,016

$5,800
−3,016
$2,784 book value

$2,784
− 2,000
$784 difference

21. Dr. Monroe purchased a new light-signaling system for the series of patient exam rooms in her new office building. The system will cost $14,780. The expected life of the system is ten years. The system will be junked at the end of its expected life. The doctor is unsure of which depreciation system to use. At the end of $3\frac{1}{2}$ years, what is the difference in the book value between the straight-line and S.O.Y.D. methods?

$$\frac{\$14{,}780}{10} \times 3.5 = \$5{,}173$$

$$\$14{,}780 \times \frac{30.5}{55} = \$8{,}196.18$$

$14,780 − $5,173 = $9,607.00
$14,780 − $8,196.18 = −6,583.82
$3,023.18 difference

22. Write a complete sentence in answer to this question. The depreciation expense for the year was $900 on a piece of equipment that has now been used for the second full year. The item is to be depreciated over a six-year period and has a salvage value of $3,000. The item is depreciated using the S.O.Y.D. method. What was the original price of the item?

$$\frac{5}{21} = \$900$$

$$\$900 \div \frac{5}{21} =$$

$3,780
+3,000
$6,780

The original price was $6,780.

23. Write a complete sentence in answer to this question. If the item in problem 22 were depreciated over a five-year period using the straight-line method, what would be the accumulated depreciation at the end of the second year?

$$\frac{\$3{,}780}{5} \times 2 = \$1{,}512$$

The accumulated depreciation would be $1,512.

★ Challenge Problem

The Bayer Corp. purchased a hydraulic press for $76,587, F.O.B. Cincinnati. Bayer Corp. is located in Wattona, South Carolina. The transportation charges were $942. The press was purchased on terms of 2/10,n/30 on July 16, 2000. It was delivered on July 23, and the invoice was paid on July 24. Trade discounts of 10 percent and 5 percent were granted by the selling firm, Gray Iron, Inc. Installation costs were $15,321. The salvage value is estimated to be $5,000 at the end of its expected life of 25 years.

Bayer uses the declining-balance method to determine the depreciation expense. The first full (fiscal) year ended on April 15, 2002.

You are asked to determine the book value of the press on April 15, 2003.

$76,587.00	*List price*
×.9	*(10% trade discount)*
$68,928.30	
×.95	*(5% trade discount)*
$65,481.89	*Selling price*
×.98	*(2% cash discount)*
$64,172.25	*Net price*
+942.00	*Transportation*
+15,321.00	*Installation*
$80,435.25	*Total cost* $\times \left(\frac{1}{25} \times 2\right) \times \frac{9}{12} = \$4{,}826.12$

$80,435.25
−4,826.12

$75,609.13 *Book value—April 15, 2002* $\times \frac{2}{25} = \$6{,}048.73$

$75,609.13
−6,048.73
$69,560.40 Book value—April 15, 2003

MASTERY TEST Chapter 17

Name ______________________ Date ______________ Score ____________

Select the best term for the explanations.

F 1. Depletion
J 2. S.O.Y.D. Method
A 3. Straight-Line Method
I 4. Declining-Balance Method
K 5. Book Value
E 6. Salvage Value
C 7. M.A.C.R.S. Method
D 8. Accumulated Depreciation

A. An equal amount of depreciation per year.
B. Made up of all costs incurred to make the item operable.
C. All properties are categorized to set the percent.
D. Depreciation per year × the number of years to date (straight-line).
E. Total cost less total depreciation: the residual value.
F. Allowance for a used portion of natural resource.
G. A method using $\frac{2}{15}$ as a fraction to multiply by salvage value.
H. A method considering the amount of operations performed. The fraction is multiplied by total cost to find book value.
I. A method using total cost or book value multiplied by a rate to accelerate the depreciation amount in the early years..
J. A method that totals the years of expected life. This number becomes the denominator in a fraction.
K. Total cost less the depreciation to date.

Calculate the missing values in each of the following problems. Use the depreciation method indicated.

	Total Cost	Salvage Value	Expected Life (years)	Year Number	Depreciation Amount	Accumulated Depreciation	Book Value	Method
9.	$21,000	$1,000	10	4	*$2,545.45*	*$12,363.64*	*$8,636.36*	S.O.Y.D
10.	35,900	1,900	8	2	*6,731.25*	*15,706.25*	*20,193.75*	Declining-Balance
11.	12,600	50	20	17	*627.50*	*10,667.50*	*1,932.50*	Straight-Line

	Total Cost	Salvage Value	Expected Life (Units)	Units this Period	Depreciation Amount	Method
12.	$42,000	$2,000	100,000	20,000	*$8,000*	Units-of-Production

13. Market Services Company purchased a new copier one year ago. The machine cost $5,800 and will be able to command a trade-in value of $500 when sold. The machine is categorized for a five-year life. What is the book value at the end of the first year using the M.A.C.R.S. method?

$5,800 × .15 = $870
$5,800
− 870
$4,930 book value

14. Hoffman Associates purchased a piece of land for $150,000 with an estimated 50,000 barrels of oil that could be produced from it. The firm pumped 2,000 barrels of oil in the third year of ownership. What is the depletion allowance for the third year?

$$\$150{,}000 \times \frac{2{,}000}{50{,}000} = \$6{,}000$$

15. A building was purchased ten years ago for $260,000. The expected life was established at thirty years. The firm decided to use a straight-line method to depreciate the property. The salvage value of the property will be $30,000. What is the book value of the property?

$(\$260{,}000.00 - \$30{,}000) \times \frac{10}{30} = \$76{,}666.67$

$260,000.00
− 76,666.67
$183,333.33

16. Flash Movers, Inc. purchased an airplane for $74,000 on October 17, 2000. The plane will have a salvage value of $3,000. The firm has decided to depreciate the property over a period of twenty years using the declining-balance method. Gwen Kimberlin, company accountant, must determine the amount of depreciation for the fiscal year ended March 31, 2003. What is the amount of depreciation for the period?

$\$74{,}000.00 \times \frac{2}{20} \times \frac{5}{12} = \$3{,}083.33$ *dep. to March 31, 2001*

− 3,083.33
$70,916.67 × .9 = $63,825.00 book value March 31, 2002
× .1
$6,382.50 depreciation March 31, 2003

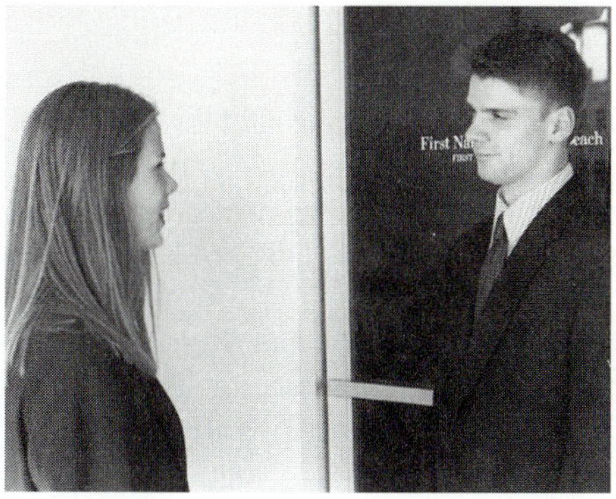

TAXES 18

Suggested Classroom Time for This Chapter

hours 1 2 3 4 5

Perceived Student Difficulty with Topical Material

easy moderate difficult

OBJECTIVES

After mastering the material in this chapter, you will be able to:

1. Calculate annual property taxes. p. 348
2. Calculate a property tax rate. p. 349
3. Calculate the customs duties on imported merchandise. p. 350
4. Calculate the personal income tax at the end of the year. p. 354
5. Describe the calculation of sales tax. p. 364
6. Understand and use the following terms and concepts:
   Assessed Valuation
   Mills
   Customs Duty
   Ad Valorem Duty
   Specific Duty
   Compound Rate
   Taxable Income
   Dependents
   Exemptions
   Deductions
   Standard Deduction

SPORT-A-MERICA

Now having two stores, Sport-A-Merica received two property tax statements from the city clerk. Kerry grimaces as she opens the statement on the old store. "It is time to think about selling the old store, Hal!" Hal responded, "I don't mind paying taxes if there is some real value Sport-A-Merica will receive in return." A review of the tax statement shows amounts for fire protection, schools, police, roads, etc. He realizes that all of those services are needed, but the amount of the tax seems so high! He can't argue with the tax rate of 31.85 mills, but the assessed valuation of $42,600 seems unreasonable. Property tax is something the business must take out of its own monies, not like the sales tax collected by Sport-A-Merica and then passed on to the government.

Then there is the tax that Hal must pay on his own income. Money is deducted from his paycheck on each payday, but that amount is only an estimate of the federal and state income taxes due at the end of the year. When he files his annual tax return, he always seems to owe a little more than what was withheld.

It seems that everything is taxed. Even the leg guards that were imported were taxed through a customs duty. That tax will add to costs, which will in turn increase the selling price. Doing the taxes for the firm and for himself is a little complicated, but it is a real challenge. Hal knows that keeping good records is a start in the right direction.

Property Taxes

Property taxes are collected by local units of government, such as cities and townships, for the operations of such services as fire protection, schools, local government, and road maintenance. This tax is levied on property owners. Church and government property are not taxed.

Tax Amount

assessed valuation

The tax amount is calculated by multiplying the taxable base by a tax rate. The taxable amount is termed the **assessed valuation.** This amount is set by a tax assessor, a representative of the taxing governmental unit. The assessed valuation is established from the apparent market value of the property.

mills

The tax rate is often expressed in terms of **mills,** or thousandths of a dollar. A 1 mill tax rate on $1 of assessed valuation would result in a tax of $.001. A tax rate of 23.6 mills would be a tax of $23.60 on each $1,000 of assessed valuation:

❍ *A mill is a very small amount—that is, until it is multiplied by the assessed valuation.*

$$\left(23.6 \text{ mills} = \frac{23.6}{1{,}000} = .0236 = 2.36 \text{ cents tax for each dollar of assessed valuation}\right)$$

The tax rate may also be expressed in *hundredths*. A tax rate of 1 hundredth would mean a $1 tax on each $100 of assessed valuation. A property taxed at 2.36 hundredths would be taxed $2.36 for each $100 of assessed valuation, or $23.60 on each $1,000.

Objective 1

Example 1

The original building Sport-A-Merica owned and still uses as an office has an assessed valuation of $42,600. The tax rate in the community is 31.85 mills. The current market value is $75,000. What is the amount of property taxes on the building?

42.6 = $1000s assessed valuation
×$31.85 = Tax per $1000 of assessed valuation (31.85 mills)
$1,356.81 = Property taxes

(42.6 × 31.85 =)

Example 2

Trizella Longoria owns a home that has a market value of $83,000 in San Benito. The assessed valuation is set at $40,000. The tax rate is $2.84 per hundred. What is the amount of the tax on her home?

400 = $100s of assessed valuation
×$2.84 = Tax per $100 of assessed valuation
$1,136 = Property taxes

(400 × 2.84 =)

Exercise A

Complete the exercise below by determining the missing value. Round off your answer to the nearest cent or hundredth of a mill.

Market Value	Assessed Valuation	Tax Rate (in Mills)	Property Taxes
1. $36,500	$22,000	13.07	$287.54 (22 × 13.07 =)
2. 169,800	90,000	25.4	***2,286.00 (90 × 25.4 =)***
3. 47,000	36,800	***21.5***	791.20 ***(791.20 ÷ 36.8 =)***
4. 19,300	***9,000***	23.5	211.50 ***(211.50 ÷ 23.5 =)***
5. 65,200	56,000	23.80	***1,332.80 (56 × 23.80 =)***

Exercise A Continued

Market Value	Assessed Valuation	Tax Rate (in Hundredths)	Property Taxes
6. $23,000	$15,600	3.15	*$491.40 (156 × 3.15 =)*
7. 48,000	30,000	2.91	*873.00 (300 × 2.91 =)*
8. 36,800	18,200	*1.42*	258.60 *(258.60 ÷ 182 =)*
9. 125,900	52,000	3.67	*1,908.40 (520 × 3.67 =)*
10. 72,800	*50,000*	4.21	2,105.00 *(2,105 ÷4.21 =)*

Determining the Tax Rate

Objective 2

The tax rate is set by the taxing unit of government. Costs for local government services are totaled and divided by the total assessed valuation within the taxing unit; that is,

$$\frac{\text{Taxes to be raised}}{\text{Total assessed valuation}} = \text{Tax rate}$$

Example 3

The Township of Kimball has established an annual budget of $746,380. The budget amount must be raised by taxing the property in the township. The total township assessed valuation is $43,904,706. What is the tax rate in mills?

$$\frac{\$746{,}380 = \text{Taxes to be raised}}{\$43{,}904{,}706 = \text{Total assessed valuation}} = .0169999 = 17 \text{ mills}$$

(746,380 ÷ 43,904,706 =)

Example 4

The City Council of Teaneck established a budget of $1,650,000 for the next fiscal year. The assessed valuation for the city is $95,000,000. What is the tax rate per hundred?

$$\frac{\$1{,}650{,}000 = \text{Taxes to be raised}}{\$95{,}000{,}000 = \text{Total assessed valuation}} = .0173684 = \$1.74 \text{ tax}$$

(1,650,000 ÷ 95,000,000 =)

● ***A local tax statement could help students understand where the tax dollars go.***

Exercise B

Determine the tax rate for each community below. Round off your answer to the nearest hundredth of a mill or cent.

Community	Total Assessed Valuation	Budget	Tax Rate (in Mills—Nearest Hundredth)
1. Watrous	$72,685,360	$1,368,419	18.83 (1,368,419 ÷ 72,685,360 =)
2. Dandby Township	14,296,000	238,300	*16.67 (238,300 ÷ 14,926,000 =)*
3. Wildwood	23,800,000	649,500	*27.29 (649,500 ÷ 23,800,000 =)*
4. Big Horn	37,567,000	721,100	*19.20 (721,100 ÷ 37,567,000 =)*
5. Portland	1,630,000	42,600	*26.13 (42,600 ÷ 1,630,000 =)*

(Continued on next page)

Exercise B Continued

Community	Total Assessed Valuation	Budget	Tax Rate (Per Hundred, Nearest Hundredth)
6. Marine City	$ 7,000,000	$ 90,000	*1.29 (90,000 ÷ 7,000,000 =)*
7. Lake Worth	30,800,000	540,000	*1.75 (540,000 ÷ 30,800,000 =)*
8. Douglas	167,900,000	3,400,000	*2.03 (3,400,000 ÷ 167,900,000 =)*
9. Refugio	6,900,000	92,000	*1.33 (92,000 ÷ 6,900,000 =)*
10. Siquine	1,360,000	23,000	*1.69 (23,000 ÷ 1,360,000 =)*

Customs Duty

Objective 3

The federal government taxes some imported goods. The tax is termed a **customs duty.** Customs duties serve many purposes, but the major purposes involve protecting American jobs, new domestic industries, and international trade agreements such as the North American Free Trade Agreement (NAFTA). For the importing firm, customs duties represent an increase in the cost of the merchandise.

The amount of customs duties for each type of product is indicated in a tariff schedule (see page 351). This document is prepared by the federal government each year, and it contains the amount of tax and type of tax. There are two types of customs duties: ad valorem and specific.

Ad Valorem Duty

An **ad valorem** customs duty is a percent of the market value of the imported item. The percent is indicated in the tariff schedule, and the market value is indicated on the sales invoice, or it is a government estimate of the actual value.

Example 5

An importer purchased 42 bolts of cloth priced at $31 a bolt. The tariff schedule indicated an ad valorem customs duty of 12 percent. What will the customs duty amount be for the importer?

42 Bolts
×$31 Per bolt
$1,302 = Market value of purchase
×.12 = Ad valorem rate
$156.24 = Customs duty

(42 × 31 = × .12 =)

Specific Duty

The tariff schedule may not indicate a percent of ad valorem customs duty but instead a specific customs duty. The **specific duty** is a set amount per unit, per measure, or per weight.

Example 6

Sport-A-Merica imported 350 leg guards at $15 each. The specific duty was $3 per unit. What is the amount of specific customs duty?

350 = Units
×$3 = Per unit specific tax
$1,050 = Specific tax

(350 × 3 =)

● ***The page from the tariff schedule shows the variation in tax rates for several items.***

HARMONIZED TARIFF SCHEDULE of the United States

Annotated for Statistical Reporting Purposes

XI
61-43

Heading/ Subheading	Stat. Suf-fix	Article Description	Units of Quantity	Rates of Duty 1 General	1 Special	2
6111 (con.)		Babies' garments and clothing accessories, knitted or crocheted (con.):				
6111.90		Of other textile materials:				
		Of artificial fibers:				
6111.90.10	00	Trousers, breeches and shorts, except those imported as parts of sets................(239)	doz....v kg	27%	Free (IL) 6% (CA) 12.6% (MX)	90%
6111.90.20	00	Blouses and shirts, except those imported as parts of sets.....(239)	doz....v kg	31.1%	Free (IL) 6.9% (CA) 13.5% (MX)	90%
6111.90.30	00	T-shirts, singlets and similar garments, except those imported as parts of sets.............(239)	doz....v kg	27.7%	Free (IL) 6.9% (CA) 13.5% (MX)	90%
6111.90.40	00	Sweaters, pullovers, sweatshirts, waistcoats (vests) and similar articles, except those imported as parts of sets.............(239)	doz....v kg	32.9% 1/	Free (IL) 6.9% (CA) 13.5% (MX)	90%
6111.90.50		Other............................	....	16.6%	Free (IL) 3.4% (CA) 8.4% (MX)	90%
	10	Sunsuits, playsuits and washsuits................(239)	doz. v kg			
	20	Sets.....................(239)	doz. v kg			
		Other:				
	30	Imported as parts of sets.............(239)	doz. v kg			
	40	Other...............(239)	doz. v kg			
6111.90.70	00	Containing 70 percent or more by weight of silk or silk waste...........	doz. v kg	5%	Free (E,IL,J) 1.2% (CA) 3.3% (MX)	60%
6111.90.90	00	Other............................(839)	doz. v kg	5.9%	Free (E*,IL) 1.2% (CA) 3.3% (MX)	60%
6112		Track suits, ski-suits and swimwear, knitted or crocheted:				
		Track suits:				
6112.11.00		Of cotton............................	....	16.6%	Free (IL) 3.4% (CA) 8.4% (MX) 2/	45%
		Garments described in heading 6101 or 6102:				
	10	Men's or boys'...........(334)	doz. v kg			
	20	Women's or girls.........(335)	doz. v kg			
		Shirts:				
	30	Men's or boys'...........(338)	doz. v kg			
	40	Women's or girls'.........(339)	doz. v kg			
		Trousers:				
	50	Men's or boys'...........(347)	doz. v kg			
	60	Women's or girls'.........(348)	doz. v kg			

1/ See subheading 9902.61.00.
2/ See subheading 9906.61.19.

Compound Rate

When the tariff schedule specifies both an ad valorem duty *and* a specific duty, it is termed a **compound rate.**

Example 7

The Far East Importers purchased 75 pieces of furniture valued at $37.50 each. The tariff schedule indicated an ad valorem duty of 43 percent and a specific duty of $4. What is the total cost of the purchase for Far East Importers?

75	= Units	75	= Units
×$37.50	= Price per unit	×$4	= Specific duty per unit
$2,812.50	= Total market value	$300	= Total specific duty
×.43	= Ad valorem rate		
$1,209.38	= Ad valorem duty		

$2,812.50	= Total market value
1,209.38	= Ad valorem duty
+ 300.00	= Total specific duty
$4,321.88	= Total cost to Far East Importers

(75 × 37.50 = × .43 = 75 × 4 =)

Exercise C

Determine the amount of customs duties in the exercises below.

	Units Purchased	Value Per Unit	Ad Valorem Rate (%)	Specific Rate/Unit	Total Customs Duties
1.	1,500 *(5,670 × .10 =)*	$3.78 *(1,500 × 1.20 =)*	10	$1.20	$2,367.00
2.	100 *(920 × .065 =)*	9.20 *(100 × .70 =)*	$6\frac{1}{2}$	.70	*129.80*
3.	144 *(144 × 11.25 =)*	46.30	—	11.25	*1,620.00*
4.	100 *(2,700 × .03 =)*	27.00 *(100 × .38 =)*	3	.38	*119.00*
5.	75 *(2,430 × .025 =))*	32.40	$2\frac{1}{2}$	—	*60.75*
6.	20 *(16,920 × .08 =)*	846.00 *(20 × .87 =)*	8	.87	*1,371.00*
7.	50 *(4,660 × .03 =)*	93.20	3	—	*139.80*
8.	100 *(450 × .02 =)*	4.50 *(100 × 3 =)*	2	3.00	*309.00*
9.	150 *(1,200 × .0425 =)*	8.00 *(150 × 16.50 =)*	$4\frac{1}{4}$	$16.50	*2,526.00*
10.	50 *(1,150 × .09 =)*	23.00	9	—	*103.50*

Federal Income Tax

Federal, state, and city income taxes are designed to tax wages, salaries, commissions, dividends, royalties, tips, and other incomes of individual wage earners. The profits of business and industry are also taxed under a similar set of laws.

At the end of a calendar year, wage earners and firms must file a report called an income tax return with the Internal Revenue Service (I.R.S.). The income tax return determines the amount of income taxes for the year. Wage earners will have paid income taxes through payroll withholdings during the year. The amount of taxes paid and gross earnings are reported to employees on a W-2 form provided by the I.R.S. and completed by the employer. The income tax for the year and the final settlement amount is worked out on the income tax return and sent to the I.R.S. The form used is known as Form 1040 or Form 1040A.

Some Americans pay no federal income tax because their income is too low or their income is not taxable (income from special investments).

a Control number	22222	Void ☐	For Official Use Only ▶ OMB No. 1545-0008
b Employer identification number	1 Wages, tips, other compensation	2 Federal income tax withheld	
c Employer's name, address, and ZIP code	3 Social security wages	4 Social security tax withheld	
	5 Medicare wages and tips	6 Medicare tax withheld	
	7 Social security tips	8 Allocated tips	
d Employee's social security number	9 Advance EIC payment	10 Dependent care benefits	
e Employee's name (first, middle initial, last)	11 Nonqualified plans	12 Benefits included in box 1	
	13 See instrs. for box 13	14 Other	
f Employee's address and ZIP code	15 Statutory employee ☐ Deceased ☐ Pension plan ☐ Legal rep. ☐ Deferred compensation ☐		

16 State Employer's state I.D. no.	17 State wages, tips, etc.	18 State income tax	19 Locality name	20 Local wages, tips, etc.	21 Local income tax

Form **W-2** **Wage and Tax Statement** 1999

Department of the Treasury—Internal Revenue Service

taxable income Income tax is based on the wage earner's ability to pay. The greater the taxable income, the higher the income tax rate. **Taxable income** is the gross earnings of an individual, less the amount of allowable expenses. The allowable expenses fall into two categories, dependents and deductions.

Dependents

Dependents are those persons who are living in the household of the wage earner and who are at least half supported by the wage earner during the year. The wage earner's spouse, children, relatives, and any person living in that household can be claimed as dependents.* For each dependent and for the wage earner, $2,750 of gross earnings are exempt from taxation. The $2,750 amount is known as an **exemption.** This amount is set aside, tax free, to cover basic needs. *exemption*

Example 8

A family earns $34,680 for the year. The family is made up of a father, mother, and three dependent children. What is the amount of tax exempt income for the family?

1 exemption—Father
1 exemption—Mother
3 exemptions—Children
5 exemptions × $2,750 = $13,750

A household with a lower income or a large number of exemptions may not need to pay income tax. In these cases, low gross earnings are balanced or overcome by the total exemption amount.

Deductions

Deductions are other allowable expenses beyond the exemption-level needs. They are also subtracted from gross earnings when determining taxable earnings. Deductions include property taxes, interest on first and second home mortgages, contributions to nonprofit organizations such as churches, dental and medical expenses over a certain amount not covered by insurance, property losses not covered by insurance, and other items.

Note: The figures used in this section are current as of the writing of this text.
*The reader should realize that the courts have established exceptions to this explanation, but the exceptions are beyond the scope of this text.

standard deduction

The wage earner may list or itemize all deductions on Form 1040 or use the **standard deduction.** The itemized deductions must be verifiable. The standard deduction amount is $7,200 for a married couple filing jointly, $4,300 for a single person, and $6,350 for head of household. These amounts apply to taxpayers under 65 years old. For those over 65 years old and/or blind, the standard deduction is a larger amount. The wage earner may select whichever method offers the greater benefits by reducing taxable income as much as possible. If the taxpayer intends to itemize deductions, Form 1040 will be used. Form 1040A will be used if the taxpayer does not plan to itemize deductions.

Gross earnings of a wage earner are adjusted by subtracting the dependent exemptions and deductions to find taxable income.

The coverage of the subject of federal income tax in this chapter is by no means complete. Students should realize that this topic is in itself a program of study and, for some, a career. The material included here on this topic is designed to give you a basic understanding, which is vital to business and individuals.

Example 9

The Paret family had gross earnings of $52,685 for the year. The family consisted of the parents and two dependent children. During the year, $3,786.40 was paid in interest on their home, $2,380 was paid for property taxes and $1,250 given in contributions. What was the family's taxable income?

Gross earnings for the year	$52,685	
Less dependent exemptions	−11,000	(4 × $2,750)
Less total itemized deductions	−7,416.40	
Taxable income	$34,268.60	

The taxable income is used to determine the amount of income tax for the year.

Objective 4
Determining the Tax

The amount of income tax is determined by using one of several tables provided by the I.R.S. The amount of taxable earnings is used to find the correct bracket on the table. The tax is determined by following the instructions on the tax table. See the table on page 355.

Example 10

The Bierston family had a taxable income of $35,332 for the year. They will file a joint tax return. What is the income tax?

$35,332 is found in the bracket $35,300 to $35,350. The tax for a married couple filing jointly is $5,299.

Exercise D

Determine the income tax from the information below. Use the tax table indicated. Compare itemized deductions to standard deduction to compute taxable earnings.

Gross Earnings	Exemptions	Itemized Deductions	Taxable Earnings	Tax Table	Income Tax
1. $55,860	5	$ 2,160	$34,910	Married-joint	$5,239
2. 55,800	2	10,410	39,890	Married-joint	5,981
3. 44,610	2	2,821	32,760	Head of household	4,916
4. 51,386	4	4,956	33,186	Married-joint	4,976
5. 43,650	1	3,660	36,600	Single	6,908
6. 66,853	6	17,647	32,706	Married-joint	4,909
7. 47,281	3	3,893	32,681	Head of household	4,901
8. 50,453	1	10,240	37,463	Single	7,146
9. 45,820	2	2,491	33,120	Married-joint	4,969
10. 49,325	1	11,490	35,085	Single	6,474

1999 Tax Table

Use if your taxable income is less than $100,000.
If $100,000 or more, use the Tax Rate Schedules.

Example. Mr. and Mrs. Brown are filing a joint return. Their taxable income on line 39 of Form 1040 is $25,300. First, they find the $25,300–25,350 income line. Next, they find the column for married filing jointly and read down the column. The amount shown where the income line and filing status column meet is $3,799. This is the tax amount they should enter on line 40 of their Form 1040.

Sample Table

At least	But less than	Single	Married filing jointly *	Married filing separately	Head of a household
		Your tax is—			
25,200	**25,250**	3,784	3,784	4,265	3,784
25,250	**25,300**	3,791	3,791	4,279	3,791
25,300	**25,350**	3,799	(3,799)	4,293	3,799
25,350	**25,400**	3,806	3,806	4,307	3,806

If line 39 (taxable income) is—		And you are—			
At least	But less than	Single	Married filing jointly *	Married filing separately	Head of a household
		Your tax is—			
32,000					
32,000	**32,050**	5,620	4,804	6,169	4,804
32,050	**32,100**	5,634	4,811	6,183	4,811
32,100	**32,150**	5,648	4,819	6,197	4,819
32,150	**32,200**	5,662	4,826	6,211	4,826
32,200	**32,250**	5,676	4,834	6,225	4,834
32,250	**32,300**	5,690	4,841	6,239	4,841
32,300	**32,350**	5,704	4,849	6,253	4,849
32,350	**32,400**	5,718	4,856	6,267	4,856
32,400	**32,450**	5,732	4,864	6,281	4,864
32,450	**32,500**	5,746	4,871	6,295	4,871
32,500	**32,550**	5,760	4,879	6,309	4,879
32,550	**32,600**	5,774	4,886	6,323	4,886
32,600	**32,650**	5,788	4,894	6,337	4,894
32,650	**32,700**	5,802	4,901	6,351	4,901
32,700	**32,750**	5,816	4,909	6,365	4,909
32,750	**32,800**	5,830	4,916	6,379	4,916
32,800	**32,850**	5,844	4,924	6,393	4,924
32,850	**32,900**	5,858	4,931	6,407	4,931
32,900	**32,950**	5,872	4,939	6,421	4,939
32,950	**33,000**	5,886	4,946	6,435	4,946
33,000					
33,000	**33,050**	5,900	4,954	6,449	4,954
33,050	**33,100**	5,914	4,961	6,463	4,961
33,100	**33,150**	5,928	4,969	6,477	4,969
33,150	**33,200**	5,942	4,976	6,491	4,976
33,200	**33,250**	5,956	4,984	6,505	4,984
33,250	**33,300**	5,970	4,991	6,519	4,991
33,300	**33,350**	5,984	4,999	6,533	4,999
33,350	**33,400**	5,998	5,006	6,547	5,006
33,400	**33,450**	6,012	5,014	6,561	5,014
33,450	**33,500**	6,026	5,021	6,575	5,021
33,500	**33,550**	6,040	5,029	6,589	5,029
33,550	**33,600**	6,054	5,036	6,603	5,036
33,600	**33,650**	6,068	5,044	6,617	5,044
33,650	**33,700**	6,082	5,051	6,631	5,051
33,700	**33,750**	6,096	5,059	6,645	5,059
33,750	**33,800**	6,110	5,066	6,659	5,066
33,800	**33,850**	6,124	5,074	6,673	5,074
33,850	**33,900**	6,138	5,081	6,687	5,081
33,900	**33,950**	6,152	5,089	6,701	5,089
33,950	**34,000**	6,166	5,096	6,715	5,096
34,000					
34,000	**34,050**	6,180	5,104	6,729	5,104
34,050	**34,100**	6,194	5,111	6,743	5,111
34,100	**34,150**	6,208	5,119	6,757	5,119
34,150	**34,200**	6,222	5,126	6,771	5,126
34,200	**34,250**	6,236	5,134	6,785	5,134
34,250	**34,300**	6,250	5,141	6,799	5,141
34,300	**34,350**	6,264	5,149	6,813	5,149
34,350	**34,400**	6,278	5,156	6,827	5,156
34,400	**34,450**	6,292	5,164	6,841	5,164
34,450	**34,500**	6,306	5,171	6,855	5,171
34,500	**34,550**	6,320	5,179	6,869	5,179
34,550	**34,600**	6,334	5,186	6,883	5,190
34,600	**34,650**	6,348	5,194	6,897	5,204
34,650	**34,700**	6,362	5,201	6,911	5,218
34,700	**34,750**	6,376	5,209	6,925	5,232
34,750	**34,800**	6,390	5,216	6,939	5,246
34,800	**34,850**	6,404	5,224	6,953	5,260
34,850	**34,900**	6,418	5,231	6,967	5,274
34,900	**34,950**	6,432	5,239	6,981	5,288
34,950	**35,000**	6,446	5,246	6,995	5,302
35,000					
35,000	**35,050**	6,460	5,254	7,009	5,316
35,050	**35,100**	6,474	5,261	7,023	5,330
35,100	**35,150**	6,488	5,269	7,037	5,344
35,150	**35,200**	6,502	5,276	7,051	5,358
35,200	**35,250**	6,516	5,284	7,065	5,372
35,250	**35,300**	6,530	5,291	7,079	5,386
35,300	**35,350**	6,544	5,299	7,093	5,400
35,350	**35,400**	6,558	5,306	7,107	5,414
35,400	**35,450**	6,572	5,314	7,121	5,428
35,450	**35,500**	6,586	5,321	7,135	5,442
35,500	**35,550**	6,600	5,329	7,149	5,456
35,550	**35,600**	6,614	5,336	7,163	5,470
35,600	**35,650**	6,628	5,344	7,177	5,484
35,650	**35,700**	6,642	5,351	7,191	5,498
35,700	**35,750**	6,656	5,359	7,205	5,512
35,750	**35,800**	6,670	5,366	7,219	5,526
35,800	**35,850**	6,684	5,374	7,233	5,540
35,850	**35,900**	6,698	5,381	7,247	5,554
35,900	**35,950**	6,712	5,389	7,261	5,568
35,950	**36,000**	6,726	5,396	7,275	5,582
36,000					
36,000	**36,050**	6,740	5,404	7,289	5,596
36,050	**36,100**	6,754	5,411	7,303	5,610
36,100	**36,150**	6,768	5,419	7,317	5,624
36,150	**36,200**	6,782	5,426	7,331	5,638
36,200	**36,250**	6,796	5,434	7,345	5,652
36,250	**36,300**	6,810	5,441	7,359	5,666
36,300	**36,350**	6,824	5,449	7,373	5,680
36,350	**36,400**	6,838	5,456	7,387	5,694
36,400	**36,450**	6,852	5,464	7,401	5,708
36,450	**36,500**	6,866	5,471	7,415	5,722
36,500	**36,550**	6,880	5,479	7,429	5,736
36,550	**36,600**	6,894	5,486	7,443	5,750
36,600	**36,650**	6,908	5,494	7,457	5,764
36,650	**36,700**	6,922	5,501	7,471	5,778
36,700	**36,750**	6,936	5,509	7,485	5,792
36,750	**36,800**	6,950	5,516	7,499	5,806
36,800	**36,850**	6,964	5,524	7,513	5,820
36,850	**36,900**	6,978	5,531	7,527	5,834
36,900	**36,950**	6,992	5,539	7,541	5,848
36,950	**37,000**	7,006	5,546	7,555	5,862
37,000					
37,000	**37,050**	7,020	5,554	7,569	5,876
37,050	**37,100**	7,034	5,561	7,583	5,890
37,100	**37,150**	7,048	5,569	7,597	5,904
37,150	**37,200**	7,062	5,576	7,611	5,918
37,200	**37,250**	7,076	5,584	7,625	5,932
37,250	**37,300**	7,090	5,591	7,639	5,946
37,300	**37,350**	7,104	5,599	7,653	5,960
37,350	**37,400**	7,118	5,606	7,667	5,974
37,400	**37,450**	7,132	5,614	7,681	5,988
37,450	**37,500**	7,146	5,621	7,695	6,002
37,500	**37,550**	7,160	5,629	7,709	6,016
37,550	**37,600**	7,174	5,636	7,723	6,030
37,600	**37,650**	7,188	5,644	7,737	6,044
37,650	**37,700**	7,202	5,651	7,751	6,058
37,700	**37,750**	7,216	5,659	7,765	6,072
37,750	**37,800**	7,230	5,666	7,779	6,086
37,800	**37,850**	7,244	5,674	7,793	6,100
37,850	**37,900**	7,258	5,681	7,807	6,114
37,900	**37,950**	7,272	5,689	7,821	6,128
37,950	**38,000**	7,286	5,696	7,835	6,142
38,000					
38,000	**38,050**	7,300	5,704	7,849	6,156
38,050	**38,100**	7,314	5,711	7,863	6,170
38,100	**38,150**	7,328	5,719	7,877	6,184
38,150	**38,200**	7,342	5,726	7,891	6,198
38,200	**38,250**	7,356	5,734	7,905	6,212
38,250	**38,300**	7,370	5,741	7,919	6,226
38,300	**38,350**	7,384	5,749	7,933	6,240
38,350	**38,400**	7,398	5,756	7,947	6,254
38,400	**38,450**	7,412	5,764	7,961	6,268
38,450	**38,500**	7,426	5,771	7,975	6,282
38,500	**38,550**	7,440	5,779	7,989	6,296
38,550	**38,600**	7,454	5,786	8,003	6,310
38,600	**38,650**	7,468	5,794	8,017	6,324
38,650	**38,700**	7,482	5,801	8,031	6,338
38,700	**38,750**	7,496	5,809	8,045	6,352
38,750	**38,800**	7,510	5,816	8,059	6,366
38,800	**38,850**	7,524	5,824	8,073	6,380
38,850	**38,900**	7,538	5,831	8,087	6,394
38,900	**38,950**	7,552	5,839	8,101	6,408
38,950	**39,000**	7,566	5,846	8,115	6,422
39,000					
39,000	**39,050**	7,580	5,854	8,129	6,436
39,050	**39,100**	7,594	5,861	8,143	6,450
39,100	**39,150**	7,608	5,869	8,157	6,464
39,150	**39,200**	7,622	5,876	8,171	6,478
39,200	**39,250**	7,636	5,884	8,185	6,492
39,250	**39,300**	7,650	5,891	8,199	6,506
39,300	**39,350**	7,664	5,899	8,213	6,520
39,350	**39,400**	7,678	5,906	8,227	6,534
39,400	**39,450**	7,692	5,914	8,241	6,548
39,450	**39,500**	7,706	5,921	8,255	6,562
39,500	**39,550**	7,720	5,929	8,269	6,576
39,550	**39,600**	7,734	5,936	8,283	6,590
39,600	**39,650**	7,748	5,944	8,297	6,604
39,650	**39,700**	7,762	5,951	8,311	6,618
39,700	**39,750**	7,776	5,959	8,325	6,632
39,750	**39,800**	7,790	5,966	8,339	6,646
39,800	**39,850**	7,804	5,974	8,353	6,660
39,850	**39,900**	7,818	5,981	8,367	6,674
39,900	**39,950**	7,832	5,989	8,381	6,688
39,950	**40,000**	7,846	5,996	8,395	6,702
40,000					
40,000	**40,050**	7,860	6,004	8,409	6,716
40,050	**40,100**	7,874	6,011	8,423	6,730
40,100	**40,150**	7,888	6,019	8,437	6,744
40,150	**40,200**	7,902	6,026	8,451	6,758
40,200	**40,250**	7,916	6,034	8,465	6,772
40,250	**40,300**	7,930	6,041	8,479	6,786
40,300	**40,350**	7,944	6,049	8,493	6,800
40,350	**40,400**	7,958	6,056	8,507	6,814
40,400	**40,450**	7,972	6,064	8,521	6,828
40,450	**40,500**	7,986	6,071	8,535	6,842
40,500	**40,550**	8,000	6,079	8,549	6,856
40,550	**40,600**	8,014	6,086	8,563	6,870
40,600	**40,650**	8,028	6,094	8,577	6,884
40,650	**40,700**	8,042	6,101	8,591	6,898
40,700	**40,750**	8,056	6,109	8,605	6,912
40,750	**40,800**	8,070	6,116	8,619	6,926
40,800	**40,850**	8,084	6,124	8,633	6,940
40,850	**40,900**	8,098	6,131	8,647	6,954
40,900	**40,950**	8,112	6,139	8,661	6,968
40,950	**41,000**	8,126	6,146	8,675	6,982

Example 11

John Talbot has provided you with the following statistics to use in preparing his 1999 federal income tax return:

Mailing label from I.R.S.	Form W-2	
SSN 456-22-3311	Wages	$45,662.19
MR JOHN A TALBOT	Federal withholding	8,788.00
4614 10TH STREET	State withholding	1,985.88
CAROL, MI 49231	City of Carol local tax	228.31
	Social Security tax	2,831.06
	Medicare tax	662.11

Form 1098

$4,526.00 home mortgage interest paid in 1999

Form 1099-MISC

$755.00 prize won in 50/50 draw

Form 1099-G

$560.00 state income tax refund from Michigan

Bills paid in 1999

1999 real estate taxes on home	$1,577.00
1998 real estate taxes on home	252.19
1999 real estate taxes on son's home	550.22
City of Carol income tax	25.89
Trinity Lutheran Church tithe	2,505.00
Income tax preparation fee	105.00
Dr. Allan Jones (for John)	800.00
Mercy Hospital (for John)	2,051.00
John's dental cleaning	50.00

Mr. Talbot is less than 65 years old, is divorced, is not blind, and lives alone. He itemized his deductions on his 1998 income tax return; his itemized deductions on that return totaled $10,656.00.

Form **1040** Department of the Treasury — Internal Revenue Service
U.S. Individual Income Tax Return 1999 (99) IRS use only — Do not write or staple in this space.

For the year Jan 1-Dec 31, 1999, or other tax year beginning , 1999, ending , OMB No. 1545-0074

Label (See instructions.)
Use the IRS label. Otherwise, please print or type.

Your First Name	MI	Last Name	Your Social Security Number
JOHN	A	TALBOT	456-22-3311
If a Joint Return, Spouse's First Name	MI	Last Name	Spouse's Social Security Number
Home Address (number and street). If You Have a P.O. Box, See Instructions.		Apartment No.	▲ **Important!** ▲ You **must** enter your social security number(s) above.
4614 10TH STREET			
City, Town or Post Office. If You Have a Foreign Address, See Instructions.	State	ZIP Code	
CAROL	MI	49231	

Presidential Election Campaign (See instructions.)

	Yes	No	
Do you want $3 to go to this fund?			**Note:** Checking 'Yes' will not change your tax or reduce your refund.
If a joint return, does your spouse want $3 to go to this fund?			

Filing Status

Check only one box.

1 [X] Single
2 [ ] Married filing joint return (even if only one had income)
3 [ ] Married filing separate return. Enter spouse's SSN above & full name here ... ►
4 [ ] Head of household (with qualifying person). (See instructions.) If the qualifying person is a child but not your dependent, enter this child's name here ... ►
5 [ ] Qualifying widow(er) with dependent child (year spouse died ► 19). (See instructions.)

Exemptions

6a [X] **Yourself.** If your parent (or someone else) can claim you as a dependent on his or her tax return, **do not** check box 6a — No. of boxes checked on 6a and 6b: 1

b [ ] **Spouse**

c **Dependents:**

(1) First name Last name	(2) Dependent's social security number	(3) Dependent's relationship to you	(4) ✓ if qualifying child for child tax credit (see instructions)

If more than six dependents, see instructions.

No. of your children on 6c who: • lived with you • did not live with you due to divorce or separation (see instructions) Dependents on 6c not entered above

d Total number of exemptions claimed — Add numbers entered on lines above ► 1

Income

Attach Copy B of your Forms W-2 and W-2G here. Also attach Form(s) 1099-R if tax was withheld.

If you did not get a W-2, see instructions.

Enclose, but do not staple, any payment. Also, please use **Form 1040-V.**

7	Wages, salaries, tips, etc. Attach Form(s) W-2		7	45,662.
8a	**Taxable** interest. Attach Schedule B if required		8a	
b	**Tax-exempt** interest. **Do not** include on line 8a	8b		
9	Ordinary dividends. Attach Schedule B if required		9	
10	Taxable refunds, credits, or offsets of state and local income taxes (see instructions)		10	560.
11	Alimony received		11	
12	Business income or (loss). Attach Schedule C or C-EZ		12	
13	Capital gain or (loss). Attach Schedule D if required. If not required, check here ► [ ]		13	
14	Other gains or (losses). Attach Form 4797		14	
15a	Total IRA distributions 15a	b Taxable amount (see instrs)	15b	
16a	Total pensions & annuities 16a	b Taxable amount (see instrs)	16b	
17	Rental real estate, royalties, partnerships, S corporations, trusts, etc. Attach Schedule E		17	
18	Farm income or (loss). Attach Schedule F		18	
19	Unemployment compensation		19	
20a	Social security benefits 20a	b Taxable amount (see instrs)	20b	
21	Other income. List type & amount (see instrs) 50/50 DRAWING PRIZE		21	755.
22	Add the amounts in the far right column for lines 7 through 21. This is your **total income** ►		22	46,977.

Adjusted Gross Income

23	IRA deduction (see instructions)	23		
24	Student loan interest deduction (see instructions)	24		
25	Medical savings account deduction. Attach Form 8853	25		
26	Moving expenses. Attach Form 3903	26		
27	One-half of self-employment tax. Attach Schedule SE	27		
28	Self-employed health insurance deduction (see instructions)	28		
29	Keogh and self-employed SEP and SIMPLE plans	29		
30	Penalty on early withdrawal of savings	30		
31a	Alimony paid **b** Recipient's SSN ►	31a		
32	Add lines 23 through 31a		32	
33	Subtract line 32 from line 22. This is your **adjusted gross income** ►		33	46,977.

BAA For Disclosure, Privacy Act, and Paperwork Reduction Act Notice, see instructions. Form **1040** (1999)

FDIA0112 11/16/99

Form **1040** (1999) JOHN A TALBOT 456-22-3311 Page **2**

Tax and Credits

34	Amount from line 33 (adjusted gross income)	34	46,977.
35a	Check if: ☐ **You** were 65/older, ☐ Blind; ☐ **Spouse** was 65/older, ☐ Blind. Add the number of boxes checked above and enter the total here ► 35a ☐		
b	If you are married filing separately and your spouse itemizes deductions or you were a dual-status alien, see instructions and check here ► 35b ☐		
36	Enter your **itemized deductions** from Schedule A, line 28, **Or standard deduction** shown on the left. **But** see instructions to find your standard deduction if you checked any box on line 35a or 35b **or** if someone can claim you as a dependent	36	11,100.
37	Subtract line 36 from line 34	37	35,877.
38	If line 34 is $94,975 or less, multiply $2,750 by the total number of exemptions claimed on line 6d. If line 34 is over $94,975, see the worksheet in the instructions for the amount to enter	38	2,750.
39	**Taxable income.** Subtract line 38 from line 37. If line 38 is more than line 37, enter -0-	39	33,127.
40	**Tax** (see instrs). Check if any tax is from a ☐ Form(s) 8814 b ☐ Form 4972 ►	40	5,928.
41	Credit for child and dependent care expenses. Attach Form 2441	41	
42	Credit for the elderly or the disabled. Attach Schedule R	42	
43	Child tax credit (see instructions)	43	
44	Education credits. Attach Form 8863	44	
45	Adoption credit. Attach Form 8839	45	
46	Foreign tax credit. Attach Form 1116 if required	46	
47	Other. Check if from a ☐ Form 3800 b ☐ Form 8396 c ☐ Form 8801 d ☐ Form (specify) ____	47	
48	Add lines 41 through 47. These are your **total credits**	48	
49	Subtract line 48 from line 40. If line 48 is more than line 40, enter -0- ►	49	5,928.

Standard Deduction for Most People

Single: $4,300

Head of household: $6,350

Married filing jointly or Qualifying widow(er): $7,200

Married filing separately: $3,600

Other Taxes

50	Self-employment tax. Attach Schedule SE	50	
51	Alternative minimum tax. Attach Form 6251	51	
52	Social security and Medicare tax on tip income not reported to employer. Attach Form 4137	52	
53	Tax on IRAs, other retirement plans, and MSAs. Attach Form 5329 if required	53	
54	Advance earned income credit payments from Form(s) W-2	54	
55	Household employment taxes. Attach Schedule H	55	
56	Add lines 49-55. This is your **total tax** ►	56	5,928.

Payments

57	Federal income tax withheld from Forms W-2 and 1099	57	8,788.
58	1999 estimated tax payments and amount applied from 1998 return	58	
59a	**Earned income credit**. Attach Schedule EIC if you have a qualifying child. b Nontaxable earned income: amount ► ____ and type ► ____	59a	
60	Additional child tax credit. Attach Form 8812	60	
61	Amount paid with request for extension to file (see instructions)	61	
62	Excess social security and RRTA tax withheld (see instrs)	62	
63	Other payments. Check if from a ☐ Form 2439 b ☐ Form 4136	63	
64	Add lines 57, 58, 59a, and 60 through 63. These are your **total payments** ►	64	8,788.

Refund

Have it directly deposited! See instructions and fill in 66b, 66c, and 66d.

65	If line 64 is more than line 56, subtract line 56 from line 64. This is the amount you **Overpaid**	65	2,860.
66a	Amount of line 65 you want **Refunded to You** ►	66a	2,860.
	► b Routing number ____ ► c Type: ☐ Checking ☐ Savings		
	► d Account number ____		
67	Amount of line 65 you want **Applied to Your 2000 Estimated Tax** ►	67	

Amount You Owe

68	If line 56 is more than line 64, subtract line 64 from line 56. This is the **Amount You Owe.** For details on how to pay, see instructions ►	68	
69	Estimated tax penalty. Also include on line 68	69	

Sign Here

Joint return? See instructions. Keep a copy for your records.

Under penalties of perjury, I declare that I have examined this return and accompanying schedules and statements, and to the best of my knowledge and belief, they are true, correct, and complete. Declaration of preparer (other than taxpayer) is based on all information of which preparer has any knowledge.

Your Signature	Date	Your Occupation	Daytime Telephone Number (optional)
►		SOFTWARE DESIGNER	
Spouse's Signature. If a Joint Return, **Both** Must Sign.	Date	Spouse's Occupation	

Paid Preparer's Use Only

Preparer's Signature ►	Date	Check if self-employed ☐	Preparer's SSN or PTIN
Firm's Name (or yours if self-employed) and Address ►	Self-prepared	EIN	
		ZIP Code	

FDIA0112 11/15/99 Form **1040** (1999)

Schedule A (Form 1040)

Department of the Treasury Internal Revenue Service (99)

Itemized Deductions

► Attach to Form 1040.
► See Instructions for Schedule A (Form 1040).

OMB No. 1545-0074

1999

07

Name(s) Shown on Form 1040: JOHN A TALBOT

Your Social Security Number: 456-22-3311

Medical and Dental Expenses		**Caution:** Do not include expenses reimbursed or paid by others.				
	1	Medical and dental expenses (see instructions)	1	2,901.		
	2	Enter amount from Form 1040, line 34 ..... 2 46,977.				
	3	Multiply line 2 above by 7.5% (.075)	3	3,523.		
	4	Subtract line 3 from line 1. If line 3 is more than line 1, enter -0-			4	0.
Taxes You Paid (See instructions.)	5	State and local income taxes	5	2,240.		
	6	Real estate taxes (see instructions)	6	1,829.		
	7	Personal property taxes	7			
	8	Other taxes. List type and amount ►	8			
	9	Add lines 5 through 8			9	4,069.
Interest You Paid (See instructions.)	10	Home mortgage interest and points reported to you on Form 1098	10	4,526.		
	11	Home mortgage interest not reported to you on Form 1098. If paid to the person from whom you bought the home, see instructions and show that person's name, identifying number, and address ►	11			
Note: Personal interest is not deductible.	12	Points not reported to you on Form 1098. See instructions for special rules	12			
	13	Investment interest. Attach Form 4952 if required. (See instructions.)	13			
	14	Add lines 10 through 13			14	4,526.
Gifts to Charity	15	Gifts by cash or check. If you made any gift of $250 or more, see instructions	15	2,505.		
If you made a gift and got a benefit for it, see instructions.	16	Other than by cash or check. If any gift of $250 or more, see instructions. You **Must** attach Form 8283 if over $500	16			
	17	Carryover from prior year	17			
	18	Add lines 15 through 17			18	2,505.
Casualty and Theft Losses	19	Casualty or theft loss(es). Attach Form 4684. (See instructions.)			19	
Job Expenses and Most Other Miscellaneous Deductions	20	Unreimbursed employee expenses – job travel, union dues, job education, etc. You **Must** attach Form 2106 or 2106-EZ if required. (See instructions.) ►	20			
	21	Tax preparation fees	21			
(See instructions for expenses to deduct here.)	22	Other expenses – investment, safe deposit box, etc. List type and amount ► TAX PREPARATION FEE 105.	22	105.		
	23	Add lines 20 through 22	23	105.		
	24	Enter amount from Form 1040, line 34 ..... 24 46,977.				
	25	Multiply line 24 above by 2% (.02)	25	940.		
	26	Subtract line 25 from line 23. If line 25 is more than line 23, enter -0-			26	0.
Other Miscellaneous Deductions	27	Other – from list in the instructions. List type and amount ►			27	
Total Itemized Deductions	28	Is Form 1040, line 34, over $126,600 (over $63,300 if married filing separately)? [X] **No.** Your deduction is not limited. Add the amounts in the far right column for lines 4 through 27. Also, enter this amount on Form 1040, line 36. [ ] **Yes.** Your deduction may be limited. See instructions for the amount to enter.			28	11,100.

BAA For Paperwork Reduction Act Notice, see separate instructions.

Schedule **A** (Form 1040) 1999

FDIA0301 10/27/99

Schedule **A & B** (Form 1040) 1999 OMB No. 1545-0074 Page **2**

Name(s) Shown on Form 1040. Do Not Enter Name and Social Security Number if Shown on Schedule A. | **Your Social Security Number**

Schedule B — Interest and Ordinary Dividends

08

Part I Interest

(See instructions for Form 1040, line 8a.)

Note: If you received a Form 1099-INT, Form 1099-OID, or substitute statement from a brokerage firm, list the firm's name as the payer and enter the total interest shown on that form.

Note: If you had over $400 in taxable interest income, you must also complete Part III.

			Amount
1	List name of payer. If any interest is from a seller-financed mortgage and the buyer used the property as a personal residence, see the instructions and list this interest first. Also, show that buyer's social security number and address . . . ►	1	
2	Add the amounts on line 1 . . .	2	
3	Excludable interest on series EE and I U.S. savings bonds issued after 1989 from Form 8815, line 14. You **Must** attach Form 8815 . . .	3	
4	Subtract line 3 from line 2. Enter the result here and on Form 1040, line 8a . . . ►	4	

Part II Ordinary Dividends

(See instructions for Form 1040, line 8a.)

Note: If you received a Form 1099-DIV, or substitute statement from a brokerage firm, list the firm's name as the payer and enter the ordinary dividends shown on that form.

Note: If you had over $400 in ordinary dividends, you must also complete Part III.

			Amount
5	List name of payer. Include only ordinary dividends. If you received any capital gain distributions, see the instructions for Form 1040, line 13 . . . ►	5	
6	Add the amounts on line 5. Enter the total here and on Form 1040, line 9 . . . ►	6	

Part III Foreign Accounts and Trusts

(See instructions.)

	You must complete this part if you **(a)** had over $400 of interest or ordinary dividends; **(b)** had a foreign account; or **(c)** received a distribution from, or were a grantor of, or a transferor to, a foreign trust.	Yes	No
7a	At any time during 1999, did you have an interest in or a signature or other authority over a financial account in a foreign country, such as a bank account, securities account, or other financial account? See instructions for exceptions and filing requirements for Form TD F 90-22.1 . . .		
b	If 'yes,' enter the name of the foreign country . . ►		
8	During 1999, did you receive a distribution from, or were you the grantor of, or transferor to, a foreign trust? If 'yes,' you may have to file Form 3520. See instructions . . .		

BAA For Paperwork Reduction Act Notice, see Form 1040 instructions. FDIA0401 10/13/99 Schedule **B** (Form 1040) 1999

Example 12

Justin A. Tenbush and his wife Taylor both work, and they want to file a joint income tax return for 1999. In May of 1999, they were blessed with a baby girl, named Jennifer. They did not have to pay a babysitter in 1999, thanks to Grandma. This couple has requested that you do their federal income tax return. They are both under 65, they are not blind, and they don't live with their parents. Their address is 813 Lincoln Ave., North Street, MI 48049. Mr. and Mrs. Tenbush brought you the following information:

Taylor Tenbush	
Social Security number	333-23-4211
Wages	$15,976.13
Federal withholding	2,395.00
Social Security tax	990.52
Medicare tax	231.66
State of Michigan tax	734.91
Federal Credit Union interest	34.52
Justin A. Tenbush	
Social Security number	441-23-9941
Wages	$28,542.14
Federal withholding	2,652.00
Social Security tax	1,769.62
Medicare tax	413.87
State of Michigan tax	1,312.94
Unemployment benefits	560.00
Jennifer Tenbush	
Social Security number	555-48-6444

The family qualifies for the child tax credit. This credit gives a family an additional $500 per child (maximum $1,500).

Form **1040A** Department of the Treasury — Internal Revenue Service
U.S. Individual Income Tax Return (99) **1999** IRS use only — Do not write or staple in this space.

OMB No. 1545-0085

Label (see the instructions.)

Use the IRS label. Otherwise, please print or type.

Your First Name and Initial		Last Name	Your Social Security Number
JUSTIN	A	TENBUSH	441-23-9941
If a Joint Return, Spouse's First Name and Initial		Last Name	Spouse's Social Security Number
TAYLOR		TENBUSH	333-23-9941
Home Address (number and street). If You Have a P.O. Box, See Instructions.		Apt Number	▲ **Important!** ▲ You **must** enter your SSN(s) above.
813 LINCOLN AVE			
City, Town or Post Office, State, and ZIP Code. If You Have a Foreign Address, See Instructions.			
NORTH STREET	MI	48049	

Presidential Election Campaign Fund (See instructions.)	Yes	No
Do you want $3 to go to this fund?		
If a joint return, does your spouse want $3 to go to this fund?		

Note: Checking 'Yes' will not change your tax or reduce your refund.

Filing status

Check only one box.

1 [ ] Single
2 [X] Married filing joint return (even if only one had income)
3 [ ] Married filing separate return. Enter spouse's social security number above and full name here ►
4 [ ] Head of household (with qualifying person). (See instructions.) If the qualifying person is a child but not your dependent, enter this child's name here ►
5 [ ] Qualifying widow(er) with dependent child (year spouse died ► 19). (See instructions.)

Exemptions

6a [X] **Yourself.** If your parent (or someone else) can claim you as a dependent on his or her tax return, **do not** check box 6a

b [X] **Spouse**

No. of boxes checked on 6a and 6b 2

c **Dependents.**

If more than seven dependents, see instructions.

(1) First name Last name	(2) Dependent's social security number	(3) Dependent's relationship to you	(4) ✓ if qualifying child for child tax credit
JENNIFER TENBUSH	555-48-6444	Daughter	X

No. of your children on 6c who:
- **lived with you** 1
- **did not live with you due to divorce or separation**

Dependents on 6c not entered above

d Total number of exemptions claimed — **Add numbers entered on lines above** 3

Income

Attach Copy B of your Form(s) W-2 here. Also attach Form(s) 1099-R if tax was withheld.

If you did not get a W-2, see instructions.

Enclose, but do not staple, any payment.

7	Wages, salaries, tips, etc. Attach Form(s) W-2		7	44,518.
8a	**Taxable** interest. Attach Schedule 1 if required		8a	35.
b	**Tax-exempt** interest. **Do not** include on line 8a	8b		
9	Ordinary dividends. Attach Schedule 1 if required		9	
10a	Total IRA distributions 10a	10b Taxable amount	10b	
11a	Total pensions and annuities 11a	11b Taxable amount	11b	
12	Unemployment compensation, qualified state tuition program earnings, and Alaska Permanent Fund dividends		12	560.
13a	Social security benefits 13a	13b Taxable amount	13b	
14	Add lines 7 through 13b (far right column). This is your **total income** ►		14	45,113.

Adjusted gross income

15	IRA deduction (see instructions)	15		
16	Student loan interest deduction (see instructions)	16		
17	Add lines 15 and 16. These are your **total adjustments**		17	
18	Subtract line 17 from line 14. This is your **adjusted gross income** ►		18	45,113.

BAA For Paperwork Reduction Act Notice, see instructions. Form **1040A** (1999)

FDIA1312 11/10/99

JUSTIN A & TAYLOR TENBUSH 441-23-9941

Form **1040A** (1999) Page **2**

Taxable income	19	Enter the amount from line 18	19	45,113.
	20a	Check if: ☐ **You** were 65 or older ☐ Blind; ☐ **Spouse** was 65 or older ☐ Blind — **Enter number of boxes checked** ► 20a ☐		
	b	If you are married filing separately and your spouse itemizes deductions, see instructions and check here ► 20b ☐		
	21	Enter the **standard deduction** for your filing status. **But** see instructions if you checked any box on line 20a or 20b **or** if someone can claim you as a dependent. • Single – $4,300 • Married filing jointly or Qualifying widow(er) – $7,200 • Head of household – $6,350 • Married filing separately – $3,600	21	7,200.
	22	Subtract line 21 from line 19. If line 21 is more than line 19, enter 0	22	37,913.
	23	Multiply $2,750 by the total number of exemptions claimed on line 6d	23	8,250.
	24	Subtract line 23 from line 22. If line 23 is more than line 22, enter 0. This is your **taxable income** ►	24	29,663.
Tax, credits, and payments	25	Find the tax on the amount on line 24 (see instructions)	25	4,451.
	26	Credit for child and dependent care expenses. Attach Schedule 2 — 26		
	27	Credit for the elderly or the disabled. Attach Schedule 3 — 27		
	28	Child tax credit (see instructions) — 28 — 500.		
	29	Education credits. Attach Form 8863 — 29		
	30	Adoption credit. Attach Form 8839 — 30		
	31	Add lines 26 through 30. These are your **total credits**	31	500.
	32	Subtract line 31 from line 25. If line 31 is more than line 25, enter 0	32	3,951.
	33	Advance earned income credit payments from Form(s) W-2	33	
	34	Add lines 32 and 33. This is your **total tax** ►	34	3,951.
	35	Total federal income tax withheld from Forms W-2 and 1099 — 35 — 5,047.		
	36	1999 estimated tax payments and amount applied from 1998 return — 36		
	37a	**Earned income credit.** Attach Schedule EIC if you have a qualifying child — 37a		
	b	Nontaxable earned income: amount ► ______ and type ► ______		
	38	Additional child tax credit. Attach Form 8812 — 38		
	39	Add lines 35, 36, 37a and 38. These are your **total payments** ►	39	5,047.
Refund	40	If line 39 is more than line 34, subtract line 34 from line 39. This is the amount you **overpaid**	40	1,096.
Have it directly deposited! See instructions and fill in 41b, 41c, and 41d.	41a	Amount of line 40 you want **refunded to you**	41a	1,096.
	b	Routing number ► ______ c Type: ☐ Checking ☐ Savings		
	d	Account number ► ______		
	42	Amount of line 40 you want **applied to your 2000 estimated tax** — 42		
Amount you owe	43	If line 34 is more than line 39, subtract line 39 from line 34. This is the **amount you owe.** For details on how to pay, see instructions	43	
	44	Estimated tax penalty (see instructions) — 44		

Sign Here

Under penalties of perjury, I declare that I have examined this return and accompanying schedules and statements, and to the best of my knowledge and belief, they are true, correct, and accurately list all amounts and sources of income I received during the tax year. Declaration of preparer (other than the taxpayer) is based on all information of which the preparer has any knowledge.

Joint return? See instructions. Keep a copy for your records.

Your Signature	Date	Your Occupation	Daytime Telephone Number (optional)
		TRUCK DRIVER	
Spouse's Signature. If Joint Return, **Both** Must Sign.	Date	Spouse's Occupation	
		BANK TELLER	

Paid Preparer's Use Only

Preparer's Signature ►	Date	Check if self-employed ☐	Preparer's SSN or PTIN
Firm's Name (or yours if self-employed) and Address ► SELF-PREPARED		EIN	
		ZIP Code	

FDIA1312 11/10/99 Form **1040A** (1999)

Sales Tax

Objective 5

Some cities and most states have a sales tax. The tax is usually a specified percent of the total value of the merchandise sold. If the retailer's cash register does not automatically calculate the tax, the state often furnishes a table to place on the cash register. This table shows the clerk how much sales tax to add to the price of the merchandise. This eliminates the necessity of calculating the sales tax for each sale. A 6 percent sales tax table is shown below.

Amount of Sale	Tax	Amount of Sale	Tax	Amount of Sale	Tax
.00 - .09	.00	8.84 - 9.09	.54	17.84 - 18.09	1.08
.10 - .16	.01	9.10 - 9.16	.55	18.10 - 18.16	1.09
.17 - .33	.02	9.17 - 9.33	.56	18.17 - 18.33	1.10
.34 - .50	.03	9.34 - 9.50	.57	18.34 - 18.50	1.11
.51 - .66	.04	9.51 - 9.66	.58	18.51 - 18.66	1.12
.67 - .83	.05	9.67 - 9.83	.59	18.67 - 18.83	1.13
.84 - 1.09	.06	9.84 - 10.09	.60	18.84 - 19.09	1.14
1.10 - 1.16	.07	10.10 - 10.16	.61	19.10 - 19.16	1.15
1.17 - 1.33	.08	10.17 - 10.33	.62	19.17 - 19.33	1.16
1.34 - 1.50	.09	10.34 - 10.50	.63	19.34 - 19.50	1.17
1.51 - 1.66	.10	10.51 - 10.66	.64	19.51 - 19.66	1.18
1.67 - 1.83	.11	10.67 - 10.83	.65	19.67 - 19.83	1.19
1.84 - 2.09	.12	10.84 - 11.09	.66	19.84 - 20.09	1.20
2.10 - 2.16	.13	11.10 - 11.16	.67	20.10 - 20.16	1.21
2.17 - 2.33	.14	11.17 - 11.33	.68	20.17 - 20.33	1.22
2.34 - 2.50	.15	11.34 - 11.50	.69	20.34 - 20.50	1.23
2.51 - 2.66	.16	11.51 - 11.66	.70	20.51 - 20.66	1.24
2.67 - 2.83	.17	11.67 - 11.83	.71	20.67 - 20.83	1.25
2.84 - 3.09	.18	11.84 - 12.09	.72	20.84 - 21.09	1.26
3.10 - 3.16	.19	12.10 - 12.16	.73	21.10 - 21.16	1.27
3.17 - 3.33	.20	12.17 - 12.33	.74	21.17 - 21.33	1.28
3.34 - 3.50	.21	12.34 - 12.50	.75	21.34 - 21.50	1.29
3.51 - 3.66	.22	12.51 - 12.66	.76	21.51 - 21.66	1.30
3.67 - 3.83	.23	12.67 - 12.83	.77	21.67 - 21.83	1.31
3.84 - 4.09	.24	12.84 - 13.09	.78	21.84 - 22.09	1.32
4.10 - 4.16	.25	13.10 - 13.16	.79	22.10 - 22.16	1.33
4.17 - 4.33	.26	13.17 - 13.33	.80	22.17 - 22.33	1.34
4.34 - 4.50	.27	13.34 - 13.50	.81	22.34 - 22.50	1.35
4.51 - 4.66	.28	13.51 - 13.66	.82	22.51 - 22.66	1.36
4.67 - 4.83	.29	13.67 - 13.83	.83	22.67 - 22.83	1.37
4.84 - 5.09	.30	13.84 - 14.09	.84	22.84 - 23.09	1.38
5.10 - 5.16	.31	14.10 - 14.16	.85	23.10 - 23.16	1.39
5.17 - 5.33	.32	14.17 - 14.33	.86	23.17 - 23.33	1.40
5.34 - 5.50	.33	14.34 - 14.50	.87	23.34 - 23.50	1.41
5.51 - 5.66	.34	14.51 - 14.66	.88	23.51 - 23.66	1.42
5.67 - 5.83	.35	14.67 - 14.83	.89	23.67 - 23.83	1.43
5.84 - 6.09	.36	14.84 - 15.09	.90	23.84 - 24.09	1.44
6.10 - 6.16	.37	15.10 - 15.16	.91	24.10 - 24.16	1.45
6.17 - 6.33	.38	15.17 - 15.33	.92	24.17 - 24.33	1.46
6.34 - 6.50	.39	15.34 - 15.50	.93	24.34 - 24.50	1.47
6.51 - 6.66	.40	15.51 - 15.66	.94	24.51 - 24.66	1.48
6.67 - 6.83	.41	15.67 - 15.83	.95	24.67 - 24.83	1.49
6.84 - 7.09	.42	15.84 - 16.09	.96	24.84 - 25.09	1.50
7.10 - 7.16	.43	16.10 - 16.16	.97	25.10 - 25.16	1.51
7.17 - 7.33	.44	16.17 - 16.33	.98	25.17 - 25.33	1.52
7.34 - 7.50	.45	16.34 - 16.50	.99	25.34 - 25.50	1.53
7.51 - 7.66	.46	16.51 - 16.66	1.00	25.51 - 25.66	1.54
7.67 - 7.83	.47	16.67 - 16.83	1.01	25.67 - 25.83	1.55
7.84 - 8.09	.48	16.84 - 17.09	1.02	25.84 - 26.09	1.56
8.10 - 8.16	.49	17.10 - 17.16	1.03	26.10 - 26.16	1.57
8.17 - 8.33	.50	17.17 - 17.33	1.04	26.17 - 26.33	1.58
8.34 - 8.50	.51	17.34 - 17.50	1.05	26.34 - 26.50	1.59
8.51 - 8.66	.52	17.51 - 17.66	1.06	26.51 - 26.66	1.60
8.67 - 8.83	.53	17.67 - 17.83	1.07	26.67 - 26.83	1.61

6% FLORIDA SALES TAX CHART

Amount of Sale	Tax	Amount of Sale	Tax	Amount of Sale	Tax
26.84 - 27.09	1.62	35.17 - 35.33	2.12	43.51 - 43.66	2.62
27.10 - 27.16	1.63	35.34 - 35.50	2.13	43.67 - 43.83	2.63
27.17 - 27.33	1.64	35.51 - 35.66	2.14	43.84 - 44.09	2.64
27.34 - 27.50	1.65	35.67 - 35.83	2.15	44.10 - 44.16	2.65
27.51 - 27.66	1.66	35.84 - 36.09	2.16	44.17 - 44.33	2.66
27.67 - 27.83	1.67	36.10 - 36.16	2.17	44.34 - 44.50	2.67
27.84 - 28.09	1.68	36.17 - 36.33	2.18	44.51 - 44.66	2.68
28.10 - 28.16	1.69	36.34 - 36.50	2.19	44.67 - 44.83	2.69
28.17 - 28.33	1.70	36.51 - 36.66	2.20	44.84 - 45.09	2.70
28.34 - 28.50	1.71	36.67 - 36.83	2.21	45.10 - 45.16	2.71
28.51 - 28.66	1.72	36.84 - 37.09	2.22	45.17 - 45.33	2.72
28.67 - 28.83	1.73	37.10 - 37.16	2.23	45.34 - 45.50	2.73
28.84 - 29.09	1.74	37.17 - 37.33	2.24	45.51 - 45.66	2.74
29.10 - 29.16	1.75	37.34 - 37.50	2.25	45.67 - 45.83	2.75
29.17 - 29.33	1.76	37.51 - 37.66	2.26	45.84 - 46.09	2.76
29.34 - 29.50	1.77	37.67 - 37.83	2.27	46.10 - 46.16	2.77
29.51 - 29.66	1.78	37.84 - 38.09	2.28	46.17 - 46.33	2.78
29.67 - 29.83	1.79	38.10 - 38.16	2.29	46.34 - 46.50	2.79
29.84 - 30.09	1.80	38.17 - 38.33	2.30	46.51 - 46.66	2.80
30.10 - 30.16	1.81	38.34 - 38.50	2.31	46.67 - 46.83	2.81
30.17 - 30.33	1.82	38.51 - 38.66	2.32	46.84 - 47.09	2.82
30.34 - 30.50	1.83	38.67 - 38.83	2.33	47.10 - 47.16	2.83
30.51 - 30.66	1.84	38.84 - 39.09	2.34	47.17 - 47.33	2.84
30.67 - 30.83	1.85	39.10 - 39.16	2.35	47.34 - 47.50	2.85
30.84 - 31.09	1.86	39.17 - 39.33	2.36	47.51 - 47.66	2.86
31.10 - 31.16	1.87	39.34 - 39.50	2.37	47.67 - 47.83	2.87
31.17 - 31.33	1.88	39.51 - 39.66	2.38	47.84 - 48.09	2.88
31.34 - 31.50	1.89	39.67 - 39.83	2.39	48.10 - 48.16	2.89
31.51 - 31.66	1.90	39.84 - 40.09	2.40	48.17 - 48.33	2.90
31.67 - 31.83	1.91	40.10 - 40.16	2.41	48.34 - 48.50	2.91
31.84 - 32.09	1.92	40.17 - 40.33	2.42	48.51 - 48.66	2.92
32.10 - 32.16	1.93	40.34 - 40.50	2.43	48.67 - 48.83	2.93
32.17 - 32.33	1.94	40.51 - 40.66	2.44	48.84 - 49.09	2.94
32.34 - 32.50	1.95	40.67 - 40.83	2.45	49.10 - 49.16	2.95
32.51 - 32.66	1.96	40.84 - 41.09	2.46	49.17 - 49.33	2.96
32.67 - 32.83	1.97	41.10 - 41.16	2.47	49.34 - 49.50	2.97
32.84 - 33.09	1.98	41.17 - 41.33	2.48	49.51 - 49.66	2.98
33.10 - 33.16	1.99	41.34 - 41.50	2.49	49.67 - 49.83	2.99
33.17 - 33.33	2.00	41.51 - 41.66	2.50	49.84 - 50.09	3.00
33.34 - 33.50	2.01	41.67 - 41.83	2.51		
33.51 - 33.66	2.02	41.84 - 42.09	2.52	55.09 -	3.30
33.67 - 33.83	2.03	42.10 - 42.16	2.53	60.09 -	3.60
33.84 - 34.09	2.04	42.17 - 42.33	2.54	65.09 -	3.90
34.10 - 34.16	2.05	42.34 - 42.50	2.55	70.09 -	4.20
34.17 - 34.33	2.06	42.51 - 42.66	2.56	75.09 -	4.50
34.34 - 34.50	2.07	42.67 - 42.83	2.57	80.09 -	4.80
34.51 - 34.66	2.08	42.84 - 43.09	2.58	85.09 -	5.10
34.67 - 34.83	2.09	43.10 - 43.16	2.59	90.09 -	5.40
34.84 - 35.09	2.10	43.17 - 43.33	2.60	95.09 -	5.70
35.10 - 35.16	2.11	43.34 - 43.50	2.61	100.09 -	6.00

6% FLORIDA SALES TAX CHART

The sales tax is found by locating the appropriate bracket of "Amount of Sale" and then reading the tax to the right. For example, the total amount of sale ($34) is found in the bracket $33.84–$34.09. The sales tax is $2.04.

ASSIGNMENT Chapter 18

Name ______________________ **Date** ______________ **Score** ____________

Skill Problems

Select the best term for the explanation.

A. Assessed Valuation B. Mills C. Customs Duty D. Ad Valorem Duty
E. Specific Duty F. Compound Rate G. Taxable Income H. Dependents
I. Exemptions J. Deductions K. Standard Deduction

1. ___D___ A type of customs duty that is a percent of the market value of the item.
2. ___B___ A means of expressing one thousandth.
3. ___E___ A customs duty that is a flat amount per unit.
4. ___K___ The amount subtracted from wage earners income that needs no verification.
5. ___G___ The amount of income left after subtracting the amount for exemptions and deductions.
6. ___A___ The base on which property taxes are computed.
7. ___C___ A tax on imported goods; imposed by the federal government.
8. ___F___ The combination of a specific rate and an ad valorem rate on imported goods.

Determine the property taxes for problems 9 through 13.

	Assessed Valuation	Tax Rate (in Mills)	Property Taxes
9.	$47,000	18.6	*$874.20 (47 × 18.6 =)*
10.	62,500	31.06	*1,941.25 (62.5 × 31.06 =)*
11.	128,000	17.3	*2,214.40 (128 × 17.3 =)*
	Assessed Valuation	**Tax Rate (in Hundredths)**	**Property Taxes**
12.	51,300	4.61	*2,364.93 (513 × 4.61 =)*
13.	216,400	5.37	*11,620.68 (2,164 × 5.37 =)*

Complete the exercises below by determining the missing value. Round off your answer to two decimals.

	Market Value	Assessed Valuation	Tax Rate	Property Taxes	
14.	$ 210,000	$125,500	*13.07* mills	$1,640	*($1,640 ÷ 125.5 =)*
15.	1,450,000	750,780	2.70 hundredths	*20,271.06*	*(7,507.8 × 2.70 =)*
16.	38,400	*20,639.81*	24.82 mills	512.28	*(512.28 ÷ 24.82 × 1,000 =)*
17.	127,500	60,700	22.64 mills	*1,374.25*	*(60.7 × 22.64 =)*
18.	734,650	*425,000*	1.46 hundredths	6,205.00	*(6,205 ÷ 1.46 × 100 =)*
19.	26,000	12,800	*7.43* hundredths	951.50	*(951.50 ÷ 128 =)*
20.	195,100	91,300	6.15 hundredths	*5,614.95*	*(913 × 6.15 =)*
21.	89,500	39,600	4.12 mills	*163.15*	*(39.6 × 4.12 =)*

Determine the tax rate in mills for each community below. Round off your answers to the nearest hundredth of a mill or cent.

	Community	Total Assessed	Budget	Tax Rate (in Mills)
22.	Corpus Christi	$462,800,000	$15,900,000	*34.36 (15,900,000 ÷ 462,800,000 =)*
23.	Marlette	8,963,800	193,500	*21.59 (193,500 ÷ 8,963,800 =)*

	Community	Total Assessed	Budget	Tax Rate (in Hundredths)
24.	Wheaton	83,000,000	1,506,000	*1.81 (1,506,000 ÷ 83,000,000 =)*
25.	Holt	72,923,244	2,151,000	*2.95 (2,151,000 ÷ 72,923,244 =)*
26.	Bradon	51,800,000	1,408,000	*2.72 (1,408,000 ÷ 51,800,000 =)*

	Community	Total Assessed Valuation	Budget	Tax Rate		
27.	Dester	$ 4,800,000	$ 121,500	*25.31*	mills	*(121,500 ÷ 4,800,000 × 1,000 =)*
28.	Ballard	675,000,000	9,078,750	*13.45*	mills	*(9,078,750 ÷ 675,000,000 × 1,000 =)*
29.	Daly Waters	6,400,000	89,000	*13.91*	mills	*(89,000 ÷ 6,400,000 × 1,000 =)*
30.	Saltillo	1,100,000	18,000	*16.36*	mills	*(18,000 ÷ 1,100,000 × 1,000 =)*
31.	Joston	14,650,000	650,000	*4.44*	hundredths	*(650,000 ÷ 14,650,000 × 100 =)*
32.	Meers	7,900,000	217,250	*2.75*	hundredths	*(217,250 ÷ 7,900,000 × 100 =)*
33.	Santa Fe	95,600,000	3,540,000	*3.70*	hundredths	*(3,540,000 ÷ 95,600,000 × 100 =)*
34.	Cotler	10,150,000	261,400	*2.58*	hundredths	*(261,400 ÷ 10,150,000 × 100 =)*

Determine the amount of customs duties in the exercises below.

	Units Purchased	Value Per Unit	Ad Valorem Rate (%)	Specific Rate/Unit	Total Customs Duties	
35.	950	$ 7.45	$1\frac{1}{2}$	$ .30	*$391.16*	*(950 × 7.45 × .015 =) + (950 × .30 =)*
36.	1,200	9.95	6	1.05	*1,976.40*	*(1,200 × 9.95 × .06) + (1,200 × 1.05 =)*
37.	800	25.20	0	.30	*240.00*	*(800 × .3 =)*
38.	355	33.45	$3\frac{1}{2}$	.25	*504.37*	*(355 × 33.45 × .035 =) + (355 × .25 =)*
39.	500	10.64	$4\frac{3}{4}$	—	*252.70*	*(500 × 10.64 × .0475 =)*
40.	50	246.80	3	—	*370.20*	*(50 × 246.8 × .03 =)*
41.	400	6.81	—	3.10	*1,240.00*	*(400 × 3.1 =)*
42.	650	49.30	8	1.25	*3,376.10*	*(650 × 49.3 × .08 =) + (650 × 1.25 =)*

Determine the total cost of the merchandise purchased, including customs duties, in problems 43 through 47.

	Quantity Purchased	Market Value/Unit	Duties Ad Valorem	Specific		Total Cost
43.	500	$ 6.90	12.3% *(3,450 × .123 =)*	—		*$3,874.35*
44.	25	85.00	— *(25 × .64 =)*	$ .64		*2,141.00*
45.	50	2.00	10 *(100 × .10 =)*	.60	*(50 × .60 =)*	*140.00*
46.	93	8.50	17 *(790.50 × .17 =)*	1.23	*(93 × 1.23 =)*	*1,039.28*
47.	600	23.65	— *(600 × 3.62 =)*	.47		*14,472.00*

Determine the income tax from the information below. Use the tax table indicated from within your chapter. Compare itemized deductions to standard deduction to compute taxable earnings.

Gross Earnings	Exemptions	Itemized Deductions	Taxable Earnings	Tax Table	Income Tax
$47,581	3	$ 3,785	48. *$32,131*	Married-joint	49. *$4,819*
55,880	2	14,921	50. *35,459*	Married-joint	51. *5,321*
40,954	1	2,910	52. *33,904*	Single	53. *6,152*
77,821	5	26,182	54. *37,889*	Married-joint	55. *5,681*
49,540	4	4,985	56. *32,190*	Head of household	57. *4,826*

Gross Earnings	Exemptions	Itemized Deductions	Taxable Earnings	Tax Table	Income Tax
$50,600	3	$ 5,450	58. *$35,150*	Married-Joint	59. *$5,276*
60,456	4	17,450	60. *32,006*	Married-Joint	61. *4,804*
75,460	1	36,640	62. *36,070*	Single	63. *6,754*
54,900	4	11,543	64. *32,357*	Single	65. *5,718*
45,500	2	2,349	66. *32,800*	Married-Joint	67. *4,924*
70,840	6	15,681	68. *38,659*	Head of household	69. *6,338*
58,650	5	2,876	70. *38,550*	Head of household	71. *6,310*
39,681	1	1,987	72. *32,631*	Single	73. *5,788*

Business Application Problems

1. Ester will import 500 silk accessories at a cost of $3.86 each. The Harmonized Tariff Schedule indicates an ad valorem rate of 8 percent and a specific duty of $.32 on these items. What is the total amount of customs duties Ester will pay on the 500 items?

$3.86 × 500 × .08 = $154.40

$.32 × 500 = $160 + $154.40 = $314.40 duty

2. What is the cost per unit in problem 1, including the customs duties?

$3.86 × .08 = $.3088

$3.86 + $.3088 + $.32 = $4.49 each

3. If the assessed valuation of an average home in a small community is $65,000 and the property tax rate is 27.04 mills, what is the annual property tax?

65 × $27.04 = $1,757.60 tax

4. If the community in problem 3 were to approve a school millage increase of 2.1 mills, what would be the additional amount of property tax per year?

65 × $2.10 = $136.50 additional

5. Seaboard Limited will either import the needed 6,000 plastic brackets at $5.21 each plus an ad valorem duty of 7 percent from Indonesia or have them made in Mexico for $5.49 each, duty free. What is the lowest total price for the 6,000 units?

$5.21 × 6,000 × .07 = $2,188.20

$5.21 × 6,000 = $31,260 + $2,188.20 = $33,448.20

$5.49 × 6,000 = $32,940 Mexico

6. How much will Seaboard Limited save in problem 5 by using the lowest total price?

$33,448.20 − $32,940 = $508.20 saved

7. Baldwin's Store operates in a state with a 5 percent sales tax. Over the period of the past month the store has collected $5,836.60 in sales taxes. What is the amount of merchandise sales for the store in the past month?

$5,836.60/.05 = $116,732. sales

8. Bullton has an annual budget of $1,089,500. This amount will be raised through property taxes. The total assessed valuation is $39,748,500. To the nearest hundredth, what is the tax rate in mills in Bullton?

$1,089,500/$39,748,500 = .027409 = 27.41 mills

9. The Meuller Grocery Store property in Bullton is assessed at $212,000. What is the amount of property tax per year on the property? (Use your answer from problem 8 in your calculations.)

212 × $27.41 = $5,810.92 tax

10. First Class Importers bought 50 walnut boxes that cost $5,310, including customs duties. The ad valorem duty was $810 of the total. What was the ad valorem tax rate?

$5,310 − $810 = $4,500 $810/$4,500 = 18% rate

11. Nancy imported 60 woven baskets from Peru at a cost of $18 each. The tariff on the baskets was a compound rate of $.625 specific and 12 percent ad valorem. If Nancy distributes the tariff over the 60 items, what will her cost be per unit?

$18 each
×60 units × $.625 = $37.50 specific
$1,080 × .12 = $129.60 ad valorem

$1,080.00
37.50
+129.60
$1,247.10 ÷ 60 = $20.79 per unit

12. You have been appointed tax clerk for your township. Your first assignment is to determine the tax rate on property. The total budget amount is $1,428,400. The market value of all of the business and residential property is $137,000,000. The assessed valuation of the property in the township is $61,330,000. What is the tax rate in mills (nearest hundredth)?

$$\frac{\$1{,}428{,}400 \text{ budget}}{\$61{,}330{,}000 \text{ assessed valuation}} = .0232903$$

= 23.29 mills

13. The tax rate in your community is 29 mills. Your property tax is $1,073. What is the assessed valuation of your property?

$1,073 ÷ .029 = $37,000 assessed valuation

14. Jeremy is single. He earned $39,880 during the year. His total itemized deductions are $826 for the year. His W-2 form stated that he had paid a total of $5,260 in federal income tax for the year. What is the amount that he still owes the I.R.S.?

$39,880
−2,750 dependent exemption
−4,300 standard deduction
$32,830 taxable income

$5,844 tax
−5,260 paid in
$584 owed

15. Tom and Nancy Grubbs are interested in estimating the property taxes on their new home. The home will cost $79,900. The city is assessing property at approximately 46 percent of market value. The tax rate in the community is 28.4 mills. What is the estimated property tax?

$79,900
×.46
$36,754 approximate assessed valuation

$36,754
×.0284
$1,043.81 estimated property taxes

16. Christina Ferrara purchased a used car that cost $9,800. She found that the car was subject to a 5 percent state sales tax and a 1 percent city sales tax. She plans to put $3,000 down on the purchase and ask the credit union for a loan on the balance. What will be the amount of her loan request?

$9,800 × .06 = $588

$9,800
+588
−3,000
$7,388 loan request

17. The Pear Tree Gift Shop of Alta must make a choice between the purchase of two items to fill their open-to-buy budget. The first item is a locally woven evening shawl, priced at $40 each. The other item is an Irish woolen jacket priced at $30. The Irish jacket must be imported from Dublin. An ad valorem duty of 20 percent and a specific duty of $5 must be added to the price of the imported jacket. There is a state sales tax of 5 percent on merchandise in Alta. What is the price to the customer, after taxes, of the lowest priced item?

Shawl
$40
+2 sales tax
$42 total

Jacket
$30
6 = 20% ad valorem
+5 = Specific
$41.00
+2.05 sales tax
$43.05 total

The shawl is the lowest priced item.

18. The Eye Center of Braton increased the space of their office building by 60 percent from the previous structure by making an addition. The property taxes are based on a tax rate of 23.1 mills. Last year the property tax was \$3,950. Estimate the assessed valuation for the completed building.

\$3,950 ÷ .0231 = \$170,995.67

\$170,995.67 last year's assessed valuation
×1.60
\$273,593.07 this year's estimated assessed valuation

19. Jim and Rose earned \$48,400 in the last year. They recently had a baby girl. They will claim \$7,240 in itemized deductions on this year's income tax return. If they file a joint return, what is the tax amount?

\$48,400
−8,250 for exemptions
−7,240 itemized deductions
\$32,910 taxable income

\$4,939 (from table) tax

20. Buelah Swarts paid \$3,985 income tax for the year. Her gross earnings were \$25,508. What was the income tax rate on the gross earnings (nearest percent)?

$$\frac{\$3{,}985}{\$25{,}508} = 16\% \text{ tax rate}$$

21. The tax on property in the community of Mahosic Notch is 32¢ per \$100 of assessed valuation. The taxable property within the community has an assessed valuation of \$82,970,570. The treasurer has found that the delinquency rate is usually $1\frac{1}{2}$ percent of the total tax revenue. What is the amount of the delinquency for the community of Mahosic Notch for the year?

829,705.7 × \$.32 = \$265,505.82
×.015
\$3,982.59 delinquent

22. Write a complete sentence in answer to this question. When Jonny received the invoice for the crate of Chinese baskets, there was a total due of \$390. The invoice price included \$70 for customs duties. The specific duties totaled \$2.80. The remainder were ad valorem duties. What was the percent of ad valorem duties?

\$390 − \$70 = \$320 amount of merchandise
\$70 − \$2.80 = \$67.20 ad valorem tax

$$\frac{\$67.20}{\$320} = 21\%$$

The ad valorem duty was 21 percent.

23. Write a complete sentence in answer to this question. Carl noted that if he bought a car in his home town he would have to pay a sales tax of 7 percent. He could travel to a nearby city and purchase the same car. The sales tax in the other city is 5 percent. The cost of the trip would be \$65. The car that he is interested in will cost \$14,800. How much will he save by making the trip?

\$14,800	*\$296*
×.02	*−65*
\$296	*\$231*

He saved \$231 by making the trip.

★ Challenge Problem

The Yankee Trader, an import shop, purchased 100 figurines from a dealer in London for \$18 apiece list price. Because it was such a large purchase, it was sold by the London dealer at a 10 percent discount, F.O.B. destination. The duty on the purchase was 25 percent ad valorem and \$3 specific. The items will be sold at a markup of 40 percent on cost. A 3 percent sales tax will be added to the selling price. What is the amount that a consumer will have to pay for a pair of figurines?

\$18.00 list price
−1.80 trade discount
\$16.20 selling price
+4.05 ad valorem
+3.00 specific
\$23.25 dealer cost per unit

\$23.25
C + M = S
100% + 40% = 140%
\$23.25 × 1.4 = \$32.55 selling price per unit
×2
\$65.10 per pair
+1.95 sales tax
\$67.05 price to customer per pair

MASTERY TEST Chapter 18

Name ____________________ **Date** ____________ **Score** ____________

Match the following terms and concepts.

Answer	Term		Concept
H	**1.** Ad Valorem Duty	A.	City- and/or state-imposed tax on retail sales
E	**2.** Taxable income	B.	Amount allowed for each wage earner and dependent
D	**3.** Mills	C.	Both a percent of the market value and a set amount
B	**4.** Exemption	D.	Thousandths of assessed valuation
G	**5.** Specific Duty	E.	Gross income less dependent exemptions and deductions
F	**6.** Dedutions	F.	Itemized allowable expenses
A	**7.** Sales Tax	G.	A set amount per unit, weight, or measure
		H.	Market value of imported item times percent
		I.	A maximum of 6 percent

Determine the amount of property tax in the problems below.

	Assessed Valuation	Tax Rate (in Mills)	Property Taxes
8.	$42,800	38.05 (In hundredths)	*$1,628.54*
9.	28,900	4.05	*1,170.45*

Determine the cost of the purchases to the retailer in the problems below.

	Quantity Purchased	Market Value/ Unit	Duties	Total Cost
10.	144	$16.00	15% ad valorem, $1.70 specific	*$2,894.40*
11.	50	17.58	10% ad valorem	*966.90*

Determine the income tax from the information below.

	Gross Earnings	Exemptions	Itemized Deductions	Tax Table	Income Tax
12.	$41,280	1	$4,953	Single	*$6,054*
13.	53,850	3	1,650	Married-Jointly	*5,764*

14. Ureel Township has established a $2,680,500 budget for the year. The total assessed valuation is $92,426,000. The township is expecting a 1 percent default on tax payments. The 1 percent amount must be added to the budget. What will be the property taxes on a home with an assessed valuation of $41,600? Compute the tax rate to the nearest hundredth of a mill.

$2,680,500 × .01 = $26,805

$2,680,500 + 26,805 = $2,707,305

$$\frac{\$2,707,305}{\$92,426,000} = .0292915 = 29.29 \text{ mills}$$

41.6 × $29.29 = $1,218.46 tax

INVESTMENTS

19

Suggested Classroom Time for This Chapter

hours 1 2 3 4 5

Perceived Student Difficulty with Topical Material

easy moderate difficult

OBJECTIVES

After mastering the material in this chapter, you will be able to:

1. Calculate the rate of return on various types of investments. p. 374
2. Calculate the dividend rate on common stock when the firm has preferred stock outstanding. p. 375
3. Calculate the rate of return on a bond. p. 378
4. Understand and use the following terms and concepts:
   Risk
   Liquidity
   Cost of Investment
   Rate of Return
   Inflation
   Common Stock
   Preferred Stock
   Dividends
   Participating
   Cumulative
   Secured Bonds
   Debentured Bonds
   Registered Bonds
   Bearer Bonds
   Government Bonds

SPORT-A-MERICA

As the president of Sport-A-Merica, Kerry is considering selling the old building they paid $26,000 for, which is now being used primarily for storage. Hal agrees that if they could sell the building for $38,000 or more, they could rent mini storage for warehouse space, and then they could invest the balance in securities that have a return of more than 9 percent. On the other hand, the value of the building has increased over the time it has been owned by Sport-A-Merica. If the firm holds on to the building, Kerry feels that its value will substantially increase over the next few years as development continues in the area.

One of Kerry's business friends needs an office building for his firm. If Kerry waits to sell the building, will there be a ready buyer for it? A valuable piece of property has only so many buyers at any given point in time. If the property is sold now, there will not be a need for a real estate agent, which would add a 7 percent commission to the sale. The commission would reduce the amount of profit and the rate of return to Sport-A-Merica. Both Kerry and Hal know they will have to make this decision very soon because the buyer will not wait much longer.

Risk, Liquidity, and Cost of Investment

The chapter opener is a typical example of the type of decisions that a business person must make. Investments require thinking!

risk

❍ *Limited risk usually means limited reward.*

Investors consist of individuals, as well as businesses. The investor expects to increase the initial value of the investment and thereby earn a profit. The **risk,** or chance of no profit—or even a loss—must also be considered. When the profit potential is high, the risk of loss is frequently equally high. When the profit potential is low, the risk of loss is usually also low.

liquidity

Another consideration for the investor is that of liquidity. **Liquidity** measures the ease with which an investment can be exchanged for cash. Stocks and bonds can easily be sold, usually on the same day the decision is made to do so, for cash payment three days later. Therefore, they have high liquidity. Real estate or equipment may take several days or even months before it is sold and therefore is considered to have lower liquidity. If an investor wishes to move from one investment into another, liquidity may be a very important factor.

cost of investment

Most investments have a fee or service charge that covers the cost of investing. The **cost of investment** may range from no cost for savings accounts in banks or credit unions, to up to 10 percent for real estate. The fee or service charge may be levied when the investment is bought, when the investment is sold, or both.

Rate of Return
Objective 1

Profit earned on the investment compared to the amount invested determines the **rate of return,** which is expressed as a percent. Savings accounts that have a 4 percent annual interest rate have a 4 percent rate of return if there is no service charge for investing. The service charges for investing must be subtracted from the profit in order to calculate the rate of return.

The Effect of Inflation

Inflation is a general increase in the cost or price of goods or services. In times of inflation, prices rise. The purchasing power of the dollar lessens in comparison to that of earlier time periods. Unless the investor's rate of return can keep up with the inflation rate, money (purchasing power) will actually be lost. However, the loss would have been greater if the money had simply been held; and therefore, the investor would have been at the mercy of the inflation rate. The investment's rate of return reduces the effect of inflation on the buying power of individuals and firms.

Stocks

● ***Appendix E is very useful to show students how to read stock prices, dividend yields, etc.***

The ownership or equity of a corporation is held by stockholders. Stockholders are individuals or other firms who have bought a portion of a corporation by purchasing shares of stock in the corporation (see appendix E, "Reading Market Quotations for Stocks, Bonds, and Mutual Funds").*

Stock owners' rights consist of variations in the basic rights listed below. The variations, if any, are spelled out in the stock prospectus.

1. The right to vote.
2. The right to share in the distribution of the corporation's earnings.
3. The right to maintain an equal share of ownership of the corporation if the firm decides to issue additional stock (termed *preemptive right*).
4. The right to share in distribution of the assets of the corporation on its dissolution.

There are two classes of stock: common and preferred. Their owners' rights differ in principle, and in many cases, by corporation.

Common Stock

Common stock owners have a vote in the corporate stockholders' annual meetings. They may have a return on investment in the form of an increase in the stock price or in the form of a dividend.

*For more information, write the Consumer Information Center, Pueblo, Colorado 81009, for a free pamphlet titled "Investigate Before You Invest."

Preferred Stock

Preferred stock owners usually do not have a vote in the corporate stockholders' meetings. The term *preferred* refers to the rights of the holder to the dividends issued, and to the assets of the corporation if it dissolves. Preferred stockholders are paid before common stockholders are paid. If there is a limited amount of assets or earnings to be distributed to stockholders, the preferred stockholders are paid first. Any amount left over after the preferred stockholders are paid is then distributed to common stockholders.

Objective 2

Dividends

Earnings of a corporation are usually split among the stockholders in the form of **dividends,** or they are held for future expansion of the firm. The manner in which the split is made is dependent on a variety of factors. Most of the factors are listed here.

1. The amount of liquid assets the firm can use to make the dividend payment.
2. The specific rights of the preferred stockholders.
3. The amount of preferred stock outstanding.
4. The amount of the preferred stock dividend (determined at the point issued).
5. The amount of common stock outstanding.

The decision to pay a dividend rests with the board of directors. The decision can be made quarterly, semiannually, or annually. The common method is quarterly. The decision may be announced at the stockholders' meeting, or it may be simply announced through the media. When the board decides to declare a dividend, the date is termed the *declaration date*. The dividend is to be paid in the future, based on the *date of record*. This date is used to determine who actually owns the stock. Those who own stock in the corporation as of the date of record will receive a dividend on the *payment date*. The corporation must compile a list of names and addresses of the current stockholders as of the date of record.

Dividends are usually paid as cash dividends. The corporation distributes the cash dividend in the form of a check to the stockholders of record.

Stock dividends are distributed in the form of an additional stock certificate. The advantage of a stock dividend is that it is not subject to income tax until it is sold.

Example 1 The Eltron Corporation has the following stock outstanding as of August 31.

$5 Preferred stock 160,000 shares ($5 dividend determined when issued)
Common stock 800,000 shares

Earnings for the year ended August 31 were $1,280,000. Determine the amount of the dividend per share on the common stock if the board decides to distribute all of the earnings.

Income for the year	$1,280,000
160,000 preferred shares at $5 dividend per share	−800,000
Amount to be paid to common stock shareholders	$480,000

$$800{,}000 \text{ shares }\overline{)\$480{,}000 \text{ for common stockholders}}\quad \$.60 \text{ dividend per share—common stock}$$

(480,000 ÷ 800,000 =)

Example 2 Three years ago, Marta and John Lopez bought 400 shares of Eltron Corporation for $24 per share. What rate of return are they earning if a $.60 dividend is paid annually?

$$\frac{\$.60 = \text{return}}{\$24 = \text{cost of investment}} = 2\frac{1}{2}\% \text{ rate of return}$$

(.6 ÷ 24 =)

Exercise A

Determine the dividend per share on common stock and the rate of return in each of the exercises below. (The price per share represents an average.) Round off your answer to the nearest cent or tenth of a percent.

	Earnings To Be Distributed	Number of Preferred Shares	Preferred Dividend	Number of Common Shares	Dividend per Share—Common	Price per Share	Rate of Return—Common
1.	$2,008,000	463,500	$3.50	860,000	$.45	$9.60	4.7%
	(463,500 × 3.50 =)		(2,008,000 − 1,622,250 =)		(385,750 ÷ 860,000 =)		(.45 ÷ 9.60 =)
2.	1,268,420	120,300	5.00	1,000,000	*.67*	10.00	*6.7*
	(120,300 × 5 =)		*(1,268,420 − 601,500 =)*		*(666,920 ÷ 1,000,000 =)*		*(.67 ÷ 10 =)*
3.	1,100,000	9,870	4.00	500,000	*2.12*	30.50	*7.0*
	(9,870 × 4 =)		*(1,100,000 − 39,480 =)*		*(1,060,520 ÷ 500,000 =)*		*(2.12 ÷ 30.50 =)*
4.	4,200,000	758,900	5.00	3,600,000	*.11*	9.75	*1.1*
	(758,900 × 5 =)		*(4,200,000 − 3,794,500 =)*		*(405,500 4 3,600,000 =)*		*(.11 ÷ 9.75 =)*
5.	865,390	160,000	2.00	750,000	*.73*	8.00	*9.1*
	(160,000 × 2 =)		*(865,390 − 320,000 =)*		*(545,390 ÷ 750,000 =)*		*(.73 ÷ 8 =)*

Participating

Preferred stockholders may have an additional right: Holders of **participating** preferred stock have the right to share in the earnings that are above the stated dividend rate; that is, the right of participation. Each firm or each issue of stock can specify the terms of the participation when the stock is sold. Therefore, the degree of the rights will differ with each stock issue. Participating preferred stock has two chances at earnings.

When the stock does not participate in earnings beyond the stated rate it is termed *nonparticipating* preferred stock.

Example 3

The Kreger Corporation has two classes of stock: A $5 preferred issue with 10,000 shares outstanding, and a common issue with 40,000 shares outstanding. The preferred stock certificate states that the stock is participating after the common stockholders receive a dollar of dividends per share. The two classes then split the remaining earnings on an equal share-for-share basis. The corporation decided to retain 40 percent of the $260,000 earnings for future expansion. The remainder will be distributed in the form of cash dividends. What is the amount of dividends per share for each share of preferred stock?

$260,000
×.60
$156,000 = Earnings to be distributed
−50,000 = Preferred dividends ($5 per share × 10,000 shares)
$106,000
−40,000 = Common dividends ($1 per share × 40,000 shares)
$66,000 = Dividends to be split between preferred and common

$\frac{\$66,000}{50,000}$ = $1.32 per share (preferred and common) shares

$5 + $1.32 = $6.32 dividend per share of preferred stock

(260,000 × .60 =)(106,000 − 40,000 =)(66,000 ÷ 50,000 =)(5 + 1.32 =)

Cumulative

If a corporation does poorly during a period, the dividend it was supposed to pay preferred stockholders may be *passed;* that is, the dividend may not be paid. In a later period, the firm may do very well and be able to pay preferred stockholders the dividend that was passed in the earlier period. The firm is not obligated to pay the passed dividend unless the preferred stock certificate states that the stock class has **cumulative** rights: the rights to all dividends *in arrears* (unpaid) before the distribution of dividends to the common stockholders. This right enables the preferred stockholder to receive the dividends over a period of time, even though receipt may be interrupted because of erratic earnings.

Preferred stock without this right is termed *noncumulative*.

Example 4

The earnings of the Watson Corporation have been poor. For the past two years, it has had losses and has not paid a dividend on the 5,000 shares of $3 preferred stock or the 30,000 shares of common stock. The preferred stock certificate states that the preferred shares have the right to cumulative dividends. The earnings this year were $310,000. The firm decided to distribute 80 percent to stockholders. What will be the amount of the preferred stock dividend? What will be the amount of the common stock dividend?

$310,000
×.80
$248,000 = Earnings to be distributed
−45,000 = Preferred dividends ($3 per share × 5,000 shares × 3 years) ($3 × 3 = $9)
$203,000 = Earnings for common stock

$$\frac{\$203{,}000}{30{,}000} = \$6.77 \text{ dividend per share of common stock}$$

(310,000 × .80 =)(248,000 − 45,000 =)(203,000 ÷ 30,000 =)

Preferred stock can be both participating and cumulative. If that were the case in example 4, the preferred stockholders would first receive the dividends that have been passed, and then be eligible to receive an additional dividend to the extent of the stipulation on the certificate.

Example 5

If the preferred stock certificate stated that the class is to receive cumulative rights and participation to the extent of sharing dividends above $5 per share of common after the dividends in arrears were paid in example 4, what would be the dividend per share of preferred? The additional dividend is to be determined on a share-for-share basis.

$203,000 Earnings after the dividends were made up to the preferred stockholders
−150,000 Common dividends ($5 per share × 30,000 shares)
$53,000 Dividends to be split between preferred and common

$$\frac{\$53{,}000}{35{,}000} = \$1.51 \text{ per share (full cent)}$$
(preferred and common) shares

$3 × 3 years = $9 $9 + $1.51 = $10.51 total preferred dividend

(203,000 − 150,000 =)(53,000 ÷ 35,000 =)(3 × 3 =)(9 + 1.51 =)

Exercise B

Determine the dividend per share on preferred stock from the information below; participation is based on the number of shares of stock in both classes.

	Earnings To Be Distributed	Number of Preferred Shares	Preferred Dividend	Number of Common Shares	Cumulative (Years Passed)	Participating (After/ Common)	Dividend per Share Preferred (Full Cent)
1.	$200,000 (4,000 × 2 =)	4,000	$2.00 (10,000 × 2 =)	10,000	No	Yes/$2	$14.28 172,000 ÷ 14,000 =)
2.	360,000	10,000	7.50	10,000	One	No	***15.00*** ***(7.50 × 2 =)***
3.	80,000	5,000	5.00	20,000	One	Yes/$3	***10.00*** ***(5.00 × 2 =)***
4.	450,000 ***(10,000 × 8 =)***	10,000	8.00 ***(50,000 × 5 =)***	50,000	No	Yes/$5	***10.00*** ***(120,000 ÷ 60,000 =)***

(Continued on next page)

Exercise B Continued

	Earnings To Be Distributed	Number of Preferred Shares	Preferred Dividend	Number of Common Shares	Cumulative (Years Passed)	Participating (After/ Common)	Dividend per Share Preferred (Full Cent)
5.	120,000	3,000	10.00	20,000	Two	Yes/$6	***30.00*** **(10 × 3 =)**
6.	75,000 ***(1,000 × 5 =)***	1,000	5.00 ***(30,000 × 1 =)***	30,000	No	Yes/$1	***6.29*** ***(40,000 ÷ 31,000 =)***
7.	100,000	3,000	6.00	40,000	One	No	***12.00*** ***(6.00 × 2 =)***
8.	1,500,000 ***(50,000 × 5 × 3 =)***	50,000	5.00 ***(65,000 × 4 =)***	65,000	Two	Yes/$4	***19.26*** ***(490,000 ÷ 115,000 =)***

Bonds

Unlike stocks, which represent ownership in a corporation, **bonds** represent debt. A firm that wishes to increase working capital may choose to sell bonds instead of stock. At a designated point in the future, the bonds will need to be paid off. Hence, bonds are a debt. Bondholders have no vote in the organization but have preference over common and preferred stockholders when the income is divided. Bond interest (yield) must be paid before any dividends are paid.

● *Appendix E provides a clear explanation of bond prices and yields.*

Bonds are usually issued in denominations of $1,000, but occasionally they are issued for $2,000, $3,000, $5,000, or even $100. Bond prices are quoted on the basis of 100. A price quote of 98 would mean that a $1,000 bond is now being sold for $980. A quote of $105\frac{1}{2}$ would mean a $1,055 current market price (see appendix E, "Reading Market Quotations for Stocks, Bonds, and Mutual Funds").

Objective 3

The current interest rate in the marketplace will determine the bond's selling price. If the current interest rate is lower than the interest rate at the time the bond was originally issued, the bond's price will be higher than its face value. If the current interest rate is higher than the interest rate at the time the bond was originally sold, the price will be lower than the face value.

Example 6

The Hubble Corporation's bonds are quoted on the exchange as selling at 98. The bonds have a stated rate of interest (yield) of 5 percent. What is the current effective rate of interest on the bonds?

$$\frac{5\% \text{ of } \$1{,}000}{98\% \times \$1{,}000} = \frac{\$50}{\$980} = 5.1\% \text{ current effective rate}$$

(.05 × 1,000 =)(.98 × 1,000 =)(50 ÷ 980 =)

Example 7

If the same bonds were quoted at 110, what would be the effective rate of interest on the bonds?

$$\frac{5\% \text{ of } \$1{,}000}{110\% \times \$1{,}000} = \frac{\$50}{\$1{,}100} = 4.5\% \text{ current effective rate}$$

(.05 × 1,000 =)(1.10 × 1,000 =)(50 ÷ 1,100 =)

Commercial Bonds

Bonds are sold by corporations for many reasons. The firm may want to expand their operations by building on to their existing physical plant, or they may need additional working capital. Usually, firms sell bonds to provide long-term financing. The long-term financing provides the cash necessary to make purchases of physical plants, equipment, and vehicles, as well as to do research on new products. The term of the bonds will usually run from five to fifty years. The quality of the bonds will depend on the ability of the corporation to repay the bonds, as well as on the annual interest on the bonds. The quality also depends on the type of bonds issued:

secured bonds

Secured bonds provide the bondholder with a claim to real property. If the corporation cannot meet its obligation to repay, the bondholder has claim to specified properties.

debentured bonds

Debentured bonds do not provide the bondholder with a claim to any particular property. The bonds are issued on the general credit of the corporation.

registered bonds

Bonds may also be classified by the way in which the interest is paid. A **registered bond** must be endorsed by the new owner when it is delivered. The corporation is notified and the interest is sent to the new bondholder.

bearer bonds

Bearer bonds, or coupon bonds, have a piece of the bond certificate predated with the amount of the interest stated on the face. The holder of the bearer bond removes the coupon and presents it at the bank for payment on the payment date. These bonds do not require the owner to notify the corporation when the bond is sold. The new bond owner simply removes the interest coupon when the interest is due.

Government and Municipal Bonds

The federal government has issued **government bonds** to finance expenditures for many years. The *Series E savings bonds* are the most familiar type. They may be purchased through a payroll deduction and come in denominations of $50 or more. Unlike other bonds, the interest is paid at maturity. A $50 bond is bought for $25. In seventeen years the bond is worth $50.

Municipal bonds are sold for the purpose of building schools, streets, sidewalks, municipal-owned utilities, etc. The interest rate on municipal bonds is often lower because of the relative safety of the principal. Some municipal bonds have interest with tax-free status. Therefore, the stated rate is less. An individual in a high income tax bracket may choose municipal bonds as an investment.

Exercise C

Determine the rate of return on the following $1,000 denomination bonds. Round off your answer to the nearest tenth of a percent.

Current Price Quotation	Stated Interest Rate	Rate of Return
1. $101\frac{1}{4}$ (1.0125 × 1,000 =)	$8\frac{1}{4}$% (.0825 × 1,000 =)	8.1% (82.5 ÷ 1,012.5 =)
2. 98 *(.98 × 1,000 =)*	*7 (.07 × 1,000 =)*	*7.1 (70 ÷ 980 =)*
3. $96\frac{1}{2}$ *(.965 × 1,000 =)*	$9\frac{1}{2}$ *(.095 × 1,000 =)*	*9.8 (95 ÷ 965 =)*
4. 108 *(1.08 × 1,000 =)*	$7\frac{1}{4}$ *(.0725 × 1,000 =)*	*6.7 (72.5 ÷ 1,080 =)*
5. $102\frac{1}{4}$ *(1.0225 × 1,000 =)*	$8\frac{3}{4}$ *(.0875 × 1,000 =)*	*8.6 (87.5 ÷ 1,022.5 =)*

There are many other investment possibilities in addition to the purchase of stocks and bonds. For example, the investor who wishes to maintain the same amount of liquidity can invest in *commodity futures*. This investment opportunity is very volatile and can make the investor great sums or lose the same amount in a very short period of time. The volatility is due to the fact that the investor need only place a 5 or 10 percent *margin* down on the investment position. A slight movement is multiplied by ten or twenty times. This type of investment is risky.

Real property, such as real estate, provides a much less volatile investment opportunity, but at the same time, it is associated with a lower chance of a high return. The cost of investment is also a consideration, in addition to the small rate of return. Liquidity is also often a negative. Property, unlike stocks and bonds, is usually not liquidated the same day it is offered for sale.

The cost of investing must be considered, no matter what the investment. Real estate brokers charge rates that reflect the difficulty of liquidating the asset. The standard rate for home property is usually 7 percent. The rate may be up to 10 percent for commercial property. The *seller* pays the real estate broker's commission, which is in reality paid by the buyer through the purchase price of the property.

Other investment vehicles include rare coins, art objects, antiques, old cars, comic books, and baseball cards. Some of these objects have a ready market of buyers and sellers. Other items have trends of popularity, limited buyers and sellers, and no middlemen or brokers.

Example 8

Sport-A-Merica had purchased their old building for $26,000 in July, 2000. In July 2001, a real estate broker had a buyer interested in paying $38,000 for the same property. A 7 percent commission rate on the selling price of the property would have had to be paid by Sport-A-Merica. What would have been the profit on the year-long investment? What would have been the rate of return?

$38,000
×.07
$2,660 Commission to the real estate broker

$38,000
−2,660
$35,340 Net proceeds from sale

$35,340
−26,000 Cost
$9,340 Profit

$$\text{Rate of return} = \frac{\$9,340}{\$26,000} = 35.9\%$$

(38,000 × .07 =)(38,000 − 2,660 =)(35,340 − 26,000 =)(9,340 ÷ 26,000 =)

Example 9

Hal purchased three uncirculated coins for $590 and placed them in a safe deposit box for three years. When Hal sold them, he had to pay a 5 percent dealer fee on the sale price of $950. What was the annual rate of return using a simple interest approach?

$950.00 Sale price
×.05
$47.50 Dealer fee

$950.00
−47.50
$902.50 From sale

$902.50 Received from sale
−590.00 Cost
$312.50 Return ÷ 3 years = $104.17 return per year

$$\frac{\$104.17 \text{ Return}}{\$590.00 \text{ Cost}} = \underline{17.7\% \text{ return}}$$

(950 × .05 =)(950 − 47.50 =)(902.50 − 590 =)(104.17 ÷ 590 =)

ASSIGNMENT Chapter 19

Name ______________________ **Date** ______________ **Score** ____________

Skill Problems

Select the best term for each statement below.

A. Risk	B. Liquidity	C. Cost of Investment	D. Rate of Return
E. Inflation	F. Common Stock	G. Preferred Stock	H. Dividends
I. Participating	J. Cumulative	K. Secured Bonds	L. Debentured Bonds
M. Registered Bonds	N. Bearer Bonds	O. Government Bonds	

N **1.** Bonds owned by whomever (physically) holds them. The interest coupon is detached for payment.

L **2.** Bonds that are issued by a firm without being tied to any physical equipment or other asset.

G **3.** A type of equity instrument that pays the owner a dividend before common stock is paid.

B **4.** Assets that are easily converted into cash.

D **5.** The amount earned on an investment expressed as a percent of the amount invested.

E **6.** An increase in the price of most goods and services.

C **7.** The fee that is paid to a middleman for facilitating the transaction.

A **8.** The chance taken by a business person who makes a business decision.

F **9.** A form of equity securities that provides a vote in the decisions of the firm.

O **10.** Investment vehicles issued by local, state, or federal agencies with a possible tax-free status.

Figure the rate of return per year for the following real estate investments. Round off to two decimal places.

	Property Purchase Price	Commission Rate Paid to Broker	Length of Time Invested	Sales Price	Annual Rate of Return (%)	
11.	$ 45,000	7%	3 years	$ 64,300	*10.96%*	*(64,300 × .07 = 4,501, 64,300 − 4,501 − 45,000 = 14,799 ÷ 3 = 4,933 ÷ 45,000 =)*
12.	15,500	5	8 years	37,890	*16.53*	*(37,890 × .05 = 1,894.5, 37,890 − 1,894.5 − 15,500 = 20,495.5, 20,495.5 ÷ 8 = 2,561.94 ÷ 15,500 =)*
13.	105,200	$6\frac{1}{2}$	10 years	200,000	*7.78*	*(200,000 × .065 = 13,000, 200,000 − 13,000 − 105,200 = 81,800, 81,800 ÷ 10 = 8,180 ÷ 105,200 =)*
14.	10,000	15	5 years, 9 months	35,400	*34.94*	*(35,400 × .15 = 5,310, 35,400 − 5,310 − 10,000 = 20,090, 20,090 ÷ 5.75 = 3,493.91 ÷ 10,000 =)*
15.	50,000	8	3 years, 6 months	70,000	*8.23*	*(70,000 × .08 = 5,600, 70,000 − 5,600 − 50,000 = 14,400, 14,400 ÷ 3.5 = 4,114.29 ÷ 50,000 =)*

Determine the rate of return on common stock in the following exercises. Round off percentage to two decimal places.

	Purchase Price per Share	Dividend Earned Per Share for the Year	Rate of Return (nearest tenth)	
16.	$ 93.25	$2.50	*2.7%*	*(2.50 ÷ 93.25 =)*
17.	127.30	5.00	*3.9*	*(5 ÷ 127.30 =)*
18.	87.75	1.25	*1.4*	*(1.25 ÷ 87.75 =)*
19.	10.00	$12\frac{1}{2}$¢	*1.3*	*(.125 ÷ 10 =)*
20.	78.50	2.00	*2.5*	*(2 ÷ 78.50 =)*

Determine the amount of the dividend per share on the common stock if the board decides to distribute all of the following earnings. Round off to the nearest cent.

	Earnings To Be Distributed	Number of Preferred Shares	Preferred Dividend	Number of Common Shares	Dividend per Share Common	
21.	$ 500,000	50,000	$ 5	80,000	*$3.13*	*(500,000 − (50,000 × 5) ÷ 80,000 =)*
22.	1,250,000	10,000	6.00	150,000	*7.93*	*(1,250,000 − (10,000 × 6) ÷ 150,000 =)*
23.	3,500,000	25,000	5.50	2,000,000	*1.68*	*(3,500,000 − (25,000 × 5.5) ÷ 2,000,000 =)*
24.	750,000	0	0	60,000	*12.50*	*(750,000 ÷ 60,000 =)*
25.	360,800	5,000	10	70,000	*4.44*	*(360,800 − (5,000 × 10) ÷ 70,000 =)*

Determine the dividend per share on common stock and the rate of return in each of the exercises below. (The price per share represents an average.) Round off your answer to the nearest hundredth of a percent.

Earnings To Be Distributed	Number of Preferred Shares	Preferred Dividend	Number of Common Shares	Dividend per Share—Common	Price per Share	Rate of Return—Common	
$3,009,000	21,600	$3.00	600,000	**26.** *$4.91*	$18.75	**27.** *26.19%*	*(3,009,000 − (21,600 × 3) ÷ 600,000 =)* *(4.91 ÷ 18.75 =)*
1,500,000	24,750	5.00	950,000	**28.** *1.45*	10.48	**29.** *13.84*	*(1,500,000 − (24,750 × 5) ÷ 950,000)* *(1.45 ÷ 10.48 =)*
954,625	146,821	1.00	600,451	**30.** *1.35*	15.75	**31.** *8.57*	*(954,625 − (146,821 × 1) ÷ 600,451 =)* *(1.35 ÷ 15.75 =)*
50,000	5,450	.50	75,420	**32.** *.63*	5.60	**33.** *11.25*	*(50,000 − (5,450 × .5) ÷ 75,420 =)* *(.63 ÷ 5.6 =)*
2,800,000	50,000	2.00	2,000,000	**34.** *1.35*	20.50	**35.** *6.59*	*(2,800,000 − (50,000 × 2) ÷ 2,000,000 =)* *(1.35 ÷ 20.5 =)*

Determine the rate of return in the following problems. The preferred stock is nonparticipating and noncumulative. Round off your answer to the nearest tenth of a percent.

	Earnings To Be Distributed	Number of Preferred Shares	Preferred Dividend	Number of Common Shares	Price per Share	Rate of Return—Common
36.	$ 8,600,000 *(300,000 × 5 =)*	300,000	$ 5.00	4,500,000	$ 47.00 *(7,100,000 ÷ 4,500,000 =)*	*3.4%* *(1.58 ÷ 47 =)*
37.	10,385,700 *(850,000 × 7.5 =)*	850,000	7.50	6,563,000	33.50 *(4,010,700 ÷ 6,563,000 =)*	*1.8* *(.61 ÷ 33.5 =)*
38.	23,700,000 *(900,000 × 7 =)*	900,000	7.00	23,460,000	91.00 *(17,400,000 ÷ 23,460,000 =)*	*.8* *(.74 ÷ 91 =)*
39.	3,680,000 *(250,000 × 6 =)*	250,000	6.00	800,000	80.00 *(2,180,000 ÷ 800,000 =)*	*3.4* *(2.73 ÷ 80 =)*
40.	9,000,500 *(600,000 × 8 =)*	600,000	8.00	1,200,000	121.25 *(4,200,500 ÷ 1,200,000 =)*	*2.9* *(3.50 ÷ 121.25 =)*
41.	19,900,000 *(1,900,000 × 4 =)*	1,900,000	4.00	6,300,000	78.38 *(12,300,000 ÷ 6,300,000 =)*	*2.5* *(1.95 ÷ 78.38 =)*
42.	1,820,000 *(16,500 × 8 =)*	16,500	8.00	498,500	227.00 *(1,688,000 ÷ 498,500 =)*	*1.5* *(3.39 ÷ 227 =)*
43.	4,500,700 *(400,000 × 7.5 =)*	400,000	7.50	750,000	162.50 *(1,500,700 ÷ 750,000 =)*	*1.2* *(2.00 ÷ 162.5 =)*
44.	2,000,000 *(135,000 × 10 =)*	135,000	10.00	400,000	61.63 *(650,000 ÷ 400,000 =)*	*2.6* *(1.63 ÷ 61.63 =)*
45.	7,363,590 *(600,000 × 8.5 =)*	600,000	8.50	1,260,000	49.00 *(2,263,590 ÷ 1,260,000 =)*	*3.7* *(1.80 ÷ 49 =)*

Determine the dividend per share of common stock in the following problems. Participation is to be based on the number of shares of stock in both classes.

	Earnings To Be Distributed	Number of Preferred Shares	Preferred Dividend	Number of Common Shares	Cumulative (Years Passed)	Participating (After/ Common)	Dividend per Share Common
46.	$ 4,500,000 *(600,000 × 5 =)*	600,000	$ 5.00	1,300,000	No	$10	*$1.15* *(1,500,000 ÷ 1,300,000 =)*
47.	985,000 *(50,000 × 7 × 3 =)*	50,000	7.00	80,000	Two	No	*0* *(—)*
48.	860,000 *(35,000 × 3.5 × 2 =)*	35,000	3.50	75,000	One *(75,000 × 5 =)*	5	*7.18* *(240,000 ÷ 110,000 =)*
49.	5,260,000 *(400,000 × 7.5 =)*	400,000	7.50	650,000	No *(650,000 × 2 =)*	2	*2.91* *(960,000 ÷ 1,050,000 =)*
50.	1,865,350 *(85,000 × 10 × 2 =)*	85,000	10.00	100,000	One	4	*1.65* *(165,350 ÷ 100,000 =)*
51.	4,360,500 *(350,000 × 8.5 =)*	350,000	8.50	500,000	No	7	*2.77* *(1,385,500 ÷ 500,000 =)*
52.	17,390,550 *(985,000 × 6 × 2 =)*	985,000	6.00	1,500,000	One	No	*3.71* *(5,570,550 ÷ 1,500,000 =)*
53.	4,185,300 *(200,000 × 9 × 3 =)*	200,000	9.00	350,000	Two	2	*0* *(—)*
54.	47,359,680 *(1,600,000 × 7.5 =)*	1,600,000	7.50	2,500,000	No *(2,500,000 × 10 =)*	10	*12.53* *(10,359,680 ÷ 4,100,000 =)*
55.	285,600 *(3,580 × 6.5 × 2 =)*	3,580	6.50	70,000	One	5	*3.42* *(239,060 ÷ 70,000 =)*

Determine the dividend per share on preferred stock from the information below; participation is based on the number of shares in both classes. Round off to the nearest cent.

	Earnings To Be Distributed	Number of Preferred Shares	Preferred Dividend	Number of Common Shares	Cumulative (Years Passed)	Participating (After/ Common)	Dividend per Share Preferred
56.	60,000	1,000	10.00	45,000	One	Yes/$1	*$20.00* (10 × 2 =)
57.	670,000	10,000	2.50	30,000	Two	No	*7.50* (2.5 × 3 =)
58.	120,500 (3,500 × 3.5 × 2 =) (120,500 − 24,500 =)	3,500	3.50	78,500 (78,500 × 1 =) (96,000 − 78,500 =)	One	Yes/$1	*7.21* (17,500 ÷ 82,000 =) (.21 + 3.5 + 3.5 =)
59.	400,000 (5,000 × 5 =) (400,000 − 25,000 =)	5,000	5.00	15,000 (15,000 × 3 =) (375,000 − 45,000 =)	No	Yes/$3	*21.50* (330,000 ÷ 20,000 =) (5 + 16.5 =)
60.	600,500 (15,000 × 4.25 × 3 =) (600,500 − 191,250 =)	15,000	4.25	75,000 (75,000 × 2 =) (409,250 − 150,000 =)	Two	Yes/$2	*15.63* (259,250 ÷ 90,000 =) (2.88 + (4.25 × 3) =)

Determine the rate of return on the following $1,000 denomination bonds. Round off your answer to the nearest tenth of a percent.

	Current Price Quotation		Stated Interest Rate		Rate of Return	
61.	109	(1.09 × 1,000 =)	$7\frac{3}{4}$%	(.0775 × 1,000 =)	7.1%	(77.5 ÷ 1,090 =)
62.	$105\frac{1}{2}$	(1.055 × 1,000 =)	$8\frac{1}{2}$	(.085 × 1,000 =)	8.1	(85 ÷ 1,055 =)
63.	$96\frac{3}{4}$	(.9675 × 1,000 =)	$9\frac{1}{4}$	(.0925 × 1,000 =)	9.6	(92.5 ÷ 967.5 =)
64.	98	(.98 × 1,000 =)	6	(.06 × 1,000 =)	6.1	(60 ÷ 980 =)
65.	102	(1.02 × 1,000 =)	11	(.11 × 1,000 =)	10.8	(110 ÷ 1,020 =)

Determine the rate of return. Round off your answer to the nearest tenth of a percent.

	Bond Denomination	Current Price Quotation	Stated Interest Rate	Rate of Return (nearest tenth %)
66.	$1,000 (.06 × 1,000 =)	$102\frac{1}{4}$ (1.0225 × 1,000 =)	6%	5.9% (60 ÷ 1,022.5 =)
67.	1,000 (.095 × 1,000 =)	99 (.99 × 1,000 =)	$9\frac{1}{2}$	9.6 (95 ÷ 990 =)
68.	5,000 (.08 × 5,000 =)	98 (.98 × 5,000 =)	8	8.2 (400 ÷ 4,900 =)
69.	1,000 (.0825 × 1,000 =)	$103\frac{1}{2}$ (1.035 × 1,000 =)	$8\frac{1}{4}$	8.0 (82.5 ÷ 1,035 =)
70.	5,000 (.0575 × 5,000 =)	96 (.96 × 5,000 =)	$5\frac{3}{4}$	6.0 (287.5 ÷ 4,800 =)
71.	1,000 (.0775 × 1,000 =)	$87\frac{1}{2}$ (.875 × 1,000 =)	$7\frac{3}{4}$	8.9 (77.5 ÷ 875 =)
72.	1,000 (.075 × 1,000 =)	95.5 (.955 × 1,000 =)	$7\frac{1}{2}$	7.9 (75 ÷ 955 =)
73.	1,000 (.095 × 1,000 =)	$100\frac{1}{2}$ (1.005 × 1,000 =)	$9\frac{1}{2}$	9.5 (95 ÷ 1,005 =)
74.	3,000 (.0825 × 3,000 =)	101.5 (1.015 × 3,000 =)	$8\frac{1}{4}$	8.1 (247.5 ÷ 3,045 =)
75.	100 (.09 × 100 =)	99.35 (.9935 × 100 =)	9	9.1 (9 ÷ 99.35 =)

Business Application Problems

Round off your answers to the nearest tenth of a percent.

1. A bond with a 7 percent stated rate of interest is quoted in the financial news at a price of 98. What is the rate of return on the bond?
$70/$980 = 7.1% return

2. If the bond in problem 1 was bought at $96\frac{1}{2}$ one year ago, what would be the rate of return for the year considering both the selling price and interest?
$980 − $965 = $15
$1,000 × .07 =
$70 + $15 = $85/$965 = 8.8% return

3. What would be the rate of return on the bond in problem 1 if the price of the bond would have been stated as 99?
$70/$990 = 7.1% return

4. Hadmire Corporation purchased a piece of land for $453,000 five years ago. A $45,000 real estate commission on the purchase was paid by the seller. The corporation plans to sell the land for $600,000 less a 10 percent commission. What would be the annual rate of return on the land over the five-year investment period?
$600,000 × .9 = $540,000 − $453,000 =
$87,000/5 = $17,400/$453,000 = 3.8% return

5. Jamie bought 400 shares of BDX one year ago at 53 1/4 plus a commission of $12. The stock was just sold at 63 7/8 less as commission of $12. How much money did Jamie make on BDX this year?
$63.875 − $53.25 = $10.625 × 400 =
$4,250 − $12 − $12 = $4,226 profit

6. What was the rate of return on Jamie's investment in problem 5?
$53.25 × 400 = $21,300 + $12 = $21,312
$4,226/$21,312 = 19.8% return

7. A common stock quoted at 19 3/4 pays a dividend of $.62 per year. What is the dividend rate of return?
$.62/$19.75 = 3.1% return

8. A gold coin from a treasure ship that sunk in 1542 was purchased for $400. The coin is valued at $530 one year later. What is the annual rate of return on the coin?
$530 − $400 = $130/$400 = 32.5% return

9. A CD pays 6.2 percent for a 12-month investment. A stock priced at $41 is expected to increase to $45 in one year. The stock also pays a dividend of $.08 per year. The cost of investing in the CD is zero. The cost of investing in the stock is 1 percent of the buying price and the selling price. What is the rate of return on the best investment?
$45 × .01 = $.45 $41 × .01 = $.41
$44.55 − $41.41 = $3.14 + $.08 = $3.22/$41.41 =
7.8% stock

10. A grain priced at $6.30 per bushel increased to $6.41 over a year. An investor had purchased 100,000 bushels at the beginning of the year. The cost of holding the grain was $.04 per bushel. The cost of buying and selling the grain was $.02 per bushel. How much money did the investor make in one year?
$6.41 − $6.30 = $.11 − $.04 − $.02 − $.02 =
$.03 × 100,000 = $3,000 profit

11. Becky Bush purchased a set of collector's stamps for $450 and sold them a year later for $560. She paid a 5 percent commission on the sale. What was the rate of return on her investment?

$560	*$560.00*	*$82 / $450 = 18.2% return*
×.05	*−28.00*	
$28	*−450.00*	
	$82.00	

12. John Cooper bought a new sports car 20 years ago. The car cost $6,300. Though he drove the car for the first three years, he decided to store the vehicle as an investment. The car is now worth $26,000 as a collector's item. On a simple interest basis, what is the rate of return per year on John's investment?
$26,000 − $6,300 = $19,700

$$\frac{\$19{,}700}{\$6{,}300} = \frac{312.7\%}{20} = 15.6\% \text{ return}$$

13. An investor is considering two investments. One is a preferred stock quoted at $39.50 paying a $3.50 annual dividend. The alternative is a commercial bond quoted at $92\frac{1}{4}$ and paying a stated $8\frac{3}{4}$ percent interest rate. Which is the best return?

$\frac{\$3.50}{\$39.50}$ = *8.9% rate of return on stock*

$\frac{\$87.50}{\$922.50}$ = *9.5% rate of return on bond*

Bond return is best.

14. A bond is quoted at $102\frac{1}{2}$ in the financial section of the newspaper. The bond pays a stated interest rate of 8 percent. What is the effective rate of return?

$\frac{\$80 \text{ interest}}{\$1{,}025 \text{ market price}}$ = *7.8% rate of return*

15. Steve is considering the purchase of a jewelry business for $128,000. The markup rate is an average 18 percent on sales, of which one-half is overhead and one-half profit. The company does $142,300 of business a year. He currently has the $128,000 invested in commercial bonds quoted at $97\frac{3}{4}$ and paying a stated $9\frac{1}{4}$ percent interest. Which investment would provide the greatest rate of return?

Bonds

$\frac{\$92.50}{\$977.50}$ = 9.5%

Jewelry Business

$142,300

C + M = S $\frac{\$12{,}807}{\$128{,}000}$ = 10%

82% + 18% = 100%

$142,300 × .18 = $25,614 ÷ 2 = $12,807 profit

Jewelry Business

16. The Berman Company purchased a future position in wheat for $3,200 on a margin of 10 percent. The price of wheat went up by $.30 on each of the 10,000 bushels purchased for the total amount of $3,200. The broker's fee was $50 to sell. What was the rate of return on the investment over the period of one year?

10,000 bu.
×$.30
$3,000 increase

$3,200 initial cost
+3,000 increase
−50 broker's fee
$6,150 net from sale

$6,150
−3,200 initial cost
$2,950 return

$\frac{\$2{,}950}{\$3{,}200}$ = *92.2% return on investment*

17. Justin bought a home for $42,000 in the spring of 1997. The home required some repairs that cost a total of $3,500. The home was sold a year later for $50,000. The brokerage firm charged a commission rate of 8 percent on the sale of the property. What was the rate of return on the investment?

$50,000 sale price
−42,000 cost
−3,500 repairs
−4,000 broker's fee ($50,000 × .08 =)
$500 return

$\frac{\$500}{\$42{,}000 + \$3{,}500}$ = *1.1% return on investment*

18. The Dillon Corporation purchased 100 shares of stock of the Deweuke Corporation at $54\frac{3}{8}$ per share. The brokerage firm charges a flat 1 percent per transaction. The Dillon Corporation held the stock for 12 months. There were four $.80 dividends paid during the time that the stock was held. The stock was sold at a price of $62\frac{1}{8}$. What was the rate of return on the investment?

$.80
×4
$3.20 dividends
×100
$320

$54.375
×100
$5,437.50
×.01
$54.38 broker's fee

$5,437.50
+54.38
$5,491.88 cost

$ 62.125
×100
$6,212.50
+320.00 dividends
−62.13 broker's fee
$6,470.37 − $5,491.88 = $978.49 return

$\frac{\$978.49}{\$5{,}491.88}$ = *17.8% rate of return*

19. Cody Allan is considering the purchase of a large amount of grain as an investment. The price per bushel is $3.60. The commission rate is $1\frac{1}{2}$ percent on the purchase and on the sale. The storage will cost Cody 1 cent per bushel per month. Cody feels that the price will rise to $3.90 per bushel in six months. What will be the net profit on 6,000 bushels?

$3.60
×6,000
$21,600 cost of grain × .015 = $324
+324 commission
$21,924 cost of purchase
+360 (6,000 bu × $.01 × 6 months)
$22,284 total cost
$3.90
×6,000
$23,400 × .015 = $351
−351 commission
$23,049 net
−22,284
$765 net profit

20. A stock was quoted at $42.50 on the local exchange. The corporation pays a dividend after it pays the preferred stockholders a $7.50 dividend per year. There are 5,000 shares of preferred stock outstanding at this point. How much would the company have to earn in order to pay each common stockholder a dividend of $4 a share if they distribute 40 percent of the earnings over the 5,000 shares of preferred and the 10,000 shares of common?

5,000 shares of preferred
×$7.50
$37,500 dividends to preferred

10,000 shares of common
×$4
$40,000 dividends to common

$37,500
+40,000
$77,500 of dividends

B =	*B = P ÷ R*
R = 40%	*B = $77,500 ÷ .4*
P = $77,500	*B = $193,750 earnings*

21. Four years ago, Kathy purchased a 400-year-old painting valued at $230,000. She expects to be able to sell the painting for $400,000 in one more year. There will be a 10 percent broker's fee on the sale. What will be the annual rate of return on the painting over the five-year period if she is successful (nearest percent)?

$400,000 × .10 = $40,000
−230,000
$170,000
−40,000
$130,000 ÷ 5 = $26,000

$$\frac{\$26{,}000}{\$230{,}000} = 11.3\% \text{ rate of return}$$

22. Write a complete sentence in answer to this question. A piece of land that was purchased for $120,000 was resold for $160,000 three years later. There were transfer costs of $690 on the sale as well as a realtor's commission of $11,200. What was the annualized rate of return on the investment?

$160,000
−120,000
−690
−11,200
$28,110

$$\frac{\$28{,}110}{\$120{,}000} = 23.4\% \div 3 = 7.8\%$$

The annualized rate of return was 7.8%.

23. Write a complete sentence in answer to this question. Alvin Perkins invested his extra \$20,000 in a certificate of deposit that paid $6\frac{1}{4}$ percent interest. The CD was held for a period of one year. Alvin could have purchased 1,000 shares of a stock for about the same amount of money. The stock has paid a quarterly dividend of 10¢ per share. The price of the stock has risen from $\$19\frac{7}{8}$ to $\$20\frac{1}{4}$. Alvin would need to pay a broker's fee of \$9. Would he have been better off to buy the stock and sell it a year later? What amount of money would he have earned on the best investment?

\$.10 × 4 = \$.40 × 1,000 = \$400 dividend
\$20.25 − \$19.875 = \$.375 × 1,000 = \$375

\$400 (dividend)
+375.00 (price change)
−9.00 (broker's fee)
\$766.00 (stock)

\$20,000
×.0625
\$1,250 (CD)

He would have had a greater return with the CD investment, with a \$1,250 return.

★ Challenge Problem

Rochelle Way purchased 400 shares of Piad Corporation common stock on June 1 for \$33 per share. There are 1,680,000 shares of common stock outstanding and 500,000 shares of \$5 preferred stock outstanding. The firm reported annual earnings of \$7.6 million on November 1. The earnings were paid out in dividends on November 7. Rochelle sold her stock on December 1 for \$40 per share. She paid a commission rate of 1 percent on the purchase and on the sale. What was the rate of return?

\$7,600,000 earnings
−2,500,000 preferred dividends (500,000 × \$5)
\$5,100,000

\$5,100,000 ÷ 1,680,000 common shares
\$3.04 common dividend

400 shares
×\$3.04 dividend
\$1,216 dividends

400 shares
×\$40 price
\$16,000 sale × .01 = \$160 commission
−160
\$15,840 from sale

400 shares
×\$33
\$13,200 cost × .01 = \$132 commission
+132
\$13,332 total cost

\$15,840 net from sale
−13,332 total cost
\$2,508 return from sale
+1,216 dividends
\$3,724 return

$$\frac{\$3{,}724 \text{ return}}{\$13{,}332 \text{ invested}} = 27.9\% \text{ return in 6 months}$$

MASTERY TEST Chapter 19

Name ______________________ **Date** ______________ **Score** ____________

True or False

(T) F **1.** Cumulative preferred stock has the right to "catch up" on dividends that have not been paid in the past.

(T) F **2.** Selling fees reduce the return on the investment to the investor.

T (F) **3.** Municipal bonds carry a lower rate of interest but have a greater risk.

T (F) **4.** In a period of inflation, the investor is better off to hold cash so as not to lose to the rate of inflation.

(T) F **5.** Common stockholders have the right to vote on corporate issues brought before the annual stockholders' meeting.

Determine the rate of return in the following problem. The preferred stock is nonparticipating and noncumulative. Round off your answer to the nearest tenth of a percent.

Earnings To Be Distributed	Number of Preferred Shares	Preferred Dividend	Number of Common Shares	Price per Share	Rate of Return
6. $6,500,000	250,000	$4	2,000,000	$38	*7.2%*

Determine the rate of return in the following problem.

Bond Denomination	Current Price Quotation	Stated Interest Rate	Rate of Return
7. $1,000	102	7%	*6.9%*

Determine the dividend per share of common stock in the following problems. Participation is to be based on the number of shares of stock in both classes.

Earnings To Be Distributed	Number of Preferred Shares	Preferred Dividend	Number of Common Shares	Cumulative (Years Passed)	Participating (After Common)	Dividend per Share (Common)
8. $1,020,000	60,000	$ 8	120,000	None	$8	*$4.50*
9. 3,400,000	240,000	12	600,000	Two	7	*0*

10. Bill Dominy purchased a piece of land for $52,000 one year ago. The land was sold to an investor for $75,000. The commission fee was 10 percent on the sale. What was the rate of return on the investment?

$75,000 × .1 = $7,500

$75,000 − 52,000 − 7,500 = $15,500

$$\frac{\$15,500}{\$52,000} = 29.8\% \text{ return}$$

Financial Statement Analysis 20

Suggested Classroom Time for This Chapter

hours 1 2 3 4 5

Perceived Student Difficulty with Topical Material

easy moderate difficult

Objectives

After mastering the material in this chapter, you will be able to:

1. Use the basic accounting equation in balance sheet problems. p. 392
2. Complete a balance sheet and income statement when key values are missing. p. 392
3. Complete a comparative balance sheet and income statement. p. 392
4. Complete a horizontal analysis on a balance sheet or income statement. p. 392
5. Calculate the following ratios:
    a. current ratio p. 398
    b. acid-test ratio p. 398
    c. net income to net sales ratio p. 398
    d. accounts receivable to net sales ratio p. 399
    e. accounts payable to net purchases ratio p. 399
6. Use and understand the following terms and concepts:
    Balance Sheet
    Assets
    Liabilities
    Proprietorship
    Comparative Balance Sheet
    Horizontal Analysis
    Vertical Analysis
    Income Statement
    Revenue
    Expenses
    Ratio

Sport-A-Merica

Some time has passed, and Sport-A-Merica is continuing to grow. The firm recently opened a new store in Eastern Mall and sold the first store they started downtown. Though the space in the mall is smaller than their main store, many people now know about Sport-A-Merica, and they visit the main store for the bigger items after visiting the mall store. Wholesale orders are filled through the rented mini warehouse space.

Now that there is more history for comparison, Hal is anxious to see how Sport-A-Merica fared in recent months. He looks over the balance sheet for the end of the year and considers the assets and liabilities of the firm. He knows that increasing sales will do little to improve the bottom line for the firm unless there has been a corresponding increase in profits. Doing volume business required more people, increased overhead, and caused a lot of headaches. But has there been an increase in profits? With the recent addition of more employees and the move to the regional mall, there is a need to analyze the firm's progress. What will the business ratios tell Hal about the firm? He would like to see if the current ratio is over 1.2 to 1 for Sport-A-Merica. Numbers, purely by themselves, do little to show how the firm is being managed. It would be helpful to be able to compare this period's performance with that during the same period last year. Other measurements could also be applied, as long as the figures can be compared to earlier, similar periods. Hal hopes the effort at the mall store with additional employees will prove to be worthwhile. He knows that providing quality products and excellent service at competitive prices is the key to success in the long run.

Business managers and owners need financial reports on which they can base decisions. These reports are termed *financial statements*. There are many types of financial statements. The most frequently used are the balance sheet and the income statement.

Balance Sheet

Objective 1

assets
liabilities
owner's equity

The **balance sheet** is a "financial photograph" of a firm. If time could be stopped, and if values could be expressed in dollars on a "financial film" report, we would develop a balance sheet. The balance sheet is dated, and therefore, it is only valid for the date that the financial photograph was taken. It shows the amount of **assets** (those things owned by the firm), the **liabilities** (that which is owed by the firm), and **owner's equity** (also termed *capital*, or *proprietorship*, the amount invested in the firm by the owners) of the firm at the date of the balance sheet.

The assets must equal the liabilities plus owner's equity on the balance sheet. This is known as the basic accounting equation: A = L + OE.

Assets are divided into two classifications. *Current assets* are those assets that will be turned into cash within one year, and *fixed assets* are those assets that will last for a year or more without being turned into cash. Examples of current assets are cash, accounts receivable, merchandise inventory, and prepaid expenses. Examples of fixed assets are land, buildings, and various types of equipment.

Liabilities are classified on the same basis as assets. *Current liabilities* are those due in a year or less, and *long-term liabilities* are those due in a year or more. Examples of current liabilities are accounts payable, notes payable, and salaries payable. Examples of long-term liabilities are mortgages payable and long-term notes payable.

In a corporation the term *capital* or *owner's equity* is used. The term *proprietorship* may be used in a sole proprietorship.

Example 1

Sport-A-Merica, Inc.
Balance Sheet
December 31, 2001

Assets		**Liabilities**	
Current Assets		**Current Liabilities**	
Cash	$3,000	Accounts Payable	$2,400
Accounts Receivable	300	Notes Payable	10,000
Merchandise Inventory	15,100	Total Current Liabilities	$12,400
Total Current Assets	$18,400	**Long-Term Liabilities**	
Fixed Assets		Mortgage, Building	$84,000
Office Equipment	$28,000	Total Liabilities	$96,400
Buildings	130,000	**Owner's Equity**	
Land	42,000	Kerry Finn, Owner	$61,000
Total Fixed Assets	$200,000	Hal Balmer, Owner	61,000
Total Assets	$218,400	Total Liabilities & Owner's Equity	$218,400

❍ *The balance sheet shows how a firm appears now in financial terms, It is like a still photograph—it exists for only a moment in time.*

Objectives 2, 3, & 4

comparative balance sheet
horizontal analysis

Just as it is interesting to compare photographs of the past with those of the present, it is also interesting to compare a balance sheet from last year with one from this year. This is termed a **comparative balance sheet.** Notice that in Example 2 the information is horizontally compared (i.e., cash of 2001 is compared to cash of 2000, a **horizontal analysis**). The comparison is always made to the earlier period or date (2000 in the following example). Decreases are always shown in parentheses.

Example 2

James Williard Co.
Comparative Balance Sheet
December 31, 2000 and 2001

Assets	2001	2000	Amount of Increase (Decrease)	% Increase (Decrease)
Current Assets				
Cash	$6,840	$7,500	($660)	(8.8)
Accounts Receivable	10,085	9,360	725	7.7
Prepaid Insurance	2,580	1,620	960	59.3
Total Current Assets	$19,505	$18,480	1,025	5.5
Fixed Assets				
Office Building	$24,000	$24,000	$0	. . .
Office Furniture	1,570	1,695	(125)	(7.4)
Total Fixed Assets	$25,570	$25,695	(125)	(.5)
Total Assets	$45,075	$44,175	900	2
Liabilities and Proprietorship				
Current Liabilities				
Accounts Payable	$7,300	$5,625	$1,675	29.8
Taxes Payable	1,580	1,580	0	. . .
Total Liabilities	$8,880	$7,205	1,675	23.2
Proprietorship				
James Williard, Capital	36,195	36,970	(775)	(2.1)
Total Liabilities and Proprietorship	$45,075	$44,175	900	2

(7,500 − 6,840 =)(660 ÷ 7,500 =)

To find the 8.8 percent decrease in cash:

$$\begin{array}{r} \$7{,}500 \\ -6{,}840 \\ \hline \$660 \end{array} = \text{Difference} \qquad \frac{\$660}{\$7{,}500} = \frac{\text{Difference}}{\text{Base year (2000)}} = 8.8\%\text{ decrease}$$

Exercise A

Determine the missing values and the percent of increase or decrease on the comparative balance sheet below.

Gordon and Associates
Comparative Balance Sheet
July 1, 2000 and 2001

Assets	2001	2000	Amount of Increase (Decrease)	% Increase (Decrease) (to Nearest Tenth)
Current Assets				
Cash (7,840 − 5,890 =)	$7,840	$5,890	$1,950	33.1 (1,950 ÷ 5,890 =)
Accounts Receivable *(1,270 − 6,200 =)*	1,270	6,200	*(4,930)*	*(79.5)* *(4,930 ÷ 6,200 =)*
Merchandise on Hand *(9,250 − 3,600 =)*	9,250	3,600	***5,650***	***156.9*** *(5,650 ÷ 3,600 =)*
Total Current Assets *(18,360 − 15,690 =)*	$18,360	$15,690	***2,670***	***17.0*** *(2,670 ÷ 15,690 =)*

(Continued on next page)

Exercise A Continued

Gordon and Associates
Comparative Balance Sheet
July 1, 2000 and 2001

Assets	2001	2000	Amount of Increase (Decrease)	% Increase (Decrease) (to Nearest Tenth)
Fixed Assets				
Delivery Trucks *(—)*	$44,260	$44,260	*$0*	*0 (—)*
Total Assets *(62,620 − 59,950 =)*	*$62,620*	*$59,950*	*2,670*	*4.5 (2,670 ÷ 59,950 =)*
Liabilities and Proprietorship				
Current Liabilities				
Accounts Payable *(4,360 − 5,680 =)*	$4,360	$5,680	*($1,320)*	*(23.2) (1,320 ÷ 5,680 =)*
Long-Term Liabilities				
Mortgage Loan Payable *(26,500 − 32,500 =)*	26,500	*32,500*	*(6,000)*	*(18.5) (6,000 ÷ 32,500 =)*
Total Liabilities *(30,860 − 38,180 =)*	*$30,860*	$38,180	*(7,320)*	*(19.2) (7,320 ÷ 38,180 =)*
Proprietorship				
J. T. Storm, Capital *(31,760 − 21,770 =)*	*31,760*	*21,770*	*9,990*	*45.9 (9,990 ÷ 21,770 =)*
Total Liabilities and Proprietorship *(62,620 − 59,950 =)*	*$62,620*	*$59,950*	*2,670*	*4.5 (2,670 ÷ 59,950 =)*

Exercise B

Determine the missing values and the percent of increase or decrease on the current assets portion of the comparative balance sheet below.

Senior Home Care Corporation
Comparative Balance Sheet
March 31, 2001 and 2002

Assets	2002	2001	Amount of Increase (Decrease)	% Increase (Decrease) (to Nearest Tenth)
Current Assets				
Cash	$12,750	$15,360 (12,750 − 15,360 =)	($2,610)	(17) (2,610 ÷ 15,360 =)
Notes Receivable	2,500	500 *(2,500 − 500 =)*	*2,000*	*400 (2,000 ÷ 500 =)*
Accounts Receivable	13,368	7,920 *(13,368 − 7,920 =)*	*5,448*	*68.8 (5,448 ÷ 7,920 =)*
Cleaning Supplies	820	1,685 *(820 − 1,685 =)*	*(865)*	*(51.3) (865 ÷ 1,685 =)*
Office Supplies	60	380 *(60 − 380 =)*	*(320)*	*(84.2) (320 ÷ 380 =)*
Prepaid Rent	1,200	1,200 *(—)*	*0*	*0 (—)*
Prepaid Insurance	380	1,898 *(380 − 1,898 =)*	*(1,518)*	*(80) (1,518 ÷ 1,898 =)*
Total Current Assets	*$31,078*	*$28,943*	*2,135*	*7.4 (2,135 ÷ 28,943 =)*

Income Statement

An **income statement** is a portion of a business firm's financial history. Whereas a balance sheet shows what exists at a point in time, an income statement shows what took place during a period of time. The heading must indicate the period covered by the statement (usually a year, six months, or a quarter), not simply a date.

revenue
expenses

The income statement shows **revenue** (the result of selling goods or services) and **expenses** (used up or expired assets). The difference between revenue and expenses is termed *net income* or *net loss*. An income is earned when revenue is larger than expenses. A loss is incurred when expenses are larger than the revenue.

An income statement shows what has happened in a firm over a period of time. It is like a video—it has a starting point and an ending point. The ending point is shown in the heading.

Baxendale Accounting, Inc.
Income Statement
Year Ended July 31, 2001

Revenue		
Accounting Fees	$147,300	
Consulting Fees	9,500	
Interest Earned	1,250	
Total Revenue		$158,050
Expenses		
Salaries Expense	$82,600	
Rent Expense	12,000	
Utilities Expense	4,600	
License Fees	400	
Telephone Expense	1,360	
Computer Rental Expense	4,800	
Transportation Expense	980	
Total Expenses		$106,740
Net Income		$51,310

Service Firm

A service firm sells only time, not merchandise. The income statement for a service firm includes their sales of service time and their expenses, such as wages, supplies, expired rent, and expired insurance.

Example 3

Chilton Business Service
Income Statement
Year Ended June 30, 2001

Revenue		
Sales of Accounting Service	$135,600	
Sales of Computer Service	117,350	
Total Revenue		$252,950
Expenses		
Wages Expense	$63,480	
Computer Rental Fees	5,800	
Utilities Expense	3,865	
Rent Expense	4,800	
Expired Insurance	720	
Supplies Used	985	
Total Expenses		$79,650
Net Income, Year Ended June 30, 2001		$173,300

Comparisons within the income statement are always made to total revenue, or net sales in the case of a merchandising firm. This type of analysis is termed a *vertical analysis*. In the previous example, "Wages Expense" was 25.1 percent of total revenue; that is,

$$\frac{\$63,480}{\$252,950} = 25.1\%$$

A vertical analysis allows a manager to compare each item to net sales regardless of the size of the firm. A vertical analysis can be made of a balance sheet by comparing each balance sheet item to total assets or the total of liabilities plus owner's equity.

A comparative income statement is similar to a comparative balance sheet. An item-by-item horizontal analysis is made, so that the difference from the earlier time period can be seen.

Exercise C

Determine the missing values, the percent of total revenue, and the percent of increase or decrease in the comparative income statement below.

Andres Perez Delivery Company
Comparative Income Statement
Year Ended September 30, 2000 and 2001

	2001	% (2001)	2000	% of Increase (Decrease) (to Nearest Tenth)
Revenue				
Revenue from Deliveries	$142,360	59.2	$148,980	(4.4)
(142,360 ÷ 240,360 =)				(6,620 ÷ 148,980 =)
Special Revenue	98,000	40.8	36,000	172.2
(98,000 ÷ 240,360 =)				(62,000 ÷ 36,000 =)
Total Revenue	$240,360	100	$184,980	29.9
				(55,380 ÷ 184,980 =)
Expenses				
Gasoline Expense	$17,300	7.2	$17,000	1.8
(17,300 ÷ 240,360 =)				(300 ÷ 17,000 =)
Wages Expense	84,780	35.3	82,900	2.3
(84,780 ÷ 240,360 =)				(1,880 ÷ 82,900 =)
Repair Expense	15,800	6.6	2,600	507.7
(15,800 ÷ 240,360 =)				(13,200 ÷ 2,600 =)
Utilities Expense	2,300	1.0	2,200	4.5
(2,300 ÷ 240,360 =)				(100 ÷ 2,200 =)
Office Supplies Used	700	.3	840	(16.7)
(700 ÷ 240,360 =)				(140 ÷ 840 =)
Miscellaneous Expense	21,400	8.9	31,380	(31.8)
(21,400 ÷ 240,360 =)				(9,980 ÷ 31,380 =)
Total Expenses	$142,280	59.2	$136,920	3.9
(142,280 ÷ 240,360 =)				(5,360 ÷ 136,920 =)
Net Income for the Year	$98,080	40.8	$48,060	104.1
(98,080 ÷ 240,360 =)				(50,020 ÷ 48,060 =)

Merchandising Firm

The revenue of a merchandising firm is mainly a result of buying and reselling merchandise at a markup. The income statement for a merchandising firm has a special category of expenses termed *cost of goods sold*. The cost of goods sold represents the cost price of purchases that were resold. This category of expenses is important to a retailer or wholesaler. A measure of business performance is found by subtracting the cost of goods sold from net sales. The result is termed *gross margin*, or *gross profit*.

A merchandising firm must deal with customer returns or after-sale allowances. These are subtracted from revenue on the income statement to find net sales.

Exercise D

Determine the missing values, the percent of net sales, and percent of increase or decrease in the partial comparative income statement below.

William Elliot Motors Company
Comparative Income Statement
Quarter Ended June 30, 2002 and June 30, 2003

	Quarter 2003	% (2003)	Quarter 2002	% Increase (Decrease) (to Nearest Tenth)
Revenue				
Sales	$487,360	100.7%	$487,480	(0)
(487,360 ÷ 484,155 =)				(120 ÷ 487,480 =)
Less: Sales Return	(3,120)	*.6*	485	*543.3*
(3,120 ÷ 484,155 =)				*(2,635 ÷ 485 =)*
Sales Allowances	(85)	*0*	226	*(62.4)*
(85 ÷ 484,155 =)				*(141 ÷ 226 =)*
Net Sales	$484,155	100%	$486,769	*(.5)*
				(2,614 ÷ 486,769 =)
Cost of Goods Sold				
(107,085 ÷ 484,155 =)				
Merchandise Inventory April 1, 2003	$107,085	*22.1*	$79,350	*35.0*
				(27,735 ÷ 79,350 =)
Merchandise Purchases	262,400	*54.2*	307,850	*(14.8)*
(262,400 ÷ 484,155 =)				*(45,450 ÷ 307,850 =)*
Freight In	5,800	*1.2*	4,200	*38.1*
(5,800 ÷ 484,155 =)				*(1,600 ÷ 4,200 =)*
Goods Available for Sale	*$375,285*	*77.5*	$391,400	*(4.1)*
(375,285 ÷ 484,155 =)				*(16,115 ÷ 391,400 =)*
Less: Merchandise Inventory June 30, 2003	98,465	*20.3*	107,085	*(8.0)*
(98,465 ÷ 484,155 =)				*(8,620 ÷ 107,085 =)*
Cost of Goods Sold	*$276,820*	*57.2*	*$284,315*	*(2.6)*
(276,820 ÷ 484,155 =)				*(7,495 ÷ 284,315 =)*
Gross Profit	*$207,335*	*42.8*	$202,454	*2.4*
(207,335 ÷ 484,155 =)				*(4,881 ÷ 202,454 =)*

Ratios
Objective 5

Although the balance sheet shows "what is" and the income statement shows "what took place," other specific measures of business performance are also used. These measures are termed *ratios* and are taken from parts of the balance sheet or income statement. A **ratio** compares one item to another.

Example 4

Reilly and Co. has 42 sales representatives who are meeting their sales quotas. Another 12 sales representatives have fallen short of their quotas. What is the ratio of success to failure?

$$\frac{42}{12} = \frac{\text{success}}{\text{failure}} = 3.5 \text{ to } 1 \text{ ratio}$$

There are 3.5 successful sales representatives to each unsuccessful sales representative.

(42 ÷ 12 =)

You may want to refer to the topic of ratios in chapter 2 for more examples.

Ratios are useful when comparing one time period to another time period. This comparison will show a trend. They are also useful when used to compare different firms within an industry, although the firms may be of different size.

Current Ratio

The *current ratio* compares current assets to current liabilities. The current ratio is a measure of the firm's ability to pay off its liabilities over a time period. It should be at least 1 to 1.

Example 5

Sport-A-Merica, Inc. has current assets of \$18,400 and current liabilities of \$12,400. What is the current ratio?

$$\frac{\$18{,}400 = \text{current assets}}{\$12{,}400 = \text{current liabilities}} = 1.5 \text{ to } 1 \text{ current ratio}$$

(18,400 ÷ 12,400 =)

Acid-Test Ratio

The *acid-test ratio* measures the firm's ability to pay off its liabilities in a brief time period. The acid-test ratio differs from the current ratio by considering only cash, receivables, and marketable securities to use to compare to current liabilities. This ratio then measures the ability of a firm to pay off its liabilities without making further sales.

Example 6

The current liabilities of the Rierdon Co. are \$19,960. The firm has \$14,000 cash, \$11,365 in receivables, and a bond that has a face value of \$1,000. What is the acid-test ratio?

$$\frac{\text{cash} + \text{receivables} + \text{securities}}{\text{total current liabilities}} =$$

$$\frac{\$14{,}000 + \$11{,}365 + \$1{,}000}{\$19{,}960} = 1.3 \text{ to } 1$$

(14,000 + 11,365 + 1,000 =)(26,365 ÷ 19,960 =)

Net Income to Net Sales Ratio

An increase in net sales doesn't necessarily mean an increase in net income. Nor does it mean a higher income percent on sales. As sales increase, expenses also tend to rise. The key is to increase sales while holding down expenses. From the income statement, we can determine net sales as well as net income. To find the *net income to net sales ratio,* divide the net income by the net sales for the period.

Example 7

What is the net income to net sales ratio for the Chilton Business Service (example 3)?

$$\frac{\$173{,}300 = \text{net income}}{\$252{,}950 = \text{net sales}} = .7 \text{ to } 1$$

(173,300 ÷ 252,950 =)

Accounts Receivable to Net Sales Ratio

Extending credit to customers may increase sales, but it may also increase the amount of credit extended if customers delay payment. Though accounts receivable is an asset, it is not cash. A cash shortage and an accounts receivable abundance can place a firm in financial difficulty. Cash is required to pay the firm's debts. The *accounts receivable to net sales ratio* measures how well credit extension is being managed. The accounts receivable to net sales ratio is found by dividing the accounts receivable at the end of a period by the net sales for the period.

Example 8

The Charles Teller Co. had net sales of $1,386,450 in 2002. Accounts receivable at the end of 2002 were $263,850. What is the accounts receivable to net sales ratio?

$$\frac{\$263{,}850}{\$1{,}386{,}450} = \frac{\text{accounts receivable}}{\text{net sales}} = .19 \text{ to } 1$$

(263,850 ÷ 1,386,450 =)

Accounts Payable to Net Purchases Ratio

The *accounts payable to net purchases ratio* is very similar to the accounts receivable to net sales ratio. This ratio indicates how well credit is being managed by the firm at the buying end instead of at the selling end. It is found by dividing accounts payable at the end of a period by net purchases during that period.

Example 9

Brooks Stevens and Co. had $13,600 of accounts payable on the balance sheet dated June 1, 2001. Net purchases for the year ended June 1, 2001, are shown on the income statement at $87,365. What is the accounts payable to net purchases ratio?

$$\frac{\$13{,}600}{\$87{,}365} = \frac{\text{accounts payable}}{\text{net purchases}} = .16 \text{ to } 1$$

(13,600 ÷ 87,365 =)

Exercise E

Donna Jean, owner of the Solid Sound Studio, has the financial statements for the year ended September 30, 2003, as shown below. Ms. Jean plans to make comparisons to last year's financial performance. Determine the ratios requested to the nearest hundredth.

Solid Sound Studio—Balance Sheet—September 30, 2003

Assets		
Current Assets		
Cash	$13,120	
Accounts Receivable	4,265	
Merchandise on Hand	10,050	
Total Current Assets		$27,435
Fixed Assets		
Building		30,000
Total Assets		$57,435
Liabilities and Proprietorship		
Current Liabilities		
Accounts Payable	$3,420	
Notes Payable	15,000	
Total Liabilities		$18,420
Proprietorship		
D. Jean, Capital		39,015
Total Liabilities and Proprietorship		$57,435

Solid Sound Studio—Income Statement—Year Ended September 30, 2003

Revenue		
Sales of Merchandise	$86,420	
Less: Sales Returns	(763)	
Net Sales		$85,657
Cost of Goods Sold		
Merchandise Inventory, September 30, 2002	$7,300	
Merchandise Purchases	41,680	
Goods Available for Sale	$48,980	
Less: Merchandise Inventory, September 30, 2003	(10,050)	
Cost of Goods Sold		38,930
Gross Profit		$46,727
Operating Expenses		
Rent Expense	$4,300	
Advertising Expense	1,200	
Salaries Expense	8,900	
Payroll Tax Expense	534	
Utilities Expense	2,620	
Total Operating Expenses		17,554
Net income, Year Ended September 30, 2003		$29,173

1. Current Ratio = 1.49 to 1
   (27,435 ÷ 18,420 =)
2. Acid-Test Ratio = *.94 to 1*
   (17,385 ÷ 18,420 =)
3. Net Income to Net Sales Ratio = *.34 to 1*
   (29,173 ÷ 85,657 =)
4. Accounts Receivable to Net Sales Ratio = *.05 to 1*
   (4,265 ÷ 85,657 =)
5. Accounts Payable to Net Purchases Ratio = *.08 to 1*
   (3,420 ÷ 41,680 =)

ASSIGNMENT Chapter 20

Name ______________________ **Date** ______________ **Score** ____________

Skill Problems

Select the best term for each statement.

A. Balance Sheet	B. Assets	C. Liabilities	D. Proprietorship
E. Horizontal Analysis	F. Income Statement	G. Revenue	H. Expenses
I. Ratio	J. Comparative Balance Sheet		K. Vertical Analysis

1. *D* The portion of the firm that is owned by the investors.
2. *F* A report that shows the financial activity of a firm over a period of time.
3. *A* A "financial photograph" of a firm.
4. *K* The comparison of each item on the income statement to net sales.
5. *J* A summary of the financial condition of a firm for two periods.
6. *I* The comparison of one value to another value.
7. *E* Comparing information on the same line. The difference is then changed into a percent.
8. *B* Items that are owned by a firm or individual.

Select the best term for each statement.

A. Current Ratio
B. Acid-Test Ratio
C. Net Income to Net Sales Ratio
D. Accounts Receivable to Net Sales Ratio
E. Accounts Payable to Net Purchases Ratio
F. Ratio

B **9.** Measures the firm's ability to pay off its liabilities in a brief time period without making additional sales.
E **10.** Measures how well credit is being managed at the buying end.
C **11.** Net money earned by firm as compared to the total revenue taken in by firm.
D **12.** Measures how well credit extension is being managed.
A **13.** Measure of the firm's ability to pay off its liabilities over a time period.

Determine the missing values and the percent of total revenue in the income statement below. Round to the nearest two decimals.

Odyssey Land Ltd.
Income Statement
Ended June 30, 2002

		2002	**% (2002)**	
14.	Sales	$150,354	*100%*	
15.	Cost of Goods Sold	48,940	*32.55*	*(48,940/150,354 =)*
16.	Gross Profit	*$101,414*	*67.45*	*(101,414/150,354 =)*
17.	Operating Expenses	$20,385	*13.56*	*(20,385/150,354 =)*
18.	Administrative Expenses	5,647	*3.76*	*(5,647/150,354 =)*
19.	Total Expenses	*$26,032*	*17.31*	*(26,032/150,354 =)*
20.	Net Income	*$75,382*	*50.14*	*(75,382/150,354 =)*

Determine the missing values and the percent of increase or decrease on the comparative balance sheet below. Round off to the nearest two decimals.

Ruth Essroc Inc.
Balance Sheet
December 31, 2003

	2003	2002	% Increase (Decrease)	
Assets				
Total Current Assets	$16,000	$17,900	21. *(10.61)*	*((16,000 − 17,900) ÷ 17,900 =)*
Total Fixed Assets	140,540	141,870	22. *(.94)*	*((140,540 − 141,870) ÷ 141,870 =)*
Total Assets	23. *$156,540*	24. *$159,770*	25. *(2.02)*	*((156,540 − 159,770) ÷ 159,770 =) (16,000 + 140,540 =), (17,900 + 141,870 =)*
Liabilities & Proprietorship				
Total Current Liabilities	$12,000	$12,200	26. *(1.64)*	*((12,000 − 12,200) ÷ 12,200 =)*
Long-Term Liabilities	27. *95,400*	74,800	28. *27.54*	*(107,400 − 12,000 =), ((95,400 − 74,800) ÷ 74,800 =)*
Total Liabilities	$107,400	29. *$87,000*	30. *23.45*	*((107,400 − 87,000) ÷ 87,000 =) (12,200 + 74,800 =)*
Proprietorship				
Ruth Essroc, Owner	49,140	72,700	31. *32.47*	*((49,140 − 72,770) ÷ 72,770 =) (107,400 + 49,140 =), (87,000 + 72,770 =)*
Total Liabilities & Proprietorship	32. *$156,540*	33. *$159,770*	34. *(2.02)*	*((156,540 − 159,770) ÷ 159,770 =)*

Use the Ruth Essroc Inc. balance sheet above for the following problems.

35. 2003 current ratio = *1.33 to one*
(16,000 ÷ 12,000 =)

36. If the current assets contain $2,000 in inventory, what is the 2003 Acid-Test Ratio? *1.17 to one*
((16,000 − 2,000)/12,000 =)

Determine the missing values and the percent of increase or decrease on the comparative income statement below. Round off to the nearest two decimals.

Ruth Essroc Inc.
Income Statement
Ended December 31, 2003

	2003	2002	% Increase (Decrease)	
Net Sales	$25,600	$19,500	37. *31.28%*	*((25,600 − 19,500) ÷ 19,500 =)*
Cost of Goods Sold	10,425	38. *7,850*	39. *32.80*	*(19,500 − 11,650 =), ((10,425 − 7,850) ÷ 7,850 =)*
Gross Profit	40. *$15,175*	$11,650	41. *30.26*	*(25,600 + 10,425 =), ((15,175 − 11,650) ÷ 11,650 =)*
Operating Expenses	42. *$5,240*	$3,255	43. *60.98*	*(6,255 − 1,015 =), ((5,240 − 3,255) ÷ 3,255 =)*
Administrative Expenses	1,015	44. *1,200*	45. *(15.42)*	*(4,455 − 3,255 =), ((1,015 − 1,200) ÷ 1,200 =)*
Total Expenses	$6,255	$4,455	46. *40.40*	*((6,255 − 4,455) ÷ 4,455 =)*
Net Income	47. *$8,920*	48. *$7,195*	49. *23.97*	*(15,175 − 6,255 =), (11,650 − 7,195 =), ((8,920 − 7,195) ÷ 7,195 =)*

Use the preceding income statement for Ruth Essroc Inc. for the following problems.

50. 2003 net income to net sales ratio = *.35 to one* *(8,920/25,600 =)*

51. 2002 net income to net sales ratio = *.37 to one* *(7,195/19,500 =)*

52. If current assets for 2003 contain $4,000 in accounts receivable, what is the accounts receivable to net sales ratio for 2003? *.16 to one* *(4,000/25,600 =)*

Use the comparative balance sheet and comparative income statement below to determine the answers to the skill problems below. Round off percents to the nearest tenth. Round off ratios to the nearest hundredth.

Albrecht's
Comparative Balance Sheet
June 1, 2001 and 2002

Assets	2002	2001	% of Increase (Decrease)
Current Assets			
Cash	$8,000	$9,300	**53.** *(14.0) (1,300 ÷ 9,300 =)*
Accounts Receivable	2,800	2,960	**54.** *(5.4) (160 ÷ 2,960 =)*
Merchandise on Hand	**55.** *17,200*	15,300	**56.** *12.4 (1,900 ÷ 15,300 =)*
Total Current Assets	$28,000	**57.** *$27,560*	**58.** *1.6 (440 ÷ 27,560 =)*
Fixed Assets			
Store Display Equipment	$15,800	$15,950	**59.** *(.9) (150 ÷ 15,950 =)*
Automobile	3,000	**60.** *3,600*	**61.** *(16.7) (600 ÷ 3,600 =)*
Total Fixed Assets	**62.** *$18,800*	$19,550	**63.** *(3.8) (750 ÷ 19,550 =)*
Total Assets	**64.** *$46,800*	**65.** *$47,110*	**66.** *(.7) (310 ÷ 47,110 =)*
Liabilities and Proprietorship			
Current Liabilities			
Accounts Payable	$4,900	$4,200	**67.** *16.7 (700 ÷ 4,200 =)*
Notes Payable—Bank	10,000	12,000	**68.** *(16.7) (2,000 ÷ 12,000 =)*
Total Current Liabilities	**69.** *$14,900*	**70.** *$16,200*	**71.** *(8.0) (1,300 ÷ 16,200 =)*
Long-Term Liabilities			
Mortgage Payable	$1,000	$0	**72.** *0 (1,000 ÷ 0 =)*
Total Liabilities	**73.** *$15,900*	**74.** *$16,200*	**75.** *(1.9) (300 ÷ 16,200 =)*
Proprietorship			
L. K. Worthy, Capital	**76.** *30,900*	**77.** *30,910*	**78.** *0.0 (10 ÷ 30,910 =)*
Total Liabilities and Proprietorship	**79.** *$46,800*	**80.** *$47,110*	**81.** *(.7) (310 ÷ 47,110 =)*

Albrecht's
Comparative Income Statement
Year Ended June 1, 2001 and 2002

	2002	2001
Revenue		
Sales	$183,975	$167,385
Less: Sales Returns	3,920	3,680
Net Sales	82. *$180,055*	83. *$163,705*
Cost of Goods Sold		
Merchandise Inventory, June 1 (Beginning)	$15,300	$16,380
Merchandise Purchases	127,100	120,850
Goods Available for Sale	84. *$142,400*	85. *$137,230*
Less: Merchandise Inventory, June 1 (Ending)	17,360	15,300
Cost of Goods Sold	86. *$125,040*	87. *$121,930*
Gross Profit	88. *$55,015*	89. *$41,775*
Operating Expenses		
Wages Expenses	$12,300	$13,600
Rent Expense	6,000	6,000
Legal Fees	3,500	2,000
Payroll Tax Expense	700	750
Miscellaneous Expense	3,000	900
Total Expenses	90. *$25,500*	91. *$23,250*
Net Income, year ended June 1	92. *$29,515*	93. *$18,525*

94. Determine the net sales increase from 2001 to 2002. *10.0* % *(16,350 ÷ 163,705 =)*

95. Determine the acid-test ratio for 2002. *.72 to 1* *(10,800 ÷ 14,900 =)*

96. Determine the percent of increase of L. K. Worthy, Capital from 2001 to 2002. *0* % *(10 ÷ 30,910 =)*

97. Determine the percent of sales returns for 2002. *2.12* % *(3,920 ÷ 180,055 =)*

98. Determine the percent of cost of goods sold for 2002. *69.4* % *(125,040 ÷ 180,055 =)*

99. Determine the total liabilities and proprietorship for 2002. *$46,800* *(46,800, total assets)*

100. Determine the accounts payable to net purchases ratio for 2002. *.04 to 1* *(4,900 ÷ 127,100 =)*

101. Determine the percent of increase of operating expenses from 2001 to 2002. *9.7* % *(2,250 ÷ 23,250 =)*

102. Determine the current ratio for 2002. *1.88 to 1* *(28,000 ÷ 14,900 =)*

103. Determine the percent of increase in net income from 2001 to 2002. *59.3* % *(10,990 ÷ 18,525 =)*

104. Determine the accounts receivable to net sales ratio for 2002. *02 to 1* *(2,800 ÷ 180,055 =)*

105. Determine the net income to net sales ratio for 2002. *.16 to 1* *(29,515 ÷ 180,055 =)*

106. Determine the percent of increase in cost of goods sold from 2001 to 2002. *2.6* % *(3,110 ÷ 121,930 =)*

107. Determine the percent of gross profit for 2001. *25.5* % *(41,775 ÷ 163,705 =)*

108. Determine the current ratio for 2001. *1.70 to 1* *(27,560 ÷ 16,200 =)*

109. Determine the percent of sales returns for 2001. *2.2* % *(3,680 ÷ 163,705 =)*

110. Determine the acid-test ratio for 2001. *.76 to 1* *(12,260 ÷ 16,200 =)*

111. Determine the percent of net income to net sales for 2001. *11.3* % *(18,525 ÷ 163,705 =)*

112. Determine the accounts receivable to net sales ratio for 2001. *.02 to 1* *(2,960 ÷ 163,705 =)*

113. Determine the net income to net sales ratio for 2001. *.11 to 1* *(18,525 ÷ 163,705 =)*

Business Application Problems

1. Current assets total \$15,000 on the 2002 annual report of Hariston Corporation. The current ratio is 1.2 to 1. What is the value of current liabilities?
\$15,000/1.2 = \$12,500 value

2. In problem 1, the acid-test ratio is .9 to 1. What is the value of merchandise on hand?
1.2 − .9 = .3 × \$12,500 = \$3,750 value

3. Brookmire Parts Company had \$47,560 in merchandise on hand, \$35,890 in cash and other liquid assets plus \$160,058 in accounts receivable on the last day of the month. The firm also had \$209,510 in current liabilities on the same date. To the nearest tenth, what is the current ratio?
\$47,560 + \$35,890 + \$160,058 =
\$243,508/\$209,510 = 1.2 to 1

4. In problem 3, the accounts receivable to net sales ratio is .11 to 1. What is the value of net sales?
\$160,058/.11 = \$1,455,072.70 value

5. In problem 3, determine the acid-test ratio to the nearest tenth.
\$160,058 + \$35,890 = \$195,948/\$209,510 = .9 to 1

6. The current assets in problem 3 are 70 percent of the total assets of the firm. The total liabilities are 40 percent of the total assets of the firm. What is the value of the capital stock of the firm?
\$243,508/.7 = \$347,868.57 × .6 = \$208,721.14 capital

7. The net sales of Holotron Company increased from \$1,690,200 to \$2,394,800 over a period of one year. In the same period the accounts receivable to net sales ratio changed from .2 to 1 to .3 to 1. What is the amount of change in accounts receivable?
\$1,690,200 × .2 = \$338,040 \$2,394,800 × .3 =
\$718,440 − \$338,040 = \$380,400 change

8. The total assets of Shardin Corporation are \$740,700, of which \$560,400 are fixed assets. Current liabilities are \$194,600. Long-term liabilities are twice the amount of current liabilities. What is the value of capital stock?
\$194,600 × 2 = \$389,200
\$740,700 − \$194,600 − \$389,200 = \$156,900 value

9. To the nearest tenth, what is the current ratio in problem 8?
\$740,700 − \$560,400 = \$180,300/\$194,600 = .9 to 1

10. The net sales at the Virginia Cutter Corporation were \$127,640 for the month. Net income for the same period was \$21,610. To the nearest tenth, what is the net income to net sales ratio?
\$21,610/\$127,640 = .2 to 1

11. Nona and Groome Associates had operating expenses of \$65,900 in 2001. The operating expenses for 2002 were \$72,865. Sales for the same period increased by 18 percent. Did the percent increase in operating expenses exceed the percent increase in sales?
\$72,865
−65,900
\$6,965
$$\frac{\$6{,}965}{\$65{,}900} = 10.6\% \quad \text{No}$$

12. The total assets of the Moloton Corporation are \$13,495,640. The total liabilities are equal to proprietorship. What are the total liabilities of the Moloton Corporation?
\$13,495,640 assets equals liabilities plus proprietorship
$$\frac{\$13{,}495{,}640}{2} = \$6{,}747{,}820 \text{ total liabilities}$$

13. Thunderbird and Hutchinson, Inc., had an acid-test ratio of 1.4 to 1. The current ratio for the same quarter was 2.3 to 1. The firm had \$61,500 in current liabilities. What is the value of the merchandise on hand if the firm did not have any prepaid assets?
\$61,500 × (2.3 − 1.4) = \$55,350 merchandise on hand

14. The accounts receivable to net sales ratio is .095 to 1. Accounts receivable are $64,920. What are net sales?

B = *B = P ÷ R*
R = 9.5% *B = $64,920 ÷ .095*
A = $64,920 *B = $683,368.42 net sales*

15. The total assets of the Levy Corporation were three times the amount of the liabilities. The owner's equity consisted of $450,000 of capital stock. What was the amount of the firm's assets?

A = L + P
1A = 3L
$450,000 ÷ 2L = $225,000
(3) $225,000 = (1) $225,000 + $450,000
A = $675,000

16. Bilby and Son, Inc., had an acid-test ratio of 1.8 to 1. The current liabilities total $421,860. The current ratio is 3.1 to 1. What is the value of merchandise on hand?

3.1
−1.8
1.3

$421,860 × 1.3 = $548,418 merchandise on hand

17. Salane Interiors has $62,460 in cash, $22,300 in receivables, and $81,415 in liabilities. What is the acid-test ratio for the firm? (Round to the nearest hundredth.)

$$\frac{\$62,460 + \$22,300}{\$81,415} = 1.04 \text{ to } 1$$

18 Write a complete sentence in answer to this question. Hollen and Cartwright had an accounts payable to net purchases ratio of 2 to 1. Their purchases for the period were $43,800, less $2,650 in purchase returns and allowances. What is the amount of the accounts payable?

$43,800
−2,650
$41,150 × 2 = $82,300
The amount of accounts payable is $82,300.

★ Challenge Problem

A comparative balance sheet shows a decrease in current liabilities from 2001 to 2002 by 14.3 percent. The current ratio in 2002 is 1.7 to 1. The acid-test ratio in 2002 is 1.5 to 1. What were current liabilities in 2001? Merchandise inventory and prepaid assets were $32,000 in 2002.

1.7 current ratio
−1.5 acid-test ratio
.2 merchandise inventory to current liabilities
$32,000 ÷ .2 = $160,000 current liabilities, 2002
B =
R = 85.7% (100% − 14.3%)
A = $160,000

B = A ÷ R
B = $160,000 ÷ .857
B = $186,697.78 current liabilities, 2001

MASTERY TEST Chapter 20

Name ______________________ **Date** ______________ **Score** ____________

1. The purpose of a comparative balance sheet or comparative income statement is to compare *the current year* to *the previous year*.
2. Assets = *Liabilities* + Owner's Equity
3. The report that gives the recent financial history of a firm is the *income statement*.
4. When the revenue of a firm is less than the expenses for the same period, the firm has incurred a *loss*.
5. Dividing the current assets—less merchandise inventory—by current liabilities will determine the *acid test* ratio.
6. The report that shows a firm's current financial status is termed the *balance sheet*.
7. The ratio that indicates how well the firm is managing the credit extended as a buyer is the *accounts payable to net purchases* ratio.
8. The ratio that measures how well our firm is collecting from our customers is the *accounts receivable to net sales* ratio.
9. The two factors that are shown on an income statement are *revenues* and *expenses*.
10. There are two general types of income statements. They are for *service* firms and *merchandising* firms.

The following information was taken from the 2002 comparative financial statements of the Owens Art Co. Use this information to solve the problems below. Round off your answers to the nearest tenth of a percent or tenth.

	2002	**2001**
Net Sales	$240,000	$220,000
Cost of Goods Sold	180,000	178,000
Merchandise Inventory	12,500	13,800
Cash	5,400	5,400
Income for the year	34,800	29,350
Accounts Payable	6,000	6,000
Accounts Receivable	7,900	8,500
Net Purchases	178,600	177,000

11. Determine the percent of increase of net sales *9.1* %
12. Determine the current ratio of 2002. *4.3 to 1*
13. Determine the acid-test ratio for 2002. *2.2 to 1*
14. Determine the percent of cost of goods sold to net sales in 2002. *75* %
15. Determine the accounts receivable to net sales ratio in 2002. *0 to 1*
16. Determine the accounts payable to net purchases ratio in 2002. *0 to 1*
17. Determine the current ratio for 2001. *4.6 to 1*
18. Determine the percent of change in cash from 2001 to 2002. *0* %
19. Determine the net income to net sales ratio for 2002. *.1 to 1*
20. Determine the percent of increase of net purchases. *.1* %

79,458	48,424	75,214	77,002	85,943	83,739
-	-	-	302,289	-	29,053
403,840	347,840	341,564	382,133	377,780	724,793
100,339	119,667	144,160	137,456	146,064	191,345
-	92,667	137,572	150,848	196,320	257,777
-	-	-	26,650	33,630	40,448
88,575	84,243	115,560	79,385	-	-
31,421	33,016	28,214	30,450	37,511	44,269
36,741	49,375	50,368	32,058	24,221	22,242
23,583	25,702	36,182	28,956	-	-
-	49,934	197,749	-	-	-
272,633	688,135	-	-	-	-
452,955	450,540	464,636	508,439	619,313	671,912
195,713	195,738	202,326	202,973	198,854	217,878
37,957	-	63,071	55,655	71,976	84,370
9,136	22,009	11,328	2,613	3,836	-
-	-	38,464	33,098	-	-
82,443	104,885	123,655	129,171	165,346	174,181
126,001	145,198	157,436	155,452	177,480	143,023
-	-	-	-	-	-

Statistics 21

Suggested Classroom Time for This Chapter

hours 1 2 3 4 5

Perceived Student Difficulty with Topical Material

easy moderate difficult

Objectives

After mastering the material in this chapter, you will be able to:

1. Calculate the mean value of a data set. p. 410
2. Calculate the median value of a data set. p. 411
3. Calculate the mode value of a data set. p. 412
4. Calculate the range in a data set. p. 412
5. Use and understand the following terms:
   Data Set
   Mean
   Median
   Mode
   Range

Sport-A-Merica

In his ongoing search for information that will assist them in improving Sport-A-Merica's sales and profits, Hal continues to examine the financial facts. He looks carefully at where Sport-A-Merica customers come from, how Sport-A-Merica advertises, and how Sport-A-Merica's capital is structured. He looks at how to use the data that is collected every day by the cash register, the current mailing list of customers, and the count of customers that come into the store each day. Hal knows that he is not fully utilizing all the information available. For example, on a recent weekend 385 people came into the store. At times, there were just a few customers to deal with. At other times, it seemed like there were five times as many people in the store judging by the crowd near the cash register.

To improve sales and keep ahead of the competition, some strategic plans must be made. Projections can be made that will provide guidelines and keep budgets and building plans in line with sales increases for Sport-A-Merica's future.

Before he makes any recommendations to Kerry, Hal wants to review what has happened in the past at Sport-A-Merica and then make plans.

Statistics are used in business problem-solving to help select the best alternative. A manager faced with a problem involving market analysis, pay rates for employees, quality control, or population effects can use statistics as a tool with which to solve the problem. Statistics are becoming more important as our population grows and as we move into a financial/information/service-oriented economy. Sophisticated information-gathering systems are being developed to deal with these changes.

data set

Statistics are used to help understand a data set. The **data set** is a series of values grouped by year, employee classification, sales by product, scrapped material by finished product, etc. That is, all the pieces of information are in some way related. They have something in common.

Quality Paint Co. Sales by Product

Year	Hightone A	Percent Increase*	Glass Cover	Percent Increase	Flat Cover	Percent Increase
1996	$185,687	—	$ 63,470	—	$39,450	—
1997	192,562	3.7	82,357	29.8	41,680	5.7
1998	192,800	3.8	96,119	51.4	40,350	2.3
1999	195,362	5.2	101,350	59.7	41,600	5.4
2000	195,020	5.0	103,280	62.7	42,965	8.9
2001	197,790	6.5	104,050	63.9	45,307	14.8

*1996 is used as a base year for comparison.

There are two sources for such data. *Primary data* is gathered by the firm from its own records or from its own survey. A firm that conducts a telephone survey of households in the community regarding a product recently introduced by the firm would be collecting primary data.

Secondary data is gathered by the government, trade associations, or public agencies. This information is published and is readily available. The census taken by the federal government is an example of a source of secondary data, because this data is available as public information.

A thorough study of statistical analysis includes a great deal of advanced math and is beyond the scope of this text. However, a basic understanding requires very basic arithmetic.

Each value in a data set has meaning and is of importance to a decision maker. There are also some useful generalizations that can be made about the data. These generalizations are termed mean, median, mode, and range. They will assist the decision maker by indicating overall general values that describe the data set.

Finding the Mean

Objective 1

The **mean** is the arithmetic average of a data set. The mean is found by dividing the sum of the values by the number of values in the data set.

Example 1

In the chart for Quality Paint Company, the mean for the sales of Hightone A is $193,203.50.

$$\frac{\text{sum of values (1996 to 2001)}}{\text{number of values (years)}} = \frac{\$1{,}159{,}221}{6} = \$193{,}203.50 = \text{mean}$$

(1,159,221 ÷ 6 =)

In this example, the actual value *$193,203.50* is not found in the data set. The mean is only a representative value.

Example 2

Find the mean in the following data set of hourly rates at Sport-A-Merica.

$7.75 $7.20
$8.00 $9.50

$$\frac{\text{Sum of values (hourly rates)}}{\text{Number of values (employees)}} = \frac{\$32.45}{4} = \$8.11/\text{hr} = \text{mean}$$

(32.45 ÷ 4 =)

Finding the Median
Objective 2

The **median** is the middle-ranked value in a data set. The values must be ordered or ranked from high to low, or from low to high. The middle value is then selected as the median. The middle-ranked value is found by taking the number of ranked values, adding one, and dividing by two $\left(\frac{n+1}{2}\right)$. The result is the location in the data set.

Example 3

● ***The process of ordering the data set is tedious, but it makes finding the median, mode, and range very easy.***

Find the median in the data set below.

147 92 46 35 81 72 65 63 81 19 64 96 163

Ranked data set:

163 147 96 92 81 81 72 65 64 63 46 35 19

There are thirteen values. The middle value is the seventh value $\left(\frac{13+1}{2} = 7\right)$.

The seventh value is 72.

(13 + 1 = 14 ÷ 2 =)

If there is an even number of values in the data set, there cannot be a middle value. In such cases, the two middle values are *averaged* to determine the median.

Example 4

Find the median in the following data set.

$72.50 $57.80 $86.80 $63.00 $130.20 $47.50

Ranked data set:

$130.20 $86.80 $72.50 $63.00 $57.80 $47.50

middle ranked values ($72.50 and $63.00)

There are 6 values. The middle two values are $72.50 and $63.00. The median is $67.75.

$$\$72.50 + \$63.00 = \frac{\$135.50}{2} = \$67.75 = \text{median}$$

(72.50 + 63.00 = 135.50 ÷ 2 =)

Finding the Mode

Objective 3

The **mode** is the value of greatest frequency. Some data sets have values that appear more than once. The repetition is a fact that a decision maker should notice. An f is used to indicate frequency.

The mode is easily found by ranking a data set as one does to find the median.

Example 5

Find the mode in the following data set.

47 91 63 81 82 52 63 37 44 64
27 63 82 88 93 59 51 37

Ranked data set:

93 91 88 82 82 81 64 63 63 63 59 52 51 47 44 37 37 27

(82 82: $f = 2$; 63 63 63: $f = 3$; 37 37: $f = 2$)

The mode is 63. Sixty-three has the greatest frequency ($f = 3$).

Occasionally, two or more values will have the greatest frequency. When two values have the greatest frequency, the data set is said to have a *bimodal distribution*. If a data set has no values with frequency greater than one, there is no mode.

Example 6

The data set below represents the customer count per hour over the two-day weekend at Sport-A-Merica. Find the mode.

12 13 27 31 51 36 24 62 36 18 18 12 18 12 15

Ranked data set:

62 51 36 36 31 27 24 18 18 18 15 13 12 12 12

(36 36: $f = 2$; 18 18 18: $f = 3$; 12 12 12: $f = 3$)

There are two modes, 18 and 12. Each has a frequency of 3.

Finding the Range

Objective 4

The measures of data have so far been generalizations about the normal or average value of the data. The **range** measures how broad the data is. It is found by subtracting the smallest value in the data set from the largest value. The difference is the range.

Example 7

Using the data in example 6, find the range of customers per hour at Sport-A-Merica over the two-day weekend.

The largest value is 62. The smallest value is 12.
The difference is 50. The range is 50.

(62 − 12 =)

❍ *Notice that the range* ***is not*** *the largest number.*

Example 8

Find the mean, median, mode, and range of the following data set.

**Rejected Parts—
Poly-Ormin**

Date		Rejects	Ranked Data Set	
June	1 =	7	23	
	2 =	1	18	
	3 =	6	16	
	4 =	14	14	
	5 =	8	13	2
	6 =	7	13	
	7 =	6	12	
	8 =	5	11	3
	9 =	18	11	
	10 =	9	11	
	11 =	4	10	2
	12 =	6	10	
	13 =	13	9	2
	14 =	8	9	
	15 =	12	8	3
	16 =	6	8	
	17 =	5	8	
	18 =	13	7	2
	19 =	3	7	
	20 =	5	6	4
	21 =	10	6	
	22 =	11	6	
	23 =	11	6	
	24 =	23	5	4
	25 =	16	5	
	26 =	9	5	
	27 =	11	5	
	28 =	10	4	
	29 =	8	3	
	30 =	5	1	
			270	= Sum

$$\text{Mean} = \frac{\text{sum of values}}{\text{number of values}} = \frac{270}{30} = 9$$

$$\text{Median} = \frac{8 + 8}{2} = 8$$

Mode = 6 and 5

Range = 23 − 1 = 22

(270 ÷ 30 =)(23 − 1 =)

Exercise A

Find the mean, median, mode, and range in the following data sets. Round off your answers to the nearest tenth of a unit or cent.

Exercise	1	2	3	4	5	6	7
	173	65	120	15	42	$4.20	$3.93
	47	23	280	29	48	6.80	4.12
	62	44	216	13	62	7.20	3.08
	49	26	381	18	57	7.20	3.63
	39	83	90	16	39	4.90	3.90
	87	53	261	21	42	7.35	3.47
	55	19	142	26	50	8.20	4.01
	23	22	288	18	51	6.57	4.20
	89	26	120	32	83	6.80	3.91
	62	30	138	36	41	6.50	3.87
	44	18	164	29	60	9.00	4.28
	162	25	93	13	86	4.75	4.60
	12	15	287	16	39		4.07
	93	82	306	28	42		4.30
	55	23	291	29	48		3.90
	62	47	222	13	56		3.98
		36	427	30			4.22
		23	230				3.68
		5					
		24					
Mean =	69.6 (1,114 ÷ 16 =)	*34.5 (689 ÷ 20 =)*	*225.3 (4,056 ÷ 18 =)*	*22.5 (382 ÷ 17 =)*	*52.9 (846 ÷ 16 =)*	*$6.62 (79.47 ÷ 12 =)*	*$3.95 (71.15 ÷ 18 =)*
Median =	58.5 (55 + 62 = ÷ 2 =)	*25.5 (25 + 26 = ÷ 2 =)*	*226 (222 + 230 = ÷ 2 =)*	*21*	*49 (48 + 50 = ÷ 2 =)*	*$6.80*	*$3.96 (3.93 + 13.98 = ÷ 2 =)*
Mode =	62	*23*	*120*	*13 & 29*	*42*	*$6.80 & $7.20*	*$3.90*
Range =	161 (173 − 12 =)	*78 (83 − 5 =)*	*337 (427 − 90 =)*	*23 (36 − 13 =)*	*47 (86 − 39 =)*	*$4.80 (9.00 − 4.20 =)*	*$1.52 (4.60 − 3.08 =)*

Exercise	8	9	10	11	12	13
	$11.20	$6.57	$4.60	$10.62	$7.00	$7
	17.95	7.02	4.58	9.88	6.50	3
	8.76	8.08	4.40	8.62	7.75	4
	13.70	7.01	4.61	8.68	6.00	7
	10.30	7.75	4.60	6.84	6.25	8
	8.50	6.83	4.91	9.30	5.50	9
	13.80	8.10	4.56	8.61	6.75	10
	17.00	8.51	4.23	8.60	7.25	8
	13.80	7.70	4.60	8.48	7.00	6
	6.90	9.00	5.01	9.32	6.75	5
	10.00	6.70	4.71	8.91	8.00	6
		6.90	4.68	10.00	5.75	9
		8.08	4.62	16.40		2
		9.23	4.60	9.30		8
		8.53	4.41			1
						4
Mean =	*$11.99* (*131.91 ÷ 11 =*)	*$7.73* (*116.01 ÷ 15 =*)	*$4.61* (*69.12 ÷ 15 =*)	*$9.54* (*133.56 ÷ 14 =*)	*$6.71* (*80.5 ÷ 12 =*)	*$6.06* (*97 ÷ 16 =*)
Median =	*$11.20*	*$7.75*	*$4.60*	*$9.11* (*8.91 + 9.30 = ÷ 2 =*)	*$6.75*	*$6.50* (*6 + 7 = ÷ 2 =*)
Mode =	*$13.80*	*$8.08*	*$4.60*	*$9.30*	*$7.00 & $6.75*	*$8.00*
Range =	*$11.05* (*17.95 − 6.90 =*)	*$2.66* (*9.23 − 6.57 =*)	*$.78* (*5.01 − 4.23 =*)	*$9.56* (*16.40 − 6.84 =*)	*$2.50* (*8.00 − 5.50 =*)	*$9.00* (*10 − 1 =*)

Data is often grouped by categories of value. In a large data set, the data may be grouped by intervals of 100 or $10, or by years. This grouping allows a frequency to be stated for each category.

Example 9

Using the data below, group the values for the daily sales at the new Sport-A-Merica store into $10 categories. Provide an indication of frequency in each category.

Data Set:

$477	$462	$392	$459	$460	$419	$446	$410
$427	$437	$432	$436	$429	$431	$436	$423
$448	$453	$386	$419	$399	$448	$407	$427

Categories by $10 Intervals	Frequency
$381 to $390	1
391 to 400	2
401 to 410	2
411 to 420	2
421 to 430	4
431 to 440	5
441 to 450	3
451 to 460	3
461 to 470	1
471 to 480	1

ASSIGNMENT Chapter 21

Name ______________________ **Date** ______________ **Score** ______________

Skill Problem

Select the best term for each statement.

A. Data Set B. Mean C. Median D. Mode E. Range

1. *C* The middle ranked value.
2. *B* The average of a data set.
3. *E* The difference between the largest and smallest value of a data set.
4. *A* A group of values that have something in common.
5. *D* The value of greatest frequency in a data set.

Solve the following problems. Round off to whole number.

Net Sales for Svene Imports

1999	$680,400
2000	900,600
2001	1,640,300
2002	1,460,800
2003	1,365,700

6. In the chart for Svene Imports, the mean for the net sales is *1,209,560*.
(680,400 + 900,600 + 1,640,300 + 1,460,800 + 1,365,700 ÷ 5 =)

Name	Week Ending: Aug. 11, 2001	Week Ending: Aug. 18, 2001
Sean Campbell	40	38
Michael Finkel	42	$38\frac{1}{2}$
Carol Ernst	38	35

7. The mean for the hours worked over the two-week period shown above is *39 hours*.
(40 + 42 + 38 + 38 + 38.5 + 35 ÷ 6 =)
8. Rank the hours above.
42, 40, $38\frac{1}{2}$, 38, 38, 35
9. The median for the hours above is *$38\frac{1}{4}$ hours*.
10. Rank the data set below.

354	687	364	242
264	843	584	368
287	945	367	764

242, 264, 287, 354, 364, 367, 368, 584, 687, 764, 843, 945

11. What is the median of the data from problem 10? *368*
(367 + 368/2 =)
12. What is the mean of the data in problem 10? *505.75*
(354 + 264 + 287 + 687 + 843 + 945 + 364 + 584 + 367 + 242 + 368 + 764 ÷ 12 =)
13. What is the mode for data in problem 7? *38*
14. What is the mode for data in problem 10? *No mode*
15. What is the range for data in problem 7? *7 hours*
(42 − 35 =)
16. What is the range for data in problem 10? *703*
(945 − 242 =)

Number of XZ50s Sold Per Week:

Jan.	Feb.	Mar.	Apr.
98	99	50	87
87	88	115	101
65	64	70	87
98	98	74	100

17. What is the mean of XZ50s sold over the four-month period shown above? *86.3 XZ50s*
(98 + 87 + 65 + 98 + 99 + 88 + 64 + 98 + 50 + 115 + 70 + 74 + 87 + 101 + 87 + 100 ÷ 16 =)
18. Rank the XZ50s.
115, 101, 100, 99, 98, 98, 98, 88, 87, 87, 87, 74, 70, 65, 64, 50
19. What is the median for XZ50s? *88*
20. What is (are) the mode(s) for XZ50s sold? *98 and 87*
21. What is the range for the XZ50s sold? *65*
(115 − 50 =)

Find the mean, median, mode, and range in the following data sets. Round off your answers to the nearest tenth of a unit or whole cent.

A.	B.	C.	D.
		$6.45	$1.56
286		5.61	1.23
642	18	6.45	4.84
873	16	6.21	2.56
941	12	6.84	3.84
365	13	6.45	4.95
187	11	5.25	7.64
365	16	7.61	9.15
384	15		3.21

22. Mean of A = *505.4*
(4,043 ÷ 8 =)

23. Median of A = *374.5*

24. Mode of A = *365*

25. Range of A = *754*
(941 − 187 =)

26. Mean of B = *14.4*
(101 ÷ 7 =)

27. Median of B = *15*

28. Mode of B = *16*

29. Range of B = *7*
(18 − 11 =)

30. Mean of C = *$6.36*
(50.87 ÷ 8 =)

31. Median of C = *$6.45*

32. Mode of C = *$6.45*

33. Range of C = *$2.36*
(7.61 − 5.25 =)

34. Mean of D = *$4.33*
(38.98 ÷ 9 =)

35. Median of D = *$3.84*

36. Mode of D = *no mode*

37. Range of D = *$7.92*
(9.15 − 1.23 =)

Using the data below, group the values for the daily sales at the new garment store into $10 categories. Provide an indication of frequency in each category.

$592	$619	$620	$608	$626
623	616	629	621	599
613	609	599	599	604

	Categories by $10 Intervals	Frequency
38.	$591 to $600	*4*
39.	601 to 610	*3*
40.	611 to 620	*4*
41.	621 to 630	*4*

Determine the mean, median, mode, and range in the following data sets. Round off your answer to the nearest tenth or cent.

Problem	**42.**	**43.**	**44.**	**45.**	**46.**	**47.**	**48.**	**49.**	**50.**	**51.**
		386	27		3	48		62	6	$1.47
	53	412	29		8	62		38	9	1.51
	77	302	31		6	37	17	47	12	1.48
	81	219	31	10	12	41	26	61	11	1.50
	79	309	28	13	14	36	15	53	14	1.48
	52	427	27	18	5	40	21	46	8	1.48
	61	403	31	13	10	38	32	42	11	1.60
	77	386	30	16	11	32	27	47	5	1.46
	59	392	27	15	7	44	16	45	3	1.47
		472	32	20	12	37	24		10	1.48
			30	11	10	47	26			1.50
			26			50				1.52
			31							1.61
Mean =	*67.4* *(539 ÷ 8 =)*	*370.8* *(3,708 ÷ 10 =)*	*29.2* *(380 ÷ 13 =)*	*14.5* *(116 ÷ 8 =)*	*8.9* *(98 ÷ 11 =)*	*42.7* *(512 ÷ 12 =)*	*22.7* *(204 ÷ 9 =)*	*49* *(441 ÷ 9 =)*	*8.9* *(89 ÷ 10 =)*	*$1.50* *(19.56 ÷ 13 =)*
Median =	*69*	*389* *(386 + 392 = ÷ 2 =)*	*30*	*14*	*10*	*40.5* *(40 + 41 = ÷2 =)*	*24*	*47*	*9.5* *(9 + 10 = ÷ 2 =)*	*$1.48*
Mode =	*77*	*386*	*31*	*13*	*10 & 12*	*37*	*26*	*47*	*11*	*$1.48*
Range =	*29* *(81 − 52 =)*	*253* *(472 − 219 =)*	*6* *(32 − 26 =)*	*10* *(20 − 10 =)*	*11* *(14 − 3 =)*	*30* *(62 − 32 =)*	*17* *(32 − 15 =)*	*24* *(62 − 38 =)*	*11* *(14 − 3 =)*	*$.15* *(1.61 − 1.46 =)*

Business Application Problems

Round off your answers to the nearest tenth or percent.

1. Larry's Lawn Service recorded the number of lawns cut over the past several weeks. Rainfall had a major effect on the amount of service required. From the data set below, what is the range of service required?

Weekly Lawn Service

23, 26, 29, 28, 32, 34, 31, 33, 39, 38, 47, 33, 28, 22

47 − 22 = 25 range

2. In problem 1, what is the mean number of lawns cut per week?

23 + 26 + 29 + 28 + 32 + 34 + 31 + 33 + 39 + 38 + 47 + 33 + 28 + 22 = 443/14 = 31.64 = 31.6 lawns

3. The Tire Shop did a price study of their tire products. They found that all of their tires were priced between $49 and $129 as shown in the table below.

Tire Prices

$52.50, $49, $89, $129, $74.95, $69.50, $99.95, $59.95

What is the mean of the prices?

$52.50 + $49 + $89 + $129 + $74.95 + $69.50 + $99.95 + $59.95 = $623.85/8 = $77.98 mean

4. What is the median in problem 3?

$49, $52.50, $59.95, $69.50, $74.95, $89, $99.95, $129 $69.50 + $74.95 = $144.45/2 = $72.23 median

5. Becky and Al are considering the purchase of a retail business located in a mall. They sat outside of the store for a whole day to audit the traffic. The table below shows the number of customers leaving the store by each hour.

Morning	Afternoon	Evening
9 = 13	12 = 37	5 = 52
10 = 14	1 = 27	6 = 54
11 = 22	2 = 22	7 = 63
	3 = 24	8 = 52
	4 = 30	

What is the range of customers?

63 − 13 = 50 range

6. In problem 5, what is the mode?

22 and 52 mode

7. In problem 5, what is the mean?
13 + 14 + 22 + 37 + 27 + 22 + 24 + 30 + 52 + 54 + 63 + 52 = 410/12 = 34.16 = 34.2 mean

8. Mac had a very successful first quarter selling commercial property. The selling price of each property is listed below.
$320,000, $265,000, $400,500, $78,500, $326,000
What is the mean value of the properties Mac sold for the quarter?
$320,000 + $265,000 + $400,500 + $78,500 + $326,000 = $1,390,000/5 = $278,000 mean

9. In problem 8, what is the range of the selling prices?
$400,500 − $78,500 = $322,000 range

10. What is the median value in problem 8?
$78,500, $265,000, $320,000, $326,000, $400,500 $320,000 median

11. Andre operated a bail bond service. He recorded the number of fees he collected in the first eight days of the year. Find the mean and median for the data.

January	1	$400	*$1,300*
	2	580	*1,200*
	3	900	*950*
	4	1,300	*890*
	5	950	*900*
	6	890	*880*
	7	1,200	*580*
	8	880	*+400*
			$7,100

890 and 900: $1,790 ÷ 2 = $895 median

$7,100 ÷ 8 = $887.50 mean

12. The Suburban Transit Authority logged the following mileage during the month of August. What was the mean, median, mode, and range for the month?

Date		Mileage	Date		Mileage
August	1	17,460	August	16	28,356
	2	20,300		17	29,041
	3	28,356		18	28,386
	4	27,462		19	28,572
	5	28,360		20	28,674
	6	26,391		21	18,330
	7	18,492		22	19,437
	8	21,005		23	26,990
	9	28,431		24	27,385
	10	27,300		25	28,104
	11	28,106		26	28,671
	12	27,998		27	18,200
	13	28,304		28	18,200
	14	18,620		29	20,465
	15	19,962		30	28,465
				31	29,231

Mean = *24,937.2*
Median = *27,462*
Mode = *18,200 & 28,356*
Range = *11,771*

13. Word processing output at the Matthais Corp. is shown below for the two shifts during the week of August 15. What is the mean, median, mode, and range of the A.M. shift and the P.M. shift?

Shift	Operator	Mon.	Tues.	Wed.	Thur.	Fri.
A.M.	1	86	68	73	79	83
	2	76	84	78	83	80
	3	82	84	83	89	86
	4	79	68	73	76	77
P.M.	1	68	64	70	69	73
	2	79	83	81	80	76
	3	86	86	86	85	80

A.M. Shift

Mean = *79.4*
Median = *79.5*
Mode = *83*
Range = *21*

P.M. Shift

Mean = *77.7*
Median = *80*
Mode = *86*
Range = *22*

14. The operators for Oliver Express, an independent trucking firm, had the following number of 40-pound boxloads of fruit for the week of September 3. What was the mean, median, mode, and range of their loads?

985	980	994	996	1,020	965
1,005	1,000	968	980	990	
960	963	968	1,004		

Mean = *985.2*
Median = *985*
Mode = *980 & 968*
Range = *60*

1,020
1,005
1,004
1,000
996
994
990
985—median
980
980 > *mode*
968
968 > *mode*
965
963
+ 960
14,778 ÷ 15 = 985.2 mean

15. The Courier recorded the following results of an effort to increase the number of subscribers through a recent circulation drive. The amount of circulation for the first ten-day period is shown below. Determine the mean, median, mode, and range for circulation from May 6 to May 15.

May	Circulation
6	42,300
7	42,360
8	43,000
9	43,050
10	43,630
11	43,700
12	43,900
13	43,900
14	43,980
15	44,250

10, 11: *87,330 ÷ 2 = 43,665*
12, 13: *f = 2*

434,070 ÷ 10 = 43,407

Mean = *43,407*
Median = *43,665*
Mode = *43,900*
Range = *1,950*

16. Determine the mean, median, and mode for your mastery tests to date.

Mean = ____________________

Median = ____________________

Mode = ____________________

17. Find the difference between the mean and the median in the number of takeoffs and landings from the international airport data below.

492 352 403 189 381 429 223

189 + 223 + 352 + 381 + 403 + 429 + 492 =
2,469 ÷ 7 = 352.7

381
−352.7
28.3 difference

18. The number of months that each of the cars in the fleet of taxi cabs went without a breakdown is shown below. What is the range in months?

20 31 18 24 19 27 31 30 12 25
31 22 8 26 29 23 16 23 20 26
26 30 20 21 26 36 18 31

36 − 8 = 28 (range)

19. Write a complete sentence in answer to this question. Human Resources has maintained records of all retired employees for ten years. The years of service have been averaged and are shown below. What is the mean of the ten-year period?

21 23 26 21 24 25 28 28 30 26

21 + 23 + 26 + 21 + 24 + 25 + 28 + 28 + 30 + 26 =
252 ÷ 10 = 25.2
The mean of the ten year period is 25.2

20. Write a complete sentence in answer to this question. The Office Manager noted the following absences during the week of July 17.

Mon.	Tue.	Wed.	Thur.	Fri.
8	6	5	7	12

Find the median for the week.

5
6
7——median
8
12
The median for the week of July 17 was 7.

★ Challenge Problem

Find the mean and median for the information below.

43 16 18 67 9 48 68 37 23 84 35 57 127
75 78 107 47 68 32 38 126 26 60 90 28
93 10 93 85 27 53 56 53 46 26 13

Mean = 1,962 ÷ 36 = 54.5

$$\text{Median} = \frac{53 + 48}{2} = 50.5$$

MASTERY TEST Chapter 21

Name ______________________ Date ______________ Score ____________

True or False

T (F) **1.** The middle-ranked value in a data set is termed the mode.

T (F) **2.** The mean is the value that occurs most often in the data set.

(T) F **3.** There may be more than one mode in a data set.

(T) F **4.** The data set is a group of related numbers.

(T) F **5.** The mean may not actually be in the data set as a number.

(T) F **6.** The median is the value that is in the middle of a data set after the values have been placed in a numerical order.

(T) F **7.** The mean, median, and mode are values that allow certain generalizations to be made about the data set.

8. The Stanley Ambulance Service Company has recorded the following calls over the past two weeks. Determine the mean, median, and mode for the period. Round off your answer to the nearest tenth.

July	Calls	July	Calls
6	5	13	10
7	14	14	6
8	8	15	7
9	13	16	13
10	6	17	2
11	7	18	14
12	12	19	13

2, 5, 6, 6, 7, 7, 8, 10, 12, 13, 13, 13, 14, + 14

130 ÷ 14 = 9.3

Mean *9.3*

Median *9.0*

Mode *13.0*

9. Spectrum Views had a sale on video cassettes for home use. The sales are recorded below. Determine the mean sale amount for the period.

$ 230
290
260
200
325
290
285
300

$2,180 ÷ 8 = $272.50

Mean *272.50*

10. The output of two employees doing word processing is shown below. Which employee had the best output as measured by the mean?

	Nancy	Gwen
Mon.	35	34
Tues.	32	35
Wed.	28	26
Thur.	32	35
Fri.	32	32
	159 ÷ 5 = 31.8	*162 ÷ 5 = 32.4*

Gwen = 32.4

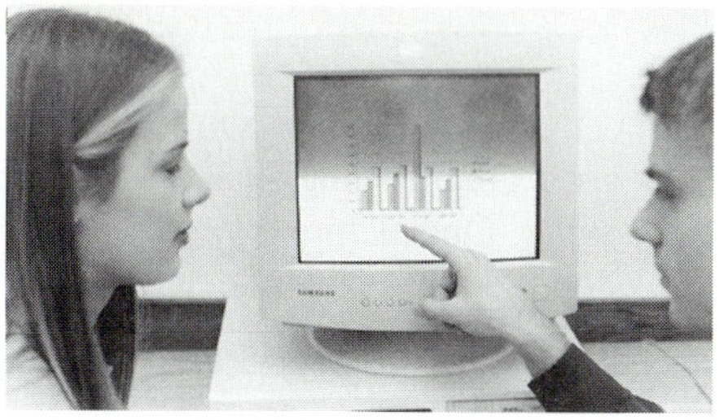

CHARTS AND GRAPHS 22

Suggested Classroom Time for This Chapter

hours 1 2 3 4 5

Perceived Student Difficulty with Topical Material

easy moderate difficult

OBJECTIVES

After mastering the material in this chapter, you will be able to:

1. Draw a circle graph, bar graph, and line graph. p. 426
2. Describe information on a bar or line graph using a base period or a direct comparison. p. 428
3. Describe two or more sets of information on one bar or line graph. p. 428
4. Use and understand the following terms:
   Circle Graph
   Bar Graph
   Line Graph

SPORT-A-MERICA

Who hasn't heard the phrase, "A picture is worth a thousand words"? In business, Hal knows that phrase is applied often. Business people appreciate the quick reference of some kind of visual representation of data. It saves them time because it communicates the information quickly and forms an overall statement of the facts. The business person can continue on and read the information that backs up the visual if he or she feels it is necessary.

Most of the data that pertain to the financial records of the firm are kept on a personal computer for reference and data manipulation. Hal knows that many times during the year the information will be used to secure a loan or locate a supplier's file. The information will also be needed for internal reports and reviews. Being able to construct and read a graph is very helpful. Hal will construct a graph to show the portion of sales that goes toward overhead and the portion that goes to profit. Selecting the correct type of graph will be important. He wants to make his point clearly to Kerry, because she has the final say in the major decisions. Though the computer has graphic capabilities, the creator of the graph will need to be able to explain the graph to the viewers.

Pictures can convey a great deal as long as the picture is constructed correctly and clearly and the message is understood.

A graph is used when a business manager wishes to show the relationship of two or more facts to each other. Though the exact figures are usually not available at a glance, the graph does allow a quick comparison of the information. The saying "One picture is worth a thousand words" is true of a graph as well. Graphs are often used to supplement actual statistics.

There are three types of graphs: circle, bar, and line.

Circle Graph
Objective 1

A **circle graph** shows the whole or base as a complete circle. The parts that make up the whole are shown as pie-shaped slices of the circle. The bigger the part is to the whole, the bigger the pie slice.

A circle is made up of 360°, a half circle equals 180°, one-quarter circle equals 90°, and so on. A protractor may be used to measure the degrees.

To determine the size of the pie slice, compare the part to the whole and then multiply by 360°, that is $\frac{\text{part}}{\text{whole}} \times 360° =$ degrees in pie slice.

The $\frac{\text{part}}{\text{whole}}$ relationship is another way of expressing $\frac{\text{amount}}{\text{base}}$, which is how a rate is determined. Therefore, you can also express the size of the slice of the circle graph with a rate expressed as a percent.

● *Students can estimate the portion of the graph for the appropriate degrees as a check on computer-generated circle graphs.*

Example 1

The Buton Company has three departments. Sales has five employees, Manufacturing has thirty-five employees, and there are ten employees in the Office. Draw a circle graph showing a comparison of each department to the total number of employees.

5 = Sales
35 = Manufacturing
+10 = Office
50 = Total Employees

$$\frac{\text{Sales}}{\text{Total}} = \frac{5}{50} \times 360° = 36° \text{ (10\%)}$$

$$\frac{\text{Manufacturing}}{\text{Total}} = \frac{35}{50} \times 360° = 252° \text{ (70\%)}$$

$$\frac{\text{Office}}{\text{Total}} = \frac{10}{50} \times 360° = 72° \text{ (20\%)}$$

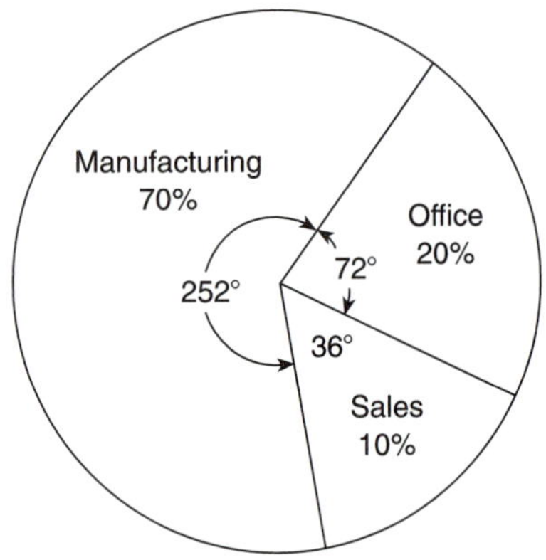

(5 + 35 + 10 =)(5 × 360 = 1,800 ÷ 50 =)(5 ÷ 50 =)

Circle graphs can be used to show a comparison of a variety of factors. The comparison is always made to the whole, to the entire circle.

Example 2

Of every dollar that Sport-A-Merica receives in sales, \$.30 is overhead, \$.55 is the cost of the merchandise for sale, and the remainder is profit. Draw a graph to show the profit, overhead, and cost.

$$\frac{\$.30}{\$1.00} \times 360° = 108° \ (30\%)$$

$$\frac{\$.55}{\$1.00} \times 360° = 198° \ (55\%)$$

$$\frac{\$.15}{\$1.00} \times 360° = 54° \ (15\%)$$

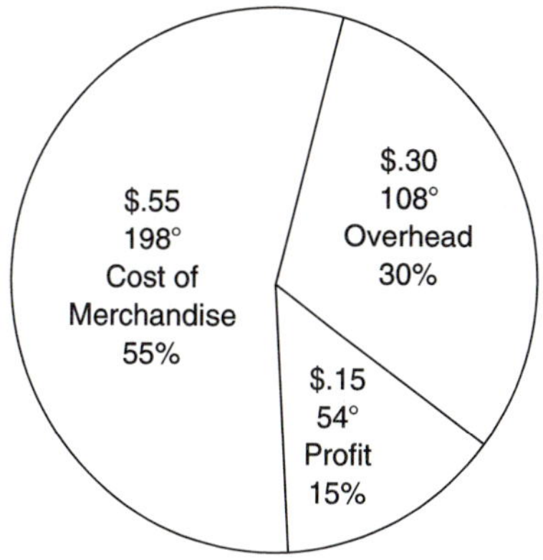

❍ *Remember:*
1/2 circle = 180°
1/4 circle = 90°
1/8 circle = 45°
How about 270°?
(3/4 circle!)

Note that in some problems the percent will need to be rounded after multiplying the 360°. The answers must then be rounded to the nearest degree and nearest percent. The sum of the degrees may therefore not total exactly 360 or the sum of the percents may not total 100. The difference is caused by rounding.

(.30 × 360)(108 ÷ 1.00 =)(.30 ÷ 1.00 =)

Bar Graph

A **bar graph** is used to show increases or decreases in an item over a period of time. There are two factors shown on a bar graph: usually time and the measurable item, such as sales, number of employees, rejected parts, or costs.

Each column or bar represents one time period. By comparing the height of each bar, a manager can see the changes.

The bar graph can be used to show only the increase or decrease from a base time period or it can show the total amount.

The bar graph is best used to show changes or trends.

Example 3

The sales for Campbell and Associates were as follows:

January = \$136,000   March = \$126,000   May = \$360,000
February = \$240,000   April = \$259,000   June = \$370,000

Draw a bar graph to show the sales for January through June.

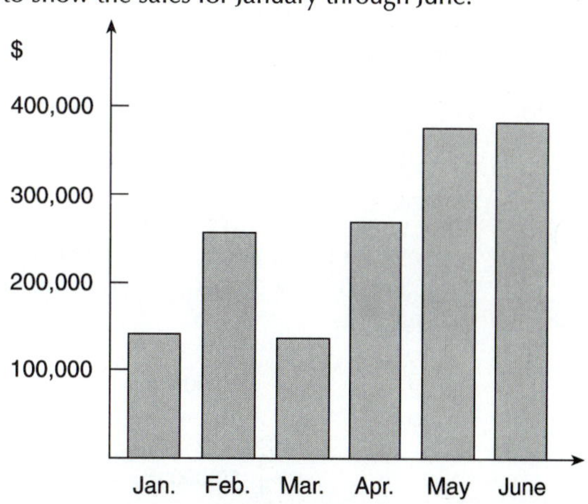

Objective 2

Example 4

Using the sales for Campbell and Associates (from example 3), draw a bar graph showing increases and decreases in sales. Use January as a base.

January sales = $136,000 = base month.

$$\text{February sales} = \frac{\$240{,}000 - \$136{,}000}{\$136{,}000} = 76\% \text{ increase}$$

$$\text{March sales} = \frac{\$126{,}000 - \$136{,}000}{\$136{,}000} = 7\% \text{ decrease}$$

$$\text{April sales} = \frac{\$259{,}000 - \$136{,}000}{\$136{,}000} = 90\% \text{ increase}$$

$$\text{May sales} = \frac{\$360{,}000 - \$136{,}000}{\$136{,}000} = 165\% \text{ increase}$$

$$\text{June sales} = \frac{\$370{,}000 - \$136{,}000}{\$136{,}000} = 172\% \text{ increase}$$

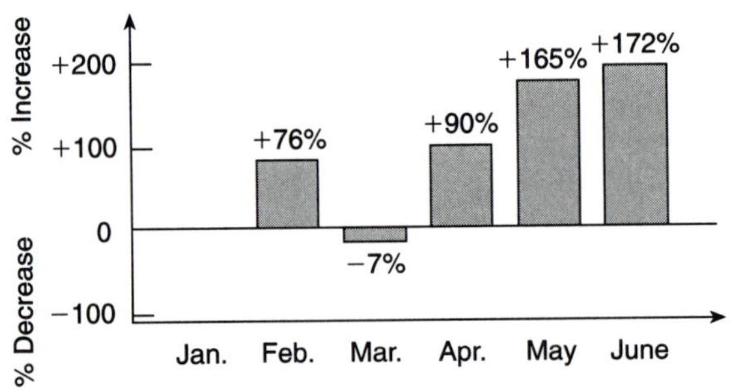

(240,000 − 136,000 = 104,000 ÷ 136,000 =)

Objective 3

By using colors or shading techniques, more than one item can be shown on a bar graph.

Care must be taken to select a representative period as a base, or else the results will not accurately reflect the comparison of periods.

Example 5

The Portland Meat Company packages beef, pork, and poultry. The output over the past five months is shown below.

	Beef (lb)	Pork (lb)	Poultry (lb)	Total (lb)
February	4,600	2,650	2,280	9,530
March	4,960	2,400	2,600	9,960
April	4,850	2,500	2,480	9,830
May	5,280	2,600	3,250	11,130
June	5,800	2,300	4,160	12,260

From this information, draw a single bar graph that will show (1) total output, (2) beef output, (3) pork output, and (4) poultry output.

- *Use examples in business publications to show graphs. U.S.A. Today uses graphs in nearly every publication.*

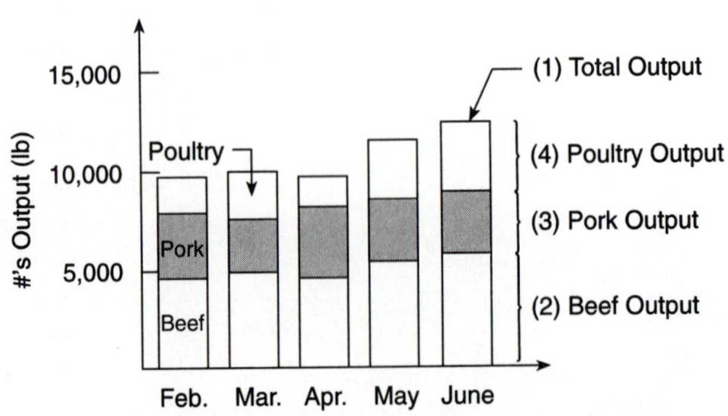

Line Graph

A **line graph** is very similar to a bar graph. The two factors are shown as on a bar graph. The bars are not shown, but instead, a point is established at the top of where the bar would have been shown. The points are then connected with straight lines. The lines show the differences between one period and others. The line graph is best used to show changes and trends.

Example 6 Draw a line graph to show the number of rejected parts in the Quality Assurance Department over the past eight weeks.

Week of	Rejected Parts
Oct 3	48
10	216
17	57
24	63
31	81
Nov 7	54
14	37
21	61

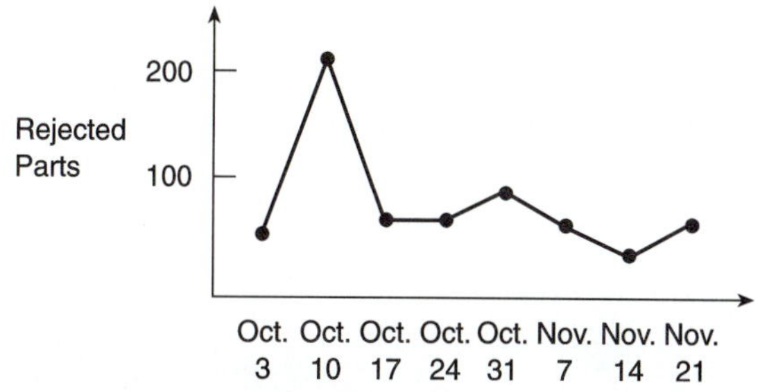

Bar graphs and line graphs can be used to show several items on a single graph.

Example 7 Draw a line graph to show selling price and cost price per unit for the Gootch Corporation over the past six years.

Year	*Selling Price*	*Cost Price*
1997	$4.63	$2.80
1998	4.82	3.35
1999	5.20	3.80
2000	5.20	3.65
2001	5.35	4.30
2002	5.38	4.30

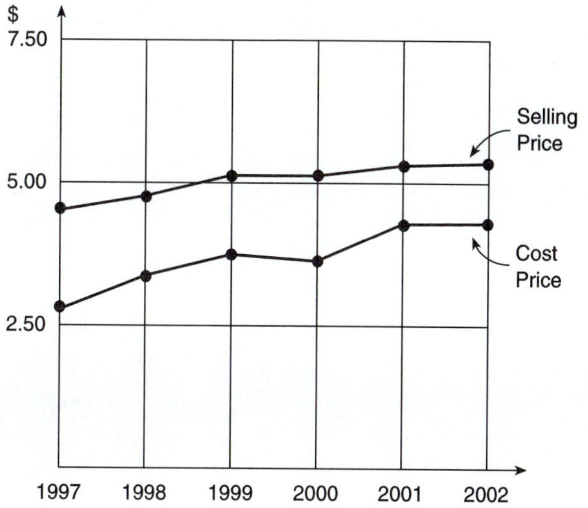

Exercise A

Draw a circle graph for each of the exercises below. Round off to the nearest percent before multiplying by 360°. Round off to the nearest degree.

1. The Plum Tree Shop has five departments. The sales of each department are shown below.

Junior Misses	$47,360 ÷ 232,420 × 360° = 73° (20%)
Sports Wear	39,400 ÷ 232,420 × 360° = 61° (17%)
Young Misses	53,980 ÷ 232,420 × 360° = 84° (23%)
Ladies Fashions	72,480 ÷ 232,420 × 360° = 112° (31%)
Accessories	+19,200 ÷ 232,420 × 360° = 30° (8%)
	$232,420

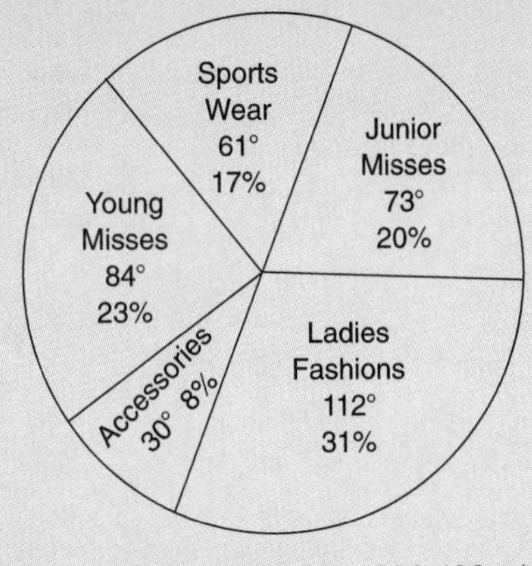

(47,360 ÷ 232,420 = × 360 =)(47,360 ÷ 232,420 =)

2. Davis Personnel found that there was a significant number of new positions in the area of technical jobs.

Administrative	430
Sales	270
Computer	310
Technical	360
Labor	+120
	1,490

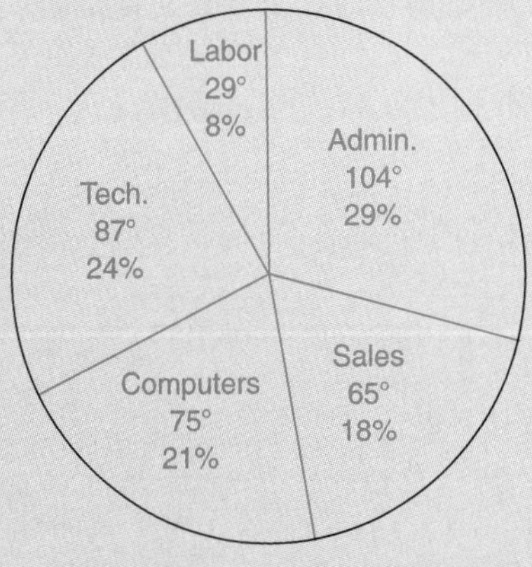

$\frac{430}{1,490}$ × 360° = 104° (29%) *$\frac{310}{1,490}$ × 360° = 75° (21%)* *(430 × 360 = 154,800 ÷ 1,490 =)*
(430 ÷ 1,490 =)

$\frac{270}{1,490}$ × 360° = 65° (18%) *$\frac{360}{1,490}$ × 360° = 87° (24%)*

$\frac{120}{1,490}$ × 360° = 29° (8%)

3. The township of Delray noted the following new housing units being built in the month of May.

Single family	108
Apartment	230
Multiunit	420

(108 × 360 = 38,880 ÷ 758 =)
(108 ÷ 758 =)

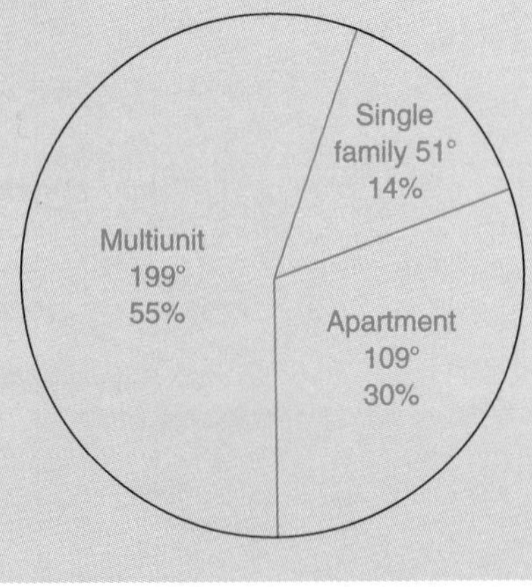

Exercise A Continued

4. The sales for Sport-A-Merica for the month were as follows:

Hiking and Camping Equipment	$42,600
Hunting and Fishing Equipment	31,850
Baseball, Football, and Soccer	12,600
Miscellaneous	8,980

(42,600 × 360 = 15,336,000 ÷ 96,030 =)
(42,600 ÷ 96,030 =)

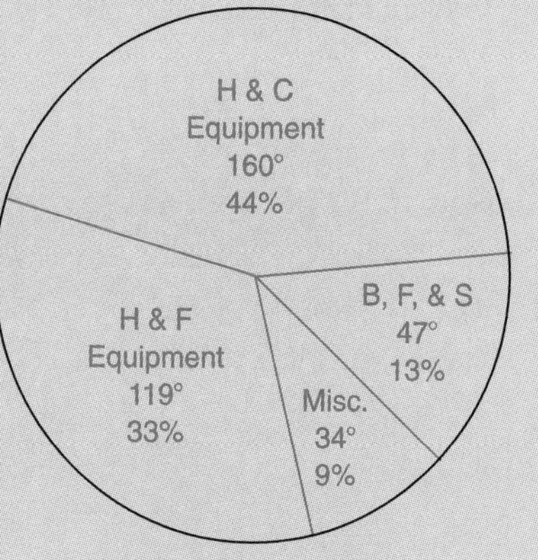

Exercise B

Draw a bar graph for each of the exercises below.

1. Monthly sales for the TASCO Company were as follows:

March	$4,300,000
April	6,059,300
May	7,300,500
June	4,800,000
July	5,693,000
August	8,360,000

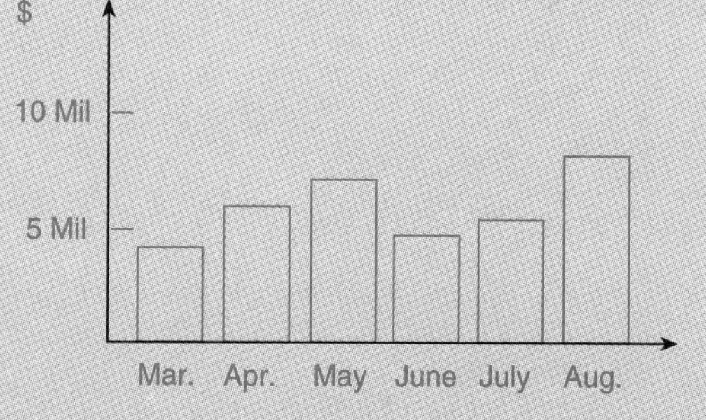

2. A traffic count at the corner of Weber and Staples indicated the following volume for the week of January 23.

Monday	2,380
Tuesday	3,200
Wednesday	1,780
Thursday	2,900
Friday	3,450

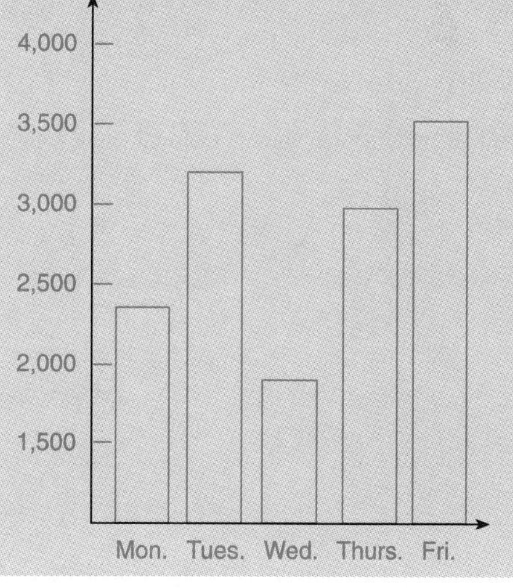

(Continued on next page)

Exercise B Continued

3. The number of employees at year-end for the Pilson & Weber Company are shown below. Show the change per year, using 1998 as a base year.

1998	4,680
1999	4,935
2000	4,320
2001	4,163
2002	4,528

(4,935 − 4,680 = 255 ÷ 4,680 =)
(4,680 − 4,320 = 360 ÷ 4,680 =)
(4,680 − 4,163 = 517 ÷ 4,680 =)
(4,680 − 4,528 = 152 ÷ 4,680 =)

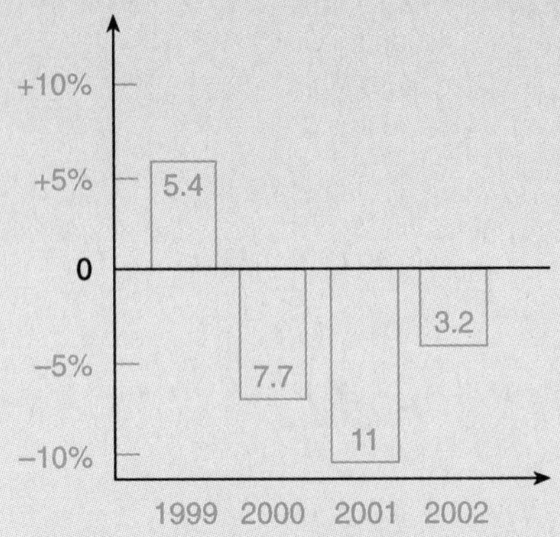

4. The total number of employees and the number of male employees are shown below for the Scott Company. Draw a bar graph to show total employees, male employees, and female employees.

Year	*Total Employees*	*Male Employees*
1998	60	39
1999	56	38
2000	84	56
2001	85	52
2002	90	63

(60 − 39 =)(56 − 38 =)(84 − 56 =)
(85 − 52 =)(90 − 63 =)

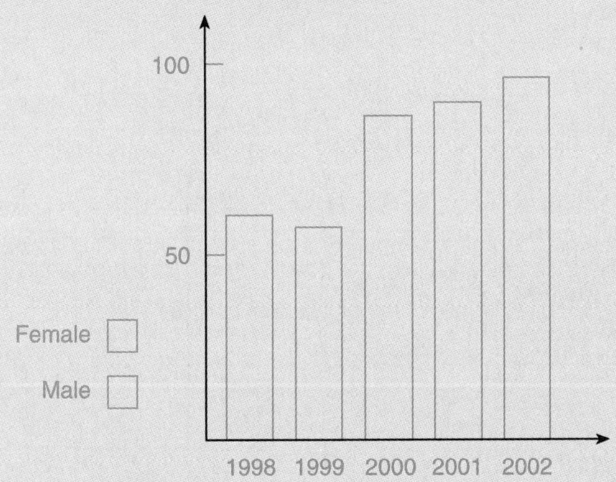

Exercise C

Draw a line graph for each of the exercises below.

1. The shipping costs over five months are shown below.

March	$4,360
April	4,823
May	4,230
June	3,986
July	5,852

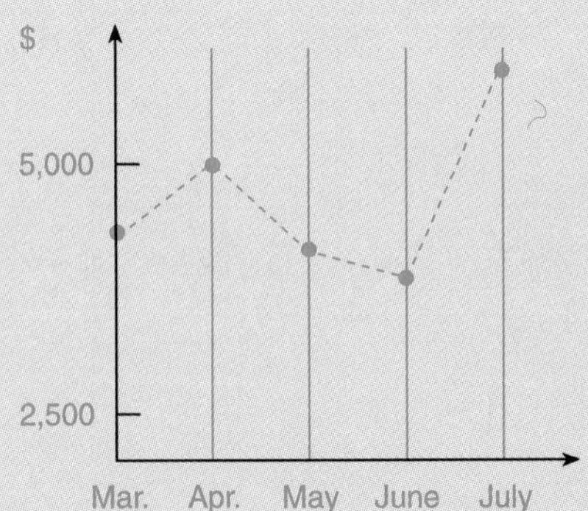

Exercise C Continued

2. Employee turnover has been as follows for the months of July through November.

July	3
August	7
September	8
October	0
November	5

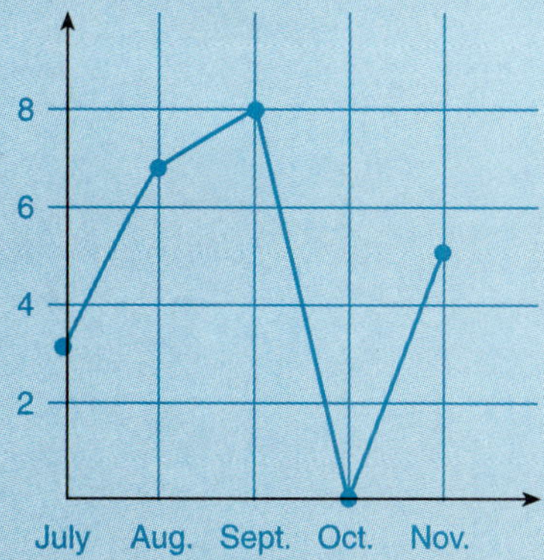

3. The sales of the Apersond Company have been steadily increasing over the six-month period of March through August. Much of the success in increased sales is due to the introduction of their quick-release valve. Draw a line graph to show both total sales and the sales of the quick-release valve.

Month	*Total Sales*	*Quick-Release Valve*
March	$32,680	$14,230
April	32,800	16,050
May	33,560	16,400
June	34,700	16,362
July	34,890	16,800
August	35,620	17,950

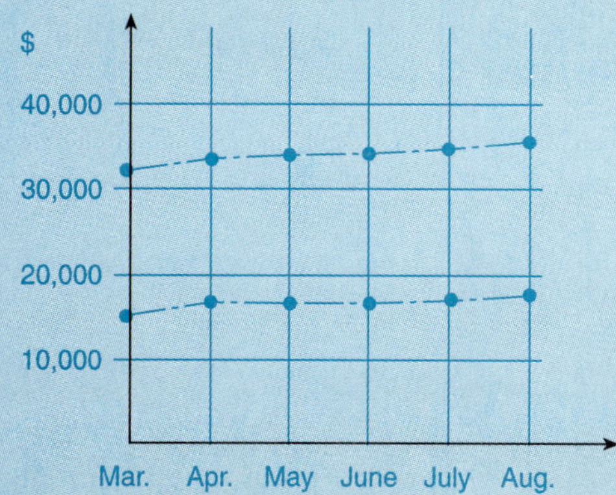

ASSIGNMENT Chapter 22

Name ______________________ **Date** ______________ **Score** ____________

Skill Problems

Match the following terms with the correct description.

A. Circle Graph B. Bar Graph C. Line Graph

B,C **1.** Best used to show changes or trends.

A **2.** Parts of a component compared to the whole.

C **3.** Points are connected to show trends.

Complete the following sentences.

4. A graph is used when a business manager wishes to show the *relationship* of two or more facts to each other.

5. In a bar graph, each *column or bar* represents one time period.

6. The circle is made up of *360* degrees.

7. In a line graph, the lines show the *differences* between one period and others.

8. The part/whole relationship is another way of expressing *percentage/base*, which is how a rate is determined.

9. The *bar graph* can show the increases or decreases from a base time period.

10. By using *colors, or shading* techniques, more than one item can be shown on a bar graph.

11. *Bar graphs* and *line graphs* can be used to show several items on a single graph.

12. A *line graph* is very similar to a bar graph.

13. In a line graph, instead of bars, a point is established, then the points are connected with *straight lines*.

Draw and use a circle graph for problems 14–26. Round off to the nearest percent before multiplying by 360 degrees. Then round off to the nearest degree. Show your figures and place percents in respective segments.

The Evergreen Waste Management Company wanted to show what was going into the land fill from each truck. Use the following figures to create the circle graph.

14. 7 yards paper *(7 ÷ 25 = 28% × 360° = 101 degrees)*

15. 6 yards plastics *(6 ÷ 25 = 24% × 360° = 86 degrees)*

16. 1 yard yard waste *(1 ÷ 25 = 4% × 360° = 14 degrees)*

17. 2 yards steel *(2 ÷ 25 = 8% × 360° = 29 degrees)*

18. 5 yards glass *(5 ÷ 25 = 20% × 360° = 72 degrees)*

19. 4 yards other *(4 ÷ 25 = 16% × 360° = 58 degrees)*

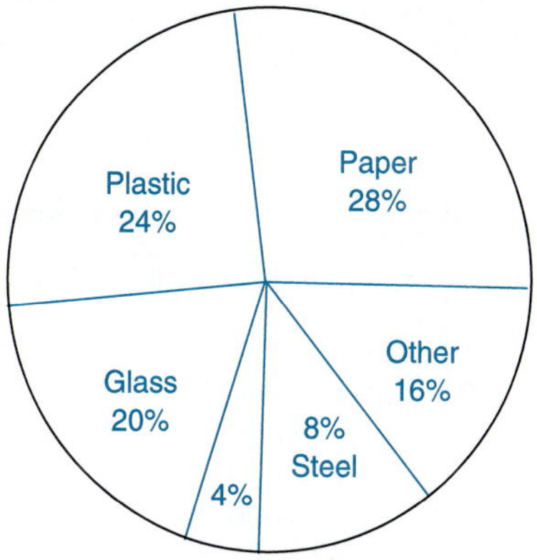

The mail room wanted to show what type of mail was being produced from the company to help control costs. Use the following figures to create a circle graph.

20. 12,000 direct mail marketing promos
(12,000 ÷ 19,000 = 63% × 360° = 227 degrees)

21. 5,000 corporate letters
(5,000 ÷ 19,000 = 26% × 360° = 94 degrees)

22. 1,000 responses to customers
(1,000 ÷ 19,000 = 5% × 360° = 18 degrees)

23. 1,000 personal letters of employees
(18 degrees)

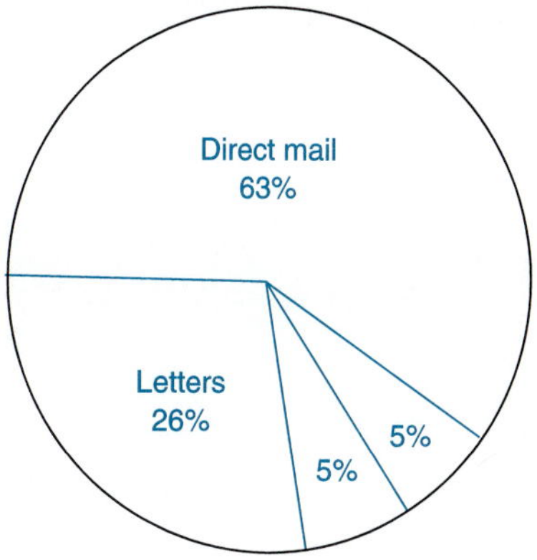

24. If the total expense for the company mailings above is $64,500, how much money is spent on direct mail marketing promos? ***$40,635*** ***(64,500 × .63 =)***

25. If each personal letter cost the company above $3.23 when time and materials are considered, how much money could the company save by eliminating the personal letters of employees? ***$3,230.00*** ***(1,000 × 3.23 =)***

26. From the information in problems 21 and 24, how much was spent on corporate letters? ***$16,770*** ***(64,500 × .26 =)***

In circle graph problems, round off to nearest percent before multiplying by 360°. Round off to nearest degree.

27. Draw a circle graph to show the following information.

Total Sales (A, B, C, and D)	$62,580
Product A	$4,350
Product B	$30,600
Product C	$4,860
Product D	$22,770

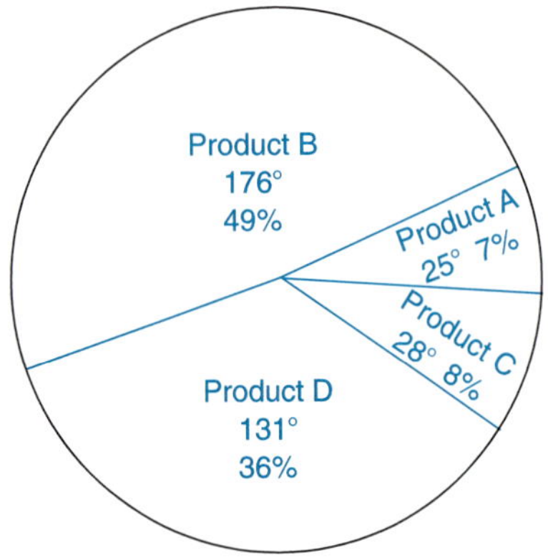

28. Of all the married couples in one state, 57 percent have two incomes. Draw a circle graph to show this relationship.
57% × 360° = 205°

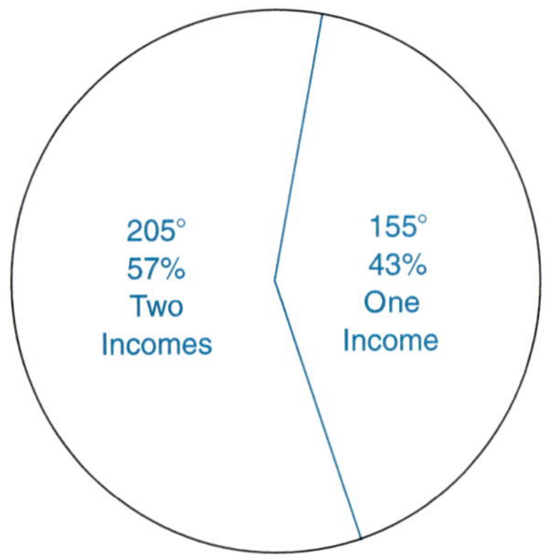

Draw a bar graph for the problems below.

29. Yearly amount of safety salt used on roads.

1997	38,500,000 tons
1998	40,800,000
1999	8,400,000
2000	25,750,000
2001	52,640,000
2002	42,670,000

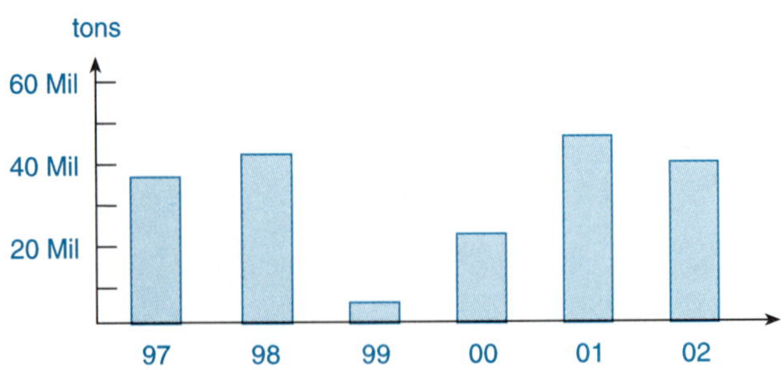

30. Indicate the volume of merchandise returned for the week of December 27.

Monday	568 items
Tuesday	742
Wednesday	820
Thursday	961
Friday	1,200
Saturday	1,500

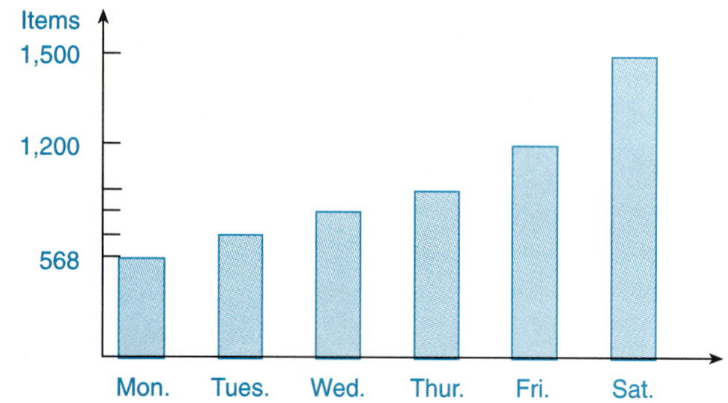

31. Draw a line graph to show the number of cars entering the company parking lot over the past six weeks.

Week of:	Cars Entered:
Dec. 6	809
13	928
20	2,105
27	701
Jan. 3	502
10	402

32. If it takes one employee working in the parking lot for every 500 cars entering, draw a line graph to show the number of employees needed to cover the same parking lot in problem 31. Round off to the nearest employee.

Week of:	*Employees Needed:*
Dec. 6	*2*
13	*2*
20	*4*
27	*2*
Jan. 3	*2*
10	*1*

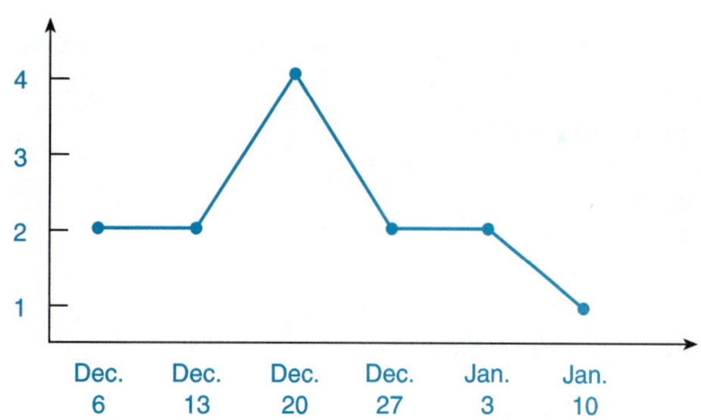

Draw a bar graph showing increases and decreases in sales. Use July sales of $6,500 as a base.

33. August $7,550 ***((7,550 − 6,500) ÷ 6,500 = 16% increase)***

34. September $3,400 ***((3,400 − 6,500) ÷ 6,500 = 48% decrease)***

35. October $5,600 ***((5,600 − 6,500) ÷ 6,500 = 14% decrease)***

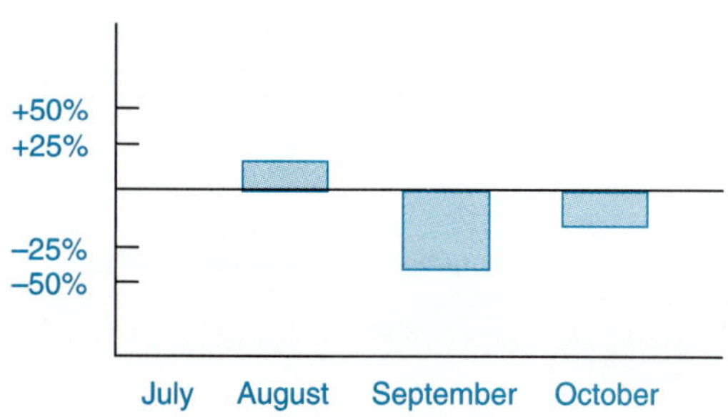

36. Draw a line graph to show the passenger miles flown by Freedom Airlines.

June	4,390
July	8,620
August	8,890
September	9,450
October	4,600
November	6,450
December	7,900

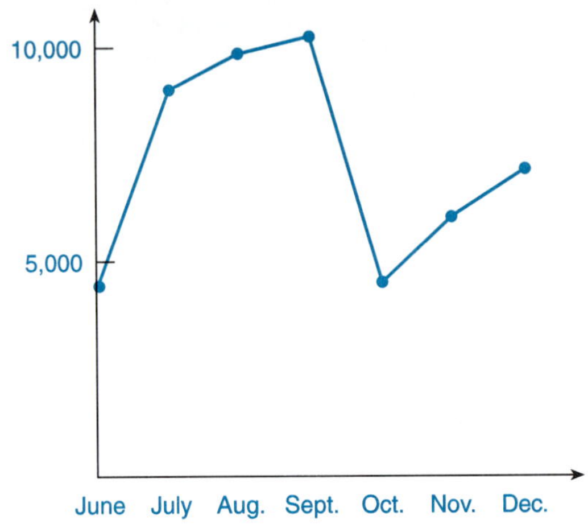

Draw a bar graph showing increases and decreases in insurance premiums. Use Company A premium of $8,750 as a base.

37. Company B $11,375 ***((11,375 − 8,750) ÷ 8,750 = 30% increase)***

38. Company C $7,435 ***((7,435 − 8,750) ÷ 8,750 = 15% decrease)***

39. Company D $10,500 ***((10,500 − 8,750) ÷ 8,750 = 20% increase)***

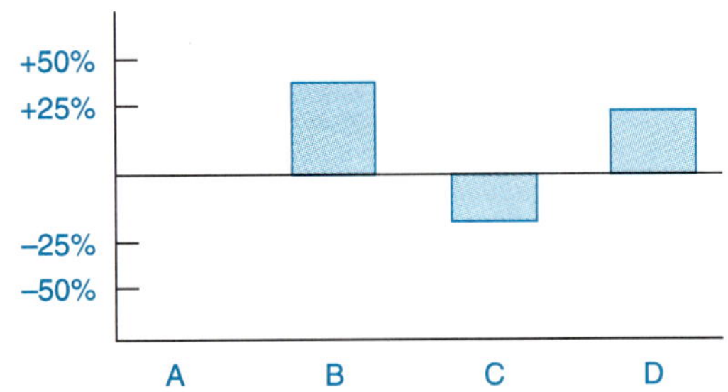

40. Draw a bar graph to show the changes in sales for the years reported below. Use 1997 as a base year.

1997	$61,190
1998	$48,650
1999	$56,890
2000	$69,400
2001	$72,350
2002	$68,870

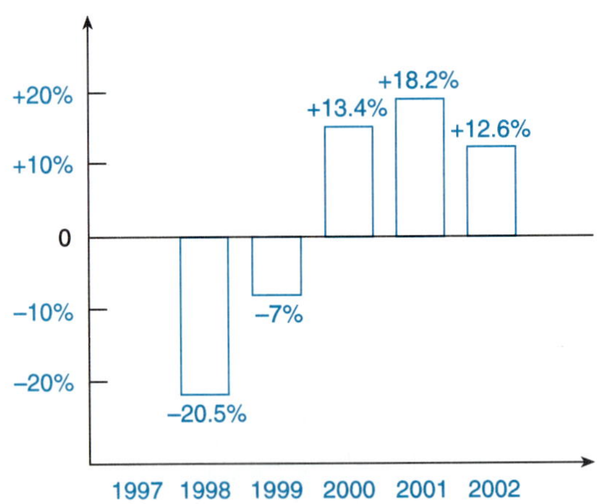

41. Draw a line graph to show your mastery test scores.

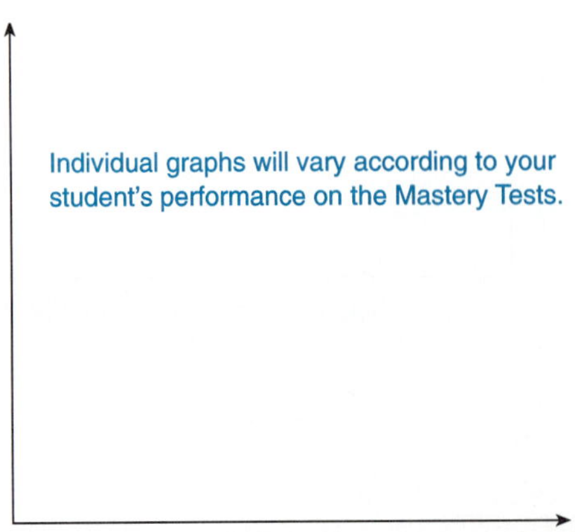

42. The telemarketing firm made three types of survey calls this week. From the information below, draw a single bar graph that will show (1) political surveys, (2) household products survey, (3) entertainment industry survey, and (4) total surveys done.

	Political	Household	Entertainment	Total Surveys Done
Monday	50	64	45	159
Tuesday	48	72	25	145
Wednesday	33	25	45	103
Thursday	65	45	33	143
Friday	48	44	62	154

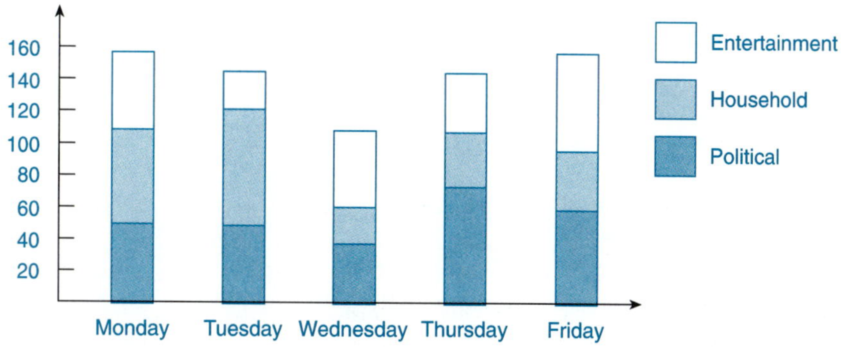

43. Draw a line graph to show net income and net sales for eight years.

Year	*Net Income*	*Net Sales*
1997	$5,000,000	$8,500,000
1998	1,000,000	12,000,000
1999	7,200,000	18,000,000
2000	8,800,000	17,000,000
2001	10,100,000	16,000,000
2002	5,900,000	17,500,000

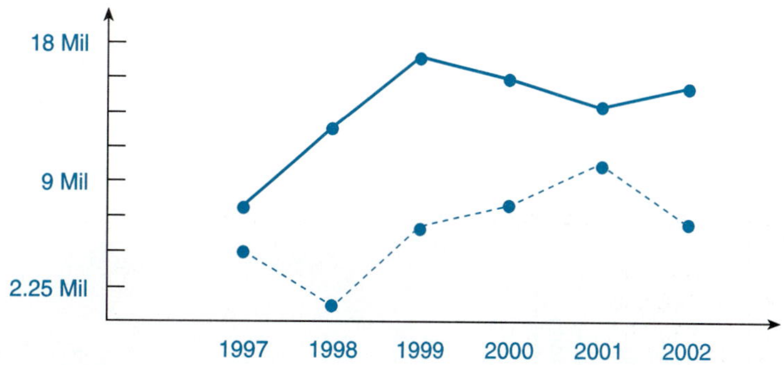

44. The total number of vehicles and the number of vehicles sold for A1 Used Car Sales is shown below. Draw a single bar graph to show the total vehicles sold, the cars, and trucks, vans, and SUVs sold.

Year	Total Vehicles Sold	Cars
1998	60	45
1999	70	30
2000	50	40
2001	70	55

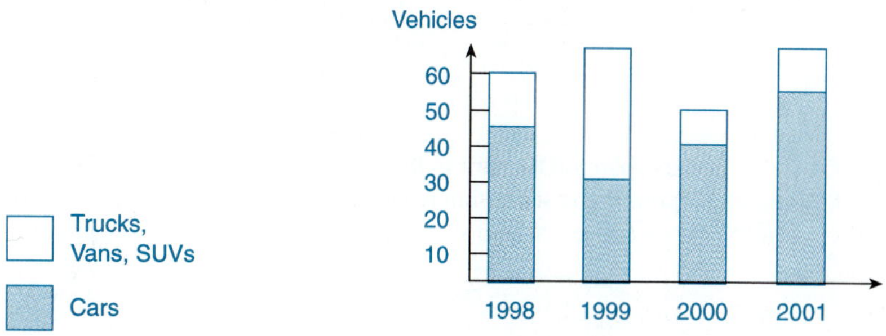

Business Application Problems

1. Sales for Bye-Gone Travel for the month were placed into five segments: Airline Travel = $452,600; Cruises = $184,000; Seaside Resorts = $634,500; Motel/Hotel Rooms = $41,300; and Auto Rentals = $84,750. Draw a circle graph to show the sales for the month.

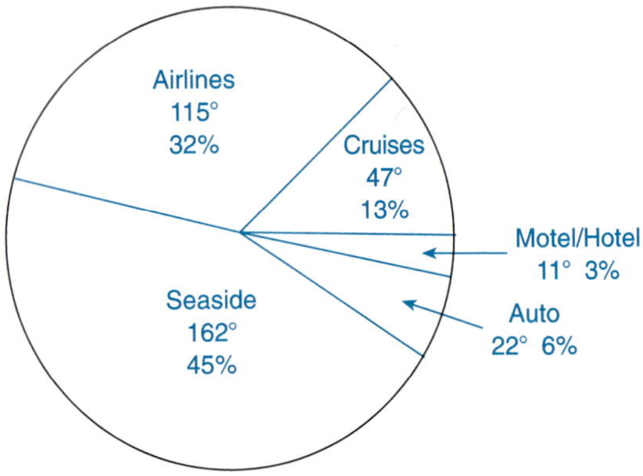

2. Draw a bar graph to represent the value of each segment of the business in problem 1.

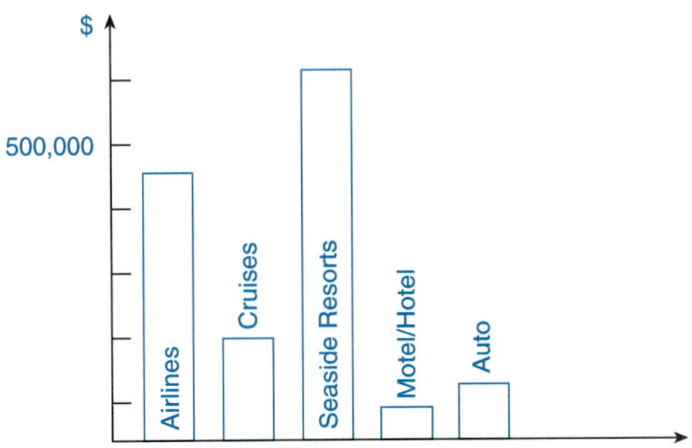

3. Of all the dollars spent on tourism in the United States, 85 percent is spent in the states that have an ocean coastline. Draw a circle graph to show this relationship.

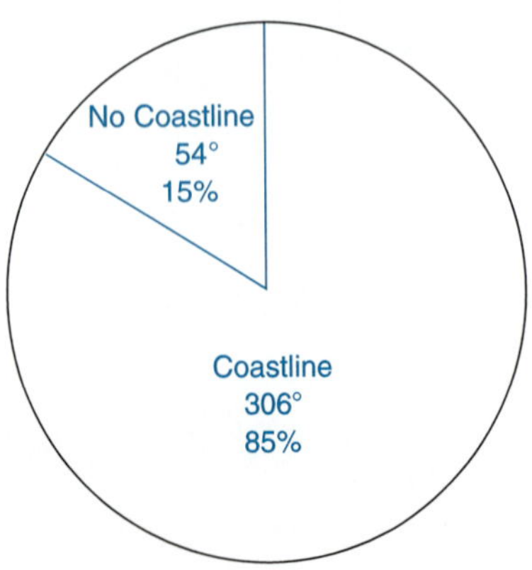

4. In one state, 14 percent of the citizens are under the poverty level, 15 percent earn above $75,000 and 3 percent earn above $100,000. The balance of the citizens earn between the poverty level and $75,000. Draw a circle graph to show these facts.

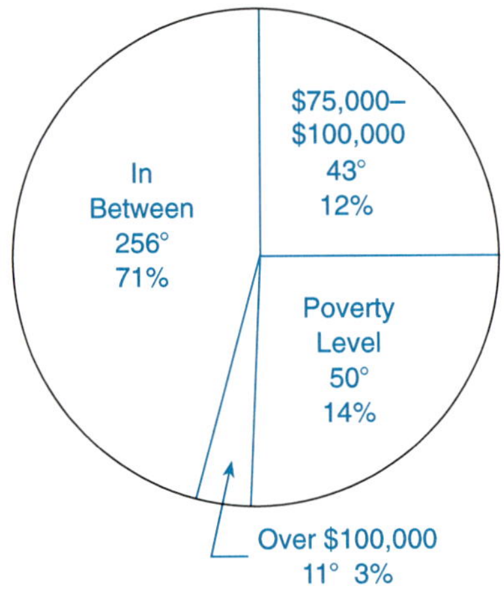

5. Draw a bar graph to show the information in problem 4.

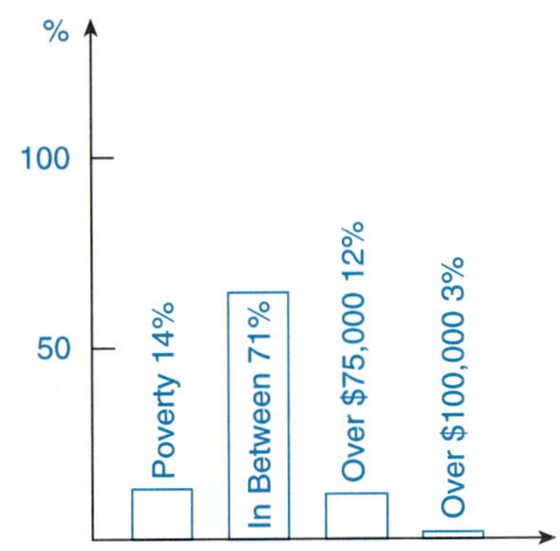

6. The number of loans in default for Sedrick Finance is listed by month for the first half of the year. Draw a bar graph to show the defaults by the month.
January 32, February 45, March 38, April 22, May 25, June 36

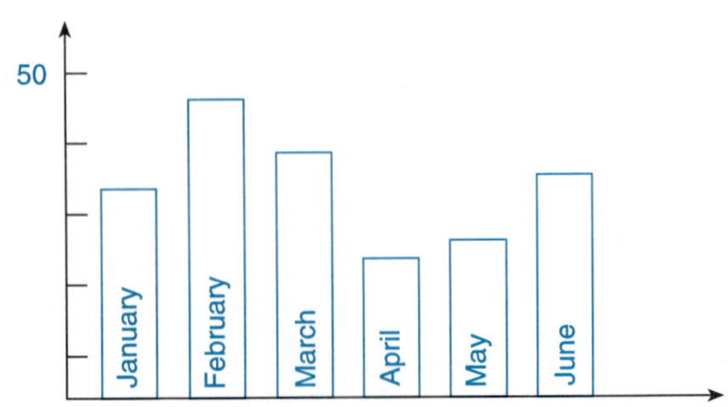

7. Using the information in problem 6, draw a bar graph to show how the number of defaults changed, based on the defaults for January.

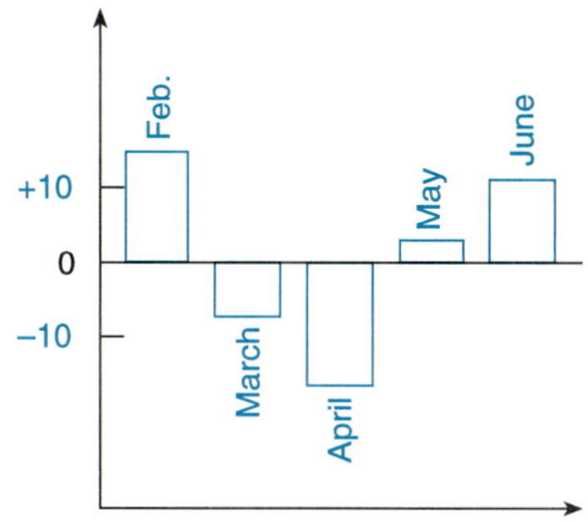

8. The miles driven by each of the service technicians on Tuesday for Circo, Inc. are shown below.
45, 133, 89, 38, 134, 96, 91, 24
Draw a bar graph to represent the miles driven by each technician.

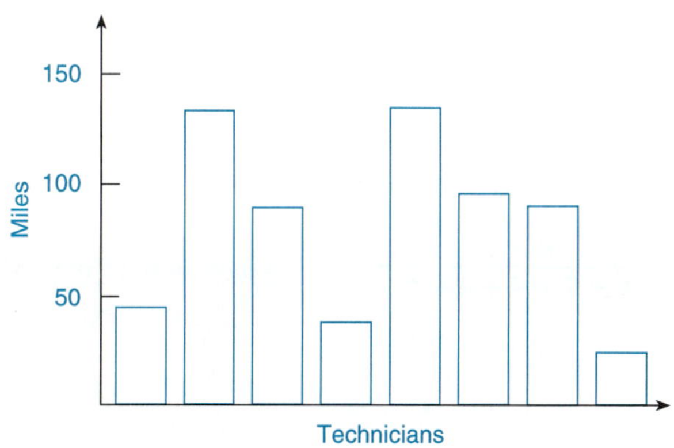

9. The total sales derived from the rural areas of the county have risen slightly over the past six years as shown below. Draw a line graph to represent the sales by year.

1997	1998	1999	2000	2001	2002	2003
$350	$340	$360	$375	$355	$389	$370

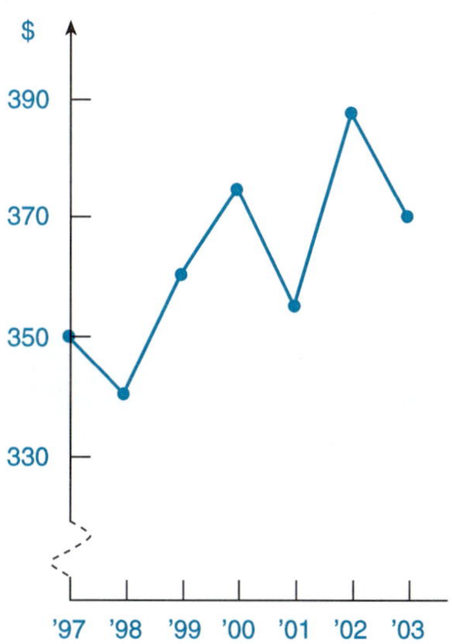

10. Using the information in problem 9, draw a line graph to represent the changes in sales based on 1997.
1998 −$10, 1999 + $10, 2000 + $25, 2001 + $5, 2002 + $39, 2003 + $20

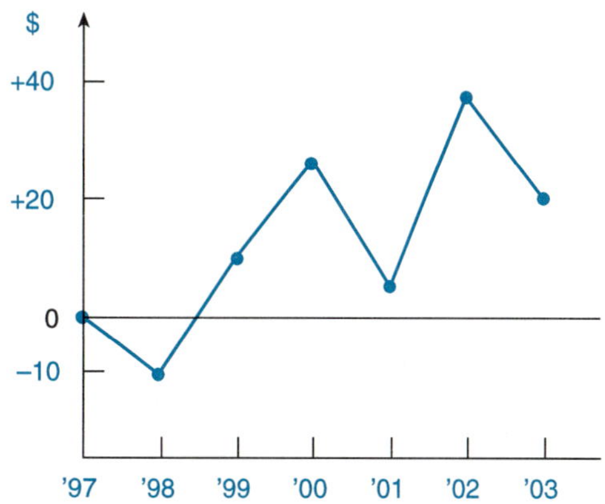

11. Draw a graph that will show the units produced and scrapped parts from the information below.

August	*Units Produced*	*Units Scrapped*
2	6,005	312
9	5,656	362
16	5,820	341
23	4,362	340
30	5,885	322

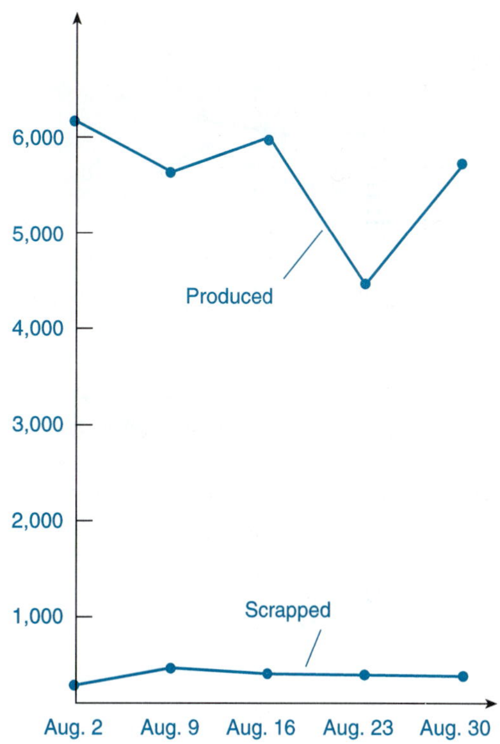

12. Draw a graph that will show the following information: Of the 42 new employees hired last month, 12 were fired, 18 were transferred, and the remainder were retained in their positions.

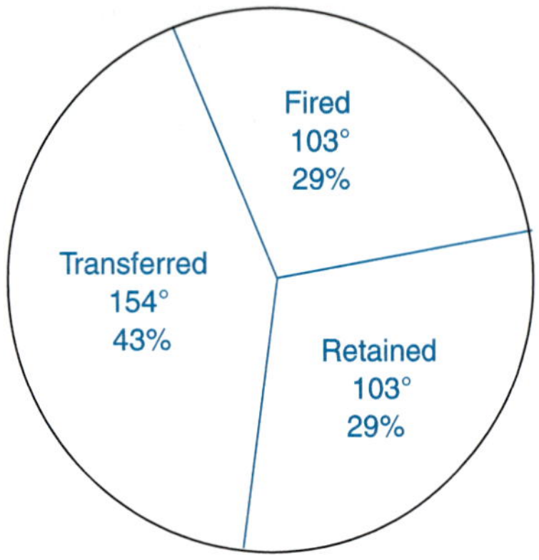

13. The prices of two of the elements of production have been recorded by the purchasing agent of a small manufacturer over the past four months. Using the data below, draw a graph to show the price changes.

	Product A	*Product B*
June	\$1.20	\$1.45
July	1.25	1.45
August	1.38	1.50
September	1.20	1.55

14. Draw a graph to show the number of available workers at a construction hiring office during each of the previous five weeks.

Week	*Workers Available*
1	42
2	30
3	30
4	18
5	25

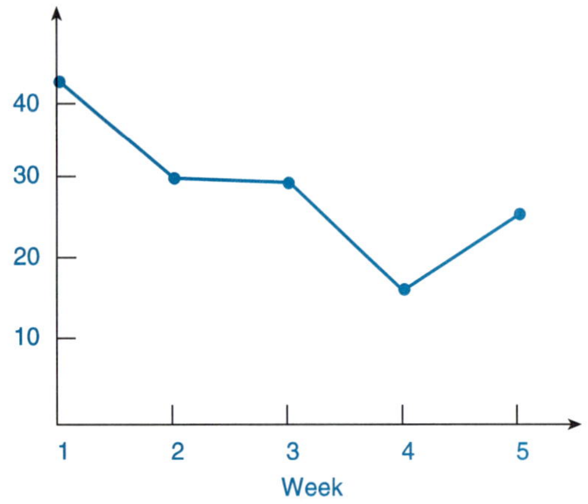

15. The number of employees that have resigned are shown below by age group. What is the percent of the total that are less than 40 years old?

$$\frac{6 + 5}{6 + 5 + 4 + 2 + 1} = \underline{61\% \text{ less than 40 years old}}$$

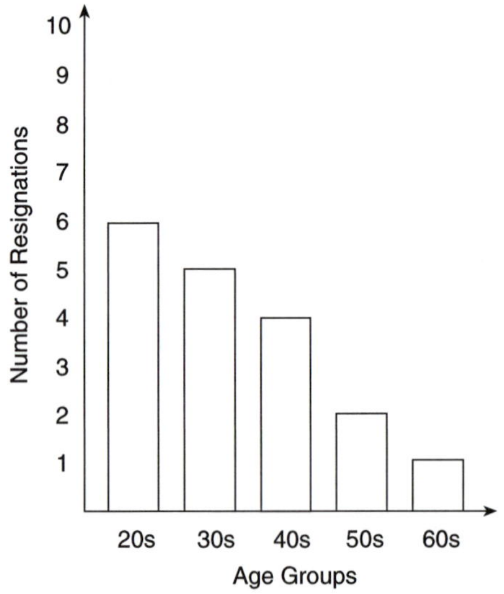

16. If the number of employees regularly employed by the firm above is 210, what is the percent of resignations for the firm?

$\frac{18}{210}$ = 8.6% resignations

17. The price of a condominium has risen by 4 percent last year and by 8 percent the previous year. Show the effect of the price increase on a condominium unit that sold for $38,000 in 2001 in a line graph.

2001 ($38,000) × 1.08 = $41,040

2001 + 1 ($41,040) × 1.04 = $42,681.60

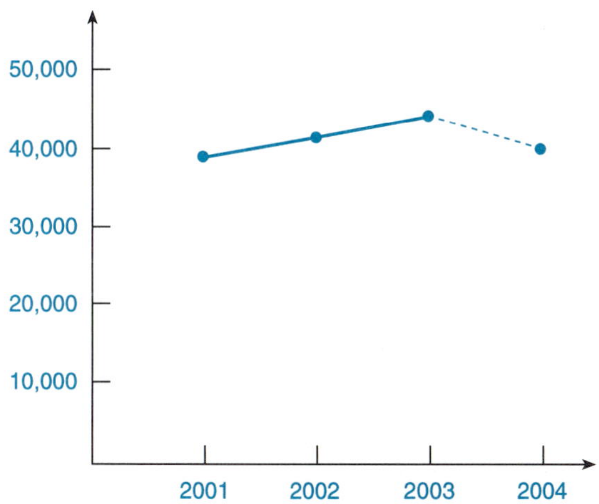

18. Write a complete sentence in answer to this question. What would the line graph in problem 17 look like if the price is projected to drop by 2 percent next year?

2001 + 2 ($42,681.60) × .98 = $41,827.97

The chart would show a drop in price to $41,827.97.

★ Challenge Problem

Chart the mean, median, and mode of the following grouped data sets.

Group A				*Group B*			
37	62	61	31	25	53	49	61
83	49	49	19	35	18	68	53
46	19	26	35	19	26	53	27
19	38	33	27	26	42	20	30

Group C			
57	49	38	18
18	81	19	31
16	26	25	27
32	53	30	62

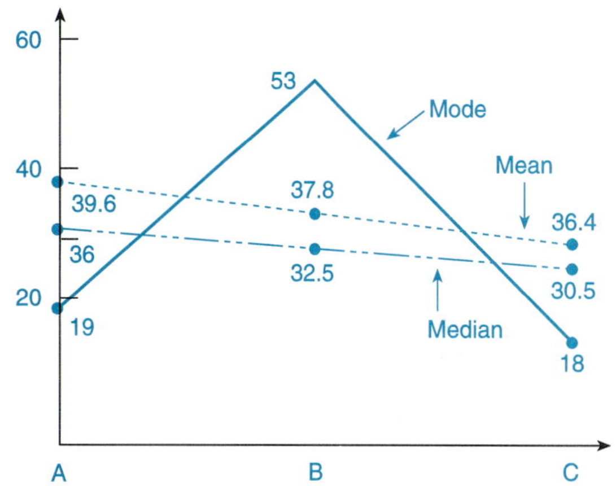

MASTERY TEST Chapter 22

Name ______________________ **Date** ______________ **Score** ____________

1. Draw a circle graph to show the following information.

   Total Revenue = \$4,500,000
   Revenue from
   Credit Sales = \$ 800,000

   $\frac{\$800,000}{\$4,500,000} \times 360° = 64°/18\%$

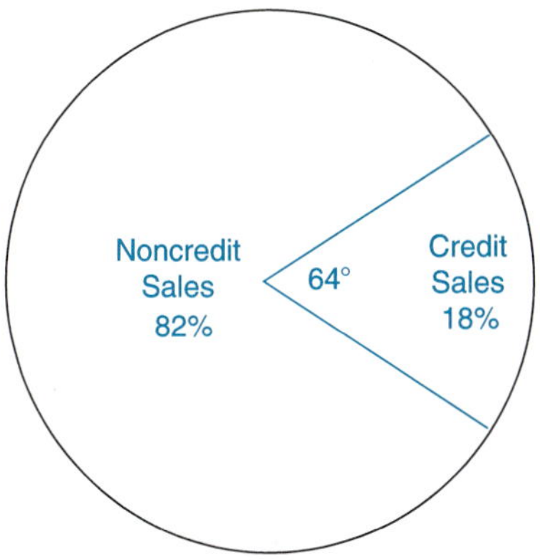

2. Draw a bar graph to show the following information for the Marlin Bank.

Year	Total Savings	Certificates of Deposit
1997	\$2,600,000	\$1,100,000
1998	3,000,000	1,080,000
1999	3,700,000	1,430,000
2000	3,800,000	1,500,000
2001	4,200,000	1,450,000
2002	4,580,000	1,700,000

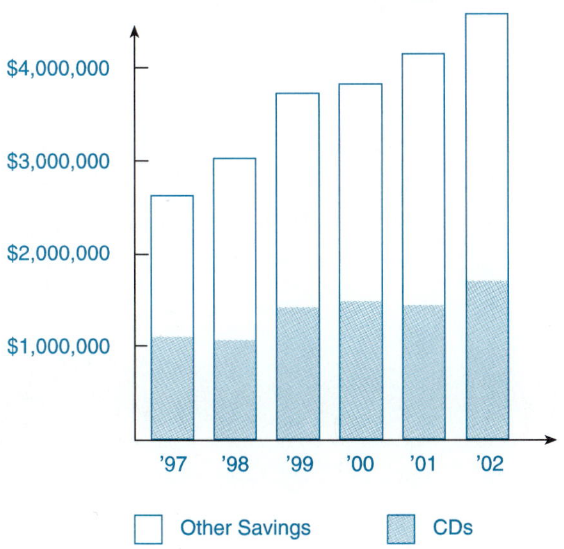

3. Draw a line graph to show the number of employees from 1998 to 2002.

Year	*Employees*
1998	390
1999	430
2000	360
2001	400
2002	420

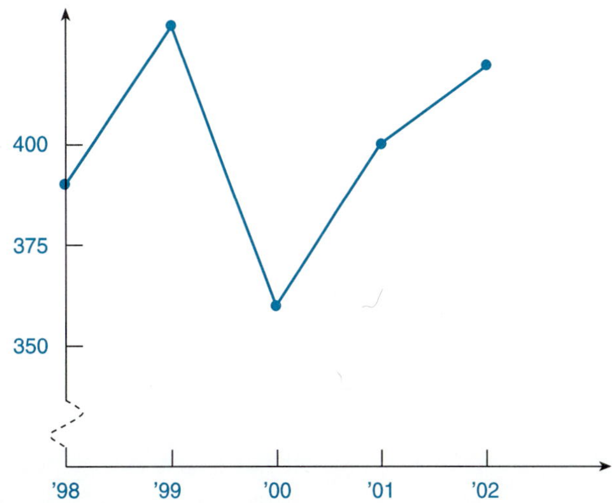

4. The result of the Green Greens Open are shown below. Prepare a graph that will show the scores of those who hit in the 30s, 40s, and 50s.

Scores			
36	56	***30s***	***f = 7***
34	41	***40s***	***f = 7***
42	37	***50s***	***f = 2***
38	40		
41	45	$\frac{7}{16} \times 360° = 158°, 44\%$	
51	33		
39	49	$\frac{2}{16} \times 360° = 45°, 12\%$	
43	38		

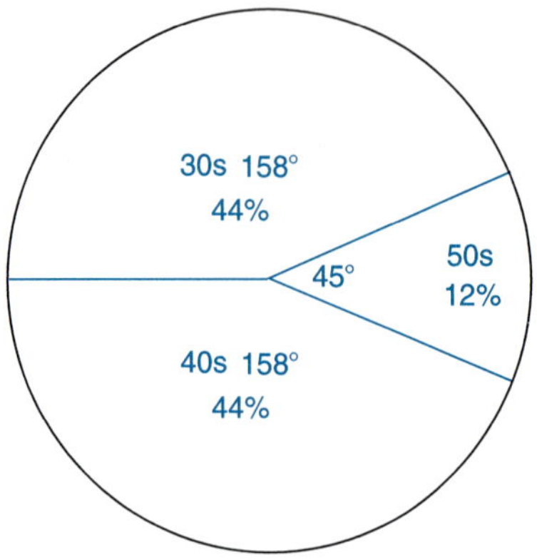

5. A graph must be prepared to be used as a visual aid at a departmental meeting. The graph must show the number of labor-hours worked and lost last year due to illness or accidents over the six-month period January 1 to June 30. Prepare a line graph based on the information below.

Month	*Hours Worked*	*Hours Lost*
January	8,900	600
February	9,100	550
March	8,750	700
April	9,400	290
May	8,050	350
June	9,400	450

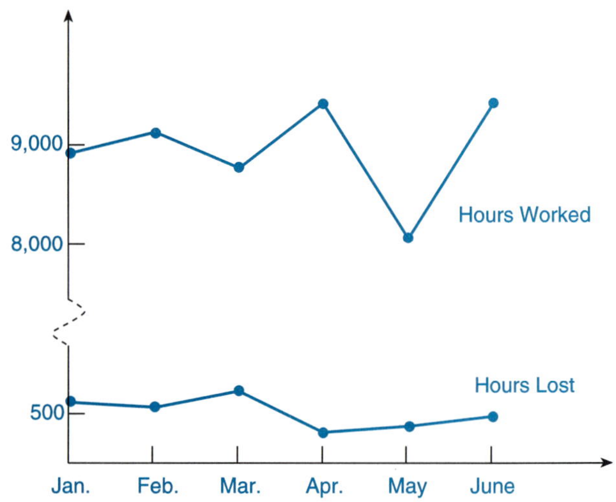

REVIEW 23

The purpose of this chapter is to review the material that you have mastered in the book's 22 chapters. This chapter may be useful in preparation for a final exam. Each problem is referenced to a chapter portion for quick review if necessary.

A review of your mastery test scores will further pinpoint areas that need extra review.

1. $92 + 8.003 + 16 + 9.627 + 8 =$ ***133.63***
(Reference: chapter 3, obj. 3)

2. $6\frac{3}{4} - 4\frac{7}{8} =$ $1\frac{7}{8}$
(Reference: chapter 2, obj. 4)

3. $21 \div \frac{3}{5} =$ ***35***
(Reference: chapter 2, obj. 4)

4. $2\frac{3}{5} \times 1\frac{2}{3} =$ $4\frac{1}{3}$
(Reference: chapter 2, obj. 4)

5. $6 - 1.306 =$ ***4.694***
(Reference: chapter 3, obj. 3)

6. $89 \times 3.2 =$ ***284.8***
(Reference: chapter 3, obj. 3)

7. $16 \div .005 =$ ***3,200***
(Reference: chapter 3, obj. 3)

8. The footing for an addition along one side of an existing tool shed is 40′ long. The footing will be 42″ deep and 18″ wide. Cement to fill the hole costs $33 a cubic yard. Using the nearest cubic yard, what will be the cost for the cement? (Reference: chapter 6, obj. 2)
40′ × 3.5′ × 1.5′ = 210 cu. ft
210 ÷ 27 = 7.77 = 8 cu. yd
×$33
$264

9. The front yard of a new home is to be sodded. The yard measures 80′ by 60′. Sod costs $8 a square yard. How much will it cost to buy the sod for the front yard? Use the nearest full yard. (Reference: chapter 6, obj. 2)
80′ × 60′ = 4,800 sq. ft ÷ 9 = 533 sq. yd
×$8
$4,264

10. Base = $800
Rate = 30%
Amount = ***$240***
(Reference: chapter 7, obj. 6)
A = B × R
A = $800 × .3
A = $240

11. Eight percent of a new store will be set aside for customer service and checkout. The new store will have 72,000 square feet of space. How much of the store will be used for customer service and checkout? (Reference: chapter 7, obj. 7)
72,000 = B
×.08 = R
5,760 sq. ft = A

12. How many square meters are there in problem 11? (Reference: chapter 6, obj. 4)
5,760 sq. ft
×.0929
535.104 sq. m

13. Sixty percent of the books sold at The Shelf were fiction. Fiction sales amounted to $45,000 for the period. What were total sales for The Shelf for the period? (Reference: chapter 7, obj. 7)
B = A ÷ R
B = $45,000 ÷ .6
B = $75,000

14. Anglebrandt & Company purchase on account. Their policy is to take advantage of all discounts available. A purchase of \$700 worth of merchandise from Stubbs, Inc., on terms of 2/10,n/60 was made on July 16. Payment is made on July 24. How much was remitted? (Reference: chapter 14, obj. 3)

\$700
−14 (2% of \$700)
\$686 remitted

15. M. Gayeski & Sons purchased \$1,700 worth of goods from Arnott Associates on August 9. Ten percent and 5 percent discounts were offered. Terms were 2/10,n/30. How much was remitted if payment was made within the discount period? (Reference: chapter 14, obj. 3)

\$1,700
−246.50 (10% & 5%)
\$1,453.50
−29.07 (2%)
\$1,424.43

16. What was the latest date that M. Gayeski & Sons could have received a cash discount in problem 15? (Reference: chapter 14, obj. 3)

August 19

17. Dick Brennan sells products on a commission formula plan. He earns a salary of \$500 a week. His commission rate is 1 percent on sales above \$60,000, and an additional $\frac{1}{2}$ percent on sales above \$100,000. His sales this week were \$117,000. What was his gross income? (Reference: chapter 5, obj. 1)

\$500 Salary
+570 (1% of \$117,000 − \$60,000) commission
+85 $\left(\frac{1}{2}\%\text{ of } \$117{,}000 - \$100{,}000\right)$ *commission*

\$1,155 income

18. Cost Price = \$300
Markup = 40% on cost
Selling Price = *\$420*
(Reference: chapter 15, obj. 1)

\$300
$C + M = S$
$100\% + 40\% = 140\%$
$A = B \times R$
$A = \$300 \times 1.4$
$A = \$420$

19. Cost Price = \$300
Markup = 40% on sales
Selling Price = *\$500*
(Reference: chapter 15, obj. 1)

\$300
$C + M = S$
$60\% + 40\% = 100\%$
$B = A \div R$
$B = \$300 \div .6$
$B = \$500$

20. Markup on cost = 25%
Markup on retail = *20%*
(Reference: chapter 15, obj. 5)

$C + M = S$
$100\% + 25\% = 125\%$ $\frac{25\%}{125\%} = 20\%$

21. Cost Price = *\$69.23*
Markup = 30% on cost
Selling Price = *\$90*
Discount = 10%
List Price = \$100
(Reference: chapter 15, obj. 2)

\$100
$S + TD = L$
$90\% + 10\% = 100\%$
$\$100 = B$
$\times .9 = R$
$\$90 = A$

\$90
$C + M = S$
$100\% + 30\% = 130\%$
$A \div R = B$
$\$90 \div 1.3 = \$69.23$

22. Dave Yost bought a light truck for \$18,700. It cost \$85 to have it delivered and another \$7,215 to have a custom body installed. The unit should last five years, at which time the salvage value is expected to be \$4,000. What will be the book value at the end of the second year using the declining-balance method? (Reference: chapter 17, obj. 2)

\$18,700
85
+7,215
\$26,000 total cost
×.6 $\left(\frac{1}{5} \times 2 = .4\right)$
$\$15{,}600 \times .6 = \$9{,}360$

23. What is the depreciation for year three of Yost's truck using the M.A.C.R.S. method if the property is categorized as a five-year item? (Reference: chapter 17, obj. 2)

\$26,000
×.192
\$4,992 depreciation

24. What is the accumulated depreciation for year four for Yost's truck using the sum-of-years'-digits method? (Reference: chapter 17, obj. 2)

$26,000
−4,000

$\$22{,}000 \times \frac{14}{15} = \underline{\$20{,}533.33}$

25. Ms. Jurk signed a 90-day promissory note on December 17, 2000, for $3,000 at 9 percent interest. What date will the note come due? (Reference: chapter 9, obj. 2)

March 17, 2001

26. How much will Ms. Jurk have to pay when the $3,000 note comes due? Use ordinary interest. (Reference: chapter 10, obj. 3)

$I = P \times R \times T$

$I = \$3{,}000 \times .09 \times \frac{90}{360}$

$I = \$67.50$

$3,000.00 face value
+67.50 interest
$3,067.50 due

27. Ken Beach made a deposit of $6,000 in his local bank, which pays 4 percent compounded semiannually. The deposit was made six years ago today. How much is the deposit worth today? (Reference: chapter 13, obj. 2)

$6,000
×1.26824179
$7,609.45 current value

28. Martin Franke would like to make a deposit in his bank that pays 5 percent compounded annually in order that he may have $8,492 two years from now. How much must he deposit today? (Reference: chapter 13, obj. 3)

$8,492
×.90702948
$7,702.49 deposit

29. Interest = $9 (ordinary)
Principal = $600
Rate = 12%
Time = *45*
(Reference: chapter 9, obj. 4)

$T = \frac{I}{P \times R} = \frac{\$9}{\$600 \times .12} = \frac{\$9}{\$72} \times 360 = 45$

30. Note face value on January 16, 2002 = $1,000
Interest on note = 8%, term of 90 days (ordinary)
Discount date = February 2, 2002
Discount rate = 10%
Proceeds = $*999.32*
(Reference: chapter 10, obj. 3)

$I = P \times R \times T$

$I = \$1{,}000 \times .08 \times \frac{90}{360}$

$I = \$20$

$I = \$1{,}020 \times .10 \times \frac{73}{360}$

$I = \$20.68$

$1,020.00
−20.68
$999.32 proceeds

31. A piece of land was purchased on July 1 for $46,000. A down payment of $10,000 was made, and the balance was financed at 14 percent. The monthly payment was determined to be $700. What is the principal balance after the second monthly payment? (Reference: chapter 11, obj. 2)

$46,000
−10,000

$\$36{,}000 \times .14 \times \frac{1}{12} = \$420$

+ 420
−700

$\$35{,}720 \times .14 \times \frac{1}{12} = \$416.73$

$35,720.00
+416.73
−700.00
$35,436.73 balance

32. A. J. Corporation received its monthly bank statement with a balance of $1,072.21. The chief accountant found that the balance in the cash account of the firm is $942.17. He also noted that the bank had not recorded two deposits of $40 each made yesterday. Also missing from the bank statement was a check for $217.62. The statement noted a service charge, but it was so blurred that it was illegible. What was the service charge? (Reference: chapter 4, obj. 2)

Bank Statement

$1,072.21
+80.00
−217.62
$934.59

$942.17
−934.59
$7.58 service charge

33. The purchase records of the Ureel Company for an item called "Soil Rich" are presented below:

Date of Purchase	*Quantity*	*Cost/Unit*
July 16, 1999	100	\$1.20
June 4, 2000	600	\$1.15
January 15, 2001	400	\$1.60
September 4, 2001	250	\$2.00

What is the value of ending inventory of 300 units using the L.I.F.O. method? (Reference: chapter 16, obj. 1)

100 × \$1.20 = \$120
200 × 1.15 = 230
\$350 value of ending inventory

34. Use the F.I.F.O. method to determine ending inventory in problem 33. (Reference: chapter 16, obj. 1)

250 × \$2.00 = \$500
50 × 1.60 = 80
\$580 value of ending inventory

35. Use the average cost method to determine ending inventory in problem 33. (Reference: chapter 16, obj. 1)

100 × \$1.20 = \$120
600 × 1.15 = 690
400 × 1.60 = 640
250 × 2.00 = 500
1,350 units \$1,950

$$\frac{\$1,950}{1,350} \times 300 = \$433.33 \text{ value}$$

36. The Social Security tax is a fixed rate on the gross income. The employer is taxed an equal percent. True or false? (Reference: chapter 5, obj. 2)

True

37. Eric earns \$6 per hour. He is paid overtime (time and one-half) for all hours in excess of 40 per week. He worked 46 hours last week. What were his gross earnings? (Reference: chapter 5, obj. 1)

40 × \$6 = \$240
6 × \$9 = \$54
\$240 + \$54 = \$294 gross earnings

38. Two weeks ago, Eric earned \$270. Federal income tax was \$63.20. Social Security tax is 6.2 percent of \$76,200. Medicare tax is 1.45 percent. Eric's accumulated earnings for the year before this check totaled \$32,200. Eric has one other deduction of a city income tax of $\frac{1}{2}$ percent. What was Eric's take-home pay? (Reference: chapter 5, obj. 2)

\$270.00 × .062 = \$16.74
270.00 × .0145 = \$3.92

270.00
−63.20
−16.74
−3.92
−1.35 (one half percent of \$270.00)
\$184.79 take home pay

39. The selling price of some merchandise is \$870. How much must be paid if there is a 4 percent sales tax on the merchandise? (Reference: chapter 18, obj. 5)

\$870.00 × .04 = \$34.80
\$870.00 + 34.80 = \$904.80

40. Sandy Baxendale imported a shipment of foreign china subject to an ad valorem tax of 10 percent and a specific tax of \$2. The customs agent determined the market value of each of the 11 items to be \$70. How much must Sandy pay for the merchandise including customs duties? (Reference: chapter 18, obj. 3)

\$70
×11 × \$2 = \$22 specific
\$770
×.10
\$77 ad valorem

\$770 merchandise
+22 specific
+77 ad valorem
\$869 total

41. Meharg, a small community in Ohio, has a tax base of \$1,600,000 (assessed valuation). The community must raise \$147,000 in taxes. To the nearest tenth of a mill, what will be the tax rate? (Reference: chapter 18, obj. 2)

$$\frac{\$147,000}{\$1,600,000} = .091875 = 91.9 \text{ mills}$$

42. Beauvais' \$47,000 house is assessed at \$25,000. The tax rate in his community is 21.3 mills. How much are the property taxes? (Reference: chapter 18, obj. 1)

25
×\$21.30
\$532.50 property taxes

43. 7, 17, 3, 1, 8, 6, 5, 3, 10.
What is the mean in the above data? (Reference: chapter 21, obj. 1)
6.7 mean

44. What is the mode in problem 43? (Reference: chapter 21, obj. 3)
3 mode

45. What is the median in problem 43? (Reference: chapter 21, obj. 2)
6 median

46. Calculate the amount of payoff on a loss of $26,000, when the coinsurance clause specification is 80 percent, and the amount of coverage is $30,000, on a $39,000 building. (Reference: chapter 8, obj. 3)

$$\frac{\$30{,}000}{\$31{,}200} \times \$26{,}000 = \$25{,}000 \text{ payoff}$$

47. Goods listed at $8,360 were shipped to a buyer in Texas F.O.B. shipping point, terms 2/10,n/35. The cost of freight was $462. What is the amount of payment the buyer sent to the seller if the invoice was paid within the discount period? (Reference: chapter 14, obj. 4)

$8,360	***$8,360.00***
×.02	***−167.20***
$167.20	***$8,192.80 payment to seller***

48. Shari Krzysiak purchased a bond at $103\frac{1}{2}$ with a stated rate of $10\frac{1}{4}$ percent interest. To the nearest tenth of a percent, what is the actual rate of return? (Reference: chapter 19, obj. 3)

$$\frac{\$102.50}{\$1{,}035} = 9.9\% \text{ rate of return}$$

49. Determine the missing figures in the partial income statement below. (Reference: chapter 20, obj. 2)

	2000	**2001**
Cost of Goods Sold		
Merchandise Inventory, January 1	$42,690	$44,620
Purchases	61,456	***71,210***
Freight In	+2,341	+1,780
Goods Available for Sale	$106,487	$117,610
Merchandise Inventory, March 31	−83,420	−81,485
Cost of Goods Sold	$23,067	***$36,125***

50. Prepare a circle graph to show the following sales. (Reference: chapter 22, obj. 1)

Products priced from $18 to $25	f = 10
25 to 50	f = 7
50 to 80	f = 18
80 to 150	f = 13
150 and up	f = 18

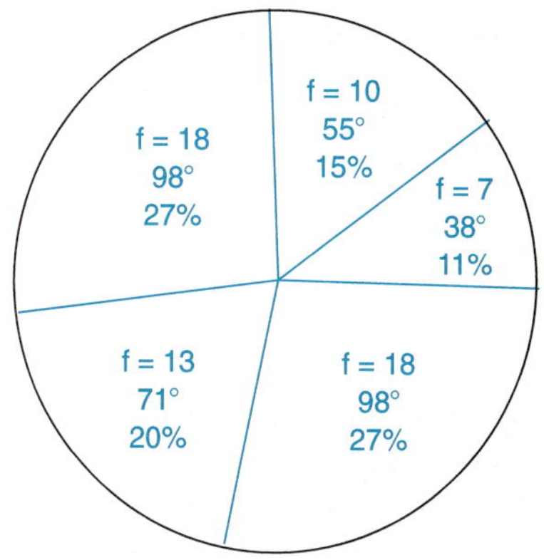

★ Review Challenge Problem

The Ristau Company imported 60 upholstery machines for use in its furniture factory. The machines were listed at $800 each in the catalog of the Kreger Company. Kreger Company shipped the merchandise F.O.B. shipping point. They also offered the Ristau Company a 10 percent trade discount because of the size of the order. Terms of 2/10,n/30 E.O.M. were also extended by the Kreger Company for early payment. The purchase was made on July 26, 2002. The delivery was made to the location of Ristau's home office on August 10, with an invoice date of August 8. The Ristau Company made the payment for the 60 machines on September 6, 2002.

When the machines entered the country, a $100 specific duty and a 5 percent ad valorem duty were imposed on the invoice amount.

The Kreger Company paid their sales representative a 7 percent commission on the sale of the machines. It was the largest single order they had ever received.

The machines were installed on August 23. This date will be used to depreciate the machines.

Each machine cost $50 to install.

The machines are estimated to have an expected life of ten years. They are also expected to have a salvage value of $100 each at the end of their expected life.

The firm uses the S.O.Y.D. method to compute depreciation. The fiscal year ends on December 1.

Determine the total depreciation on the 60 machines for the second fiscal year ending December 1, 2003.

$800
×60
$48,000 × .1 = $4,800
−4,800 trade discount
$43,200 × .02 = $864
−864 cash discount
$42,336 net price
+2,160 ad valorem
+6,000 specific duty
$50,496 total cost of machines

($43,200 × .05 = $2,160 ad valorem duty
60 × $100 = $6,000 specific duty
60 × $50 = $3,000 installation cost)

$50,496 total cost of machines
+3,000 installation costs
$53,496 total depreciable costs
−6,000 salvage value (60 × $100)
$47,496 total depreciation over life of 10 years

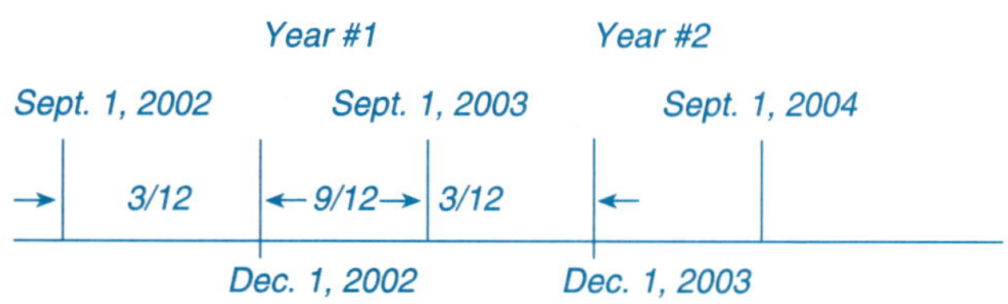

$$\$47{,}496 \times \frac{10}{55} \times \frac{9}{12} = \$6{,}476.73$$

$$\$47{,}496 \times \frac{9}{55} \times \frac{3}{12} = \underline{\$1{,}943.02}$$

$8,419.75 depreciation on the 60 machines

Calculator Selection and Operation

A

Thirty years ago most students and business people solved business math problems by using the skills they had learned and by putting their problems on paper. The only other option was an expensive, large, semiportable desk-model calculator. Today, the calculator, a small, highly portable battery-operated device, usually priced under $5, is available. This offers the student and business person a much better means to solve business math problems. Through the advances of technology, the student is able to take a large amount of the long and tedious efforts out of the "arithmetic" aspect of business math. More effort can be placed on understanding the "why" of business math.

The calculator has earned a place in the business world and is being used more in the classroom. Some cautions should be stated, however. Calculators should not be used for problems in the first three chapters of this book. The skills covered in these chapters are necessary for an understanding of the operation of the calculator and for working problems when a calculator is not available.

Selection

When selecting a calculator for business math, the student should look for a model that has (1) the four mathematical functions, (2) a floating decimal point, and (3) an eight-place readout. Other features that may interest the student are memory functions, especially the accumulating memory. This device can be purchased for $5 or less.

Operation

The fundamental operations on the calculator are performed just as you would state the problem. Press the keys on the calculator as you say the problem.

3 plus 9 equals 12

3 + 9 = 12

84 divided by 2 equals 42

8 4 ÷ 2 = 42

With the floating decimal feature, the decimal point is automatically placed by the machine.

42.07 times .36 equals 15.1452

4 2 . 0 7 × . 3 6 = 15.1452

54.06 minus 18.561 equals 35.499

5 4 . 0 6 − 1 8 . 5 6 1 = 35.499

Fractions can be added and subtracted by finding the decimal equivalent of each fraction and then performing the addition or subtraction.

$\frac{3}{4}$ minus $\frac{5}{8}$ equals

[3] [÷] [4] [=] .75

[5] [÷] [8] [=] .625

[.] [7] [5] [−] [.] [6] [2] [5] [=] .125

Fractions in problems with whole numbers can be done by using the following example.

$$147 \times \frac{3}{8} = 147 \times 3 = 441 \div 8 = 55.125$$

[1] [4] [7] [×] [3] [=] 441 [÷] [8] [=] 55.125

Multiple-function problems are also performed the way they are read.

$$\frac{\left(3.18 \times \frac{1}{4}\right) + 27}{1.6} =$$

[3] [.] [1] [8] [×] [1] [=] 3.18 [÷] [4] [=] .795 [+] [2] [7] [=]
27.795 [÷] [1] [.] [6] [=] 17.371875

Memory

A memory feature can allow the operator to do a part of a problem, store the answer, and then retrieve the stored answer later. This is helpful when solved information is needed in conjunction with another part of the problem. The memory portion of the calculator usually has four keys: an M+ key for adding a value to memory, an M− key for subtracting a value from memory, an RM key for memory recall, and a CM key for clearing the memory. The following problems will show the advantages of the memory feature.

$$\frac{47 \times 16.5}{8.3 + 9.6} =$$

[8] [.] [3] [+] [9] [.] [6] [=] 17.9 [M+]

[4] [7] [×] [1] [6] [.] [5] [=] 775.5 [÷] [MR] [=] 43.324022

The memory function saves writing down partial answers, errors in copying, and time!

$(75.67 - 8.06) - (85 \times .1687)$

[8] [5] [×] [.] [1] [6] [8] [7] [=] 14.3395 [M+]

[7] [5] [.] [6] [7] [−] [8] [.] [0] [6] [=] 67.61 [−] [RM] [=] 53.2705

The operation of calculators differs slightly from one to another. You should consult the instruction pamphlet accompanying your unit or the instructions on the back panel of the unit.

Use the problems above to check the operation of your own calculator.

MATH SHORTCUTS B

Interest-Time Shortcuts

When the interest rate and the time factor are considered together, a shortcut method for computing interest can be used. This method is often termed the 6 percent/60-day or 4 percent/90-day method. It is especially useful when a calculator is not available.

If a promissory note is negotiated for $1,868 at 6 percent for 60 days, the rate and time factors $(.06 \times \frac{60}{360})$ can be reduced to .01 $\left(\text{i.e., } \frac{\overset{.01}{\cancel{.06}}}{1} \times \frac{\overset{1}{\cancel{60}}}{\underset{\underset{1}{\cancel{6}}}{\cancel{360}}} = .01\right)$. This factor multiplied by the principal will yield the interest amount. In this problem, the interest would be $18.68. By moving the decimal point in the principal amount two places to the left, we have determined the interest amount.

The same approach is used with a problem involving 4 percent and 90 days.

$$I = P \times R \times T$$

$$I = \$16{,}520 \times \overset{.01}{\cancel{.04}} \times \frac{\overset{1}{\cancel{90}}}{\underset{\underset{1}{\cancel{4}}}{\cancel{360}}} = \$165.20 \text{ interest}$$

Not all promissory notes are at the stated rates or time periods as those in the previous problems. The shortcut can still be used with some adjustments.

Problem: $I = P \times R \times T$

$$I = \$1{,}568 \times .06 \times \frac{90}{360}$$

$\left(.06 \times \frac{60}{360}\right) = .01$ 60 days = $15.68 interest
30 days = <u>7.84</u> interest one-half the time period = one-half the interest
90 days = $23.52 total interest

The same type of thinking can be applied to similar problems.

Problem: $I = P \times R \times T$

$$I = \$450 \times .08 \times \frac{90}{360}$$

$$\left(.04 \times \frac{90}{360}\right) = .01$$

(4%/90 days) = $4.50 —interest
+ (4%/90 days) = 4.50 —interest
Interest (8%/90 days) = $9.00

The application can become even more interesting, as well as challenging to the problem solver.

Problem: I = P × R × T

$$I = \$4{,}840 \times .075 \times \frac{90}{360}$$

(6%/60 days) interest	\$48.40
(6%/30 days) interest	+24.20
(6%/90 days) interest	\$72.60
$\left(1\frac{1}{2}\%/90 \text{ days} = \frac{1}{4} \text{ of } 6\%/90 \text{ days}\right)$	+18.15
Interest $\left(7\frac{1}{2}\%/90 \text{ days}\right)$	\$90.75

Problem: I = P × R × T

$$I = \$1{,}570 \times .08 \times \frac{135}{360}$$

(4%/90 days) interest	\$15.70
(4%/90 days) interest	+15.70
(8%/90 days) interest	\$31.40
$\left(8\%/45 \text{ days} = \frac{1}{2} \text{ of } 8\%/90 \text{ days}\right)$	+15.70
Interest (8%/135 days)	\$47.10

Aliquot Parts

Another shortcut method used in business problems is termed *aliquot parts*. This method makes use of the fractional parts of 100. Knowledge of fractions and their decimal equivalents is very useful here. The table below may serve as a refresher.

Common Fraction	Decimal Equivalent	Common Fraction	Decimal Equivalent
$\frac{1}{25}$	.04	$\frac{1}{6}$	.166
$\frac{1}{20}$	.05	$\frac{1}{5}$	.2
$\frac{1}{12}$	.0833	$\frac{1}{4}$	.25
$\frac{1}{11}$	.091	$\frac{1}{2}$	.5
$\frac{1}{10}$	.1	$\frac{5}{6}$	.833
$\frac{1}{9}$	.111	$\frac{3}{8}$	.375
$\frac{1}{8}$	.125	$\frac{5}{8}$	.625
$\frac{1}{7}$	.14285	$\frac{7}{8}$	.875

The clear advantage of aliquot parts becomes evident when the student can visualize the shortcut's use and then do the problem in his or her head.

Problem: What is the total price of 680 units of a product to be sold at 25¢ each?

25¢ is equal to $\frac{1}{4}$ of \$1.00; $\frac{1}{4}$ of 680 = 170 or \$170

Problem: What is the cost per unit of 750 units produced at a total cost of \$1,860?

750 units equals $\frac{3}{4}$ of 1,000; $\frac{3}{4}$ of \$1,860 = $\frac{\$1{,}395}{1{,}000}$ or \$1.40 each

Problem: A product is priced at \$28. Twenty percent of the product price is packaging cost. What is the packaging cost?

20% = $\frac{1}{5}$ of \$28 = \$5.60

MATH OF COMPUTERS C

Throughout this text, we have been using a numbering system termed the decimal system. This system has a base of 10. It uses the digits 0 through 9. Once ten units have been counted, the decimal system records that ten by placing a one in the column to the left. Each place value counts ten units of the position to the right. This system works well for most quantitative problem-solving methods, but not for the computer. The computer operates using only two digits, 0 and 1, because it has only "on-off" or "yes-no" capability. Hence, there is the need for only two digits. This number system is termed the binary system because of the use of two (bi) digits.

Below is a conversion table of decimal numbers to binary numbers.

Decimal Numbers (Base 10)	Binary Numbers (Base 2)	
0	0	
1	1	(2^0 power = 1)
2	10	(2^1 power = 2)
3	11	(2^1 power + 2^0 power = 3)
4	100	(2^2 power = 4)
5	101	(2^2 power + 2^0 power = 5)
6	110	
7	111	
8	1000	
9	1001	
10	1010	
11	1011	
12	1100	
13	1101	
14	1110	
15	1111	
16	10000	
17	10001	
18	10010	
19	10011	
20	10100	

Notice that each one is a power of two. Each position to the left increases the power by one.

Converting binary to decimal

Problem: 100000 = 2^5 power = 32

Problem: 10100 = 2^4 power (16) + 2^2 power (4) = 20

Problem: 1011000 = 2^6 power (64) + 2^4 power (16) + 2^3 power (8) = 88

Converting decimal to binary numbers

Problem: 23 = 10111 binary; that is, 2^4 power (16) + 2^2 power (4) + 2^1 power (2) + 2^0 power (1)

Problem: 17 = 10001 binary; that is, 2^4 power (16) + 2^0 power (1)

Problem: 67 = 1000011 binary; that is, 2^6 power (64) + 2^1 power (2) + 2^0 power (1)

Decimal numbers can be converted to binary numbers by dividing the decimal number by two until the dividend is zero. The remainder, placed in reverse order, is the binary number.

Problem: Convert 259 decimal into a binary number.

	Remainder
2)259	
2)129	1
2)64	1
2)32	0
2)16	0
2)8	0
2)4	0
2)2	0
2)1	0
0	1

259 decimal equals 100000011 binary.

Problem: Convert 123 decimal into binary.

	Remainder
2)123	
2)61	1
2)30	1
2)15	0
2)7	1
2)3	1
2)1	1
0	1

123 decimal equals 1111011 binary.

Employment Math Test D

Business firms often use a mathematics test as a screening device in personnel selection. The skill of mathematics as well as logic ability is measured in a mathematics test. Some of the types of problems used on employment tests are shown below.

Arithmetic Progression

This type of problem tests logic more than it tests mathematical skills. The applicant is tested on his or her ability to see an arithmetic pattern and extend that pattern to the blank space.

Problem: 2, 4, 8, 16, 32, ________

The pattern in this problem is to double the previous number. The blank would be filled with a 64.

Problem: 5, 10, 50, 100, 500, 1,000, ________

The pattern in this problem is to multiply the first value by 2, the next value by 5, and then repeat this pattern. The blank should be filled with 5,000.

Problem: 60, 30, 15, 20, 10, 5, 10, ________

The pattern is ÷ by 2, ÷ by 2, add 5. The blank should be filled with a 5 (÷ by 2).

Problems:

1. 72, 70, 66, 64, 60, 58, 54, 52, ________
2. 80, 70, 61, 53, 46, 40, ________
3. 360, 180, 60, 30, ________
4. 4,000, 800, 200, 40, 10, ________
5. 6,720, 840, 120, 20, 4, ________
6. 68, 61, 63, 56, 58, 51, 53, ________
7. 600, 300, 100, 25, ________
8. 72, 76, 73, 75, 79, 76, 78, 82, 79, ________
9. 8, 40, 10, 30, 15, ________
10. 3, 9, 27, 108, 432, ________

Find the answers below.

1.	48	**6.**	46
2.	35	**7.**	5
3.	10	**8.**	81
4.	2	**9.**	15
5.	1	**10.**	1,296

A Sample Employment Math Test

An employer may ask you to take a test to determine your math skills. This is legal if math will be part of your actual job performance (bona fide occupational skill). The following problems are a sampling of the type you may encounter on such a test. Test your skill. (See your instructor for answers to these problems.)

1. Add $\frac{1}{7}, \frac{1}{8}, \frac{1}{21}, \frac{1}{14}$
(a) $\frac{57}{83}$ (b) $\frac{65}{168}$ (c) $\frac{15}{32}$ (d) $1\frac{13}{97}$ *(b)*

2. A photocopy machine can run 390 copies in one-half hour. How many copies can be run in $3\frac{1}{4}$ hours?
(a) 2,535 (b) 1,268 (c) 1,860 (d) 2,350 *(b)*

3. To drive from Billerston to Dillworth takes $15\frac{1}{2}$ hours. The same trip by bus takes $27\frac{3}{4}$ hours. How much time is saved by driving?
(a) $3\frac{3}{4}$ (b) $43\frac{1}{4}$ (c) $12\frac{3}{4}$ (d) $12\frac{1}{4}$ *(d)*

4. Sarah plans to save 10 percent of her salary each week. When she has saved enough she will buy a bond for $375. Her salary is $130 weekly. How many weeks will she need to work before she can buy her first bond?
(a) 27 (b) 4,875 (c) 26 (d) 29 *(d)*

5. A women's apparel shop has 4,368 units in stock. Half of the stock is slacks, $\frac{1}{8}$ is sweaters, $\frac{1}{16}$ is belts and scarves. How many units are not included above?
(a) 3,003 (b) 1,365 (c) 1,638 (d) 2,965 *(b)*

6. The receiving department uses 43 inspection forms each day. How many days will a six-gross inventory last?
(a) 20 (b) 18 (c) 0 (d) 36 *(a)*

7. $\frac{1}{2}$ of $\frac{3}{8}$ of a stock of pens have been used in the past ten days. The total inventory was 862. What is the daily usage rate?
(a) 162 (b) 32 (c) 46 (d) 16 *(d)*

8. $3\frac{5}{8} \times 17\frac{1}{4} =$
(a) $62\frac{17}{32}$ (b) $13\frac{5}{8}$ (c) $50\frac{11}{32}$ (d) $73\frac{7}{16}$ *(a)*

9. 280 of the employees are in the painting department. They represent 40 percent of the total labor force. How many are in the labor force?
(a) 112 (b) 11,200 (c) 700 (d) 168 *(c)*

10. 140 percent of x is 1,344. What is x?
(a) 1,882 (b) 538 (c) 188 (d) 960 *(d)*

11. The Wilkum Company bought $630 of merchandise. They were able to deduct a 10 percent discount. What was the net amount due?
(a) $693 (b) $63 (c) $567 (d) $630 *(c)*

12. A train averages 63 miles per hour. The departure time was 9:20 P.M. How far had it gone by midnight?
(a) 168 miles (b) 924 miles (c) 147 miles (d) 146 miles *(a)*

13. A merchant purchased a lamp at $17 and marked it up by 30 percent over the cost. What was the amount of the markup?
(a) $11.90 (b) $22.10 (c) $5.10 (d) $10.20 *(b)*

14. An employee, Jack Terril, earns a commission of $3\frac{1}{4}$ percent on sales. His sales were $43,680. What was the commission?
(a) $1,419.60 (b) $1,310.40 (c) $1,683 (d) $15,387.68 *(a)*

15. If the interest on a 30-day promissory note for $500 is $6.67, what was the interest rate using ordinary interest?
(a) 10.3% (b) 5.8% (c) 16% (d) 12% *(c)*

16. Bill earned a commission of $463 this week. This was a 30 percent increase over last week. How much did Bill earn last week?
(a) $601.90 (b) $356.15 (c) $324.10 (d) $138.90 *(b)*

17. David word-processed 14 more letters today than did Karen. This represented a 20 percent increase over Karen's output. What was Karen's output?
(a) 56 (b) 70 (c) 84 (d) 280 *(b)*

18. A farmer plowed a field at the rate of $2\frac{1}{4}$ acres per hour. How many acres will he plow in $3\frac{1}{2}$ hours?
(a) $5\frac{3}{4}$ (b) $7\frac{7}{8}$ (c) $8\frac{3}{4}$ (d) $9\frac{1}{8}$ *(b)*

19. Of the 465 students enrolled in a school, 93 were ill with the flu. What percent were ill?
(a) 20% (b) 93% (c) 50% (d) 30% *(a)*

20. If a 20 percent discount on an invoice saved a firm $79, what was the original amount of the invoice?
(a) $158 (b) $15.80 (c) $395 (d) $98.75 *(c)*

21. One hundred sixteen people registered for a workshop on energy conservation. One-fourth were late registrants. The early registrants showed a two-to-one male-to-female relationship. How many of the early registrants were males?
(a) 15 (b) 47 (c) 58 (d) 37 *(c)*

22. An $8.95 gallon of paint will cover 500 square feet. How much will it cost to cover the side walls of a room that measures 15 feet by 10 feet by 7 feet 6 inches high?
(a) $17.90 (b) $8.95 (c) $26.85 (d) $89.50 *(b)*

23. Multiply and round off to the nearest tenth: 8.63×5.0473.
(a) 43.6 (b) 43.5 (c) 13.7 (d) 37.8 *(a)*

24. A pool measuring 20 feet by 40 feet by 5 feet deep must be filled with water. How many hours will be required if it will fill at the rate of 900 gallons an hour? There are $7\frac{1}{2}$ gallons in a cubic foot.
(a) 33 (b) 4 (c) 23 (d) 31 *(a)*

25. Jim processes 40 reports a day, which is 25 percent more than Tom's output. How many does Tom process?
(a) 65 (b) 50 (c) 32 (d) 30 *(c)*

26. Mark had $3,957 in his savings account. He bought a new motorcycle for $1,680. What percent of his savings was left after the purchase?
(a) 58% (b) 47% (c) 42% (d) 43% *(a)*

27. If wheat sold for $3.60 a year ago, and it sells for $4.25 today, what is the percent of increase?
(a) 118% (b) 18% (c) 85% (d) 15% *(b)*

Complete the following using arithmetic progression.

28. 60, 30, 15, 20, 10, 5, 10, ______
(a) 10 (b) 0 (c) 5 (d) 15 *(c)*

29. 5, 10, 30, 120, ______
(a) 600 (b) 240 (c) 60 (d) 480 *(d)*

30. 72, 76, 73, 75, 79, 76, 78, 82, 79, ______
(a) 76 (b) 83 (c) 81 (d) 77 *(c)*

READING MARKET QUOTATIONS

E

The price of stocks and bonds are printed in the business section of most newspapers. The price, as well as its activity in the market, are of interest to investors.

A sample of a stock exchange report found in a newspaper is shown below.

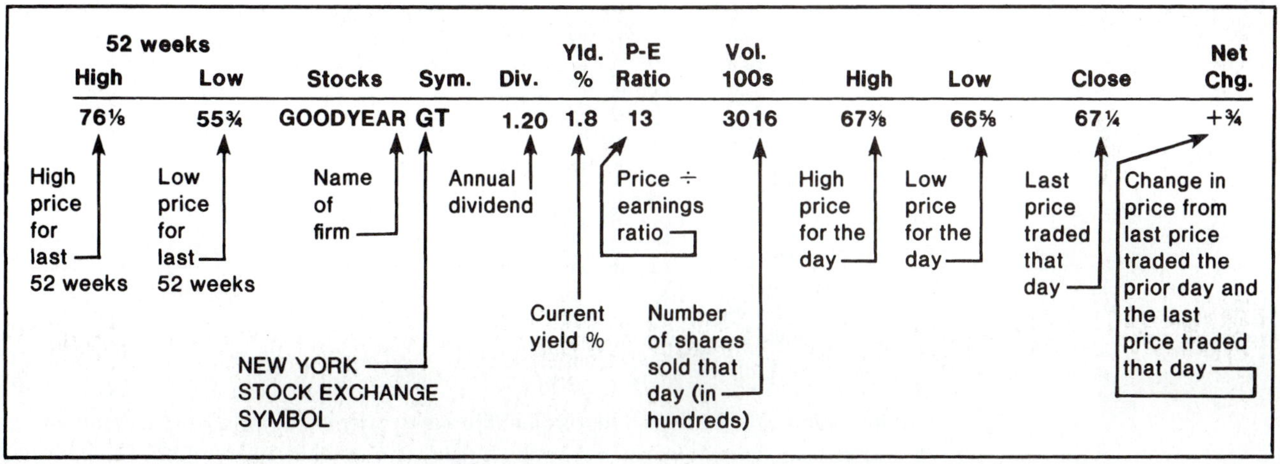

	52 Weeks Hi	Lo	Stock	Sym	Div	Yld %	PE	Vol 100s	Hi	Lo	Close	Net Chg
	76⅛	55¾	**Goodyear**	GT	1.20f	1.8	13	3016	67⅜	66⅝	67¼	+ ¾
	22	**7½**	**Gottschks**	**GOT**		...	**dd**	**485**	**8½**	**8⅛**	**8⅛**	**– ⅝**
	45	32	GraceWR	GRA	1.40	3.8	dd	2403	37¾	36¾	36⅞	– ⅝
	27¾	23	Graco	GGG	.76	3.0	18	12	25¾	25½	25⅝	– ⅛
	10⅜	5⅛	GrahamFld	GFI		...	23	137	6	5¾	5⅞	
	61⅛	39	Grainger	GWW	.66	1.2	21	572	55¼	54½	55⅛	+ ⅜
	37¼	24⅜	GrandMet	GRM	1.05e	4.0	...	880	26¾	26⅛	26½	+ ¼
	35¼	21⅜	GtAtlPac	GAP	.80	3.4	dd	447	24	23¾	23¾	– ⅛
	79	50¼	GtLakesChm	GLK	.32	.4	23	1552	76	74⅞	75¾	+ ¾
	77½	60	GtNorIron	GNI	5.75e	8.6	...	21	66½	65½	66½	+ ¾
	20	12½	GtWestFnl	GWF	.92	5.3	58	4230	17¾	17⅛	17⅜	– ⅛
	60	51	GtWestFnl pf		4.38	7.6	...	26	58	57¾	57¾	– ¼
n	24⅞	22⅜	GtWestFnl pfA			...	...	158	24⅞	24¾	24⅞	
n	15⅛	11½	GreatChina	GCH			...	203	13⅜	13	13⅜	+ ⅜
▲	34½	29	GreenMtPwr	GMP	2.10	6.0	14	43	35	34½	35	+ ½
s▲	**32**	**15⅛**	**GreenTree**	**GNT**	**.33**	**1.0**	**14**	**6045**	**33⅞**	**31½**	**33⅝**	**+1⅝**

Bond prices are quoted in a similar fashion in the newspaper each business day. A sample of a **bond** exchange report is shown below.

bond

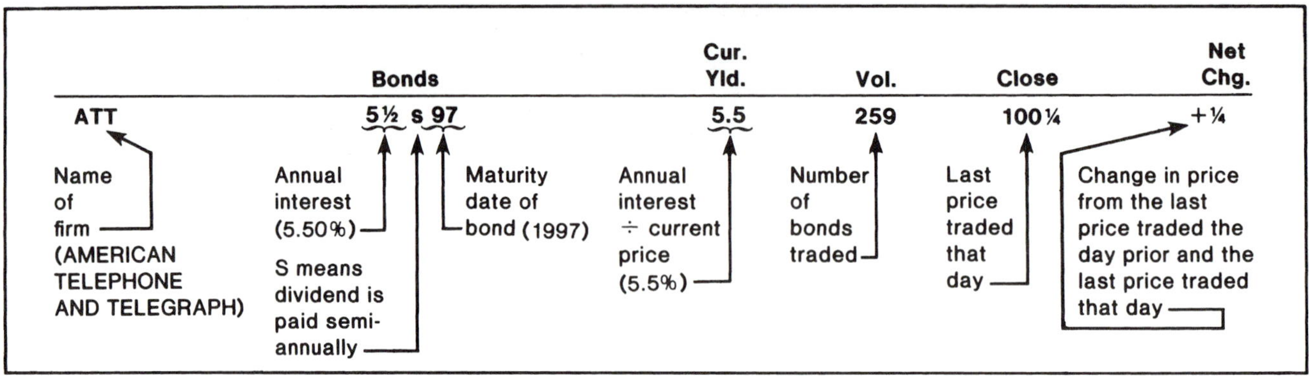

Bonds	Cur Yld	Vol	Close		Net Chg
ATT 5½97	5.5	259	100¼	+	¼
ATT 6s00	6.0	208	99¾	+	¾
ATT 5⅛01	5.6	70	91⅞		...
ATT 7s01	6.9	315	100⅞	+	⅛
ATT 7⅛03	7.0	934	101¾	+	⅝
ATT 8⅝26	8.2	140	105⅝		...
ATT 8⅝31	7.8	37	110	+	¼
ATT 7⅛02	6.9	100	104	−	½
ATT 8⅛22	7.7	190	105⅞	+	⅛
ATT 8⅛24	7.7	135	106	+	1
Amoco 8⅝16	7.8	18	110		...
AmocoCda 7⅜13	6.2	39	119	−	1
Andarko 6¼14	5.7	5	109	+	½
Anhr 8s96	7.7	10	103¾	+	1¾
Anhr 8⅝16	8.1	10	105⅞	−	¾
Apache 7½00	cv	29	132	+	3
ArizP 7.45s02	7.3	20	101⅜		...
Arml 8.7s95	8.7	3	100⅛	+	⅛
Arml 9.2s00	9.4	13	97½	+	1⅞
Arml 8½01	9.2	5	92⅞	−	1
AshO 8.8s00	8.3	30	106	+	1¼
AshO 8.2s02	8.1	25	101		...
AshO 6¾14	cv	117	89¼	+	¾
ARch 10⅜95	9.3	1	112⅛	−	⅛
ARch 10⅞05	8.4	10	129¼		...
ARch 9⅞16	8.0	100	124	+	3⅞

Mutual fund values are also reported in the newspaper each business day. A mutual fund represents ownership of several bonds, stocks, or a combination of both. The daily report lists each organization followed by their group of funds. The report below shows the Capital Appreciation Fund of Fidelity Investments.

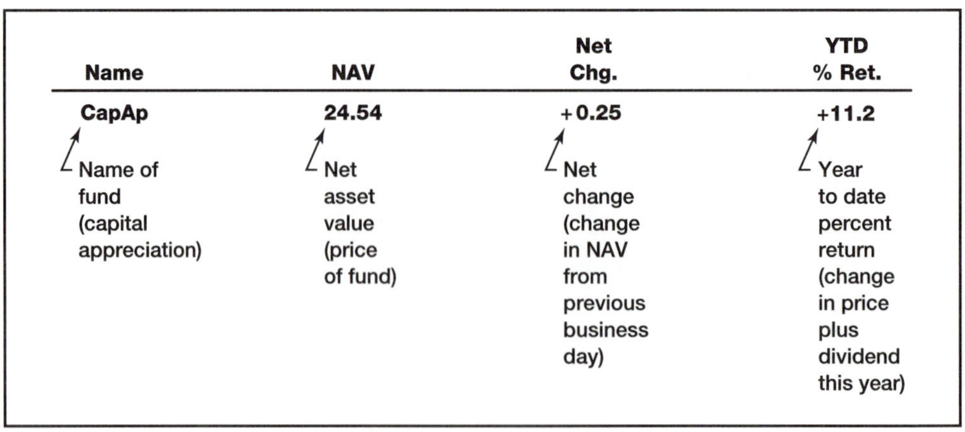

Name	NAV	Net Chg	YTD %ret
Fidelity Invest:			
A Mgr.	17.65	+0.13	+ 2.4
AggrGr	38.01	+0.80	+22.3
AMgrGr	19.24	+0.22	+ 3.0
AMgrIn	12.28	+0.04	+ 1.3
Balanc	17.68	+0.11	+ 8.7
BluCh	51.54	+0.75	+ 2.3
Canad r	14.73	+0.15	+ 5.3
CapAp	24.54	+0.25	+11.2
CpInc r	10.05	+0.01	+11.4
Cngs	393.32	+3.29	+ 4.8
Contra	59.55	+0.53	+ 5.8

Glossary

A

Accumulated Depreciation The amount that an asset has depreciated to date.

Ad Valorem Duty A type of customs duty that is a percent of the market value of the item.

Adjusted Balance The actual amount available in the checking account at a point in time.

Amortizing Paying off a loan by making monthly payments.

Amount The result of taking a number of hundredths of the base.

Annual Percentage Rate (APR) The interest rate on a loan to a consumer, stated as prescribed by Regulation Z of the Truth-in-Lending Act.

Annuity A savings instrument that usually includes a series of investments that earn interest and accumulate to a desired amount by a specific future date.

Annuity Due A type of annuity that requires that the investments be made at the beginning of the investment period.

Assessed Valuation The base on which property taxes are computed.

Assets Items that are owned by a firm or individual.

Automatic Teller Machine (ATM) A machine available on a 24-hour basis to accept deposits and provide cash withdrawals from deposit accounts.

Average Cost Method An accounting approach used to determine the value of inventory.

B

Balance Sheet A summary of the financial condition of a firm at a particular point in time.

Balloon Payment A single payment that is due on a loan (usually a home mortgage) at the end of one to five years, although the amount of regular payments is based on a longer term.

Bank Card A type of credit card. This type of card is the most versatile.

Bank Statement An analysis by the banking institution of the transactions and current balance of an account based on available information.

Bar Graph A visual method of displaying numerical information in a two-dimensional form.

Base The basis of comparison; the number of which so many hundredths is taken.

Bearer Bonds Bonds owned by whoever (physically) holds them. The interest coupon is detached for payment.

Beginning Inventory The value of the merchandise a firm has at the beginning of an accounting period such as a month or a year.

Beneficiary The person or organization that will receive the financial benefits of a life insurance policy.

Blank Endorsement The simplest form of endorsement. This endorsement requires only the endorser's signature; the endorser places no restrictions on the document.

Bond A debt instrument of a firm or a government organization.

Book Value The total cost of an asset less the accumulated depreciation.

Borrowing Used in the subtraction function. Using a larger place value to provide enough to subtract from.

C

Calendar The number of days in each month of a year.

Cash Discount A reduction of the price of the merchandise offered to the buyer for early payment.

Cash Surrender Value The money that will be received if the insurance policy holder gives up the claims of the policy.

Check A demand by a depositor on his or her bank to pay the payee the amount specified.

Check Register A record of the checking account balance from the viewpoint of the depositor.

Check Stub A record of information about a check that is completed by the depositor of a checking account.

Circle Graph A visual method to display numerical information. A circle is divided from the center into the portions that represent the data.

Closing Costs The amount, paid at the closing meeting, for escrow, points, and fees.

C.O.D. (Cash on Delivery) The merchandise delivered by this method must be paid for when it is received.

Coinsurance The policy holder and the insurance company share in the financial loss of property damage.

Commission Formula A combination of ever-increasing rates of pay for sales designed to motivate an employee to sell more.

Common Denominator A value that can be used to establish a common label for the fractions under consideration, for purposes of addition or subtraction.

Common Fraction A type of fraction in which the numerator is smaller than the denominator.

Common Stock A form of equity securities that provides a vote in the decisions of the firm.

Comparative Balance Sheet A balance sheet that shows figures for two periods. A percent of change is shown and is determined by comparing the second period to the first period.

Compound Interest A method of computing interest in which the interest earned in one period is added to the principal; interest for the next period is calculated on both interest and principal from the previous period.

Compound Rate The combination of a specific rate and an ad valorem rate on imported goods; imposed by the federal government.

Compounded Amount The amount returned to an investor at the end of the term of investment.

Cost of Goods Sold The price that was paid for the merchandise that was sold during the period.

Cost of Investment The fee that is paid to a middleman for facilitating the transaction.

Cost Price The price that the retailer pays for goods.

Credit Card A purchase device issued to those with a good credit history.

Cumulative A stipulation on a preferred stock that allows the holder to receive dividends that were not paid in the past.

Currency Exchange The rate at which money from one country trades for money from another country.

Customs Duty A tax on imported goods; imposed by the federal government.

D

Data Set A group of values that have something in common.

Debentured Bonds Bonds that are issued by a firm without being tied to any physical equipment or other asset.

Decimal Point Placement Rule The answer in a multiplication problem should contain as many decimal places as the two values in the problem had.

Declining Balance Method An accelerated depreciation method that does not consider salvage value in the initial calculations; depreciation occurs at double the straight-line rate.

Deductions, Income Tax Filing Amounts that can be subtracted from the gross income amount for such items as interest expense on a home, contributions, and medical expenses.

Deductions, Payroll Amounts subtracted from gross pay for such items as taxes, contributions, and savings.

Denominator The number on the bottom of a fraction.

Dependents Those persons that are being supported by the wage earner and that can be claimed on the income tax return.

Deposit A sum of money that adds to the balance of a checking or savings account.

Deposit Slip A record device to verify a deposit.

Depositor The entity (an individual or a firm) that holds a checking or savings account.

Deposits in Transit Deposits of money, checks, and collections that have not caught up with the records of the bank.

Depreciation An adjustment in the value of an asset due to its decline in market value. An estimate.

Difference The result of subtraction.

Direct Method A fast method for computing the effect of trade discounts on the list price of an item.

Discount Amount The dollar amount of service charge that the purchaser of a note subtracts from the maturity value.

Discount Rate The percent of the note for service charge.

Dividend The number to be divided.

Dividends Profits distributed to shareholder owners of a corporation.

Divisor The number that is used to divide by. In a fraction, the number on the bottom.

Down Payment The amount that will be paid first and in addition to the amount financed in the purchase of an item.

E

Ending Inventory The value of the merchandise a firm has at the end of an accounting period such as a month or a year.

Endorsement The signature of the payee on the back of the check. There are several types of endorsements.

Endowment Life Policy A type of life insurance in which a large portion of the premium is directed toward a savings plan that may be used for college, retirement, etc.

E.O.M. (End of Month) One of the terms of trade for a cash discount; the buyer computes the days of discount beginning at the end of the month in which the purchase was made.

Equivalent Trade Discount A single percent equal to a series or chain of trade discounts.

Escrow Account An account kept by the lending institution into which the borrower pays in order to accumulate enough money to cover property taxes and property insurance when they come due.

Exact Interest An interest rate that uses a 365-day year. Used for personal loans only.

Exemption, Income Tax Filing The amount of earnings that the federal government allows a taxpayer to claim as untaxable income when filing the year-end tax return. This amount is based on the number of dependents that the taxpayer claims.

Exemptions, Payroll Legal dependents and the employee; can be claimed by the employee in order to reduce the amount of income tax withheld.

Expected Life The period of time or the number of cycles that an asset normally functions before it is disposed of.

Expenses Used up or expired assets.

F

Face Value The amount stated on a promissory note. The amount borrowed.

F.I.F.O (First In–First Out) A method of determining the value of ending inventory or cost of goods sold.

Fixed Payment Amount A payment amount on a home mortgage that is a set (fixed) amount for the term of the loan.

Fixed Rate An interest rate that remains the same (fixed) over the term of the loan. Usually on a home mortgage.

F.O.B. (Free on Board) Destination Transportation costs to the specified buyer's location are included in the price of the item.

F.O.B. (Free on Board) Shipping Point Transportation costs from the seller's location are to be paid by the buyer.

Form W-2 A federal government form that the employer completes indicating the earnings and deductions for the year.

Form W-4 A federal government form that the employee fills out to claim exemptions.

G

Gasoline Card A credit card issued by an oil company. Used primarily for credit purchases of gasoline.

Goods Available for Sale Beginning inventory plus net purchases. The cost of goods sold comes from this category.

Government Bonds Investment vehicles issued by local, state, or federal government agencies with a possible tax-free status.

Gross Earnings The amount earned by the employee in a pay period.

Gross Sales All the sales of a firm for a time period.

H

Home Mortgage An installment loan on a home. Loan may carry a fixed or variable rate, a fixed or variable payment, or a balloon payment.

Horizontal Analysis Comparing information on the same line. The difference is then changed into a percent.

I

$I = P \times R \times T$ The basic interest formula.

Imprinting Checks The process of personalizing check forms with the name and account number of the depositor.

Improper Fraction The type of fraction that has a numerator larger than the denominator.

Income Statement A report that shows the financial activity of a firm over a period of time.

Income Tax Return A form that must be completed by every individual and firm that has income. On this form, the actual amount of tax owed is determined through calculations.

Inflation An increase in the price of most goods and services.

Insurance A way to share the risk of loss with another firm that specializes in estimating the probability of loss.

Interest A financial return to the lender for the risks undertaken during the term of a loan.
Interest Formula A method used to determine the amount of money that may be charged when money is borrowed; Interest = Principal × Rate × Time. (*See* I = P × R × T.)
Interest-bearing Note A promissory note that has an interest provision. The maturity value will be larger than the face value.
Internal Revenue Service An agency of the federal government that is responsible for collecting the income taxes due from persons and firms.

L

Leap Year A year that is evenly divisible by four. February has 29 days in a leap year.
Liabilities Amounts due others; financed obligations.
L.I.F.O. (Last In–First Out) A method of determining the value of ending inventory or cost of goods sold.
Limited Pay Life Policy A type of life insurance that has a savings factor and that involves a limited number of payments before the policy is paid up.
Line Graph A type of graph that uses a line to depict information.
Linear Measure Measurement in only one direction.
Liquidity Assets that are easily converted into cash are considered to have high liquidity.
List Price The price shown in a catalog. This price is often subject to trade discounts.
Lowest Common Denominator (LCD) The smallest value that will serve as a denominator for all of the fractions being considered.

M

Maker The person or organization that borrows money on a promissory note.
Markdown An amount or percent subtracted from the selling price.
Market Value The amount of money that an asset can be sold for at a point in time.
Markup An amount or percent added to the cost price of merchandise in order to cover overhead and profit. Can be based on cost or selling price.
Maturity Value The amount of money that the maker of a promissory note must pay at the end of the note term.
Mean The average of a data set.
Median The middle ranked value of a data set.
Metric Measurement System A system of measure that is used in nearly all of the world. The system uses a base of ten to convert measurements.
Mills A means of expressing one thousandth.
Minimum Payment Amount The smallest amount that can be paid on a credit card balance to maintain a good credit rating.
Minuend The larger of two numbers that are being subtracted.
Mixed Number A type of a fraction consisting of a whole number and a common fraction.
Mode The value of greatest frequency in a data set.
Modified Accelerated Cost Recovery System (M.A.C.R.S.) A depreciation method for tax use. For items placed into service after 1986.
Monthly Statement A record of the account balance from the viewpoint of the issuer of the credit card. This record is sent each month.
Multiplicand The top number of two numbers that are to be multiplied.
Multiplier The bottom number of two numbers that are to be multiplied.

N

Net Pay The amount due a person after deductions have been subtracted from gross pay.
Net Price The amount to be paid after all discounts have been subtracted.
Net Sales The dollar value of sales after sales returns and sales allowances have been subtracted from gross sales.
Noninterest-bearing Note A promissory note that has a maturity value that is the same as the face value.
Numerator The number on top in a fraction.

O

Ordinary Annuity A type of annuity that requires that the investments be made at the end of the investment period.
Ordinary Interest An interest rate that uses a 360-day year; used for commercial loans only.
Outstanding Checks Checks that have been written by the depositor, but that have not been received by the bank.
Owner's Equity The portion of the firm owned by the investors. Part of the basic accounting formula: Assets = Liabilities + Owner's Equity.

P

Participating A stipulation on a preferred stock that allows the stockholder to receive additional dividends after the holders of common stock have received a stated amount.
Pay Systems The methods that are available to determine gross earnings for an employee, including hourly, salary, piece-rate, and commission methods.
Payee, Check The person or organization to whom the check is made out.
Payee, Promissory Note The lender of the money on a promissory note.
Payment Amount A monthly payment on an installment loan. This payment will pay off the interest and principal over the term of the loan.
Payroll Record A form showing gross earnings, deductions, and net pay for all employees.
Percent A means of expressing some part of 100.
Place Values The names of the locations of digits in a number.
Points A point is one percent of the face value of the loan; points are usually paid by the buyer at closing.
Policy A contractual document explaining the financial commitment of an insurance company to the policy holder.
Preferred Stock A type of equity instrument that pays the owner a dividend before common stock is paid.
Premiums Payments for insurance protection.
Present Value The current value of an investment due some date in the future. The present value is found by discounting back the future value at prevailing interest rates.
Prime Number A number that is divisible only by itself and one.
Principal The amount borrowed.
Principal Balance The amount remaining to be paid on an installment loan.
Proceeds The amount due the payee when a note is discounted.
Product The result of multiplication; the answer.
Promissory Note A legally enforceable written contract; an "IOU."
Proprietorship The portion of the firm that is owned by the investors.
Purchases The merchandise that is bought during the period.

Q

Quotient The number found as a result of division.

R

Range The difference between the largest and smallest value of a data set.

Rate A percent.

Rate, Interest The annual charge for the use of money.

Rate of Return The amount earned on an investment expressed as a percent of the amount invested.

Ratio The comparison of one value to another value.

Ratio, Financial A set of comparisons of one item to another, from either a balance sheet or an income statement.

Registered Bonds A type of bond that requires that a new owner provide the firm with his or her name, address, etc., so that the firm may communicate with the new owner.

Regulation Z The portion of the Truth-in-Lending Act that specifies that interest rates must be calculated using a set method and the rate be provided to consumer borrowers.

Remainder The value that is left over after division of whole numbers.

Repayment Schedule A record of payments to be made; includes the interest amounts and the principal balance.

Restrictive Endorsement A type of endorsement of a check that determines the check's final use.

Retail Inventory Estimating Method A method of providing an approximate value of the merchandise without a physical inventory.

Revenue The result of selling goods or services.

Risk The chance taken by a business person who makes a business decision.

R.O.G. (Receipt of Goods) One of the terms of trade for a cash discount; the buyer computes the days of discount beginning with the date that the merchandise was actually received by the buyer.

Round Off Directions that determine where an answer should be terminated.

Rounding Rule Used to determine the final number for an answer. The number to the right must be 5 or larger to round up.

Rule of 78s A formula used to determine the payoff amount on a short-term loan.

S

Sales Allowances Adjustments made in the price of the merchandise in order to compensate the buyer for keeping damaged or incorrectly shipped merchandise.

Sales Returns Merchandise returned to the seller because of damage, errors in ordering, etc.

Sales Tax A tax that is determined by taking a percent of the selling price of items sold.

Salvage Value The estimated market value of an asset at the end of its life.

Secured Bonds Bonds that have certain properties pledged to the owner of the bond.

Selling Price The price that the buyer (consumer) pays for the goods.

Service Charges Expenses charged to the depositor by the bank for use of the account, imprinting of checks, etc.

Sinking Fund The accumulation of payments into an interest-paying investment; a desired amount will then be available to the investor at a future date.

Special Endorsement An endorsement of a check that limits what can be done with it in the next transaction.

Specific Duty A customs duty that is a flat amount per unit.

Specific Identification Method An inventory method in which each item is identified through a physical inventory.

Store Card A credit card that is used in only one store or chain of stores.

Straight Commission A percent of the dollar value of the sale paid to the sales person.

Straight Life Policy A type of life insurance that has a small surplus factor built into the premium.

Straight-line Method A type of depreciation that provides an equal amount of expense per year.

Subtrahend The smaller of two numbers that are being subtracted.

Sum The amount resulting from the addition of two or more numbers.

Sum-of-Years'-Digits (S.O.Y.D.) Method A type of depreciation that is accelerated and that uses the years of expected life to determine the fraction of total depreciation for that year.

Surface Measure Found by multiplying length by width.

T

Tax Rate The percent that is used to find the Social Security and Medicare tax; set by Congress.

Taxable Base The maximum amount of earnings that can be taxed for Social Security purposes for that year.

Taxable Income The amount of income left after subtracting the amount for exemptions and deductions. This amount is used to find the tax from the tax table.

Term Life Policy A type of life insurance that has no savings factor built into the premium. A best buy for the money if the face value needs to be large.

Term of Discount The period of time that the promissory note is discounted by the third party.

Term of Note The number of days between the date of the note and the due date.

Terms of Trade A numerical statement of the percent discounts that are available and the time period for qualification for cash discounts.

Time Used to compute interest or discounts. Usually a fraction of 360 days (ordinary interest) or 365 days (exact interest).

Time Card An item of control that allows a payroll department to determine the number of hours worked by employees.

Time Line A line that enables the asset life to be compared with the fiscal year to determine the proper amount of depreciation expense.

Total Cost All of the costs associated with placing the item into service. Used in determining depreciation expense.

Total Depreciation The amount that the asset will depreciate over its expected life.

Trade Discount A reduction of the list price to the buyer based on large quantity purchases or on the assumption of marketing functions being performed by the buyer.

Travel or Entertainment Card A credit card that is accepted primarily by firms in the travel or entertainment business.

Turnover The number of times that the average inventory is sold. A measure of success for the merchandiser.

U

Units-of-Production Method A method of depreciation that uses the amount of times an asset is used instead of the number of years it is used.

Universal Life Policy A type of life insurance that allows for changes in premium and coverage amounts. Surplus is invested in fixed-rate vehicles.

V

Variable Life Policy A type of life insurance that allows for changes in premium and coverage amounts. Surplus is invested in stocks and funds.

Variable Payment Amount Payment on a home mortgage that begins as a small amount and then increases with time. May cause loan balance to increase in early years.

Variable Rate Interest rate that changes; changes are based on an index of interest indicators—usually government securities.

Volume Measure Found by multiplying length by width by depth.

W

W-2 Form (*See* Form W-2.)

W-4 Form (*See* Form W-4.)

Workers' Compensation A type of insurance that provides coverage to workers who have been injured or who become ill because of work-related activities.

Answers to the Student Edition

Chapter One (Pages 1–20)
Exercise A
1. 8 **2.** 2 **3.** 3 **4.** 0 **5.** 1
Exercise B
1. thousands **2.** hundred thousands
3. tens **4.** hundreds **5.** ten thousands
Exercise C
Check number 1269. Five thousand, eight hundred forty-three and no/100
Check number 1270. One hundred sixty-five and no/100
Exercise D
1. 98,000 **2.** 2,640 **3.** 278,900
Exercise E
1. $1,900,000 **2.** $630,000
3. $320,000,000
Exercise F
A. 1. 124 **2.** 53 **3.** 112 **4.** 219
5. 90 **6.** 106 **7.** 99
B. 1. 772 **2.** 3,493 **3.** 15,215
4. 2,029 **5.** 10,655
C. 1. $8,116 **2.** $92,131 **3.** $16,417
4. $76,901 **5.** $89,140
D. 1. $114,635 **2.** $11,565
3. $84,744 **4.** $27,935
5. $23,980
Exercise G
Breakfast $13
Lunch $6
Dinner $49
Travel $353
Lodging $114
Tips $11
Misc. $9
Monday $245
Tuesday $91
Wednesday $70
Thursday $149
Week $555 (owe)
Exercise H
1269—$906, $286
1270—$286, $186
1271—$186, $117
1272—$117, $70
1273—$590, $526
Yes, pay Cindi.
Exercise I
A. 1. 19 **2.** 68 **3.** 24 **4.** 42 **5.** 28
6. 16 **7.** 62
B. 1. $17 **2.** $36 **3.** $160 **4.** $138
5. $86 **6.** $47
C. 1. $174 **2.** $98 **3.** $437 **4.** $155
5. $64 **6.** $544
Exercise J
A. 1. 67,300 **2.** 89,200 **3.** 5,460
4. 891,000 **5.** 8,940
B. 1. 8,190 **2.** 96,100 **3.** 5,700
4. 46,900 **5.** 162,000
C. 1. $134,937 **2.** $574,864
3. $228,690 **4.** $240,875
D. 1. $152,405 **2.** $251,805
3. $700,104 **4.** $138,678
Exercise K
$94, $279, $534, $46, $108, $27, $102, $120, $144, $192, $82, $104, $42, $100, $70, $60, $64, $126, $50, $20, $12, $48, $36, $68, $2,528
Exercise L
1. 157 **2.** 1,334 **3.** 118 **4.** 3,489 r32
5. 421 **6.** 2,033 r33 **7.** 224 r364
8. 19 r11
Exercise M
1. 7,930 **2.** 470 **3.** 906 **4.** 891 **5.** 60,050

Chapter One
Assignment
Skill Problems
1. 4 **3.** 7 **5.** billions **7.** hundred thousands **9.** One hundred forty-eight and 00/100 **11.** Two hundred fifty thousand and no/100 **13.** 5,450 **15.** D **17.** G **19.** I **21.** 10 **23.** 599 **25.** 13 **27.** 352 **29.** 119 **31.** 906 **33.** 9,650 **35.** 1,028 **37.** $5,562 **39.** 13 r41 **41.** $27,636 **43.** 23 r18 **45.** $69,971 **47.** 80 **49.** $32,400 **51.** 3,824 **53.** 47,280 **55.** $142,968 **57.** $32,571 **59.** 31 r1
Business Application Problems
1. 116 **3.** 70 **5.** $11,898 **7.** 544 **9.** $16,740 **11.** $16,498 **13.** $1,200 **15.** 1,071 **17.** College **19.** $2,100 **21.** $325,300 **23.** 5 **25.** 48,028 **27.** 591 **29.** The cost of 23 chairs will be $2,921.

Chapter One
Mastery Test
1. 54,400 (Objective 3)
($64 \times 850 =$)
2. $10,873 (Objective 3)
($1,457 + 5,800 + 3,575 + 41 =$)
3. $830,079 (Objective 3)
($42,803 + 787,276 =$)
4. 4,214 r7 (Objective 3)
($75,859 \div 18 =$)
5. $10,773 (Objectives 3,4)
($64,638 \div 6 =$)
6. 102,663 (Objective 3)
($549 \times 187 =$)
7. 19 r8 (Objective 3)
($768 \div 40 =$)
8. $6,249,488 (Objectives 3, 4)
($6,185,005 + 54,980 + 7,681 + 1,822 =$)
9. 680 (Objectives 1, 2, 3, 4)
($891 - 147 - 63 =$)
10. $326,723 (Objectives 1, 3, 4)
($85,605 + 200,038 + 41,080 =$)
11. $197,217 (Objective 3)
($387 \times 651 = 456 \times 120 = 251,937 - 54,720 =$)
12. Forty-three thousand, six hundred seventy-nine dollars (Objective 1)
13. $367,000 (Objectives 2, 3)
($459,073 - 91,752 =$)
14. $11,760 (Objective 3)
($40 \div 2 = 20 \div 2 = 10 \times 98 = 980 \times 12 =$)
15. 685,200 (Objective 2)

Chapter Two (Pages 21–44)
Exercise A
1. common **2.** improper **3.** improper
4. improper **5.** mixed **6.** common
7. improper **8.** common **9.** common
10. common **11.** mixed **12.** mixed
Exercise B
1. $1\frac{5}{8}$ **2.** $5\frac{7}{12}$ **3.** $2\frac{4}{5}$ **4.** $3\frac{5}{14}$ **5.** $3\frac{1}{6}$
6. $2\frac{1}{2}$ **7.** $2\frac{1}{9}$ **8.** $9\frac{3}{10}$ **9.** $4\frac{3}{10}$ **10.** $1\frac{1}{8}$
11. 4 **12.** $7\frac{1}{8}$
Exercise C
1. $\frac{27}{4}$ **2.** $\frac{59}{6}$ **3.** $\frac{39}{5}$ **4.** $\frac{15}{8}$ **5.** $\frac{551}{5}$
6. $\frac{7}{2}$ **7.** $\frac{147}{8}$ **8.** $\frac{20}{3}$ **9.** $\frac{19}{4}$ **10.** $\frac{3}{2}$
11. $\frac{87}{16}$ **12.** $\frac{21}{2}$
Exercise D
1. $\frac{1}{2}$ **2.** $\frac{3}{4}$ **3.** $\frac{7}{10}$ **4.** $\frac{2}{5}$ **5.** $\frac{1}{2}$ **6.** $\frac{7}{8}$
7. $\frac{11}{25}$ **8.** $\frac{1}{2}$ **9.** $\frac{3}{4}$ **10.** $\frac{3}{5}$ **11.** $\frac{8}{11}$
12. $\frac{3}{8}$
Exercise E
1. 15 **2.** 33 **3.** 90 **4.** 28 **5.** 15 **6.** 27 **7.** 16
8. 5 **9.** 410 **10.** 12 **11.** 15 **12.** 5
Exercise F
1. $1\frac{1}{12}$ **2.** $2\frac{1}{8}$ **3.** $1\frac{16}{105}$ **4.** $2\frac{3}{8}$ **5.** $1\frac{1}{35}$
6. $1\frac{11}{40}$ **7.** $1\frac{11}{60}$ **8.** $1\frac{37}{80}$ **9.** $1\frac{1}{16}$ **10.** $1\frac{13}{14}$
Exercise G
1. $\frac{5}{8}$ **2.** $\frac{3}{8}$ **3.** $\frac{7}{30}$ **4.** $\frac{1}{20}$ **5.** $\frac{113}{180}$
6. $\frac{11}{24}$ **7.** $\frac{7}{36}$ **8.** $\frac{5}{12}$ **9.** $\frac{51}{80}$ **10.** $\frac{5}{16}$
Exercise H
1. $15\frac{1}{4}$ **2.** $14\frac{7}{24}$ **3.** $5\frac{1}{16}$ **4.** $10\frac{23}{24}$
5. $11\frac{4}{5}$ **6.** $8\frac{67}{72}$ **7.** $24\frac{81}{100}$ **8.** $8\frac{17}{24}$
9. $15\frac{11}{12}$ **10.** $17\frac{19}{24}$

Exercise I

1. $7\frac{1}{3}$ **2.** $2\frac{1}{4}$ **3.** $5\frac{11}{40}$ **4.** $52\frac{23}{24}$
5. $2\frac{11}{30}$ **6.** $5\frac{1}{40}$ **7.** $6\frac{7}{12}$ **8.** $173\frac{1}{6}$
9. $3\frac{1}{12}$ **10.** $6\frac{11}{18}$

Exercise J

1. $\frac{1}{6}$ **2.** $\frac{77}{80}$ **3.** $\frac{5}{8}$ **4.** $\frac{5}{42}$ **5.** $\frac{1}{2}$ **6.** $\frac{7}{18}$
7. $3\frac{1}{4}$ **8.** $19\frac{19}{24}$ **9.** $\frac{25}{84}$ **10.** $\frac{13}{224}$

Exercise K

1. $1\frac{5}{7}$ **2.** $1\frac{1}{2}$ **3.** $1\frac{5}{7}$ **4.** $2\frac{1}{2}$ **5.** $1\frac{11}{16}$
6. $1\frac{2}{7}$ **7.** $3\frac{1}{2}$ **8.** 6 **9.** $4\frac{2}{7}$ **10.** $1\frac{1}{4}$

Exercise L

1. $106\frac{47}{256}$ **2.** $4\frac{5}{7}$ **3.** $4\frac{2}{13}$ **4.** 4
5. $1\frac{71}{84}$ **6.** $2\frac{53}{76}$ **7.** $10\frac{19}{35}$ **8.** $1\frac{83}{162}$

Exercise M

1. 3:2 **2.** 1:10 **3.** 4:1 **4.** 5:3 **5.** 5:2
6. 1:4

Chapter Two
Assignment
Skill Problems

1. improper **3.** mixed **5.** $8\frac{1}{2}$ **7.** $12\frac{1}{2}$
9. $\frac{16}{3}$ **11.** $\frac{48}{7}$ **13.** $\frac{3}{4}$ **15.** $\frac{2}{5}$ **17.** $1\frac{1}{4}$
19. $2\frac{2}{5}$ **21.** 12 **23.** 18 **25.** A **27.** H
29. $9\frac{1}{4}$ hours **31.** $10\frac{1}{4}$ hours **33.** 3:2
35. 3:2 **37.** $\frac{27}{143}$ **39.** $\frac{24}{35}$ **41.** $1\frac{5}{12}$
43. $1\frac{11}{24}$ **45.** $\frac{1}{30}$ **47.** $2\frac{9}{40}$ **49.** $3\frac{27}{40}$
51. $1\frac{3}{8}$ **53.** $\frac{21}{32}$ **55.** $1\frac{11}{30}$ **57.** $\frac{1}{2}$
59. $1\frac{1}{5}$ **61.** $\frac{8}{75}$ **63.** $2\frac{1}{12}$

Business Application Problems

1. 3,871 **3.** $6\frac{5}{12}$ **5.** 240 **7.** $603
9. 1:136 **11.** $21 **13.** 2 **15.** 7:1
17. $1,400,000 **19.** 8:1 **21.** $18
23. $500 **25.** 30,488 **27.** $227\frac{7}{10}$
29. There are 130 units of the condominium that are unsold.

Chapter Two
Mastery Test

1. Whole number (8), numerator (4), denominator (7) (Objective 6)
2. 1. B
   2. A
   3. C
   4. A
   5. B
   6. C (Objective 1)
3. $1\frac{11}{20}$ (Objective 4) $\left(\frac{15}{20} + \frac{16}{20} =\right)$
4. $\frac{5}{6}$ (Objective 4) $\left(1\frac{3}{6} - \frac{4}{6} =\right)$
5. 50 (Objective 4) $\left(15 \times \frac{10}{3} =\right)$
6. $4\frac{3}{8}$ (Objective 4) $\left(\frac{7}{8} \times \frac{5}{1} =\right)$
7. $5\frac{1}{3}$ (Objective 3)
8. $147,720 (Objective 4) (98,480 × 3 =, 295,440 ÷ 2 =)
9. 9:1 (Objective 5)
10. $5\frac{2}{5}$ (Objective 4) $\left(8\frac{5}{10} - 3\frac{1}{10} =\right)$
11. $11,592 (Objective 4) (57,960 ÷ 5 =)
12. $104 (Objective 4) (8 × 13 =, 104 ÷ 2 =, 52 × 2 =)
13. 270 (Objective 4) $\left(800 \times \frac{1}{2} =, 400 \times \frac{3}{4} =, 300 \times \frac{9}{10} =\right)$

Chapter Three (Pages 45–62)

Exercise A

1. 7 **2.** 8 **3.** 5 **4.** 3 **5.** 2 **6.** 9

Exercise B

1. hundred thousandths **2.** hundredths
3. millionths **4.** ten thousandths
5. tenths **6.** thousandths

Exercise C

Reading

Exercise D

Reading

Exercise E

1. Eight hundred twenty-three dollars and sixty-seven cents
2. Two thousand, one hundred five dollars and forty-six cents
3. Eight hundred sixty-seven dollars and twenty-eight cents
4. Two hundred thirty-four dollars and eighty-seven cents
5. Three thousand, eight hundred sixty-nine dollars and sixty-one cents

Exercise F

1. $935.49 **2.** $63.26 **3.** $572.21
4. $3,018.07 **5.** $51,307.36

Exercise G

1. .6 **2.** .47 **3.** .0 **4.** .546 **5.** 1

Exercise H

1. $605.84 **2.** $57.65 **3.** $486.94
4. $68,640.27 **5.** $1,754.11

Exercise I

1. 607.80 **2.** 800 **3.** 759.9
4. 550,000 **5.** 14.80

Exercise J

1. 15,576.1975 **2.** 889.1357
3. 745.22646 **4.** 1,165.7941
5. 840.48988

Exercise K

1. 612,026.0913 **2.** 346.2974
3. 79.8762 **4.** 502.2 **5.** 158,463.0286

Exercise L

1. 7.530 **2.** 16.433 **3.** .4383
4. 56.043 **5.** 41.863

Exercise M

1. $835.49 **2.** $530.05 **3.** $48.17
4. $596.99 **5.** $278.92

Exercise N

1. 8,445.2 **2.** 2.278 **3.** 49.53
4. 2.28015 **5.** 2.754

Exercise O

1. 35.7 **2.** .4 **3.** 308.7 **4.** 1.7 **5.** 0

Exercise P

1. .375 **2.** .5625 **3.** 1.6 **4.** 2.4 **5.** 2.1875

Chapter Three
Assignment
Skill Problems

1. 7 **3.** 3 **5.** 0 **7.** Eight hundred fifty-six dollars and forty-five cents **9.** .7
11. 548,658.459 **13.** .461 **15.** 52.0027
17. $10,319.65 **19.** $14,725.60
21. .75 **23.** 2.6 **25.** B **27.** $840
29. 92.964 **31.** 3.334 **33.** 6,175.8
35. 22.31 **37.** .02 **39.** .7 **41.** .1512
43. 1,458.6557 **45.** 4.35 **47.** 48.8
49. 72.1589 **51.** 1.17 **53.** .04644
55. 1,462.6193

Business Application Problems

1. $37.15 **3.** $32,145.12
5. $112,070.50 **7.** $3,234 **9.** $85
11. $219 **13.** $112.18 **15.** $74.12
17. 175 **19.** $33,513.99
21. $125,819.25 **23.** $8.20 **25.** $129.70
27. $12,286.30 **29.** The total cost of the small flower arrangements and cards will be $2,227.50 for the month.

Chapter Three
Mastery Test

1. **a.** 10
   b. .1
   c. 600
   d. 64.097
   e. 43.9 (Objective 2)
2. Three hundred eighty-five dollars and seventy cents (Objective 1)
3. 755.167 (Objective 3) (733.042 + 8.367 + 13.758 =)
4. 47.674 (Objective 3) (57.079 − 9.405 =)
5. 1 (Objective 3) (1.2 × .56 =)
6. $.47 (Objectives 2, 3) (3.72 ÷ 8 =)
7. **a.** .625 **b.** .3 **c.** 4.2 (Objective 5) (5 ÷ 8 =, 3 ÷ 10 =, 1 ÷ 5 =)
8. 19¢ (Objective 3) (1,560 ÷ 6,000 =)

9. $844,875.20 (Objective 3) (324,952 × 2.6 =)
10. 10,835 (Objectives 2, 3) (15,479 × .7 =)

Chapter Four (Pages 63–82)
Exercise A
1. restrictive **2.** special **3.** blank **4.** blank **5.** restrictive **6.** restrictive **7.** blank **8.** special **9.** blank **10.** blank
Exercise B
1. $775.44 **2.** $159.80 **3.** $3,895.07 **4.** $1,563
Exercise C
1. $789 **2.** $1,178 **3.** $1,576 **4.** $728.88 **5.** $3,540

Chapter Four Assignment
Skill Problems
1. blank **3.** restrictive **5.** blank **7.** E **9.** C **11.** depositor **13.** deposit slip **15.** deposits in transit **17.** service charges **19.** $1,166.03 **21.** $1,699 **23.** $1,663.33 **25.** 214.00 **27.** 2,936.49 **29.** 2,962.04 **31.** 545.01 **33.** 2,607.12 **35.** 2,592.77 **37.** 2,580.33 **39.** $1,663.33
Business Application Problems
1. $3,559.50 **3.** $1,527.50 **5.** $25.00 **7.** $4,724.95 **9.** $7,258 **11.** $1,919.80 **13.** $124.30 **15.** $11.50 **17.** $331.28 **19.** $425 **21.** $429.40 **23.** The bank information indicates a $5,256 balance. **25.** The real balance is an overdraw of $338.

Chapter Four Mastery Test
1. a. For deposit only, G. S. Student
b. Gerry S. Student
c. Payable to B. College G. S. Student (Objective 1)
2. $1,197.55 (Objective 2) (1,088 + 555.97 − 447.42 =, 1,063.08 − 100 + 234.47 =)
3. $10.98 (Objective 2) (47 + 25 − 37.80 =, 92.61 − 47.43 =, 45.18 − 34.20 =)
4. $7.40 (Objective 2) (89.70 − 97.10 =)
5. $754.54 (Objective 2) (2,726.34 − 1,962.80 − 9 =)

Chapter Five (Pages 83–106)
Exercise A
1. $400, $45, $445 **2.** $380, $28.50, $408.50 **3.** $320, $12, $332 **4.** $880, $66, $946 **5.** $320, $0, $320 **6.** $684, $0, $684 **7.** $360, $13.50, $373.50
Exercise B
1. $312.00, $252.00, $564.00
2. $295.20, $0, $295.20 **3.** $307.80, $328.05, $635.85 **4.** $487.50, $37.50, $525.00 **5.** $576.00, $48.00, $624.00
6. $361.00, $299.25, $660.25
7. $331.50, $229.50, $561.00
8. $390.00, $165.00, $555.00
Exercise C
1. $57,048 **2.** $1,052.04 **3.** $19,994.96 **4.** $396,541.50 **5.** $14,875
Exercise D
1. $2,504.70 **2.** $2,061 **3.** $2,641.80 **4.** $3,818 **5.** $2,943.50
Exercise E
1. $1,380 **2.** $1,840 **3.** $3,219.50 **4.** $1,255 **5.** $274.50
Exercise F
1. $8.70, $37.20, $467.10
2. $18.32, $78.33, $869.02
3. $13.02, $55.68, $564.42
4. $10.97, $46.92, $580.78
5. $10.07, $43.04, $509.37
Exercise G
1. $30, $6.31, $26.97, $371.72
2. $44, $7.55, $32.30, $437.15
3. $46, $7.06, $30.19, $403.75
4. $17, $7.35, $31.43, $451.22
5. $56, $7.19, $30.75, $402.06
6. $43, $6.70, $28.64, $383.66
7. $54, $8.63, $0, $532.37
8. $43, $6.67, $28.52, $381.80
9. $29, $6.19, $26.47, $365.34
10. $20, $6.95, $0, $452.05
Exercise H
1. $606.36 **2.** 1,593.40 **3.** 431.27 **4.** 2,926.40 **5.** 1,106.08 **6.** $141.81 **7.** 372.65 **8.** 100.86 **9.** 684.40 **10.** 258.68 **11.** $39.12 **12.** 102.80 **13.** 27.82 **14.** 188.80 **15.** 71.36 **16.** $264.06 **17.** 693.90 **18.** 187.81 **19.** 1,274.40 **20.** 481.68

Chapter Five Assignment
Skill Problems
1. C **3.** I **5.** D **7.** J **9.** P **11.** H **13.** $563.50 **15.** $287. **17.** $467.65 **19.** $12,361.92 **21.** $56,010.19 **23.** $2,527.20 **25.** $990 **27.** $19,304.20 **29.** $3,013.40 **31.** $960.25 **33.** $444.90 **35.** $332.01 **37.** $3,153.09 **39.** $2,561.31 **41.** $2,897.10 **43.** $103 **45.** $66.87 **47.** $176.90 **49.** $19.26
Business Application Problems
1. $50 **3.** $4,823.50 **5.** $992 **7.** $2,273.58 **9.** $4,095 **11.** $281.55 **13.** $7,347.15 **15.** $4,724.40 **17.** $3,648 **19.** $403.84 **21.** $9,360 **23.** $268.96 **25.** The difference will be $17. **27.** The additional take home pay will be $3,273.50.

Chapter Five Mastery Test
1. F (Objective 4)
2. T (Objective 4)
3. F (Objective 4)
4. F (Objective 4)
5. F (Objective 4)
6. T (Objective 4)
7. $453.10 (Objective 1) (394 × 1.15 =)
8. $45 (Objective 2)
9. $1,823.14 (Objective 3)
10. $407.05 (Objective 2) (40 × 9.50 =, 7 × $1\frac{1}{2}$ × 9.50 =, 479.75 × .062 =, 479.75 × .0145 =)
11. $396 (Objective 1) (40 × 9 =, 2 × 18 =)
12. $324.15 (Objective 1) (39.8 × 7 =, 584 × .078 =)

Chapter Six (Pages 107–126)
Exercise A
1. 55 yd, 1 ft, 4 in.
2. 1 lb, $6\frac{1}{3}$ oz.
3. 33 min, 5 sec
4. 424 lb, 13 oz
5. 1 hr, 26 min, 47 sec
Exercise B
1. 27 sq. yd, 1 sq ft, 7 sq. in.
2. 23 sq. yd, 6 sq. ft
3. 6 sq. yd, $7\frac{3}{4}$ sq. ft
4. 21.76 sq. mi
5. 1 sq. yd, 7 sq. ft
Exercise C
1. 5 cu. yd
2. 8 cu. yd, 24 cu. ft
3. 10 cu. yd, 24 cu. ft
4. 20 cu. yd, 4 cu. ft
5. 4,000 cu. ft
Exercise D
1. 9.632 **2.** 8.56 **3.** .806329 **4.** 76 **5.** 473,570
Exercise E
1. 3.8354 **2.** 4.4196 **3.** 39.47896 **4.** 15.1408 **5.** 8.4539 **6.** 18.123 **7.** 3.1096 **8.** .1135749 **9.** 1.7577 **10.** 8.37748
Exercise F
1. 7,050.88 **2.** 126,044.07 **3.** 3,456,950 **4.** 339,020 **5.** 4,369.5
Exercise G
1. $96 **2.** $163.10 **3.** $87.65 **4.** $7,757.44 **5.** $1,315

Chapter Six Assignment
Skill Problems
1. 3 yd, 1 ft, 11 in. **3.** 54 min, 5 sec **5.** 1 hr, 42 min, 36 sec **7.** 15 sq. yd, 3 sq. ft **9.** 39.84 sq. mi **11.** 6 cu. yd **13.** 19 cu. yd, 6 cu. ft **15.** 7 yd, 2 ft, 4 in. **17.** 10 hr, 8 min, 40 sec **19.** 55 cu. yd,

11 cu. ft **21.** 43,620 **23.** 7.865 **25.** .807604 **27.** 374,800 **29.** 7.5438 **31.** 5.1816 **33.** 51.3992 **35.** 44.4761 **37.** 8.8255 **39.** 11,206.372 **41.** 316.86 **43.** 9,909.68 **45.** 726.31 **47.** 4,534.90 **49.** 607,600
Business Application Problems
1. $14,285.71 **3.** $208 **5.** $50,025 **7.** $36,727.68 **9.** 1,173 **11.** 4 hr, 14 min **13.** 10 **15.** 200 sq. yd **17.** $133\frac{1}{3}$ hr **19.** The cost of the material will be $1,028.40.

Chapter Six
Mastery Test

1. 1 mile, 885 yd (Objective 2) (4 ÷ 3 =, 2,655 ÷ 3 =)
2. 486 (Objective 1) (60 × 5 =, 300 × 60 =, 18,000 ÷ 37 =)
3. 3 (Objective 2) (30 × 8 =, 42 × 8 =, 1,152 ÷ 500 =)
4. 56 (Objective 2) (400 × 2.5 × 1.5 =, 1,500 ÷ 27 =)
5. 33.558355 (Objective 4) (53 × 1.196 =, 63.388 × 9 =, 570.492 ÷ 17 =)
6. 5.81096 (Objective 4) (7.6 × .7646 =)
7. 583.2 (Objective 4) (90 × .4536 =, 40,824 ÷ 70 =)
8. 3 yd, 10 in. (Objective 2) (8 − 5 =)
9. 14,300 (Objective 3)
10. $5,109.84 (Objective 5) (42,300 × .1208 =)
11. no (Objective 2) (60 × 120 × 12 =)
12. 36,270,000 (Objective 5) (3,900 × 9,300 =)

Chapter Seven (Page 127–142)
Exercise A
1. .17 **2.** .56 **3.** .11 **4.** .09 **5.** 1.56
Exercise B
1. 41,.41 **2.** 62,62 **3.** 127,127 **4.** 10, 10 **5.** 21, .21
Exercise C
1. 8.2, .082 **2.** 1.375, .01375 **3.** .6, .006 **4.** 52.9, .529 **5.** 29.5, 29.5
Exercise D
1. 36 **2.** 11 **3.** 59 **4.** 6 **5.** 320 **6.** 2 **7.** 6 **8.** 10 **9.** 25 **10.** 64
Exercise E
1. 100 **2.** 86% **3.** 26 **4.** 69 **5.** 11% **6.** 500
Exercise F
1. $7.14 **2.** $180 **3.** 512% **4.** $23.20 **5.** $7.08 **6.** 43%
Exercise G
1. 25 **2.** 10 **3.** 14 **4.** 3 **5.** 11 **6.** 935 **7.** $400,000 **8.** $115,384.61 **9.** 20 **10.** 9

Chapter Seven
Assignment
Skill Problems
1. B **3.** 18 **5.** 41 **7.** .7 **9.** 175.9 **11.** 4.58 **13.** 30 **15.** .47 **17.** .17 **19.** 26 **21.** .102 **23.** 3.6 **25.** 254 **27.** 91 **29.** $64.80 **31.** $4.80 **33.** 150 **35.** $11.03 **37.** 84 **39.** 11 **41.** 14% **43.** 35 **45.** 6 **47.** 2 **49.** 2 **51.** 108 **53.** 22 **55.** $8,184
Business Application Problems
1. $4,248.10 **3.** 17% **5.** 10% **7.** $51,239.90 **9.** 66% **11.** 68 **13.** 17% **15.** $200,000 **17.** $24,000 **19.** 13% **21.** $786.36 **23.** 120 **25.** $500 **27.** 30% **29.** The sales in excess of quota were $325,000 for the month.

Chapter Seven
Mastery Test

1. .09 (Objective 2)
2. 87.5 (Objective 2)
3. 21 (Objective 2)
4. 9.5 (Objective 3)
5. T (Objective 5)
6. F (Objective 5)
7. F (Objective 6)
8. $150 (Objective 7) (600 × .25 =)
9. 2% (Objective 7) (12,856 ÷ 642,800 =)
10. $1.88 (Objective 7) (24 ÷ 1.085 =)
11. $240 (Objective 7) (1,500 × .16 =)
12. 595 (Objective 7) (85,000 × .007 =)
13. 8.9% (Objective 7) (7,000 ÷ 79,000 =)
14. $27,200,000 (Objective 7) (68,000,000 × .4 =)
15. 1.2% (Objective 7) (816,000 ÷ 68,000,000 =)

Chapter Eight (Pages 143–160)
Exercise A
1. $111.60 **2.** $95.04 **3.** $194.04 **4.** $169.32 **5.** $66.83 **6.** $250.60 **7.** $132.30 **8.** $159 **9.** $138.13 **10.** $73.27
Exercise B
1. $11,739.13 **2.** $20,000 **3.** $5,783.13 **4.** $33,333.33 **5.** $27,692.31 **6.** $10,000 **7.** $50,000 **8.** $12,000 **9.** $22,500 **10.** $23,750
Exercise C
1. $299.60 **2.** $249.15 **3.** $2,202 **4.** $96 **5.** $1,219.80 **6.** $178.10 **7.** $538.40 **8.** $774.75 **9.** $93.40 **10.** $4,414.80
Exercise D
1. $6,208 **2.** $123,750 **3.** $4,500 **4.** 0 **5.** $134,600
Exercise E
1. $156,950 **2.** $80,400 **3.** $2,968,880 **4.** $4,895,100 **5.** $3,487,128
Exercise F
1. $1.83 **2.** $.60 **3.** $1.18 **4.** $9.45 **5.** $4.69

Chapter Eight
Assignment
Skill Problems
1. E **3.** C **5.** A **7.** $410 **9.** $1,050 **11.** $132.84 **13.** $199.20 **15.** $325 **17.** $193.20 **19.** $1,302 **21.** $226.50 **23.** $9,375 **25.** $1,010.40 **27.** $300 **29.** $232,050 **31.** $1,444,800 **33.** $24,500 **35.** $7,513,560 **37.** $1,222,650 **39.** $13,461.54 **41.** $35,000 **43.** $8,163.27 **45.** $1,011.49 **47.** $49,188.38 **49.** $20,294.12 **51.** $10,158.73 **53.** $4,088.12 **55.** $13,427.49 **57.** $31,250 **59.** $80,000
Business Application Problems
1. $5,778 **3.** $790 **5.** $160,000 **7.** $1,576.40 **9.** $1,951.60 **11.** $1,923.75 **13.** $78,688.52 **15.** $75,000 **17.** $38,000 **19.** $360,000 **21.** $39,375 **23.** $440,000 **25.** The premium cost is $6 per hour.

Chapter Eight
Mastery Test

1. T (Objective 5)
2. T (Objective 5)
3. F (Objective 5)
4. T (Objective 5)
5. T (Objective 5)
6. T (Objective 5)
7. T (Objective 5)
8. F (Objective 5)
9. $136.80 (Objective 2) (18 × 7.6 =)
10. $169,000 (Objective 2) (70 × 12 =, 840 ÷ 4.98 =)
11. $27,200 (Objective 3) (34,000 × .75 =, 34,000 × .6 =, 20,400/25,500 × 34,000 =)

Chapter Nine (Pages 161–178)
Exercise A
1. 35 **2.** 44 **3.** 83 **4.** 78 **5.** 42
Exercise B
1. March 8 **2.** November 26 **3.** August 2 **4.** December 28 **5.** March 23
Exercise C
1. $5.60 **2.** $9.72 **3.** $245 **4.** $13.57 **5.** $9.71 **6.** $93.33
Exercise D
1. $5.92 **2.** $181.23 **3.** $7.80 **4.** $11.32 **5.** $9.86 **6.** $12.95
Exercise E
1. $720 **2.** $30.56 **3.** 14% **4.** 27 **5.** $750 **6.** 12% **7.** $212.50 **8.** $1,080 **9.** 7 **10.** 13% **11.** $4,132.09 **12.** $6,000.02 **13.** 24 **14.** 12% **15.** 251 **16.** $1,155 **17.** 10% **18.** $4,615.38 **19.** 59 **20.** $1,757.15

Chapter Nine
Assignment
Skill Problems
1. E **3.** C **5.** G **7.** 211 **9.** 451 **11.** 135 **13.** January 23, 2003

15. August 6, 2004 **17.** 53 **19.** 64 **21.** 96 **23.** October 1 **25.** April 15 **27.** $40. **29.** $2,396.24 **31.** $10.79 **33.** $7.89 **35.** $9.64 **37.** $15.90 **39.** $20.83 **41.** $826.67 **43.** $1,496.25 **45.** $110. **47.** 20 **49.** 8% **51.** $780 **53.** 7% **55.** $60,000 **57.** 10% **59.** 10% **61.** $3,500

Business Application Problems

1. $4,000 **3.** May 11th **5.** 10% **7.** $164,133.33 **9.** 15% **11.** $350 **13.** $414.25 **15.** Mr. Severance **17.** $2,024.56 **19.** $12,954. **21.** $184.38 **23.** $3.77 **25.** The amount of payoff on the loan is $1,542.50.

Chapter Nine
Mastery Test

1. c (Objective 1)
2. 148 (Objective 2) (25 + 30 + 31 + 30 + 31 + 1 =)
3. April 21, 2002 (Objective 2) (8 + 31 + 20 =)
4. $666.67 (Objective 2) (50,000 × .16 × 30 ÷ 360 =)
5. $657.53 (Objective 3) (50,000 × .16 × 30 ÷ 365 =)
6. June 26 (Objective 2) (31 − 12 + 26 =)
7. $13.44 (Objective 3) (1,000 × .1075 × 45 ÷ 360 =)
8. $1,013.44 (Objective 3) (1,000 + 13.44 =)
9. $46.50 (Objective 3) (1,800 × .155 × 60 ÷ 360 =)
10. $5,290.27 (Objective 3) (5,000 × .13 × 163 ÷ 365 =, 5,000 + 290.27 =)
11. $5,037.26 (Objective 3) (5,000 × .17 × 16 ÷ 365 =, 5,000 + 37.26 =)
12. $21.30 (Objective 3) (900 × .12 × 71 ÷ 360 =)

Chapter Ten (Pages 179–194)

Exercise A

1. $6,662.50 **2.** $4,080 **3.** $13,985.33 **4.** $25,746.53 **5.** $9,000 **6.** $7,303.33 **7.** $8,623.96 **8.** $5,000 **9.** $61,703.33 **10.** $3,540

Exercise B

1. 23 **2.** 23 **3.** 69 **4.** 21 **5.** 26 **6.** 75 **7.** 22 **8.** 30

Exercise C

1. $453.60 **2.** $5,048.63 **3.** $5,120.81 **4.** $6,086.85 **5.** $1,849.20 **6.** $170.90 **7.** $807.81 **8.** $1,017.06 **9.** $1,020.90 **10.** $3,533.95

Chapter Ten
Assignment

Skill Problems

1. F **3.** T **5.** F **7.** E **9.** $20,466.67 **11.** $460,800 **13.** $8,806.94 **15.** $3,050 **17.** $964.68 **19.** $10.61 **21.** $259 **23.** 131 **25.** $7,408.17 **27.** $10,100 **29.** $84.17 **31.** $263.01 **33.** 10 **35.** $5,237.06 **37.** $1,140 **39.** $1.25 **41.** $65,878.94 **43.** $4,518.66 **45.** $40,361.88 **47.** $6,996.09 **49.** $413.92

Business Application Problems

1. $16,520 **3.** $16,038.17 **5.** Balentine and Company is the maker. **7.** August 10 due **9.** $61,407.75 **11.** $3,953.33 **13.** $4,064.92 **15.** $12,980.24 **17.** $19,753.89 **19.** $1,221.10 **21.** $268.49 **23.** The proceeds were $40,425.84.

Chapter Ten
Mastery Test

1. B (Objective 4)
2. E (Objective 4)
3. D (Objective 4)
4. K (Objective 4)
5. C (Objective 4)
6. A (Objective 4)
7. 62 (Objective 3) (21 + 7 =, 90 − 28 =)
8. $95.72 (Objective 3) (3,580 × .14 × 90 ÷ 360 =, 3,705.30 × .15 × 62 ÷ 360 =)
9. $3,609.58 (Objective 3) (3,705.30 − 95.72 =)
10. $3,705.30 (Objective 2) (3,580 × .14 × 90 ÷ 360 =)
11. $893 (Objective 3) (900 × .14 × 20 ÷ 360 =, 900 − 7 =)

Chapter Eleven (Pages 195–216)

Exercise A

1. $209.99 **2.** $430.44 **3.** $444.50 **4.** $517.50 **5.** $373.15 **6.** $39.56 **7.** $29.51 **8.** $279.47

Exercise B

1. $381.98 **2.** $892.58 **3.** $344.82 **4.** $680.01 **5.** $1,140.22 **6.** $108.75 **7.** $1,297.26

Exercise C

1. $29,346.13 **2.** $13,815.86 **3.** $105,866.19 **4.** $9,549.58 **5.** $14,220.04

Exercise D

1. $123.82 **2.** $931.43 **3.** $259.37 **4.** $263.85 **5.** $216.14

Chapter Eleven
Assignment

Skill Problems

1. $672.66 **3.** $235.04 **5.** $279.86 **7.** $156.75 **9.** $417.13 **11.** $244.44 **13.** $1,850.48 **15.** $33.19 **17.** $1,445.68 **19.** $1,294.17 **21.** $4.17 **23.** $418.39 **25.** $82.29 **27.** $2.80 **29.** $253.12 **31.** $83.67 **33.** $1.41 **35.** $85.08 **37.** $85.08 **39.** $37.05 **41.** $11.03 **43.** $50.77 **45.** $7,799 **47.** $25,602.36 **49.** $25,703.99 **51.** $95.67 **53.** $9,199.23 **55.** $48,730.90 **57.** $284.57 **59.** $381.43 **61.** $911.93

Business Application Problems

1. $852.23 **3.** $82,307.16 **5.** $145.40 **7.** $5,490 **9.** $1,500 **11.** $1,207.50 **13.** $318.67 **15.** $444.30 **17.** $74,651.40 **19.** The difference over the term of the mortgage is $33,408

Chapter Eleven
Mastery Test

1. True (Objective 5)
2. False (Objective 5)
3. $22.48 (Objective 1) (8.99 × 2.5 =)
4. $810.48 (Objective 1) (86 × 6.43 =, 1,200 ÷ 12 =, 1,890 ÷ 12 =)
5. $27,819.33 (Objective 2) (28,000 × .09 × $\frac{1}{12}$ =, 27,910 × .09 × $\frac{1}{12}$ =)
6. $156.34 (Objective 4) (158.87 × 6 =, 953.22 − 900 =, 158.87 − 2.53 =)
7. $4,997.31 (Objective 4) (4,907.35 × .10 × $\frac{1}{12}$ =, 4,907.35 × .01 =)

Chapter Twelve (Pages 217–234)

Exercise A

1.	**2.**	**3.**
$38.68	$171.87	$16.87
$64.07	$153.59	$53.14
$77.31	$175.13	$55.47
		$46.02

4.	**5.**
$53	$36.66
$109.03	$208.53
$100.12	$212.12
	$184.24

Exercise B

1. 15.5 **2.** 23.0 **3.** $35.87 **4.** 14.25 **5.** 21.0

Exercise C

1. $24, $456 **2.** $85.50, $1,814.50 **3.** $12.01, $588.64 **4.** $36, $1,164 **5.** $3.35, $85.85

Chapter Twelve
Assignment

Skill Problems

1. C **3.** B **5.** F **7.** C **9.** $.37 **11.** $.48 **13.** $.39 **15.** $.24 **17.** $.09 **19.** $.55 **21.** $14.67 **23.** $14.14 **25.** $14.13 **27.** $0 **29.** $1.44 **31.** $2.10 **33.** $1.60 **35.** $2.19 **37.** $10.06 **39.** $4.85 **41.** $4.31 **43.** $.79 **45.** 17% **47.** 17% **49.** 21% **51.** 14.25% **53.** 17% **55.** 23% **57.** $1,241.60 **59.** $82.87 **61.** $81,499.40 **63.** $1,797.41 **65.** $1,532.70

Business Application Problems

1. $211.28 **3.** $20.47 **5.** $15.42 **7.** $1,684.30 **9.** $133.33 **11.** $287.63 **13.** $126.74 **15.** $9.20 **17.** $67.44 **19.** The total amount of the deposit is $839.56.

Chapter Twelve
Mastery Test

1. T (Objective 5)
2. T (Objective 5)

3. F (Objective 5)
4. F (Objective 5)
5. $1,122.91 (Objective 4) (824.8 × .04 =, 327 + 4.10 + 791.81 =)
6. $44.13 (Objective 2) (160 − 20 =, 140 × .015 =, 92.10 × .015 =, 43.48 × .015 =)
7. 17.5% (Objective 3) (100 × 8 =, 50 ÷ 7.5 =)

Chapter Thirteen (Pages 235–262)
Exercise A
1. $1,792.12 **2.** $7,234.90 **3.** $7,550.32 **4.** $10,494.58 **5.** $43,822.46 **6.** $8,704.06 **7.** $27,057.50 **8.** $6,348.67 **9.** $12,033.10 **10.** $12,854.22
Exercise B
1. $1,365.58 **2.** $1,672.77 **3.** $3,905.99 **4.** $1,531.70 **5.** $624.60 **6.** $9,970.67 **7.** $5,435.70 **8.** $6,015.68 **9.** $6,736.25 **10.** $10,053.29
Exercise C
1. $42,576.09 **2.** $6,603.39 **3.** $15,449.34 **4.** $22,337.43 **5.** $7,423.10 **6.** $9,702.61 **7.** $11,687.27 **8.** $73,088.95 **9.** $9,581.01 **10.** $24,984.23
Exercise D
1. $2,046.15 **2.** $1,439.30 **3.** $7,664.70 **4.** $3,168.78 **5.** $2,419.02 **6.** $8,088.26 **7.** $8,795.56 **8.** $4,695.14 **9.** $14,384.73 **10.** $1,670.23

Chapter Thirteen
Assignment
Skill Problems
1. F **3.** F **5.** T **7.** F **9.** T **11.** 3% **13.** $6,047.62 **15.** 30 **17.** 5% **19.** $78,663.61 **21.** 12 **23.** 3/4% **25.** $7,178.48 **27.** 14 **29.** $50 **31.** $4,050 **33.** $4,100.63 **35.** $51.26 **37.** $4,151.88 **39.** $4,203.78 **41.** $52.55 **43.** $4,256.33 **45.** $4,309.53 **47.** 3, 10, $40,553.12 **49.** 4, 30, $6,166.37 **51.** 1.5, 24, $11,192.70 **53.** $52,059.12 **55.** $16,159.49 **57.** $6,565.97 **59.** $3,683.48 **61.** $71,527.29 **63.** 1.5, 30, $24,024.76 **65.** 2, 17, $12,433.89 **67.** 5, 13, $158,773.32 **69.** $98,553.91 **71.** $879.56 **73.** $7,806.16 **75.** $13,783.21 **77.** $4,138.10 **79.** $3,409.65 **81.** $17,486.88 **83.** $478.41
Business Application Problems
1. $120,482.96 **3.** $5,232.73 **5.** $4,381.55 **7.** $5,837.11 **9.** $171,979 **11.** $6,937.82 **13.** $14,215.79 **15.** $2,867.25 **17.** $44,705.34 **19.** $16,082.17 **21.** $431,448.57 **23.** The profit was $2,019,600 for one year.

Chapter Thirteen
Mastery Test
1. $2,745.57 (Objective 1) (2,000 × 1.37278571 =)
2. $2,981.34 (Objective 3) (5,000 × .59626733 =)
3. $37,618.70 (Objective 1) (19,000 × 1.9799316 =)
4. $7,490.68 (Objective 3) (9,500 × .78849318 =)
5. $4,011.39 (Objective 4) (400 × (11.0284738 − 1) =)
6. $2,498.73 (Objective 5) (30,000 × .08329094 =)
7. $5,968.92 (Objective 4) (375 × 15.91712652 =)
8. $24,271.84 (Objective 5) (50,000 × .48543689 =)
9. $374.66 (Objective 2) (300 × 1.24886297 =)
10. $8,233.49 (Objective 4) (750 × (11.97798875 − 1) =)

Chapter Fourteen (Pages 263–284)
Exercise A
1. $5,120 **2.** $40,800 **3.** $812.43 **4.** $4,320 **5.** $2,041.20 **6.** $432 **7.** $15,552 **8.** $684 **9.** $449.55 **10.** $4,968
Exercise B
1. $7,866 **2.** $4,590 **3.** $522.50 **4.** $54,432 **5.** $59,280 **6.** $6,361.20 **7.** $13,608 **8.** $1,988.35 **9.** $1,923.75 **10.** $266.12
Exercise C
1. $654.75 **2.** $441 **3.** $200 **4.** $64.35 **5.** $175 **6.** $1,400 **7.** $69.30 **8.** $490 **9.** $118.80 **10.** $88.11
Exercise D
1. $190 **2.** $150 **3.** $59.40 **4.** $735 **5.** $600 **6.** $252.20 **7.** $550 **8.** $1,514.70 **9.** $31.20 **10.** $693
Exercise E
1. $601.40 **2.** $83.30 **3.** $120 **4.** $450 **5.** $323.73 **6.** $693 **7.** $171.50 **8.** $76.44 **9.** $321.75 **10.** $802
Exercise F
1. $492.63 **2.** $1,339.20 **3.** $233.77 **4.** $704.95 **5.** $2,245.32
Exercise G
1. $2,067.41 **2.** $8.50 **3.** $2,619 **4.** $180 **5.** 30
Exercise H
1. $2,257.70 **2.** $3,559 **3.** $23,659.51 **4.** 150 **5.** $76

Chapter Fourteen
Assignment
Skill Problems
1. $10,965 **3.** $6,925.86 **5.** $26,732.50 **7.** $6,739.20 **9.** $1,574.63 **11.** $253,634.80 **13.** $10,771.14 **15.** $30,723 **17.** 21.5%, $1,306,774.74 **19.** 41.6%, $592.18 **21.** July 10 **23.** February 10 **25.** November 16 **27.** April 14 **29.** $929.61 **31.** $425.65 **33.** $1,930 **35.** $252.23 **37.** $259.58 **39.** $881.10 **41.** $117.60 **43.** $2,787.30 **45.** $2,018.55 **47.** $2,088.06 **49.** $208.17 **51.** $3,376.48 **53.** $6,111.99 **55.** $453.99 **57.** $1,345.96 **59.** $56,517.45 **61.** $444.72 **63.** $1,681.02
Business Application Problems
1. $41,055 **3.** $212.50 **5.** $1,016.75 **7.** $54,850.50 **9.** $182.51 **11.** $4,048 **13.** $850 **15.** $1,453.50 **17.** $86 **19.** $1,764 **21.** $4,888.80 **23.** The list price was $700. **25.** The company will save $747.77.

Chapter Fourteen
Mastery Test
1. L (Objective 5)
2. J (Objective 5)
3. E (Objective 5)
4. H (Objective 5)
5. O (Objective 5)
6. C (Objective 5)
7. D (Objective 5)
8. M (Objective 5)
9. N (Objective 5)
10. G (Objective 5)
11. I (Objective 5)
12. A (Objective 5)
13. K (Objective 5)
14. F (Objective 5)
15. $492 (Objective 1) (600 × .82 =)
16. $640.80 (Objective 1) (890 × .9 × .8 =)
17. $891 (Objective 4) (900 × .99 =)
18. $733.04 (Objective 4) (748 × .98 =)
19. $564.48 (Objective 3) (800 × .9 × .8 × .98 =)
20. $812.25 (Objective 3) (1,000 × .95 × .95 × .9 =)
21. $501.06 (Objective 3) (650 × .95 × .9 × .92 × .98 =)
22. $739.31 (Objective 3) (92 × 10 × .82 × .98 =)
23. $470.45 (Objective 3) (600 × .9 × .88 × .99 =)

Chapter Fifteen (Pages 285–302)
Exercise A
1. $20, $60 **2.** $12, $72 **3.** $35, $135 **4.** $50, $175 **5.** $15.64, $83.64
Exercise B
1. $37.50, $187.50 **2.** $107.14, $357.14 **3.** $26.67, $106.67 **4.** $112, $160 **5.** $10.64, $23.64
Exercise C
1. $99 **2.** $42 **3.** $213.33 **4.** $365 **5.** $24 **6.** $20.30 **7.** $10.40 **8.** $1,131 **9.** $25.33 **10.** $153.33
Exercise D
1. $128, $32 **2.** $35, $15 **3.** $68, $28 **4.** $111.60, $18.60 **5.** $29.90, $16.10 **6.** $33.25, $14.25 **7.** 82, $17.95 **8.** 31, $20.50 **9.** $171, $228 **10.** $228.57,

$308.57 **11.** 140, $28 **12.** 25, $12
13. 37, $74 **14.** $44.80, $11.20
15. $189, $147

Exercise E
1. $59.50 **2.** $10.20 **3.** $26 **4.** $14.36
5. 25% **6.** 43% **7.** 15% **8.** 31%

Exercise F
1. 29% **2.** 33% **3.** 67% **4.** 100% **5.** 41%
6. 23% **7.** 67% **8.** 17% **9.** 31% **10.** 25%

Exercise G
1. $117.65 **2.** $42.74 **3.** $355.12
4. $95.19 **5.** $149.40 **6.** $102.78
7. $146.04 **8.** $10,000 **9.** $119.47
10. $180.72

Exercise H
1. $166.01, 6% **2.** $720.56, 16%
3. $501.96, 16% **4.** $166.52, 20%
5. $942.31, 6%

Chapter Fifteen
Assignment

Skill Problems
1. $52.25 **3.** $245 **5.** $27.70
7. $401.70 **9.** $34.30 **11.** $97.78
13. $64.62 **15.** $55.40 **17.** $521.68
19. $113.17 **21.** $660 **23.** $169.20
25. $578.03 **27.** $55.35 **29.** $17.50
31. $208.13 **33.** $1,512.28 **35.** $15
37. $49 **39.** $14.50 **41.** $25
43. $31.33 **45.** $210 **47.** $240
49. $108 **51.** $11 **53.** $74.40
55. $6.98 **57.** 22.57% **59.** 21.45%
61. 19.50% **63.** 51.00% **65.** 21.25%
67. 30% **69.** 150% **71.** $50
73. $71.65 **75.** $59.80 **77.** $36.43
79. $38.40 **81.** $140.63 **83.** $758.51
85. $135.04 **87.** $128.02
89. $1,120.40 **91.** $57.27
93. $44.58 **95.** $253.39

Business Application Problems
1. $362.60 **3.** $431.67 **5.** 31%
7. $206.75 **9.** 56% **11.** 56%
13. 57% **15.** $6,251.20 **17.** 89
19. $428.57 **21.** $90.13 **23.** The store must sell $1,866,666.67 per month.
25. They will receive $1,094,846.10 in revenue.

Chapter Fifteen
Mastery Test

1. $65 (Objective 1) (50 × 1.3 =)
2. $95 (Objective 1) (38 ÷ .4 =)
3. $142.86 (Objective 2) (200 ÷ 1.4 =)
4. $23.50 (Objective 2) (47 × .5 =)
5. 140% (Objective 3) (60 − 25 =, 35 ÷ 25 =)
6. 43% (Objective 3) (35 − 20 =, 15 ÷ 35 =)
7. 67% (Objective 6) (40 ÷ 60 =)
8. 33% (Objective 5) (50 ÷ 150 =)
9. $746.67 (Objective 1) (80 ÷ .12 =, 666.67 + 80 =)
10. $345 (Objective 2) (30 ÷ .08 =, 375 − 30 =)
11. $258.06 (Objective 7) (120 ÷ .62 =, 193.55 ÷ .75 =)
12. $188.96 (Objective 2) (380 × .85 =, 323 × .78 =, 251.94 × .75 =)
13. $392 (Objective 1) (280 × 1.4 =)
14. $1.91 (Objective 2) (2.29 ÷ 1.2 =)
15. 15% (Objective 4) (16 ÷ 105 =)

Chapter Sixteen (Pages 303–322)

Exercise A

1. 1,728	**2.** 1,150
1,428	800
$383.62	$1,743.75
$1,855.14	$3,882.25
3. 370	**4.** 2,005
190	1,273
$6,992	$7,456.60
$6,732.50	$14,245.65

Exercise B

1. $6,720.45	**2.** $264.25
$11,415.55	$22,196.75
3. $1,597.12	**4.** $7,528.51
$8,784.13	$36,701.49

Exercise C
1. $8,608, $17,669
$1,820.41, $3,738
2. $46,018, $54,351
$15,910.05, $18,784
3. $80,683, $117,595
$9,394.77, $13,695
4. $136,062, $221,256
$122,891.76, $199,824

Exercise D

1. $346,650.41	**2.** $90,330.77
$52,117.25	$49,925
6.7	1.8
3. $103,488	**4.** $17,777.78
$45,650	$3,616.67
2.3	4.9
5. $15,840	
$5,042	
3.1	

Chapter Sixteen
Assignment

Skill Problems
1. A **3.** K **5.** H **7.** D **9.** $40,928.57
11. $10,044.44 **13.** $43,737.93
15. 2,075 **17.** $723.75 **19.** 1,727
21. $7,947 **23.** $357.70 **25.** $348.70
27. $343.70 **29.** $26,588
31. $26,530.40 **33.** $26,541.99
35. $1,492.60 **37.** $1,509.52
39. $1,498.10 **41.** $233,310.28
43. $233,627.06 **45.** $233,065.50
47. $8,017.20 **49.** $417 **51.** $1,311.70
53. $200.20 **55.** $163,896 **57.** $86,402
59. $95,080 **61.** $54,840
63. $11,474.90 **65.** $16,064.86
67. $2,093.25 **69.** $2,616.56
71. $12,906.29 **73.** 3.0 **75.** $70,334
77. 2.8 **79.** 2.6

Business Application Problems
1. 6.3 **3.** $71,768 **5.** .8 **7.** $302,187
9. $5,599.72 **11.** $35,760
13. $98,678.45 **15.** $27,600.22
17. $916 **19.** The turnover is 1.8 for the month.

Chapter Sixteen
Mastery Test

1. F (Objective 4)
2. F (Objective 4)
3. T (Objective 4)
4. T (Objective 4)
5. $185,000 (Objective 2) (42,000 + 175,000 − 32,000 =)
6. $158,455 (Objective 2) (345,600 + 56,835 − 243,980 =)
7. $17,800 (Objective 2) (39,800 + 156,000 − 178,000 =)
8. $39,786 (Objective 1) (890 × 41 =, 80 × 41.20 =)
9. $40,189 (Objective 1) (650 × 41.50 =, 300 × 41.30 =, 20 × 41.20 =)
10. $39,979.77 (Objective 1)
11. 2.1 (Objective 3) (36,490 + 39,786 ÷ 2 =, 79,329 ÷ 38,138 =)
12. $192,250, $385,830 $62,215.14, $124,930 (Objective 1) (38,500 + 153,750 =, 84,600 + 301,230 =)
13. 1.0 (Objective 3) (96,800 + 42,690 ÷ 2 =, 120,970 × .6 =, 72,582 ÷ 69,745 =)
14. $230,885 (Objective 1) (34,980 × .6 =, 251,873 − 20,988 =)

Chapter Seventeen (Pages 323–346)

Exercise A

1. $2,318.33	**2.** $650
$9,273.32	$4,550
$21,756.68	$16,207
3. $100	**4.** $3,000
$300	$6,000
$600	$46,980
5. $100	
$300	
$825	

Exercise B
1. $55,000 **2.** 7 **3.** 10 **4.** $5,000
5. $20,000

Exercise C
1. $586.93, $32,351.40
2. $227.50, $393.25
3. $2,625, $8,980
4. $198.91, $3,001.19
5. $562.07, $7,296.49

Exercise D
1. 105, $\frac{12}{105}$ **2.** 210, $\frac{19}{210}$ **3.** 91, $\frac{10}{91}$
4. 120, $\frac{10}{120}$ **5.** 55, $\frac{1}{55}$

Exercise E

1. $2,330.95	**2.** $2,137.14
$32,845.24	$9,714.29
$15,654.76	$11,285.71

3. $10,458.33 **4.** $287.27
$78,437.50 $7,756.36
$57,062.50 $1,143.64
5. $211.43
$6,871.43
$728.57
Exercise F
1. $950 **2.** $10,800
$2,850 $28,800
$950 $16,200
3. $2,376.56 **4.** $988.15
$9,770.31 $8,020.74
$7,129.69 $1,979.26
5. $6,995.97
$57,416.13
$27,983.87
Exercise G
1. $8,352 **2.** $1,748
$24,592 $1,748
$33,408 $6,992
3. $825.37 **4.** $18,857.50
$825.37 $28,782.50
$32,018.63 $169,717.50
5. $1,733.55
$3,033.42
$866.58
Exercise H
1. $96,606.44 **2.** $1,760 **3.** $90,644 **4.** $36,800 **5.** $183,450
Exercise I
1. $215 **2.** $208.20 **3.** $80 **4.** $70 **5.** $196.88 **6.** $125 **7.** $118.78 **8.** $652 **9.** $1,312.50 **10.** $81.25

Chapter Seventeen
Assignment
Skill Problems
1. $66,000 **3.** $14,990 **5.** $15,476 **7.** $1,446 **9.** $4,338 **11.** $375.52 **13.** $1,712.50 **15.** $15,425 **17.** $790.50 **19.** $81.43 **21.** $230.77 **23.** $933.33 **25.** $31,976.13 **27.** $119.92 **29.** $3,152 **31.** 13/136 **33.** 8/120 **35.** 22/903 **37.** 6/78 **39.** $48,297.44 **41.** $1,190.34 **43.** $12,808.22 **45.** $51,024.76 **47.** $900 **49.** $900 **51.** $3,072 **53.** $1,387.97 **55.** $4,163.90 **57.** $4,129.97 **59.** $9,748.48 **61.** $38,993.92 **63.** $2,600 **65.** $1,111.11 **67.** $2,822.22 **69.** $12,433.80 **71.** $101.25 **73.** $911.25 **75.** $2,247.50 **77.** $80 **79.** $710 **81.** $3,306.24 **83.** $50 **85.** $345 **87.** $600 **89.** $86.67 **91.** $576 **93.** $8,332.50 **95.** $16,667.50 **97.** $11,112.50 **99.** $5,555 **101.** $3,702.50 **103.** $1,852.50 **105.** $8,841.01 **107.** $1,654.90 **109.** $4,137.73 **111.** $3,926.75 **113.** $67.40 **115.** $1,280.60 **117.** $7,000.20 **119.** $3,052.80 **121.** $4,579.20 **123.** $8,500 **125.** $89,684 **127.** $64,417.50 **129.** $665 **131.** $1,041.67 **133.** $1,216.95 **135.** $10.50
Business Application Problems
1. $1,200 **3.** $1,248 **5.** $800 **7.** $1,275 **9.** $4,992 **11.** $5,950 **13.** $5,385.15 **15.** $2,866.67 **17.** $27,000 **19.** $2,136 **21.** $3,023.18 **23.** The accumulated depreciation would be $1,512.

Chapter Seventeen
Mastery Test
1. F (Objective 1)
2. J (Objective 1)
3. A (Objective 1)
4. I (Objective 1)
5. K (Objective 1)
6. E (Objective 1)
7. C (Objective 1)
8. D (Objective 1)
9. $2,545.45
$12,363.64
$8,636.36 (Objective 2) (21,000 − 1,000 =, 20,000 × 7 ÷ 55 =)
10. $6,731.25
$15,706.25
$20,193.75 (Objective 2)
(35,900 × (1 ÷ 8 × 2 =) =)
11. $627.50
$10,667.50
$1,932.50 (Objective 2)
(12,600 − 50 =, 12,550 ÷ 20 =)
12. $8,000 (Objective 1)
(42,000 − 2,000 =,
40,000 × 20,000 ÷ 100,000 =)
13. $4,930 (Objective 1)
(5,800 × .15 =, 5,800 − 870 =)
14. $6,000 (Objective 1)
(150,000 × 2,000 ÷ 50,000 =)
15. $183,333.33 (Objective 2)
(260,000 − 30,000 =,
230,000 × 10 ÷ 30 =)
16. $6,382.50 (Objective 4) (74,000 × 2 ÷ 20 =, 7,400 × 5 ÷ 12 =, 70,916.67 × .9 =, 63,825 × .1 =)

Chapter Eighteen (Pages 347–372)
Exercise A
1. $287.54 **2.** $2,286 **3.** 21.5 **4.** $9,000 **5.** $1,332.80 **6.** $491.40 **7.** $873 **8.** 1.42 **9.** $1,908.40 **10.** $50,000
Exercise B
1. 18.83 **2.** 16.67 **3.** 27.29 **4.** 19.20 **5.** 26.13 **16.** 1.29 **7.** 1.75 **8.** 2.03 **9.** 1.33 **10.** 1.69
Exercise C
1. $2,367 **2.** $129.80 **3.** $1,620 **4.** $119 **5.** $60.75 **6.** $1,371 **7.** $139.80 **8.** $309 **9.** $2,526 **10.** $103.50
Exercise D
1. $34,910, $5,239 **2.** $39,890, $5,981 **3.** $32,760, $4,916 **4.** $33,186, $4,976 **5.** $36,600, $6,908 **6.** $32,706, $4,909 **7.** $32,681, $4,901 **8.** $37,463, $7,146 **9.** $33,120, $4,969 **10.** $35,085, $6,474

Chapter Eighteen
Assignment
Skill Problems
1. D **3.** E **5.** G **7.** C **9.** $874.20 **11.** $2,214.40 **13.** $11,620.68 **15.** $20,271.06 **17.** $1,374.25 **19.** 7.43 **21.** $163.15 **23.** 21.59 **25.** 2.95 **27.** 25.31 **29.** 13.91 **31.** 4.44 **33.** 3.70 **35.** $391.16 **37.** $240 **39.** $252.70 **41.** $1,240 **43.** $3,874.35 **45.** $140 **47.** $14,472 **49.** $4,819 **51.** $5,321 **53.** $6,152 **55.** $5,681 **57.** $4,826 **59.** $5,276 **61.** $4,804 **63.** $6,754 **65.** $5,718 **67.** $4,924 **69.** $6,388 **71.** $6,310 **73.** $5,788
Business Application Problems
1. $314.40 **3.** $1,757.60 **5.** $32,940 **7.** $116,732 **9.** $5,810.92 **11.** $20.79 **13.** $37,000 **15.** $1,043.81 **17.** Shawl—$42 **19.** $4,939 **21.** $3,982.59 **23.** He saved $231 by making the trip.

Chapter Eighteen
Mastery Test
1. H (Objective 6)
2. E (Objective 6)
3. D (Objective 6)
4. B (Objective 6)
5. G (Objective 6)
6. F (Objective 6)
7. A (Objective 6)
8. $1,628.54 (Objective 1)
(42.8 × 38.05 =)
9. $1,170.45 (Objective 1)
(289 × 4.05 =)
10. $2,894.40 (Objective 3)
(144 × 16 × .15 =, 144 × 1.70 =)
11. $966.90 (Objective 3)
(50 × 17.58 × .10 =)
12. $6,054 (Objective 4) (−$2,750, −$4,953)
13. $5,764 (Objective 4) (−$8,250, −$7,200)
14. $1,218.46 (Objectives 1, 2)
(2,680,500 × .01 =, 2,707,305 ÷ 92,426,000 =, 41.6 × 29.29 =)

Chapter Nineteen (Pages 373–390)
Exercise A
1. 45¢, 4.7% **2.** 67¢, 6.7% **3.** $2.12, 7.0% **4.** 11¢, 1.1% **5.** 73¢, 9.1%
Exercise B
1. $14.28 **2.** $15 **3.** $10 **4.** $10 **5.** $30 **6.** $6.29 **7.** $12 **8.** $19.26
Exercise C
1. 8.1% **2.** 7.1% **3.** 9.8% **4.** 6.7% **5.** 8.6%

Chapter Nineteen
Assignment
Skill Problems
1. N **3.** G **5.** D **7.** C **9.** F **11.** 10.96% **13.** 7.78% **15.** 8.23% **17.** 3.9%

19. 1.3% **21.** $3.13 **23.** $1.68
25. $4.44 **27.** 26.19% **29.** 13.84%
31. 8.57% **33.** 11.25% **35.** 6.59%
37. 1.8% **39.** 3.4% **41.** 2.5%
43. 1.2% **45.** 3.7% **47.** 0 **49.** $2.91
51. $2.77 **53.** 0 **55.** $3.42 **57.** $7.50
59. $21.50 **61.** 7.1% **63.** 9.6%
65. 10.8% **67.** 9.6% **69.** 8.0%
71. 8.9% **73.** 9.5% **75.** 9.1%
Business Application Problems
1. 7.1% **3.** 7.1% **5.** $4,226 **7.** 3.1%
9. 7.8% **11.** 18.2% **13.** Bond return 9.5% **15.** Jewelry Business **17.** 1.1%
19. $765 **21.** 11.3% **23.** Yes. He would have earned $1,250 on the C.D. investment.

Chapter Nineteen
Mastery Test

1. T (Objective 4)
2. T (Objective 4)
3. F (Objective 4)
4. F (Objective 4)
5. T (Objective 4)
6. 7.2% (Objective 3) (250,000 × 4 =, 5,500,000 ÷ 2,000,000 =, 2.75 ÷ 38 =)
7. 6.9% (Objective 3) (70 ÷ 1,020 =)
8. $4.50 (Objective 2) (60,000 × 8 =, 1,020,000 − 480,000 =, 540,000 ÷ 120,000 =)
9. 0 (Objective 2) (240,000 × 12 =)
10. 29.8% (Objective 1) (75,000 × .1 =, 75,000 − 7,500 − 52,000 =, 15,500 ÷ 52,000 =)

Chapter Twenty (Pages 391–408)
Exercise A
Accounts Receivable
($4,930), (79.5%)
Exercise B
Notes Receivable
$2,000, 400%
Accounts Receivable
$5,448, 68.8%
Exercise C
Special Revenue
40.8%, 172.2%
Exercise D
Net Sales
(.5%)
Exercise E
1. 1.49 to 1 **2.** .94 to 1 **3.** .34 to 1
4. .05 to 1 **5.** .08 to 1

Chapter Twenty
Assignment
Skill Problems
1. D **3.** A **5.** J **7.** E **9.** B **11.** C
13. A **15.** 32.55% **17.** 13.56%
19. $26,032; 17.31% **21.** (10.61)%
23. $156,540 **25.** (2.02)% **27.** $95,400
29. $87,000 **31.** 32.47% **33.** $159,770
35. 1.33 to 1 **37.** 31.28% **39.** 32.80%
41. 30.26% **43.** 60.98%
45. (15.42)% **47.** $8,920 **49.** 23.97%
51. .37 to 1 **53.** (14.0)% **55.** $17,200
57. $27,560 **59.** (.9)% **61.** (16.7)%
63. (3.8)% **65.** $47,110 **67.** 16.7%
69. $14,900 **71.** (8.0)% **73.** $15,900
75. (1.9)% **77.** $30,910 **79.** $46,800
81. (.7)% **83.** $163,705
85. $137,230 **87.** $121,930
89. $41,775 **91.** $23,250 **93.** $18,525
95. .72 to 1 **97.** 2.12% **99.** $46,800
101. 9.7% **103.** 59.3% **105.** .16 to 1
107. 25.5% **109.** 2.2% **111.** 11.3%
113. .11 to 1
Business Application Problems
1. $12,500 **3.** 1.2 to 1 **5.** .9 to 1
7. $380,400 **9.** .9 to 1 **11.** No
13. $55,350 **15.** $675,000
17. 1.04 to 1

Chapter Twenty
Mastery Test

1. current year, last year (Objective 3)
2. liabilities (Objective 1)
3. Income Statement (Objective 6)
4. loss (Objective 2)
5. acid test (Objective 5)
6. balance sheet (Objective 6)
7. accounts payable to net purchases ratio (Objective 5)
8. accounts receivable to net sales ratio (Objective 5)
9. revenue, expenses (Objective 2)
10. service, merchandising (Objective 6)
11. 9.1 (Objective 3) (240,000 − 220,000 =, 20,000 ÷ 220,000 =)
12. 4.3 (Objective 5) (5,400 + 7,900 + 12,500 =, 25,800 ÷ 6,000 =)
13. 2.2 (Objective 5) (5,400 + 7,900 =, 13,300 ÷ 6,000 =)
14. 75 (Objective 5) (180,000 ÷ 240,000 =)
15. .03 or 0 (Objective 5) (7,900 ÷ 240,000 =)
16. .03 or 0 (Objective 5) (6,000 ÷ 178,600 =)
17. 4.6 (Objective 5) (5,400 + 8,500 + 13,800 =, 27,700 ÷ 6,000 =)
18. 0 (Objective 5)
19. .1 (Objective 5) (34,800 ÷ 240,000 =)
20. 1 (Objective 5) (178,600 − 177,000 =, 1,600 − 177,000 =)

Chapter Twenty-One (Pages 409–424)
Exercise A

1. 69.6, 58.5, 62, 161
2. 34.5, 25.5, 23, 78
3. 225.3, 226, 120, 337
4. 22.5, 21, 13 & 29, 23
5. 52.9, 49, 42, 47
6. $6.62, $6.80, $6.80 & $7.20, $4.80
7. $3.95, $3.96, $3.90, $1.52
8. $11.99, $11.20, $13.80, $11.05
9. $7.73, $7.75, $8.08, $2.66
10. $4.61, $4.60, $4.60, $.78
11. $9.54, $9.11, $9.30, $9.56
12. $6.71, $6.75, $7.00 & $6.75, $2.50
13. $6.06, $6.50, $8.00, $9.00

Chapter Twenty-One
Assignment
Skill Problems
1. C **3.** E **5.** D **7.** 39 hours
9. 38¼ hours **11.** 368 **13.** 38
15. 7 **17.** 86.3 **19.** 88 **21.** 65
23. 374.5 **25.** 754 **27.** 15 **29.** 7
31. $6.45 **33.** $2.36 **35.** $3.84
37. $7.92 **39.** 3 **41.** 4 **43.** 370.8, 389, 386, 253 **45.** 14.5, 14, 13, 10
47. 42.7, 40.5, 37, 30 **49.** 49, 47, 47, 24 **51.** $1.50, $1.48, $1.48, $.15
Business Application Problems
1. 25 **3.** $77.98 **5.** 50 **7.** 34.2
9. $322,000 **11.** $887.50, $895
13. 79.4, 79.5, 83, 21; 77.7, 80, 86, 22
15. 43,407, 43,665, 43,900, 1,950
17. 28.3 **19.** The mean of the ten year period is 25.2.

Chapter Twenty-One
Mastery Test

1. F (Objective 5)
2. F (Objective 5)
3. T (Objective 5)
4. T (Objective 5)
5. T (Objective 5)
6. T (Objective 5)
7. T (Objective 5)
8. 9.3, 9, 13 (Objectives 1, 2, 3) (130 ÷ 14 =)
9. $272.50 (Objective 1) (2,180 ÷ 8 =)
10. Gwen (Objective 1) (159 ÷ 5 =, 162 ÷ 5 =)

Chapter Twenty-Two (Pages 425–446)
Exercise A

1. Junior Misses 73°, 20%
2. Administrative 104°, 29%
3. Single family 51°, 14%
4. Hiking and Camping Equipment 160°, 44%

Exercise B
3. 1999, 5.4
Exercise C
Graphs

Chapter Twenty-Two
Assignment
Skill Problems
1. B, C **3.** C **5.** column or bar
7. differences **9.** bar graph **11.** bar graphs, line graphs **13.** straight lines
15. 24% **17.** 8% **19.** 16% **21.** 26%
23. 5% **25.** $3,230 **27.** product A, 25°, 7% **33.** 16% increase **35.** 14% decrease **37.** 30% increase **39.** 20% increase
Business Application Problems
1. Airlines 32%, 115° **3.** Coastline 85%, 306°

5.

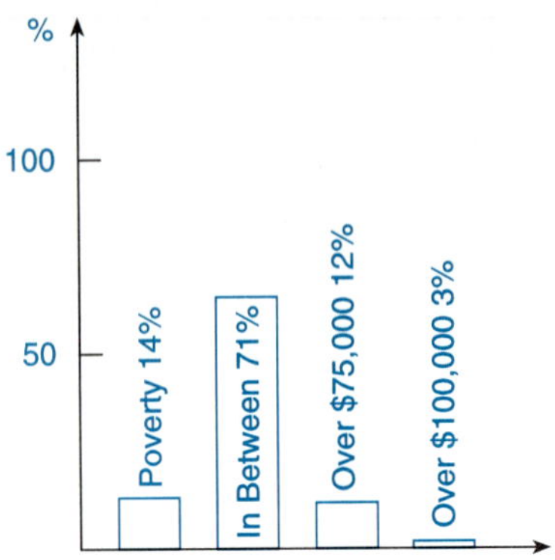

7.

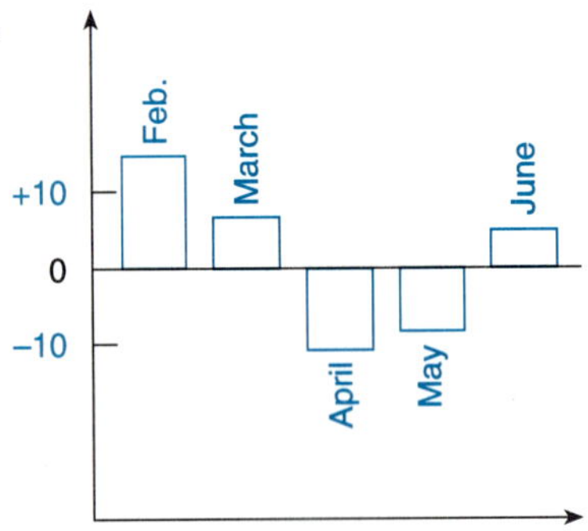

9.

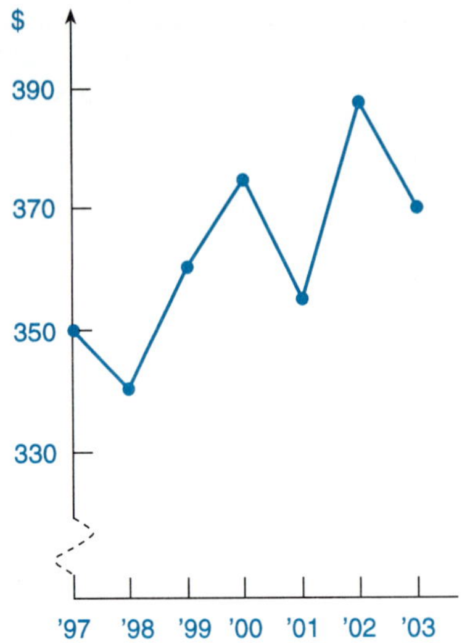

11. Line graph **13.** Line graph
15. 61% **17.** $41,040, $42,681.60

Chapter Twenty-Two
Mastery Test
1. credit sales: 64°, 18%
4. 30s: 158°, 44%

Chapter Twenty-Three (Pages 447–452)

1. 133.63 **2.** $1\frac{7}{8}$ **3.** 35 **4.** $4\frac{1}{3}$
5. 4.694 **6.** 284.8 **7.** 3,200 **8.** $264
9. $4,264 **10.** $240 **11.** 5,760 sq. ft
12. 535.104 sq. meters **13.** $75,000
14. $686 **15.** $1,424.43 **16.** August 19
17. $1,155 **18.** $420 **19.** $500
20. 20% **21.** $69.23, $90 **22.** $9,360
23. $4,992 **24.** $20,533.33
25. March 17, 2001 **26.** $3,067.50
27. $7,609.45 **28.** $7,702.49 **29.** 45
30. $999.32 **31.** $35,436.73 **32.** $7.58
33. $350 **34.** $580 **35.** $433.33
36. T **37.** $294 **38.** $184.79
39. $904.80 **40.** $869 **41.** 91.9
42. $532.50 **43.** 6.7 **44.** 3 **45.** 6
46. $25,000 **47.** $8,192.80 **48.** 9.9%
49. $71,210, $36,125
50. $f = 10$, 55°, 15%

Index

Record of Performance on Mastery Tests

Chapter	*Mastery Test Score*
1. Whole Numbers	______
2. Fractions	______
3 Decimals	______
4. Banking Records	______
5. Payroll	______
6. Business Measurements	______
7. Percent	______
8. Insurance	______
9. Simple Interest	______
10. Promissory Notes	______
11. Installment Loans	______
12. Consumer Credit	______
13. Compound Interest and Present Value	______
14. Discounts	______
15. Markup	______
16. Inventory and Turnover	______
17. Depreciation	______
18. Taxes	______
19. Investments	______
20. Financial Statement Analysis	______
21. Statistics	______
22. Charts and Graphs	______

Introduction to Instructor's Resource Section

Recognizing that your time is very valuable, this Instructor's Resource Section of the Annotated Instructor's Edition has been prepared to assist the new or adjunct instructor with day-to-day teaching preparations while using *Practical Business Math: An Applications Approach.*

Among the materials contained in this section are:

- a flowchart of chapter interdependencies (page 490),
- four sample course outlines (pages 491 and 492).

The flowchart will enable you to easily arrange the text's self-contained chapters according to your course needs. The sample course outlines will show you the flexibility of the text to different business math emphases.

Additionally, there are:

- copy/transparency masters (pages 493–507),
- three mastery tests (A, B, and C) per chapter (pages 508–573), followed by the answers (pages 575–582),
- a final exam (pages 583–587), followed by the answers (page 588).

This text is accompanied by a computerized Test Bank in addition to the materials contained in the Instructor's Resource Section. This separate document contains 100 questions per chapter. Question types include multiple choice, problem solving (skill and business application types), and questions requiring an essay response. Questions are graded by level of difficulty and identified by chapter objective. All 2,200 questions are new to this edition.

Business students are often exposed to a math test in employment situations. A sample employment math test with the answers is found on pages 462 and 463. The test can be administered at any point in your course to demonstrate the importance of understanding math.

The material in this Instructor's Resource Section is designed to help you reduce your preparation time and provide classroom resource materials and alternative testing materials. We are sure that this material will be valuable to you.

Best wishes for a successful course!

Flowchart of Chapter Interdependencies

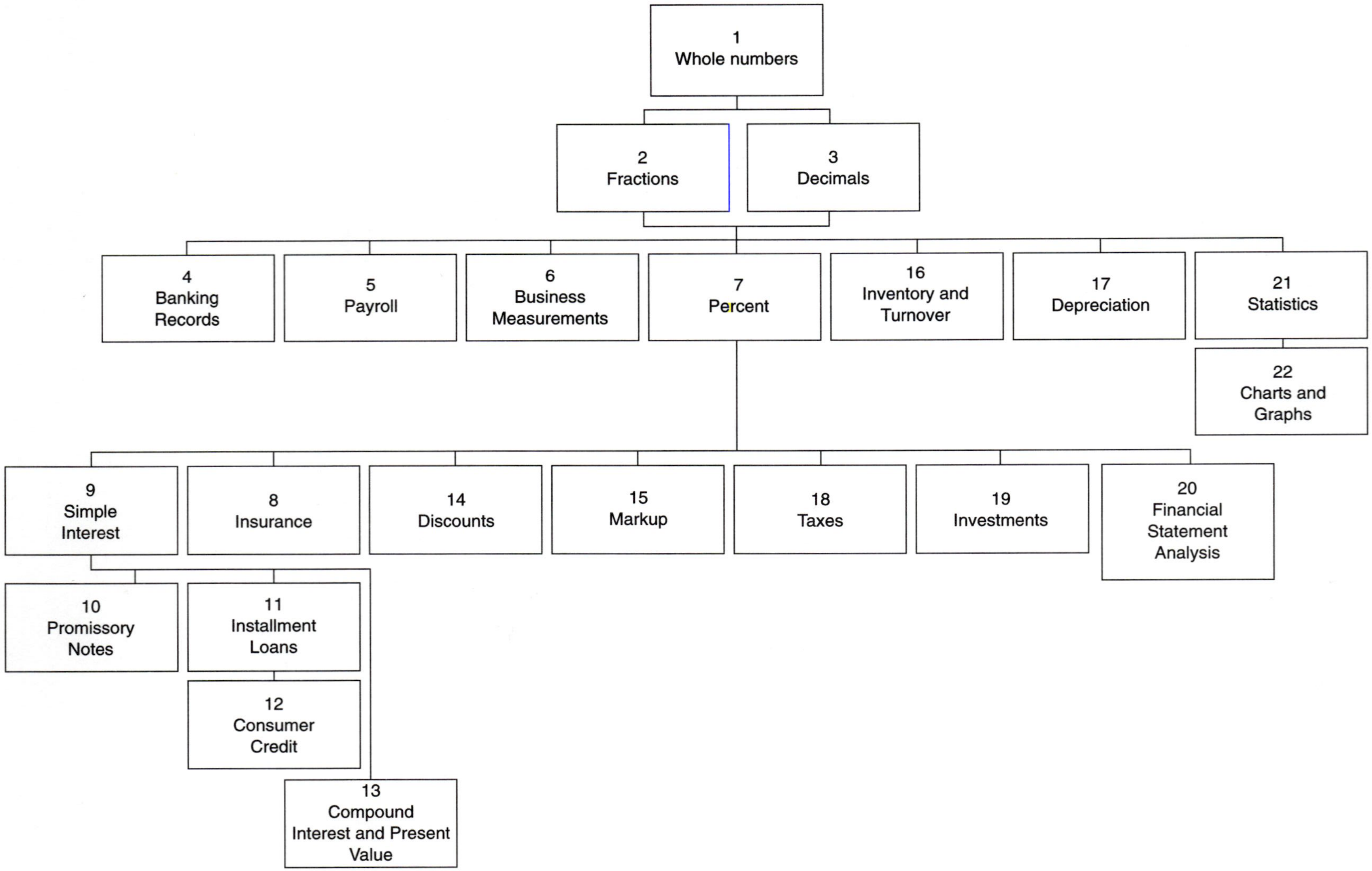

Four Suggested Course Outlines

Course Emphasis on Office Occupations

Chapter One*
Working with Whole Numbers

Chapter Two*
Working with Fractions

Chapter Three*
Working with Decimal Numbers

Chapter Four
Banking Records

Chapter Five
Payroll

Chapter Six
Business Measurements

Chapter Seven*
Percent

Chapter Eight
Insurance

Chapter Nine*
Simple Interest

Chapter Ten
Promissory Notes

Chapter Eleven
Installment Loans

Chapter Twelve
Consumer Credit

Chapter Fourteen
Discounts

Chapter Sixteen
Inventory and Turnover

Chapter Eighteen
Taxes

Chapter Twenty-two
Charts and Graphs

Chapter Twenty-three
Review

Short Course Emphasis

Chapter One*
Working with Whole Numbers

Chapter Two*
Working with Fractions

Chapter Three*
Working with Decimal Numbers

Chapter Four
Banking Records

Chapter Five
Payroll

Chapter Seven*
Percent

Chapter Nine*
Simple Interest

Chapter Ten
Promissory Notes

Chapter Eleven
Installment Loans

Chapter Twelve
Consumer Credit

Chapter Fourteen
Discounts

Chapter Fifteen
Markup

Chapter Seventeen
Depreciation

Chapter Eighteen
Taxes

Chapter Twenty
Financial Statement Analysis

*Instructor may use the Mastery Test as a pre-test to determine if students have sufficient skills to bypass these chapters.

Course Emphasis on Accounting or Business Administration

Chapter One*
Working with Whole Numbers

Chapter Two*
Working with Fractions

Chapter Three*
Working with Decimal Numbers

Chapter Four
Banking Records

Chapter Five
Payroll

Chapter Six
Business Measurements

Chapter Seven*
Percent

Chapter Eight
Insurance

Chapter Nine*
Simple Interest

Chapter Ten
Promissory Notes

Chapter Eleven
Installment Loans

Chapter Twelve
Consumer Credit

Chapter Thirteen
Compound Interest and Present Value

Chapter Fourteen
Discounts

Chapter Sixteen
Inventory and Turnover

Chapter Seventeen
Depreciation

Chapter Eighteen
Taxes

Chapter Nineteen
Investments

Chapter Twenty
Financial Statement Analysis

Chapter Twenty-one
Statistics

Chapter Twenty-three
Review

Course Emphasis on Retailing, Marketing, or Fashion Merchandising

Chapter One*
Working with Whole Numbers

Chapter Two
Working with Fractions

Chapter Three*
Working with Decimal Numbers

Chapter Four
Banking Records

Chapter Five
Payroll

Chapter Six
Business Measurements

Chapter Seven*
Percent

Chapter Nine*
Simple Interest

Chapter Ten
Promissory Notes

Chapter Eleven
Installment Loans

Chapter Twelve
Consumer Credit

Chapter Fourteen
Discounts

Chapter Fifteen
Markup

Chapter Sixteen
Inventory and Turnover

Chapter Eighteen
Taxes

Chapter Twenty
Financial Statement Analysis

Chapter Twenty-two
Charts and Graphs

Chapter Twenty-three
Review

*Instructor may use the Mastery Test as a pre-test to determine if students have sufficient skills to bypass these chapters.